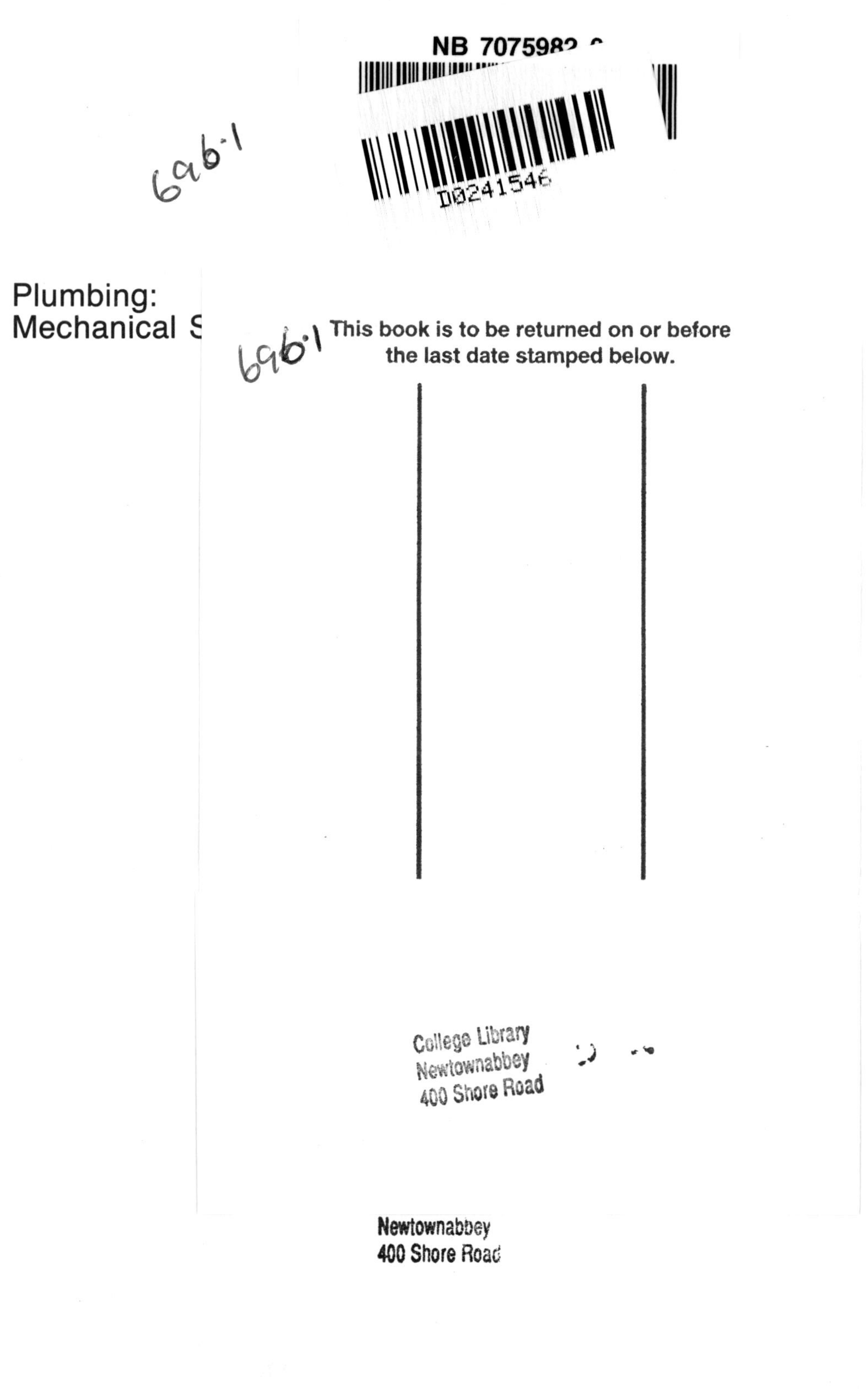
NB 7075982
D0241546
696·1
Plumbing:
Mechanical S
696·1
This book is to be returned on or before
the last date stamped below.
College Library
Newtownabbey
400 Shore Road
Newtownabbey
400 Shore Road

Plumbing: Mechanical Services

Book 1

Third Edition

G.J. Blower
Eng Tech (CEI), MIP, LCGI, Technical Teachers Cert.

Formerly Senior Lecturer, Plumbing Mechanical
Services Section, College of North East London.
Currently Visiting Lecturer at Barking College
and an NVQ Assessor

Addison Wesley Longman Limited
Edinburgh Gate, Harlow
Essex CM20 2JE, England
and Associated Companies throughout the world

First published 1982
Reprinted by Longman Scientific & Technical 1987
Second edition 1989
Fourth impression 1993
Third edition 1995
Second impression 1998

British Library Cataloguing in Publication Data
A catalogue entry for this title is available from the British Library

ISBN 0-582-23659-2

Set by 4 in 10/12 pt Compugraphic Times
Produced through Longman Malaysia, PP

Contents

Foreword

It is said that every man is a builder by instinct and, while this may be true to a degree, the best way to become a craftsman is first to be an apprentice.

To be an apprentice or trainee means to learn a trade, to learn skills and have knowledge which will both be a means of earning a living and provide an invaluable expertise for life. But whereas this used to be the main purpose and advantage of serving an apprenticeship, nowadays many people in high positions in industry have made their way to the top after beginning with an apprenticeship.

The once leisurely pace of learning a trade has now been replaced by a much more concentrated period of learning because of the shortened length of training. In addition, there is an ever-increasing number of new materials and techniques being introduced which have to be understood and assimilated into the craftsman's daily workload. There is therefore a large and growing body of knowledge which will always be essential to the craftsman, and the aim of this craft series of books is to provide this fundamental knowledge in a manner which is simple, direct and easy to understand.

As most apprentices and trainees nowadays have the advantage of attending a college of further education to help them learn their craft, the publishers of these books have chosen their authors from experienced craftsmen who are also experienced teachers and who understand the requirements of craft training and education. The learning objectives and self-testing questions associated with each chapter will be most useful to students and also to college lecturers who may well wish to integrate the books into their teaching programme.

The needs have also to be kept in view of the increasing numbers of late entrants to the crafts who are entering a trade as adults, probably under a government-sponsored or other similar scheme. Such students will find the practical, down-to-earth style of these books to be an enormous help to them in reaching craftsman status.

L. Jaques
Head of Department
Leeds College of Technology

Preface

The primary purpose of this series of books has always been to supplement the technical knowledge of craft apprentices and trainees attending a college or training centre. It is also hoped they will be of use to practising plumbers who wish to gain further information on aspects of the plumbing craft with which they may not be familiar. The reader should be aware that the NVQ courses in plumbing which have supplemented the traditional City & Guilds courses emphasise competence, but it must be stressed that competence must be based on knowledge and experience. To be experienced takes time but knowledge of the underlying technology in any craft can be acquired at any one period of one's career. Modern plumbing work embraces a multitude of skills and it should be appreciated that the British Plumbing Employers Council, who developed the scheme for the NVQs and is responsible for training in the industry, is well aware of this. The Council has, therefore, produced a series of learning modules which cover the wide range of activities with which the plumber is involved. The one exception to this is the module for the external weathering of buildings, which currently deals with sheet lead only, but modules are likely to be developed in copper, aluminium and zinc sheet work. It is hoped the sections relating to these metals will prove helpful to practising plumbers involved with sheet weathering, whether or not the aim is a recognised qualification.

The author wishes to record his thanks and appreciation to the manufacturers of tools and equipment who have assisted in the production of this book. Special thanks are also accorded to Mr K.A.E. Green, formerly Head of Department of Building Crafts, College of N.E. London, and Mr L. Jaques, formerly Head of Department of Building, Leeds College of Technology.

1994 G.J. Blower

1 Safety on site and in the workshop

After completion of this chapter the reader should understand the basic principles of general site safety and legislation relating to Health and Safety in construction.

1. Employers' and employees' responsibilities relating to safe working conditions and maintenance of safety equipment. Cause of accidents and their prevention.
2. Recognise potential hazards with all types of working platforms.
3. Understand the need for using personnel protective clothing and safety equipment.
4. Safety with electrical tools and appliances. Treatment for electric shock.
5. Causes of fire and methods of fire control. Correct selection of fire extinguishers.
6. Safe use of grinding wheels.
7. Safe use of hand tools.
8. Safe manual-lifting techniques.
9. Safety signs.
10. Safety when handling dangerous and toxic materials.
11. The principles of mechanical handling. Safe working loads.
12. Understanding the basic principles underlying the use of simple levers and pulleys.

Health and safety at work

The Factories Act of 1833 was introduced to safeguard the health, working conditions and general well-being of children employed in the textile industry, and it is from this Act that all industrial legislation relating to health, safety and welfare can be traced.

Since this Act there have been a number of other Factories Acts, each of which has sought to improve the conditions of the worker in respect of health and safety, the last and probably the most far reaching of these being the Health and Safety at Work etc. Act 1974. Under the terms of this Act severe penalties can be imposed by the Courts if the various requirements are not observed. One of the main features is to make employees, as well as employers, jointly responsible for safety at places of work and for the protection of the public.

It is mandatory for companies employing five or more employees to produce a written statement setting out their safety policy, the arrangements they make to put it into effect and make all employees aware of it. Obviously there will be a wide variation in these policies and the risks and dangers vary with the type of industry and the nature of materials used. The important thing is that every employee becomes conversant with the content of such documents.

General regulations relating to health and safety are also set out in the Factories Acts, but due to the nature of building and its hazards, special regulations are made for those employed on construction sites, these being entitled, The Construction (Health and Welfare) Regulations 1966. It is outside the range of this book to cover in depth all the aspects concerned with safety and health legislation, and the reader is advised to contact the organisations mentioned at the end of the chapter in order to become familiar with the detailed implications of this important subject.

Accidents are nearly always caused by

carelessness or lack of thought, failing to take proper precautions, ignoring instructions and failure to use safety equipment provided. In many cases accidents are caused by taking risks to get the job done quickly; such thinking is fatal. It should also be borne in mind that in many cases one's own actions may not only result in injury to oneself but also serious or even fatal injury to one's own workmates. One of the surest ways of reducing accidents is to plan and think ahead of any dangers there might be in carrying out specific tasks. A responsible person at work should always be on the look-out for possible hazards and develop a safety conscience.

It must be remembered that safety is the responsibility of employers and employees. It is the responsibility of employers to provide safety equipment, and to ensure that it is in good order and is well maintained. It is the responsibility of the employee to make sure he or she uses the equipment provided and to see that it is properly cared for whilst in use. To ignore the foregoing is in fact breaking the law.

Control of Substances Hazardous to Health (COSHH)

These regulations came into effect in October 1989 and apply to all potentially dangerous substances where they are produced or used. They also apply to the production of fumes or dust in a work environment that may not in themselves be dangerous, but when workers are exposed to these substances over long periods, they may constitute a health hazard. There are some exceptions to these regulations where very stringent legislation is already in existence, typical examples being The Control of Lead at Work Regulations 1980 and The Control of Asbestos at Work Regulations 1987.

Under the COSHH Regulations employers are required to carry out on a regular basis an assessment of the workplace, environment, and the materials used or produced in order to identify possible hazards to employees. Where hazards are identified measures must be taken either to prevent or minimise the risk of exposure to employees. This may take the form of providing regular health checks, special safety equipment and clothing. It also means, for example, that all such equipment and machinery must be in good working order and regularly maintained. Records of all such maintenance must also be kept. Where health checks are required the employer must ensure that they are carried out. He must also make sure sufficient information, instruction and, where necessary, training relating to health hazards is available to employees so that they are fully aware of any risks to their health.

General requirements on site

To ensure that the correct precautions are taken and procedures adopted, the Health and Safety at Work etc. Act 1974 requires that where twenty men or more are employed a safety officer must be appointed. For such small numbers, this is not necessarily a full-time position, but sufficient time must be made available to the officer to enable him to perform his duties. Among his many responsibilities is the periodic inspection of the scaffolding, plant and stores to ensure there are no obvious hazards. He is also responsible for recording any accident, and where necessary, reporting the matter to HM Factory Inspectorate. (This latter procedure is required where the accident involves a death or an injury which requires such treatment as will prevent the person returning to work for three days or more.)

On all sites when five or more men are employed a first-aid box should be provided by the employer, and this should contain sufficient and suitable equipment for the treatment of minor injuries. It should be available to all personnel and be used for first-aid requirements only, not, as sometimes is the case, a container or depository for all sorts of odds and ends from timesheets to sandwiches. On larger sites where more than fifty men are employed a qualified first-aider is required, and in many instances the safety officer undertakes this role.

The Construction (Health and Welfare) Regulations also specify minimum first-aid facilities, but it is only where more than fifty workers are employed on a contract that there is any requirement to appoint a qualified first-aider.

If serious injury occurs and medical aid is required urgently on site, the recommended course of action is to summon a doctor, ambulance or other medical assistance. Unfortunately, mechanical services tradesmen frequently work alone on construction sites when no other personnel are present or where means of telephone communication is not available. This will create difficulty in summoning assistance and could place the person requiring assistance in danger. It is therefore essential for all site workers to be familiar with basic first-aid procedures, as by prompt attention the danger of death or permanent injury can be greatly reduced. For simple and easy to read details of first-aid treatment the publications listed at the end of the chapter are recommended.

Remember that first-aid treatment is only preliminary help given to the patient prior to assistance being available from trained medical personnel.

As it is essential for the craftsman to work safely it is necessary to consider some of the more common hazards that face site workers. Every year many men are killed and injured in accidents involving the use of scaffolds or ladders and the vast majority of these accidents could have been prevented by common sense and the application of a few simple precautions.

Scaffolding

While the erection of scaffolding is regarded as a specialist activity to be undertaken only by trained operatives, it is important for anyone using a scaffold to be able to recognise its type and whether or not it is safe to work on.

Scaffolds are basically a temporary working platform and those used for building purposes are of two main types:

(a) Scaffolds used in the construction of a new building where one side of the scaffold is supported by putlogs resting on the building fabric as the building is constructed. It is tied to the building through windows and doorways and is sometimes referred to as putlog scaffolding.
(b) Independent scaffolds are those which are more commonly used on existing buildings for maintenance or refurbishment work, and can be used without actually being attached to the structure. Obviously some method of temporary tie to the structure will be necessary on any but low-rise buildings. Figures 1.1(a) and (b) show sections of both types of scaffold.

Some of the main safety features about scaffolding are as follows (see Fig. 1.2)

(a) All scaffolds must have a secure base as shown in Fig. 1.2(a) with a steel base plate to spread the weight of the scaffold and prevent the tube ends sinking into the ground.
(b) Scaffolds should be securely tied to the buildings as shown in Fig. 1.2(b), this being intended to prevent the scaffold falling away from the building.
(c) The cross brace shown in Fig. 1.2(c) forms a triangle (a recognised safe form of structure) and prevents the scaffold collapsing sideways. Note also the braces shown in Fig. 1.1 which serve the same purpose.
(d) The main requirements regarding working platforms and toe boards are shown in Fig. 1.3(a), the toe boards being necessary to prevent tools, materials or equipment sliding off the edge of the working platform. The guard rail and toe board are to prevent operatives stepping or falling off the working platform.
(e) It is necessary to space the putlogs carefully to avoid creating a 'trap', or an excessive overhang of the walkboards, as shown at B in Fig. 1.3(b). The permissible distance that a board may overhang a putlog is 4 × its thickness. For example, if boards of 40 mm are used they may overhang the putlog by not more than 4 × 40 = 160 mm. The distance between putlogs shown at A on Fig. 1.3(b) depends on the thickness of the board. Table 1.1 gives some examples.

Figure 1.3(c) shows the recommended way to join walk boards.

Table 1.1 Distance between putlogs.

Thickness of plank	*Distance between putlogs*
32 mm	1 m
40 mm	1.5 m
51 mm	2.6 m

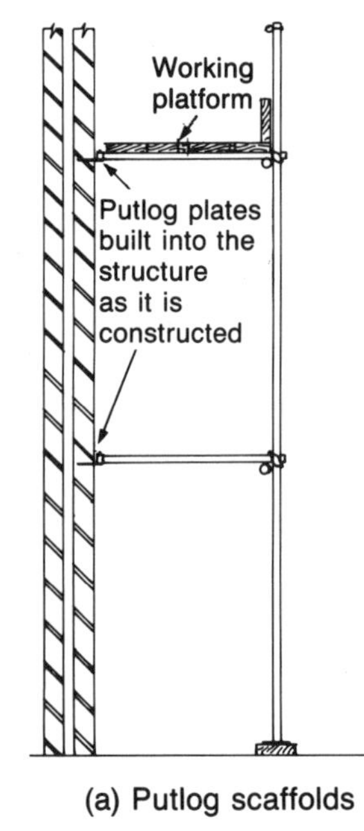

(a) Putlog scaffolds

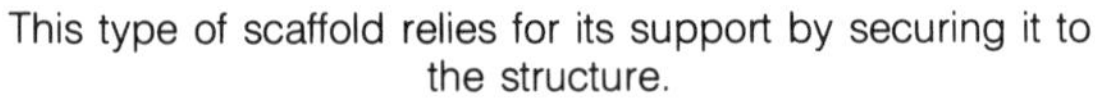
This type of scaffold relies for its support by securing it to the structure.

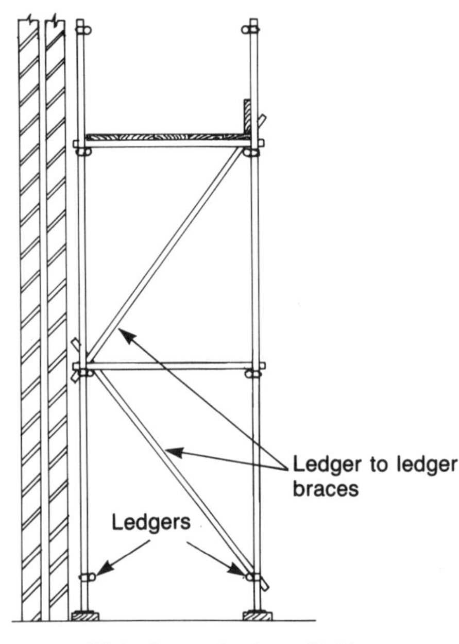

(b) Independent scaffolds

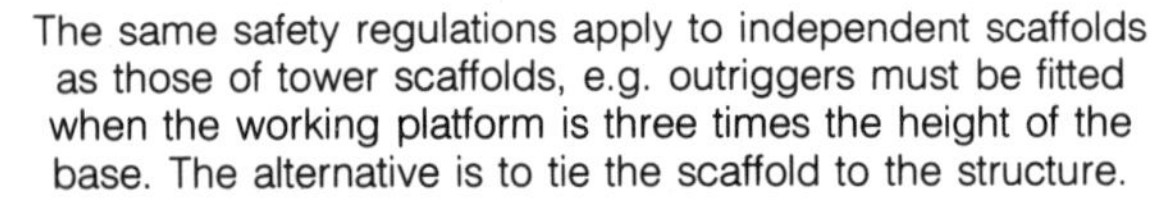
The same safety regulations apply to independent scaffolds as those of tower scaffolds, e.g. outriggers must be fitted when the working platform is three times the height of the base. The alternative is to tie the scaffold to the structure.

Fig. 1.1 Scaffolds

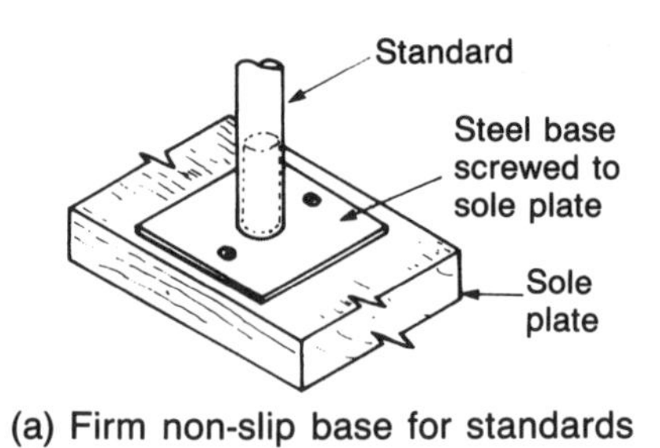

(a) Firm non-slip base for standards

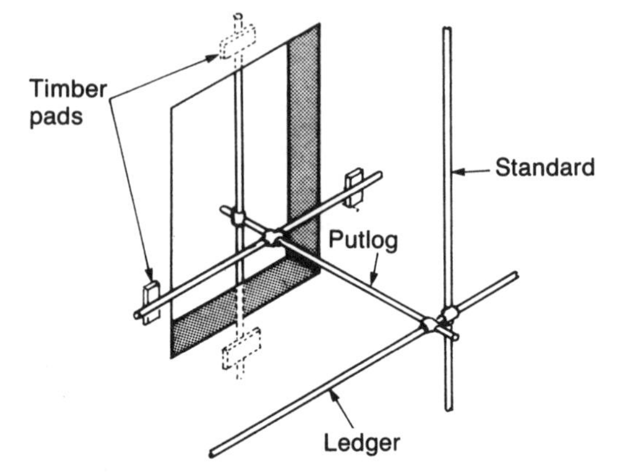

(b) Method of tying scaffolds to a building through window or door openings

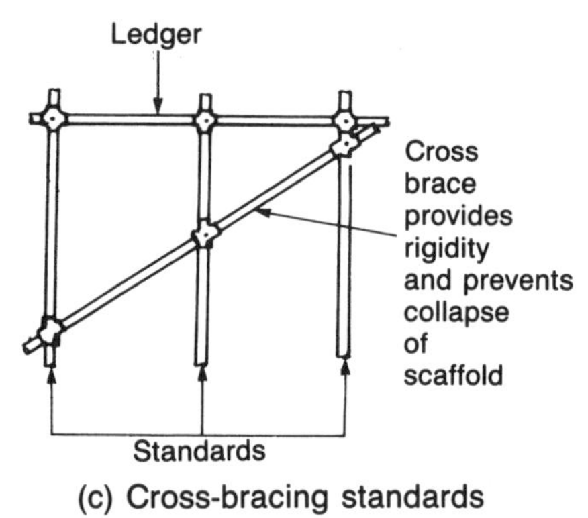

(c) Cross-bracing standards

Fig. 1.2 Securing scaffolding

The same general requirements also apply to mobile scaffolds as illustrated in Fig. 1.4, but there are also additional regulations which include the following.

(a) They should be used only on level surfaces.
(b) They should be moved only by pushing at the base.
(c) Care must be taken not to foul any overhead cables or wires when the scaffold is in motion.

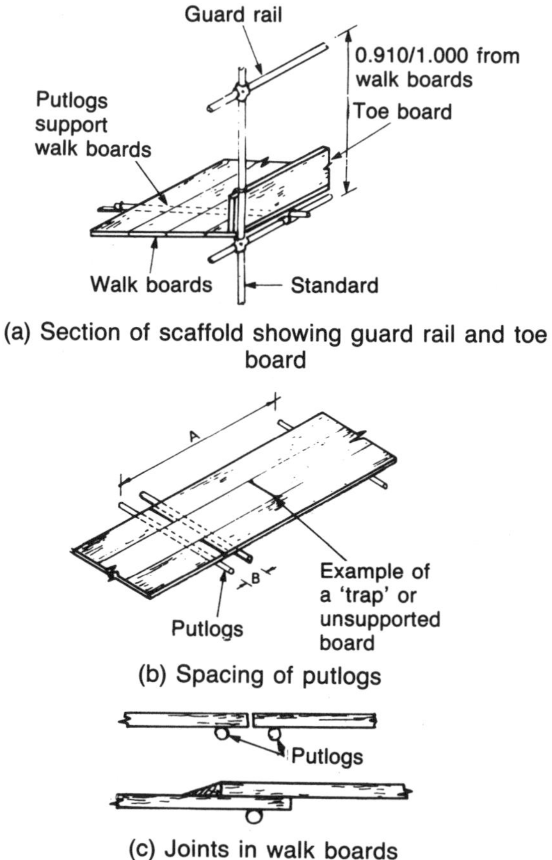

Fig. 1.3 Details of working platforms

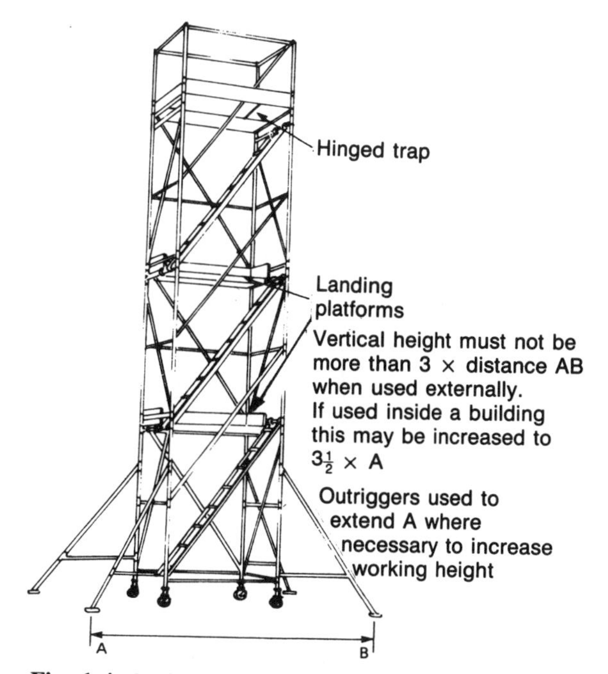

Fig. 1.4 Mobile tower scaffold

The legal provisions relating to scaffolds and the means of access to them are contained in detail in the Construction Regulations 1966.

Trestle scaffolds
Figure 1.5 illustrates two typical scaffolds of this type, both of which are useful for relatively low level work. It is essential to use them within the guidelines shown, as they are entirely free standing and do not have the rigidity of a proper scaffold. Check the base if there is any sign of 'wobble' and if necessary a length of cord attaching the trestle to a convenient fixing will doubly ensure its safety.

Ladders
Ladders should be protected by the application of several coats of clear varnish, paint must not be used as it will cover any defects such as cracks that may be present. They should never be stored in such a way as to cause permanent bowing or twisting, the best method being to stand them upright or place them in racks which provide full support for the stiles. The best ladders have a steel tie rod under each rung, these rods having two functions:

(a) Should the rung break the rod will still support the user as a temporary measure and prevent a fall.
(b) They ensure the stiles do not fall away from the rungs.

The recommendations relating to the slope of ladders are illustrated in Fig. 1.5(a) and must be observed to ensure safety.

When in use, ladders should be well secured by lashing the top to a ledger on a scaffold or some equally good fixing, and using one of the two methods shown to ensure the bottom of the ladder does not slip.

Typical examples of the misuse of ladders are shown in Fig. 1.6(c) and (d). Detail (c) shows a

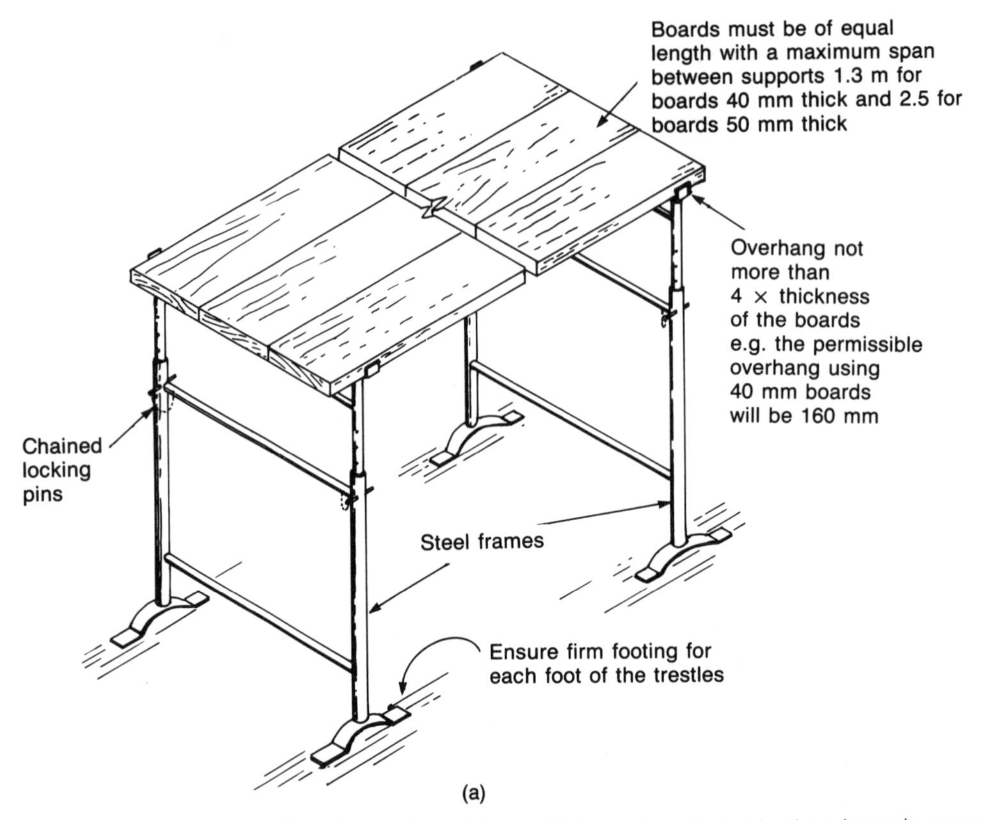

(a)

This type of trestle gives a good working platform the height of which can be adjusted by the telescopic supports which are pinned through the main trestle.

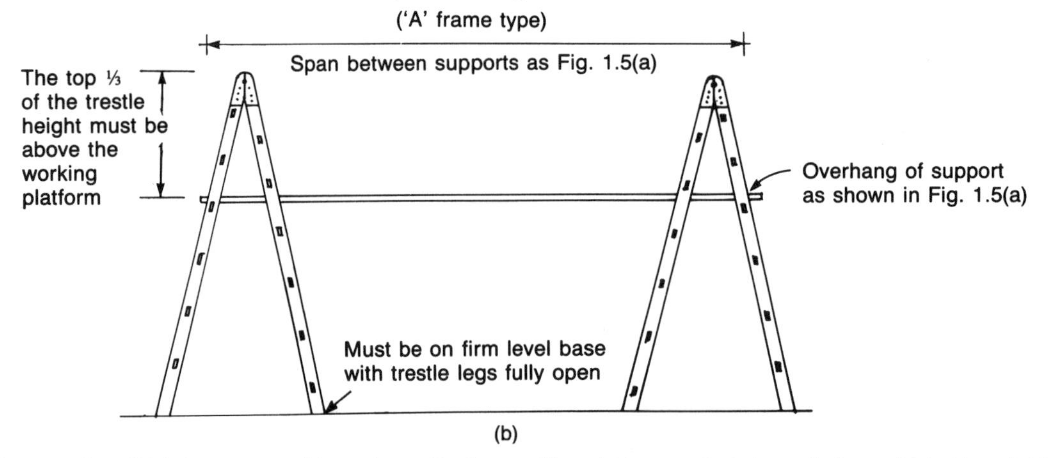

(b)

This type of scaffold does not provide such an effective working platform as that shown in figure (a) but is suitable for access.

General:

1. Trestles more than 3.6 m high must be tied to the structure, they must not be used where a person could fall more than 4.5 m. It is not permissible to use trestles on a platform scaffold unless it is firmly secured to avoid collapse.
2. Maximum height of working platforms is 2 m unless toe boards and guard rails are provided and access is by means of a separate ladder.

Fig. 1.5 Trestle scaffolds

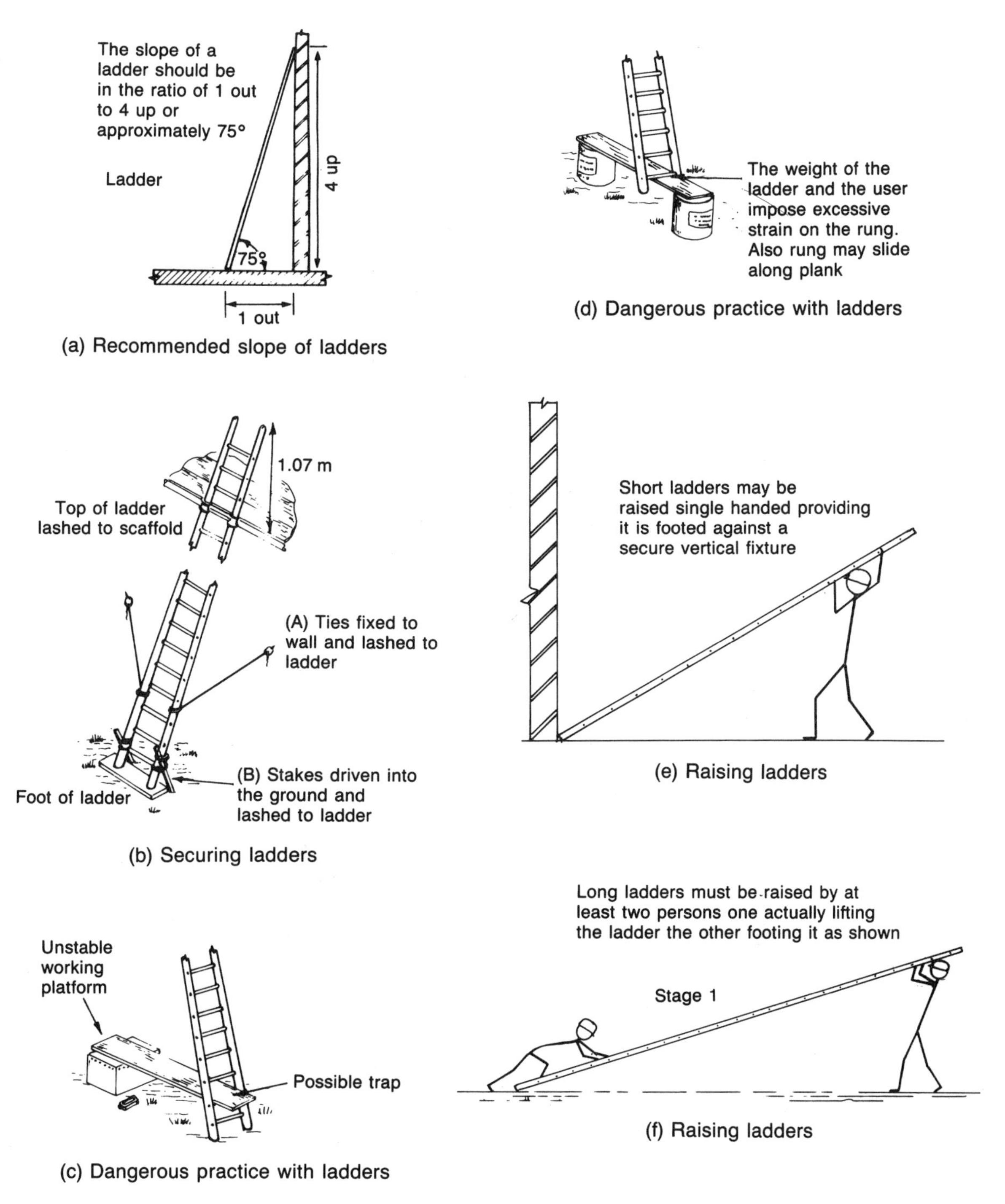

Fig. 1.6 Safety when using ladders

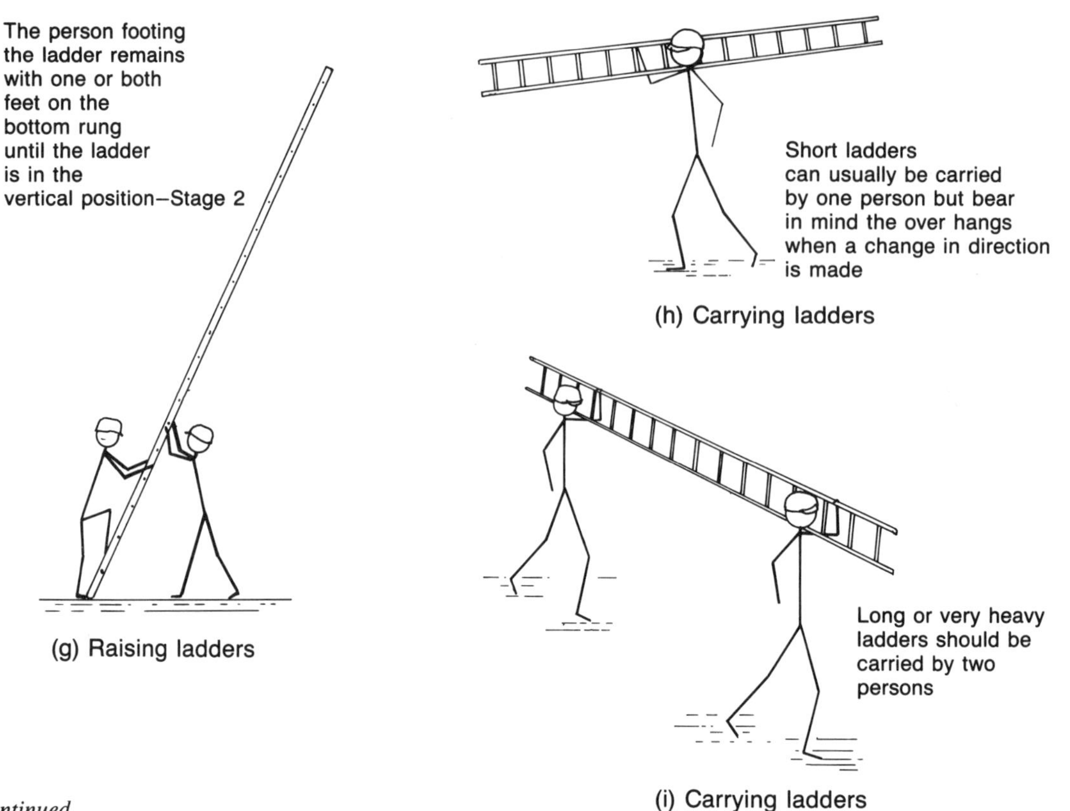

(g) Raising ladders

(h) Carrying ladders

(i) Carrying ladders

Fig. 1.6 *continued*

board forming a working platform, resting on a rung of the ladder. Quite apart from being generally dangerous this will subject the rung to undue strain. Detail (d) shows a ladder that is too short for the job and the dangerous way in which an attempt has been made to extend it. Do *not* use two short ladders tied together *always* use a ladder long enough for the job.

Handling ladders

Quite apart from making ladders secure when in use, great care must be taken when moving them about and when raising them ready for use. Some very serious injuries are on record due to accidents involving the raising or lowering of ladders. Short, light ladders can, with care, be handled by one person, but long, heavy ladders will require at least two persons, one of whom must always be at the foot of the ladder, having one or both feet on the bottom rung to prevent it sliding when raised.

Before raising a ladder always look out for any overhead obstructions – they can be a trap for the unwary. Overhead cables are a common hazard, especially on existing buildings. When carrying ladders from one place to another, never be tempted to put the arm through the rungs to make it easier to carry. It might be a little more comfortable this way but if one stumbles or trips over it could result in a broken arm! Figure 2.6(e)–(i) show the correct ways of carrying and raising ladders. Remember, too, the same procedures must also be carefully followed when lowering ladders.

Excavations

This term relates to trenches; pits or holes below ground level. Regulations in respect of excavations require certain precautions to be taken intended to prevent persons falling in, and also to ensure that effective support is provided to guard against

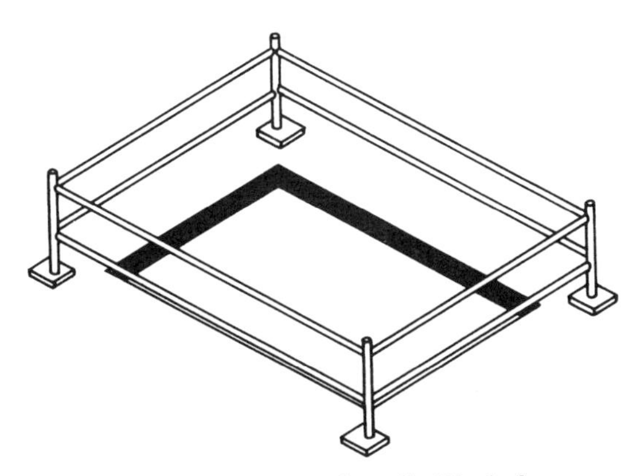

Fig. 1.7 Guard rail round stair well, lift shaft or excavation

collapse of the walls. Fencing must be erected around any excavation into which persons could fall more than two metres, but clearly it is common sense as well as legally compulsory under the Health and Safety Act for a place of work to be made safe for both workers and the general public. Similar fencing is also necessary for lift shafts or stair wells as shown in Fig. 1.7.

Excavated material should be deposited at a sufficient distance back from the edges of a trench or hole to prevent excessive mass causing the sides to collapse. Suitable materials such as stout timbers or steel plates should be used to shore up the inside of the excavation, the extent to which this is necessary depending upon the nature of the soil and the depth of the hole or trench. Extensive and deep excavations must be inspected daily and thoroughly examined every seven days by a competent person, as a certain amount of experience is required to detect the first indication of any danger.

Electric power tools and temporary lighting

It is sound advice to read and thoroughly understand before use the instructions provided with power tools, especially those tools with which the user is not familiar. The following six points however, are a general guide for safe use.

(a) The supply voltage should always be within the range for which the tool is designed.

(b) For work on building sites voltages of 110 should be used as this is far safer than the mains voltage of 240, a shock from the latter often proving fatal. A portable step-down transformer can be used to reduce the voltage from 240 to 110 volts, care being exercised in handling the transformer as these are fragile and easily damaged. As an alternative, or if no mains supply of electricity is available, a small portable generating plant may be used.

(c) Portable power tools are perfectly safe providing they are treated with care and used only for work for which they have been designed. It is important that they are serviced regularly by a competent electrician.

(d) All electric power tools must be earthed unless they are double insulated. Double insulated equipment is indicated by the mark shown in Fig. 1.8.

(e) Flexible electrical cables used for lighting or tools should never be allowed to trail on the floor where they are liable to be damaged; instead they should be tied or suspended at high level out of harm's way. Only tough rubber or PVC cables should be used, and all temporary connections should be both waterproof and unbreakable.

(f) Portable electric hand lamps must be adequately earthed and should not be used with a socket plugged into a light pendant which has only live and neutral connections. These lamps should have an electrically insulated handle and be fitted with a wire guard around the bulb. Most of the lamps are fitted with a clamp to enable them to be secured to a suitable point thus leaving both hands free.

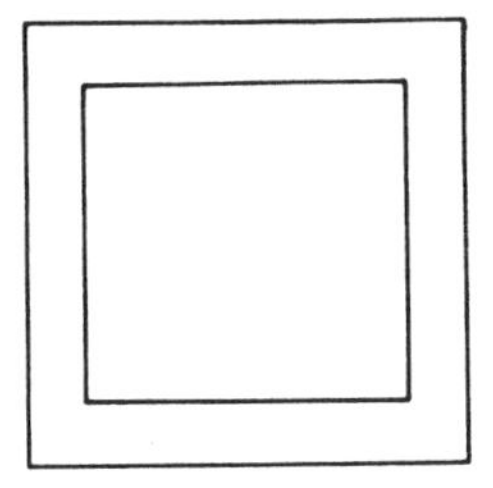

Fig. 1.8 Double insulation mark on power tools

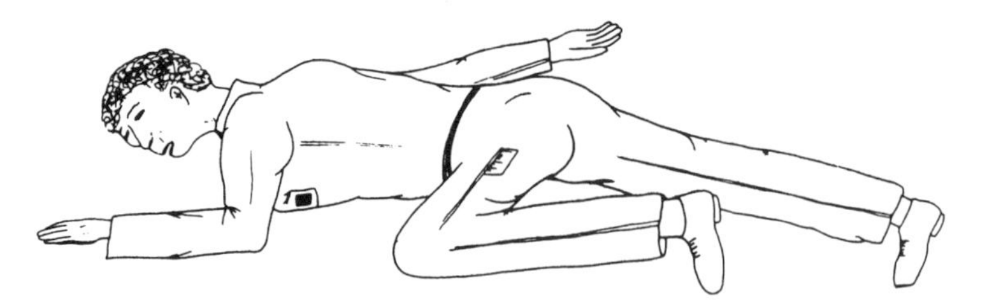

Fig. 1.9 Recovery position. After resuscitation from a severe electric shock the patient should be placed in the recovery position and kept warm until examined by a doctor or taken to hospital.

Persons suffering electric shock

An electric shock can be dangerous and in some cases can cause death if immediate action is not taken. The casualty must be removed from the power source and if the heart has stopped resuscitation procedures must be applied. The severity of shock depends on the voltage and length of time the casualty has been in contact with the current. The effects of electric shock will be severe pain due to muscle contraction, possible minor burns at the point of contact and the casualty may be thrown off his or her feet causing further injuries. The muscle contraction may prevent the casualty from releasing contact with the conductor. The first step to be taken when this occurs is to switch the current off, or if this is not possible, pull or push the casualty clear using a non-conductive material such as a piece of timber or a broom handle. Use anything that is handy providing it is not made of metal and is dry. At the same time call for help and if possible get someone to telephone for an ambulance as it is important that the casualty is treated as quickly as possible by a qualified person. It is recommended that a study is made of the St John's Ambulance Association's first-aid manual relating to treatment for electric shock and the following is given for general guidance only. If the casualty is breathing but unconscious, loosen the clothing about the neck and waist and lay the casualty in the recovery position (see Fig. 1.9), this prevents the casualty from inhaling fluid or vomit. Keep the casualty warm and dry and constantly check on breathing and pulse rate until help arrives. If the casualty is not breathing rapid action must be taken by administering mouth-to-mouth resuscitation, or if this is not possible seal the casualty's lips with your thumb and seal your lips about the casualty's nostrils to force breath into the casualty. Watch the chest, it should rise as the lungs are filled with air. When breathing is restored, the casualty should be placed in the recovery position.

If the heart has stopped beating this is usually indicated by the lips turning blue, the pupils of the eyes being dilated and absence of the carotid pulse in the neck. The following procedure should then be adopted: lay casualty on his/her back and kneel alongside. Locate the lower part of the breastbone. Place the heel of one hand on the centre of the lower part of the breastbone and cover with your other hand, locking the fingers to keep them free of the rib cage. Keeping the arms straight press sharply down on the base of the breast bone then release the pressure. This should be repeated fifteen times at the rate of 80 compressions per minute. The pulse should then be checked and two more breaths of mouth-to-mouth resuscitation given. Repeat this process until the heart beat returns and continue the mouth-to-mouth resuscitation until breathing is normal. Only when heart beats and breathing are normal should the casualty be placed in the recovery position. Never leave a patient who has suffered an electric shock, or indeed any other type of accident for that matter, until expert help has arrived.

It should be mentioned here that some have expressed concern relating to the possibility of being exposed to contagious diseases when performing mouth to mouth resuscitation. There are commercially available plastic shields which can be used to protect anyone against this possible risk. They are cheap, fold up easily and can fit comfortably into a first-aid box. They can be obtained from Laerdal Medical Ltd, Laerdal House, Goodmead Road, Orpington, Kent BR6 0HX. Tel.: 0689 876634.

Grinding tools

While the plumber generally does not use machine tools, it is quite common to find grinding equipment provided in plumbing workshops for the sharpening of cutting tools, and if these are misused they can be very dangerous. The worst accident that can occur with grinding equipment is probably the explosion or shattering of the abrasive wheel, with disastrous consequences to anyone in the vicinity. Wheel explosion is usually due to incorrect fitting of the wheel, which should only be changed by a trained and qualified person.

To ensure safety in the use of grinders, the following observations may be helpful to those with limited experience of these tools, but it must be stressed again that the fitting or changing of wheels should only be done by a person properly trained to the requirements of the standards of the Abrasive Wheel Regulations.

The wheel should be fitted on the spindle with two recessed metal discs, which take a bearing against two cardboard washers. A check is made before using the grinder to ensure these cardboard washers are in place, and the wheel is secure on the spindle (see Fig. 1.10(a)).

The wheel must be provided with a suitable guard, but some variation of the amount of the exposure of the wheel is permissible, depending upon its use. The wheel exposure shown in Fig. 1.10(b) is suitable for general-purpose work. The face of the wheel should be at right angles to its sides as shown in Fig. 1.10(a) and the tool rest should always be adjusted so that its front edge is not more than 3 mm away from the face of the wheel. The sides of the wheel should not be used for grinding as stresses may be set up causing it to explode. When the face of the wheel becomes worn it should be redressed (to remove any irregularities) by a trained operative. Do not forget that dressing will slightly reduce the diameter of the wheel and it will be necessary to re-adjust the tool rest to the correct gap.

Eye protection is essential when grinding is being carried out and some machines are fitted with a clear shatter-proof glass window which must always be in position when the machine is in use, as in Fig. 1.10(c). Safety goggles of the correct specification must also be worn, whether or not the machine is equipped with a safety window.

The following are some common faults found with grinding wheels:

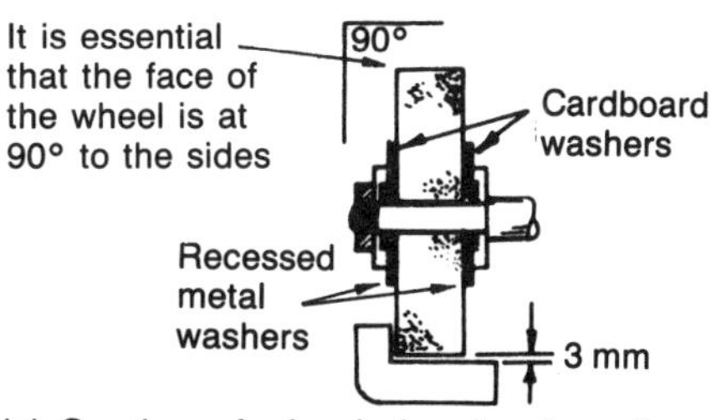

(a) Section of wheel showing how the wheel is fitted on the shaft and tolerances for tool rest

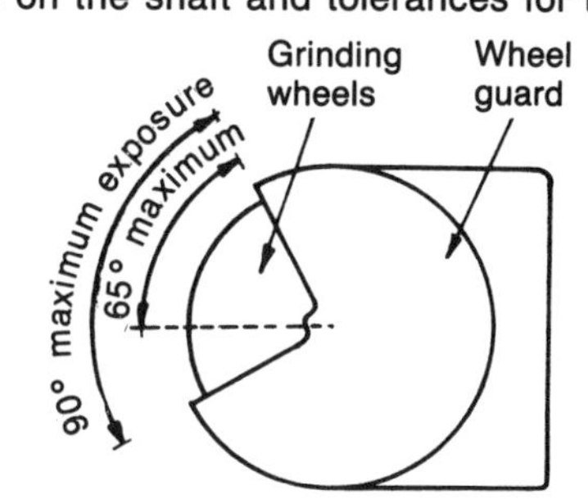

(b) Exposure of wheel for normal operation

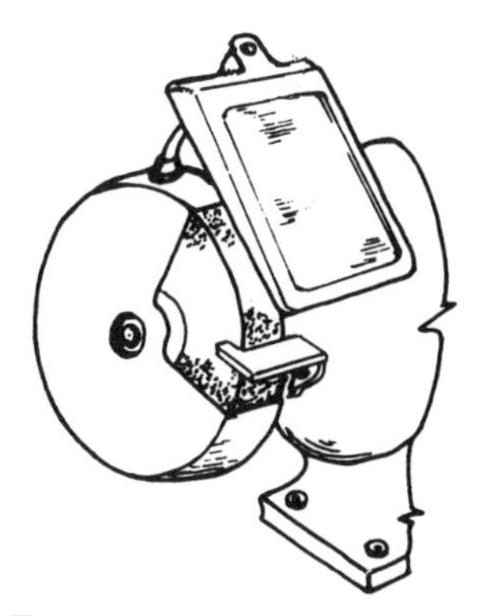

(c) Eye protection on grinding wheels

Shatter-proof glass window must be in position and goggles worn by the operator when wheel is in use

Fig. 1.10 Grinding wheels

(a) Tool rest is incorrectly adjusted, i.e. too far from the wheel or at the wrong height.
(b) Face of tool is grooved and requires redressing.
(c) Absence of adequate eye protection.

Do not use the machine if any of these apply. Also ensure the operator can stand comfortably on a clean dry floor.

Safety with hand tools

Most building craftsmen do a great deal of their work with hand tools and many quite serious accidents are caused when these are defective or badly maintained. It is not only the user who is endangered, but also his workmates who may be in the vicinity. The use of unsafe tools is an offence under the Health and Safety at Work Act.

A common case of using an unsafe tool is an improperly fixed hammer head flying off its handle. Some of the more common dangers which occur with hand tools are shown in the following illustrations.

Figure 1.11(a) shows what a plumber's hammer should *not* look like. Damaged handles should always be promptly replaced, not tied up with string or wire. The head should be securely wedged to the handle using both wooden and barbed metal wedges, as in Fig. 1.11(b). Nails are not suitable as wedges.

When using a spanner, make sure it fits the nut. *Never* use one that 'nearly' fits. Discard old spanners (Fig. 1.12) and pipe wrenches which may slip in use and result in severe hand or wrist injuries.

Cold chisels, caulking irons, punches and socket-forming tools are all used with a hammer and continuous use causes the top of these tools to turn over, commonly called 'mushrooming' as illustrated in Fig. 1.13(a). The mushroomed parts often fly off when struck with a hammer and can cause severe eye injuries so keep the heads of all tools of this nature ground as shown in Fig. 1.13(b).

The edge of cutting tools should be kept sharp and protected when not in use. Never carry sharp tools in a pocket, and always work with the hands behind the cutting edge. Although a screwdriver is

(a) A hammer like this is a danger to both the user and his workmates

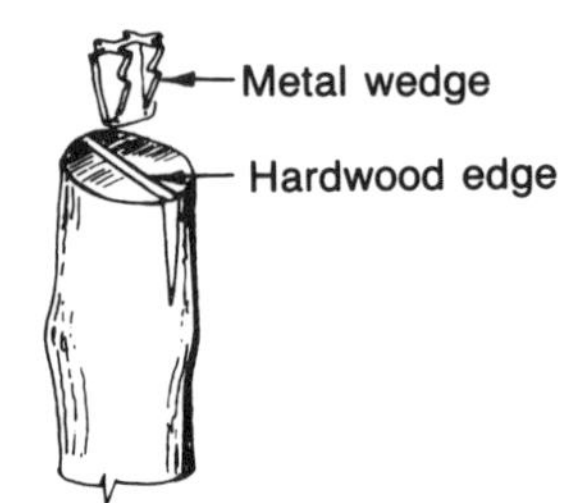

(b) Fixing hammer heads correctly with wooden and metal wedges

Fig. 1.11 Fixing hammer head

Fig. 1.12 Worn and opened spanners can slip resulting in hand injuries

(a) 'Mushroom' headed tools can result in loss of sight or serious eye injury.

(b) Correct form of grinding for the striking of chisels, points, caulking tools and socket formers

Fig. 1.13 Maintenance of tools used with a hammer

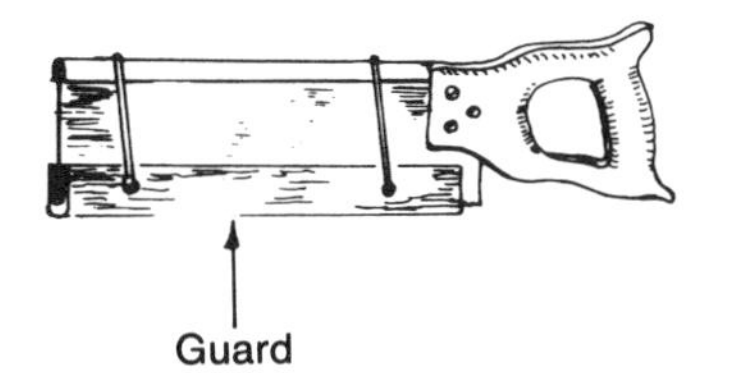

Fig. 1.14 Guards should be fitted to all sharp cutting tools when not in use.

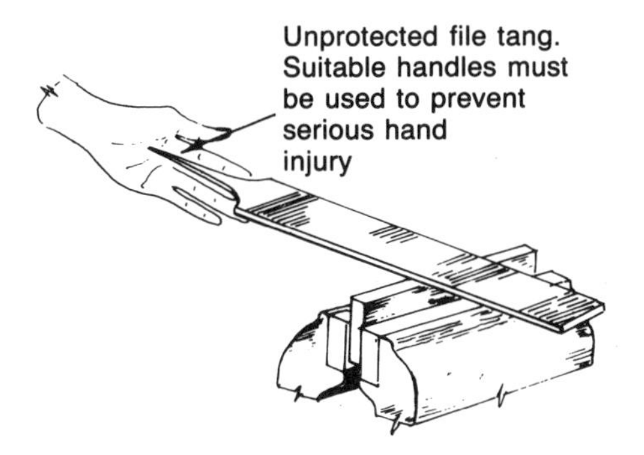

Fig. 1.15 Use of files

not normally classified as a cutting tool, it is unwise to use one on work held in the hand, use a suitable vice instead. A simple guard on a saw (Fig. 1.14) makes it safe when not in use and also protects the cutting edge.

The use of a file without a handle can be very dangerous as the sharp tang can pierce the hand as shown in Fig. 1.15. Make sure the work is securely held in the clamp or vice when filing operations are being carried out.

General site safety

One of the first rules of safety is tidyness. Building debris should never be left lying about. Nails protruding from pieces of timber as shown in Fig. 1.16 is one of the most common causes of foot injury; such nails should always be removed or hammered over. Pieces of pipe or tube can roll if trodden on and cause a fall which could be serious, especially if the person falling is carrying something such as a ladder or gas cylinder.

Remember also that while work is being carried out which does not necessarily constitute a serious danger to oneself, this may not be the case for

Fig. 1.16 One of the most common hazards on site. Never leave nails like this in timber.

people in the vicinity. Be extra careful when anyone else is working below, and in all cases display proper warning notices and cordon off possible danger areas.

Protective footwear should always be worn on site and in the workshop, boots or shoes which are strong and hard wearing, preferably with steel reinforced toe caps to protect the foot against the type of hazard shown in Fig. 1.17.

Eye injuries are painful and can result in loss of sight, so always wear the appropriate type of safety goggles or glasses when exposed to dust or flying chippings as in Fig. 1.18. Goggles are made with a variety of lenses of different impact resistance for various operations. The correct type must be worn or they may not provide the required protection.

Protection of the skin by using suitable gloves may be necessary in some cases when handling abrasive materials or those materials containing strong acids or alkalis such as cement or lime mortars. The use of barrier cream can prevent skin disease such as dermatitis, and also prevents the entry of grease and dirt into the pores of the skin.

Fig. 1.17 Protective footwear. Falling objects are extremely dangerous especially to the head and feet.

Fig. 1.18 Eye protection. Goggles are essential when cutting away brickwork.

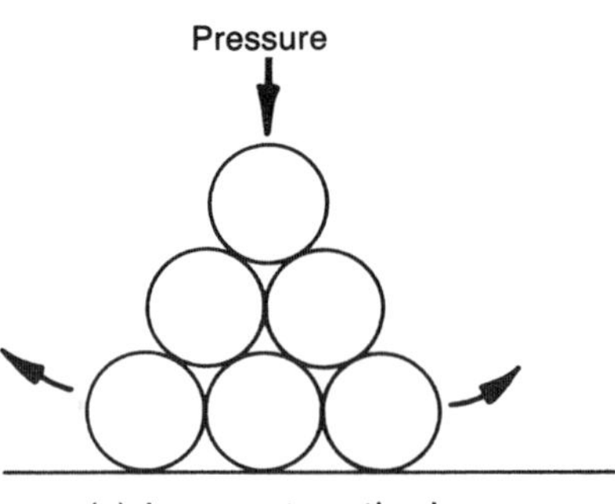

(a) Incorrect method

Pressure on the top of a stack of pipes or cylinders can force out those below. Foot or hand injury may be the result of incorrect stacking

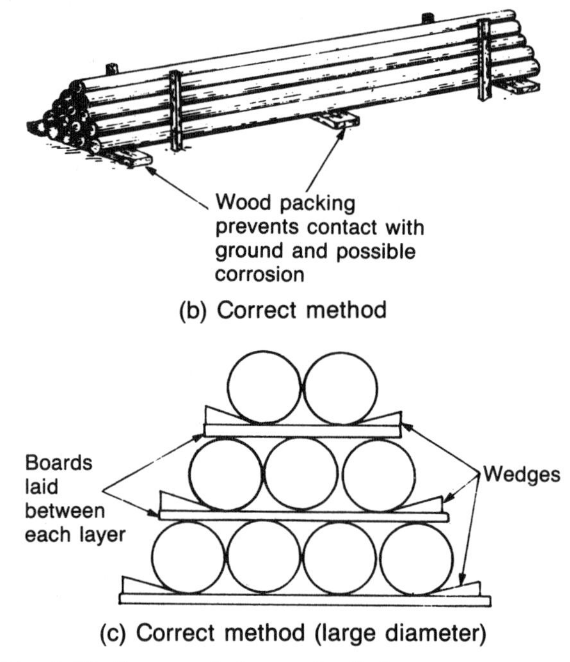

Fig. 1.19 Storage of pipes and cylinders

Hearing protection

In comparatively recent years it has been found that the hearing of a person may be seriously affected if he is subjected to persistent noise. Many power tools now in use on site such as cartridge firing tools, large hammer drills and percussion tools, generally have a very high noise level. Operatives using such tools or working in close proximity to them should protect their hearing by using suitable ear muffs.

Knee pads

Plumbers spend much of their working lives on their knees, often on cold, hard or damp surfaces, which apart from the discomfort, can be the cause of many ailments, such as rheumatism and a common complaint called 'housemaids knees'. Many of the older generation of plumbers fabricated a form of knee pad, using pieces of discarded carpet with leather straps sewn on to secure them. It is possible to purchase proper pads made of leather which give a good degree of flexibililty and are comfortable and safe to use.

Storage of pipes and cylinders

The incorrect storage of pipes or gas cylinders can be a possible site hazard. Figure 1.19(a) shows how stored circular members can roll outwards due to the pressure of those on top, and to prevent this occurring wedges or stakes should be used as shown in Fig. 1.19(b).

If the site is a large one and large quantities of pipe are held in store, a properly constructed tube rack is a better arrangement (Fig. 1.19(c)). Not only can differing sizes be separately stored, but protection from corrosion can be provided by keeping the tubes clear of the ground and covering the whole rack with waterproof sheeting.

Safety signs

Prominent notices should be displayed on site to warn of possible hazards; to indicate what protective clothing should be worn; and to provide general information. Some typical examples are

Fig. 1.20 Typical notices warning of potential hazards

shown in Fig. 1.20.

Safety signs fall into four main categories which can be recognised by their shape and colour. In some cases additional information is provided specifying the nature of the danger.

Prohibition signs

Shape	Circular
Colour	Red border and cross bar Black symbol on white background
Meaning	Shows what must *not* be done
Example	No smoking

Mandatory signs

Shape	Circular
Colour	White symbol on blue background
Meaning	Shows what *must* be done
Example	Wear hand protection

Warning signs

Shape	Triangular
Colour	Yellow background Black border and symbol
Meaning	Warns of hazard or danger
Example	Caution, risk of electric shock

Information signs

Shape	Square or oblong
Colour	White symbols on green background
Meaning	Indicates or gives information of safety provision
Example	First-aid point

Protective clothing

Construction work is usually associated with dirt, dust, dampness and mud, and additionally for the plumber oil and grease, therefore suitable overalls will protect the user's clothing against these hazards. Quite apart from this the use of protective clothing in cold weather conditions enables the wearer to keep warm. It is not easy to concentrate on a job if one is feeling cold and wet. In the event of plumbers handling lead, the use of overalls is essential so that at the end of the working day they can be removed and the possibility of contaminating public or personal vehicles, or indeed one's home, is minimised. It is recommended that clothes that have been in contact with lead should be washed separately to avoid contaminating other clothing. Protective clothing should be comfortable and such that one's movement is not constrained. They should be worn properly, not as is sometimes seen with boiler suits tied around the waist by the arms; this practice has been known to cause accidents.

Many employers sensibly insist on the use of safety helmets on their sites, as employees are less likely to sustain serious head injuries in the event of an accident. It is compulsory to wear a safety helmet for certain types of work which include the following:

1. Where overhead work is in progress.
2. On demolition sites.
3. When loading or unloading vehicles by means of mechanical equipment such as cranes or lifts.
4. In excavations more than 1.2 m deep.
5. When piling work is in progress.

It is important that the inner head band is adjusted correctly to suit the wearer as they can cause both discomfort in use and possibly render the helmet useless in the event of an accident. For identification purposes it should be clearly marked with the wearer's name.

Toxic and dangerous substances and materials

Plumbers frequently handle potentially dangerous substances and it has always been accepted in the trade that lead falls into this category. Sensible plumbers have always recognised the dangers of lead and taken the necessary precautions to avoid these, some of which are listed here.

Molten lead gives off fumes containing lead and operations involving lead welding or melting lead for caulked joints should always be carried out in a well-ventilated area. In extreme cases a suitable respirator should be worn. Protective clothing must be worn but should be removed for meal breaks and before leaving the site. Personal cleanliness is essential, and before commencing work a good barrier cream should be applied to the hands. Before a meal is consumed, and on completion of work, hands and finger nails must be thoroughly washed and scrubbed with a suitable nail brush.

Remember lead can be absorbed into the body via the mouth, through inhalation when it is vaporised, and in some cases through the skin.

Solvents and adhesives

Fumes from many of these substances are both toxic and addictive and great care should be taken to avoid inhaling them. In most cases they also constitute a very real fire hazard and great care must be taken when, for example, making solvent-welded joints on plastic pipes. When not in use all caps and stoppers on the containers of these materials should be replaced, not only for safety reasons, but also for economy. If stoppers are not replaced after use these materials will quickly become unsuitable for the purpose for which they were intended, due to the evaporation that takes place.

Asbestos

This material has been used for a long time in the construction industry for a wide variety of purposes. It has been found to be extremely dangerous in both fibre and powder form and the manufacture of asbestos and asbestos products has ceased. Its principal use in mechanical services was for insulation and in the form of asbestos cement products such as corrugated roofing sheet and flue pipes. Where work on asbestos cement products has to be carried out, possibly for maintenance purposes the principal danger is when cutting the material with a saw or drilling, thus producing asbestos dust. Providing this is done in a well-ventilated area and a suitable face mask is worn, there is minimum danger to the installer. Asbestos products are most dangerous when used for heat insulation and due to its hazardous nature it is, in many cases, being removed. This can only be done by specialist firms using the correct equipment and taking precautions to ensure no contamination of the surrounding area occurs. No one under the age of eighteen is permitted to undertake work of this nature and when such material is encountered, no attempt should be made to remove it. It is the subject of the Asbestos Regulations 1969.

Prevention of fire risks on site

Many materials used in the construction industry are highly flammable and care should be taken to ensure that fire risks are reduced to a minimum. One such risk is the blowlamp, an essential part of the plumber's tool kit. Most modern types of lamp use liquefied petroleum gas (LPG) as a fuel, and on large sites where a considerable quantity of this gas is stored, proper facilities are necessary. Notices indicating the storage of explosive gases and no-smoking signs should be prominently displayed both inside and outside the store.

The most common causes of fire when using LPG blowlamps are defective hoses, or loose connections from the hose to the cylinder or lamp. Both hoses and connections should be periodically checked for leakage, using a solution of soapy water or leak detection fluid.

Another fire hazard with which plumbers need to be concerned is the solvent used for making joints on synthetic plastic materials. Many of these solvents give off heavy flammable vapour while in use, and it is dangerous to smoke or use a flame in the vicinity where these joints are being made.

Fire caused by carelessness with the blowlamps can be dangerous and costly. Be especially careful when making soldered capillary joints in awkward corners, under suspended floors and in roof spaces. Figure 1.21(a) shows how convection currents can draw the flame from a blowlamp through a hole in the ceiling and start a fire in a sub-floor or roof space. Certain types of pipe insulation are very flammable and such insulation should be temporarily removed to give plenty of clearance for the flame when making soldered joints (see Fig. 1.21(b)).

Suitable fire-fighting equipment must always be available and ready for use, not only on building sites but also when small maintenance jobs are being carried out. Detailed information on fire control will be found in the Health and Safety at Work booklets obtainable from HMSO listed at the

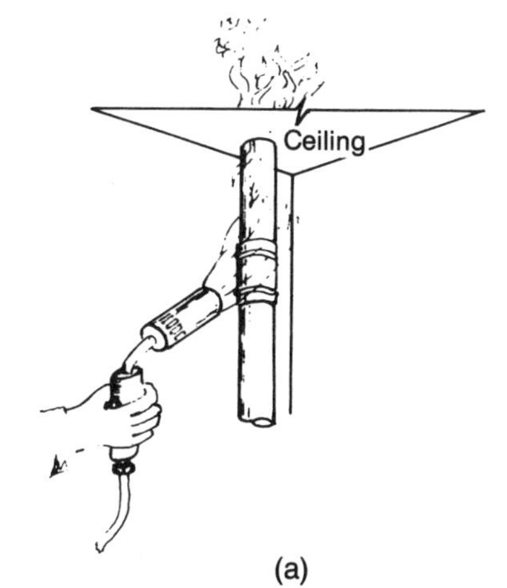

(a)
Heat from blow lamp can be drawn through a hole in the ceiling causing flammable materials to ignite

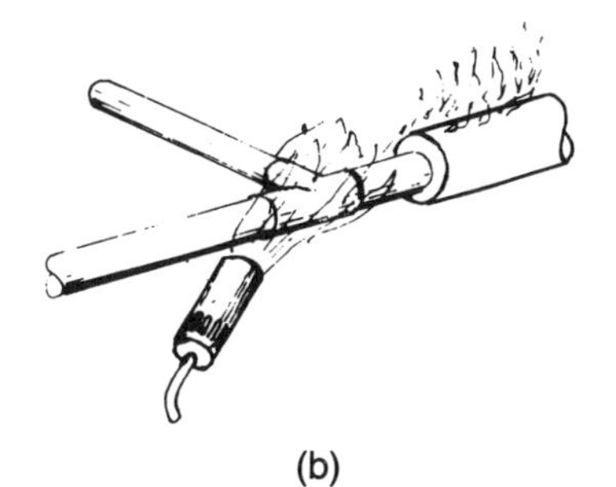

(b)
Make sure that any flammable insulation is removed before attempting to sweat a soldered joint

Fig. 1.21 Fire risks when using a blowlamp

Fig. 1.22 The three essentials of fire. For a fire to be maintained, heat, oxygen and fuel must all be present. Therefore to put out a fire one or more of these should be removed.

end of this chapter, but the following information will be helpful in extinguishing or controlling fires until the arrival of the fire brigade.

Broadly speaking there are three essentials to every fire and these can be shown as the three sides of a triangle as in Fig. 1.22. If any one part of the triangle is removed, combustion cannot continue. Fire extinguishing is based on this very simple principle:

(a) *Starving* – removing the fuel.
(b) *Smothering* – preventing the access of oxygen.
(c) *Cooling* – removing the source of heat.

Fire risks are classified in three categories as follows:

(a) *Class A risks* – mainly extinguished by cooling, usually with water.
(b) *Class B risks* – most efficiently extinguished by smothering and preventing oxygen joining with fuel.
(c) *Class C risks* – a non-conductive smothering agent must be used.

Table 1.2 gives details of extinguishing agents and types of fire for which they are suitable.

It is important to recognise that not all extinguishing agents are suitable for all types of fires. Using the wrong extinguisher can be dangerous. For example, to throw water on an electrical fire would give the fire-fighter an electric shock and in the case of a petrol or oil fire, water would easily spread the blaze. It is therefore important to know the various types of extinguishing agents and the types of fire for which they are suitable. It should be noted that BCF extinguishers are especially useful when dealing with expensive electronic equipment as it does little physical damage. It is, however, derived

Table 1.2 Extinguisher selection chart types and uses.

Class of fire ref. BS 4547 1970	*Extinguishing principles*	*A.B.C. all purpose powder*	*Metal powder*	*CO_2 gas*	*Foam*	*Water*	*B.C.F.*
Class A. Fires involving solid materials usually of an organic nature. Examples, wood, paper, textiles, etc.	Water cooling or exclusion of air	Yes Excellent	No	No	Yes	Yes Excellent	Yes
Class B. Fires involving liquids or liquefiable solids. Example, oil, fat, paint, etc.	Flame inhibiting or surface blanketing and cooling	Yes Excellent	No	Yes	Yes Excellent	No	Yes
Class C. Fire involving gases	Flame inhibiting	Yes	No	Yes	Yes	No	Yes
Class D. Fires involving metals magnesium, sodium, titanium, etc.	Exclusion of oxygen and cooling	No	Yes Excellent	No	No	No	No
Fires involving electrical hazards	Flame inhibiting	Yes	No	Yes Excellent	No	No	Yes

from the chemicals bromine, chlorine and fluorine and special care is necessary with its use as these substances give off dangerous fumes when these extinguishers are used.

Lifting and handling materials

Manual lifting and carrying

Many thousands of accidents caused in the handling and carrying of goods are reported annually to HM Inspectors of Factories. Building is among those industries having the highest accident rate and one of the major contributory factors can clearly be identified as the lack of training provided in the 'know-how' of lifting. Brute force and ignorance are of little use in lifting and can lead to strain or injury. It is more effective to 'think out' the lifting operation and then to rely on the skilful application of the right muscles. A load can be raised with less effort by using the strong leg and thigh muscles as opposed to those in the abdomen and back. This procedure is illustrated in Fig. 1.23.

The factors to be considered in the lifting operation are:

(a) maintain the back in a straight, though not necessarily vertical, position
(b) tuck the chin well in and do not drop the head forward or backward
(c) keep the arms straight and close to the body
(d) place feet about 200–300 mm apart (no wider than the hips) with one foot slightly forward and pointing in the direction of travel
(e) grip the object firmly with the whole hand (not just the finger tips) using gloves for protection where necessary
(f) bend the knees and lift by straightening the legs
(g) lift in easy stages – floor to knee – knee to carrying position
(h) ensure the load does not obscure vision
(i) move off in direction of advanced foot, keeping the load close to the body
(j) when setting down the object, reverse the lifting procedure

It is impossible to specify accurately the maximum load which can be lifted by an

(a) Correct

Legs are straightened to lift the load. Back straight

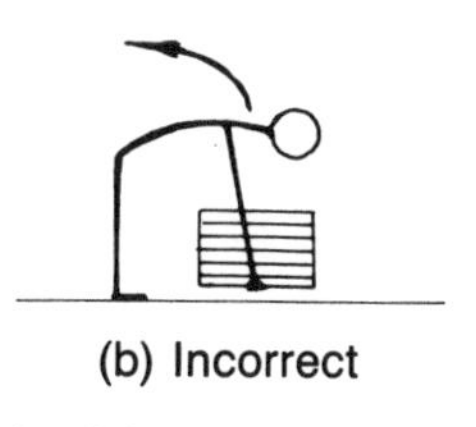

(b) Incorrect

Weight of the load taken by back. Leg muscles not being used

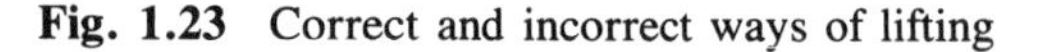

Fig. 1.23 Correct and incorrect ways of lifting

individual as this depends on the physical characteristics and the age of the person. Section 72(1) of the Factories Act 1961 stipulates that no person shall be required to lift, carry or move any load that is so heavy as to be likely to cause injury. If an object is too heavy, too large or of an awkward shape, do not attempt to lift it alone – seek assistance or use a mechanical lifting appliance.

A lot of lifting can be avoided by the old fashioned method of using rollers. Short pieces of steel pipe placed under a suitable support carrying the load will enable it to be moved very easily on flat hard surfaces. Use of rollers on sloping surfaces is, however, not recommended, as a load can get out of control and be a possible cause of accident.

Mechanical lifting and handling

The range of mechanical handling equipment on modern building sites is extensive and embraces everything from the giant tower crane to the simple lever. For normal site practice, the plumber is only required to use simple appliances such as the lever and pulley block. The term 'machine' is given to these appliances and denotes they are able to overcome a load at one position by the application of force at another.

Although there are two fundamental types of

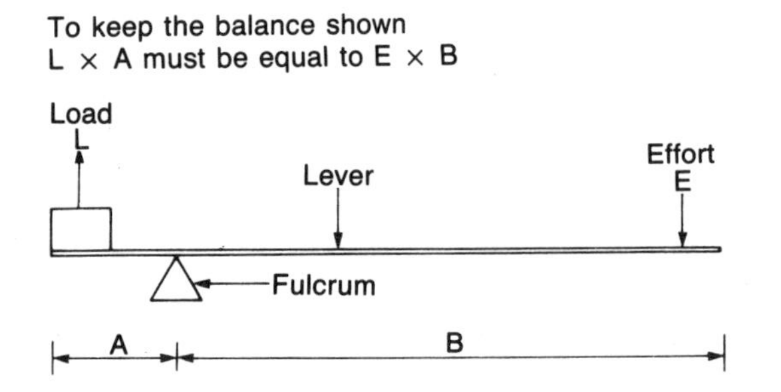

(a) The first order of levers

The longer the distance B, the less effort will be required to balance the load L. If L is increased then either B or E will have to be increased

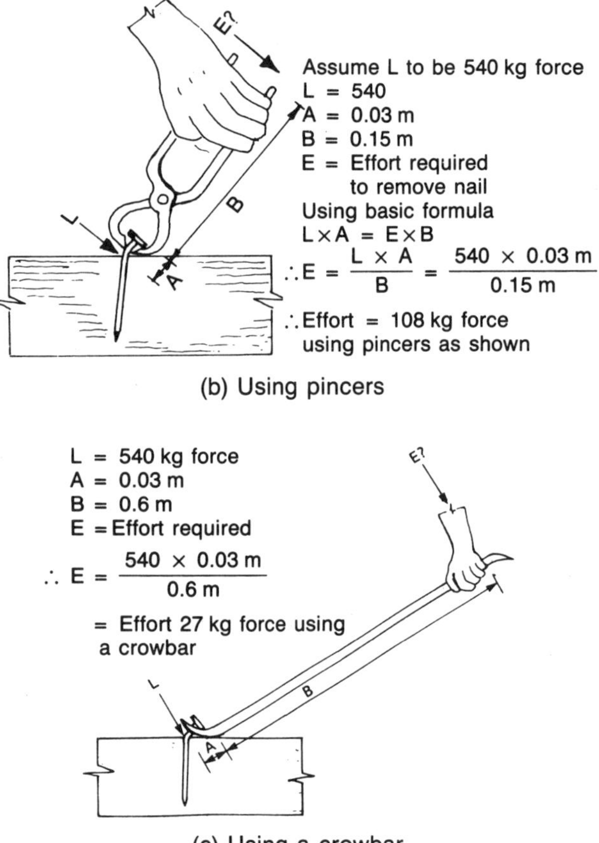

(b) Using pincers

(c) Using a crowbar

Fig. 1.24 Illustrating how levers reduce effort

simple machine, the *lever* and the *inclined plane*, only the former will be considered, as it has an everyday use on site.

The principle of leverage has a wide application,

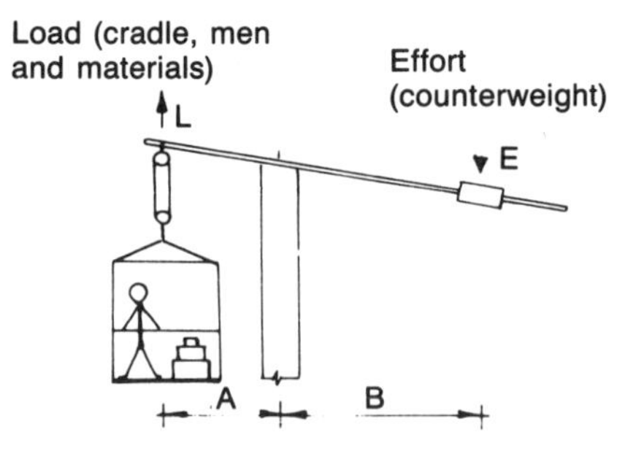

Fig. 1.25 Application of leverage to supporting suspended scaffold or cradle. Merely to keep cradle balanced then L × A = E × B, but for safety E × B should not be less than three times L × A.

the lever consisting essentially of a rigid bar which is pivoted in such a way as to allow a large load to be moved by a smaller effort. This is shown diagrammatically in Fig. 1.24 where it can be seen that to keep the lever at a state of rest, less effort E is needed in comparison with the load L because of their relative differences in distance from the pivot or point of balance called the fulcrum.

In each example L × A = E × B, where L is the symbol for the load. A practical example may serve to illustrate this point. The difficulty in removing a nail well driven home into a piece of timber is well known, and to remove it with a pair of pincers (Fig. 1.24(b)) may be difficult, if not impossible, due to the insufficiency of force afforded by the ratio L × A in relation to that of E × B. By using a crowbar or claw hammer, (Fig. 1.24(c)) the length of B will be increased, and the nail successfully withdrawn as shown.

A further example of leverage is shown in Fig. 1.25 where a cradle is used over the parapet of a building. If L × A was equal to E × B, the cage or cradle would remain stationary, so to lift it, E × B must be greater than L × A. As a safety measure, E × B must be three times greater than L × A.

The pulley, being a wheel, is a continuous lever. With the single fixed wheel pulley the only mechanical advantage is a change in the direction of the effort E to a more convenient position, i.e. pulling downwards is easier than lifting upwards. In order to lift a load using a pulley wheel, an effort of at least the same magnitude as the load must be applied. Such a machine is said to have a mechanical advantage (MA) of one, which can be expressed as a formula:

$$\text{Mechanical advantage} = \frac{\text{Load}}{\text{Effort}}$$

A gin wheel is an example of a single fixed wheel pulley and is illustrated in Fig. 1.26.

The single movable pulley shown in Fig. 1.27 will give a mechanical advantage of two. With this arrangement one end of the rope is fixed and the pulley itself is cradled in the rope. The load is suspended from the pulley block and the effort is used to pull downward on the free end of the rope. Because the load is suspended from two ropes each taking half the strain, the effort required is only 50 per cent of the load.

The common pulley tackle shown in Fig. 1.27 has only single wheels in each block, whereas Fig. 1.28 illustrates pulley blocks having more than one wheel, the number of wheels determining the mechanical advantage. It may appear at first sight that using a multiple wheel pulley or a lever is a method of obtaining something for nothing, i.e. a small effort to raise a greater load, but this

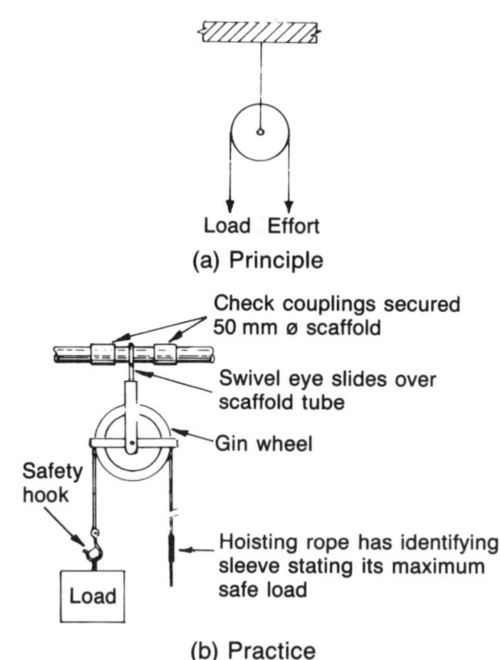

Fig. 1.26 Simple pulley or gin wheel

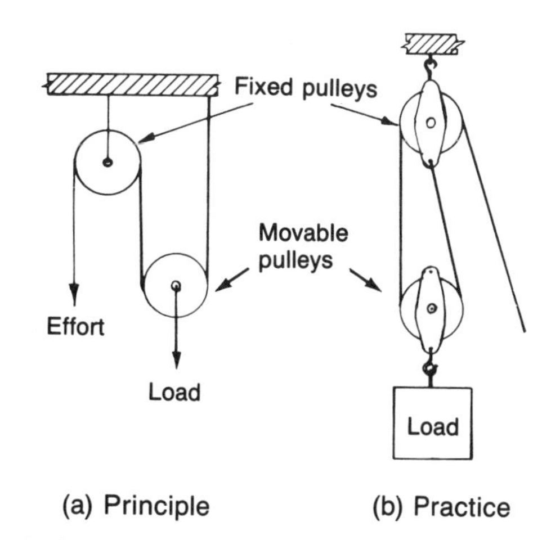

Sketch shows the working action of a single movable pulley.

Fig. 1.27 Double pulley hoists

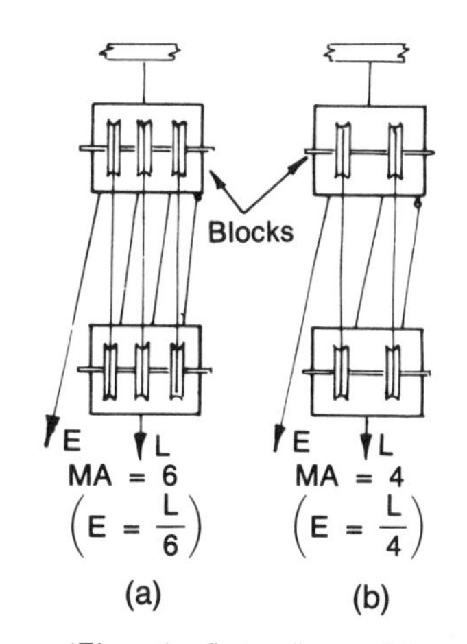

By pulling rope 'E' on the first pulley a distance of 1 m, the load will be raised by approximately 166 mm. In the case of pulley (b) a greater effort is required but the load will be raised by 250 mm

Fig. 1.28 Multi-pulley systems enable heavy loads to be lifted with little effort

conclusion is not correct. These machines cannot create additional energy and the work done by the effort is never less than the work done on the load. What in fact happens is that the load is moved through a smaller distance than the effort, i.e. by pulling the rope a distance of one metre, the load is lifted perhaps 100 mm.

With the arrangement shown in Fig. 1.27 the effort moves twice as far as the load, so the pulley is said to have a velocity ratio (VR) of two.

Velocity ratio can be expressed as follows:

$$\text{Velocity ratio} = \frac{\text{Distance moved by effort}}{\text{Distance moved by load}}$$

If a machine were 100 per cent efficient then the velocity ratio would be identical to the mechanical advantage. In practice, friction losses must be considered in addition to the mass of the pulley blocks, while the actual mechanical advantage requires the efficiency (%) of the machine to be taken into account.

Mechanical advantage can also be expressed as the formula:

$$\text{Mechanical advantage} = \text{\% Efficiency} \times \text{Velocity ratio}$$

Mechanical handling involving lifting is usually accompanied by means of a sling made of chain or rope of steel wire, fibre or composition construction. Reference to any catalogue of lifting appliances will indicate a wide range of slings made for specific lifting operations, some common types which may be used on the building site being shown in Fig. 1.29.

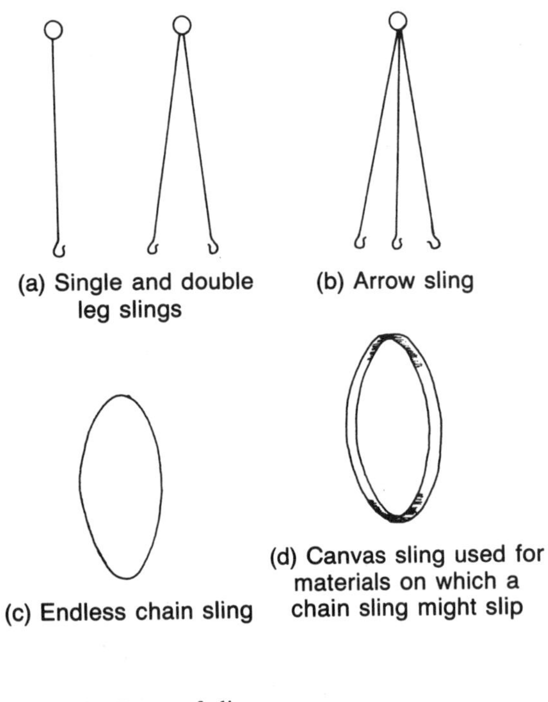

Fig. 1.29 Types of sling

Table 1.3 Two-legged slings at different angles.

Approx. angles only					
Angle of sling	0°	30°	60°	90°	120°
% of SWL (approx.)	100	96	86	69	50

Adequate strength for a specific lifting task is of vital importance. This can be found by referring to the Safe Working Load (SWL) that is either indicated on the sling or is recorded in the test report which must be provided for every lifting appliance used on site or in the workshop. An SWL of 1,500 kg means the sling should not be used for loads in excess of that figure.

Even if the lifting capacity of the sling is satisfactory, careless slinging can cause the SWL to be exceeded. The lifting capacity of any sling *decreases* as the angle between the legs *increases* – the effect can be seen from Table 1.3.

For example, a two legged sling with a SWL of 1 tonne (1,000 kg) used at an angle of 30° can carry a maximum load of

$$1{,}000 \times \frac{96}{100} = 960 \text{ kg}$$

Used at 90° the same sling will safely carry only

$$1{,}000 \times \frac{69}{100} = 690 \text{ kg.}$$

The following are typical examples of lifting operations encountered by the plumber which could be hazardous.

(a) Awkward objects such as boiler sections will swing about if the sling is 'off centre': lift slowly until load is off the ground.
(b) Compressed gas cylinders should not be lifted with chain slings. A fibre or composition rope sling may be used to lift one cylinder – never more than one at a time – provided it is correctly adjusted to prevent slipping.
(c) Long lengths of large diameter pipes may bend (deflect) or even slip from the sling. Slings should be positioned at one quarter of the pipe length from each end, see Fig. 1.30.

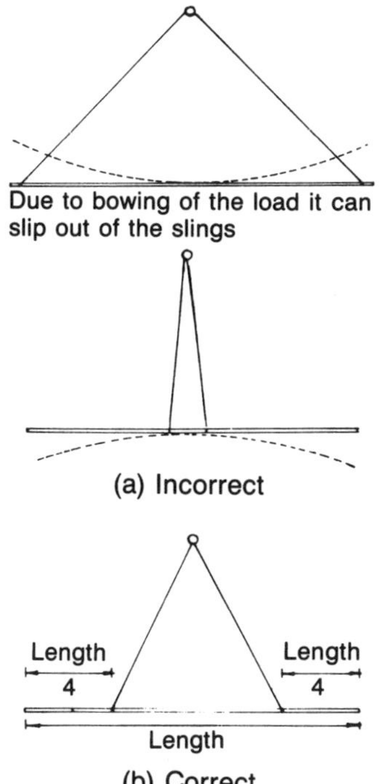

Fig. 1.30 Correct and incorrect methods of using slings

(d) Pulleys must never be attached to any element of the building, e.g. the roof truss, without checking that it will take the load.

Further reading

The following bodies are important sources of further reading and information on health and safety at work:

Health and Safety Executive, Baynards House, 1 Chepstow Place, West Bourne Place, London W2 4TF.

Royal Society for the Prevention of Accidents (RoSPA), Cannon House, The Priory, Queensway, Birmingham B4 6BS.

British Red Cross Society, 9 Grosvenor Crescent, London SW1X 7EJ.

Hazards at work: T.U.C. Guide to Health and Safety, obtainable from T.U.C. Publications Dept, Congress House, Great Russel Street, London, WC1B 3LS.

Fire Protection Association, Aldermary House, Queen Street, London EC4N

HMSO, 49 High Holborn, London WC1

Electrical Electronic Telecommunication and Plumbing Union (EEPTU), Hayes Court, Bromley, Kent.

Self-testing questions

1. Describe briefly the purpose of (a) a guard rail, and (b) toe boards, when used on scaffolding.
2. State the reason why ladders should not be painted and what methods are used for their protection and storage.
3. State three faults commonly found in connection with grinding wheels, and name the most common injury caused by ignoring the safety regulations relating to them.
4. List the precautions to be taken when using a blowlamp in a roof space or under suspended wooden floors.
5. State the three essential conditions which must be present before a fire can be started, and name the extinguishing agent recommended in the following cases:
   (a) a petrol fire
   (b) burning timber
   (c) an electrical fire
6. State the recommended supply voltage for the safe use of electrical power tools and lighting on site.
7. Describe briefly the main purposes of the Health and Safety at Work etc. Act and state the main responsibilities of employers and employees under the Act.
8. What type of sling is recommended to hoist gas cylinders? What is the maximum number of cylinders which can be lifted in a sling?
9. What is the meaning of the abbreviation 'SWL' and what is its importance?
10. If a load of 6 kg is to be lifted using a crowbar 900 mm long and the fulcrum is placed 100 mm from the load, what effort will be required to raise it using the formula
    $$E = \frac{L \times A}{B}$$
11. Make a simple sketch showing the symbol for double insulated power tools.
12. List the correct first-aid procedures for treating a person suffering from electric shock.

2 Hand tools

After completing this chapter the student should be able to:

1. Recognise and name the common tools used by the plumber.
2. Name the tools required to be supplied by the plumber.
3. State the use of each tool.
4. Identify faults in hand tools.
5. State maintenance requirements for hand tools.
6. Select and list appropriate tools for a given job.

Introduction

A saying much favoured by plumbing craftsmen is that 'Good tools are required to do a good job'. Good tools are, however, very expensive and in order to keep costs to a minimum the following points must be considered.

Selection

It must be realised that low cost items are not always a bargain. To obtain the best value for money, high quality tools which will last should be purchased.

Maintenance

To ensure a long, effective and safe working life, it is essential to keep tools in a good state of repair.

Range of tools

Basically these are the tools listed in the current tools list of the Joint Industry Board for Plumbing Mechanical Engineering Services (JIB), but the plumber usually possesses far more tools than these for his everyday use. In addition, various special tools are required, such as those used in the forming of sheet roofing materials, some of these being made by the plumber himself.

The JIB list does not include such items as welding equipment, stocks and dies, pipe vices and bending machines. These and many other expensive items of equipment are provided by the employer.

Basic tool kit

By agreement between employers and craftsmen the plumber is required to provide the basic tool kit detailed opposite. This range of tools is sufficient to allow the craftsman to undertake the normal range of plumbing work.

When the apprentice or trainee achieves the status of a trained plumber he is paid a tool allowance by his employer. This enables him to maintain the tools in good condition and to replace those which have become worn.

The reader should note that a flooring chisel is identical to a bolster except it has a much thinner blade. The basin key listed relates to the tools shown in Fig. 2.30.

Some might question the requirement of a wiping cloth in a modern tool list in view of the fact that wiped solder joints are no longer permissible for water services. It should be borne in mind, however, that a small wiping cloth of the type formerly used for wiping branch joints on lead pipe is necessary for effecting a temporary repair on existing lead pipes and defects in lead gutters. It should be noted that only plumber's wiping solder is suitable for these purposes, see page 224, Chapter 10, 'Soft solders'.

Tool list (sizes are metric or Imperial equivalent)

Allen keys – 1 set
Adjustable spanner – up to 6″
Basin key
Bending spring – 15 mm
Bending spring – 22 mm
Blow torch and nozzle – complete (similar to Primus Type B)
Bolster, chisel – 2½″
Bossing mallet
Bossing stick
Bradawl
Chisel – brick up to 20″
Chisel – wood
Dresser for lead
Flooring chisel
Footprints 9″
Gas or adjustable pliers
Glass cutter*
Hacking knife*
Hack-saw frame
Hammer – large
Hammer – small
Hand drill brace
Junior hacksaw/saw
Mole wrench
Padsaw/compass saw
Pipewrench – Stilsons up to type 14″
Pliers – insulated general type
Pocket knife (Stanley type)
Putty knife*
Rasp
Rule – 3 m tape
Screwdriver – large
Screwdriver – small
Shave hook
Snips
Spirit level – 600 mm
Tank cutter
Tool bag
Trowel
Tube cutter
Wiping cloth (one)

* Only if glazing is normally done by plumbers in the district.
Sizes are metric or Imperial equivalent.

Classification of hand tools

It is not practicable to describe within the scope of this book every tool the plumber uses, but it is necessary to briefly consider the more common items so they can be recognised and their uses understood. The application of the various tools to particular activities is more fully detailed in later chapters.

The hand tools used by the plumber can best be grouped under the following general practical operations.

Measuring and setting out

Compasses, *dividers*, *try squares*, *straight edge*, *flexible steel tape* (marked in metric), *chalk line*, *spirit level* (those shaped like a canoe are termed 'boat levels') and the *plumb bob* are tools so widely used as to require no further explanation.

Cutting (sheet and pipe)

For sheet materials such as lead, copper and aluminium, *tinman's snips* having either straight or curved blades are used. The recommended handle is the 'open ended' or 'no nip' pattern which avoids pinching the hand when they are closed.

Sheet lead, being a soft material, can also be cut using a knife (Fig. 2.1) that 'scores' the metal, which is then separated along the scored line.

Holes are cut in metal vessels, e.g. water storage tanks, by *hole saws* or *hole cutters* (see Fig. 2.2). Both tools are rotated in a clockwise direction, the saw by hand, breast or power drill, and the cutter by means of the tommy bar. The hole cutter cannot be used unless it is possible to position the pad from the inside of the vessel, e.g. it could not be used on a cylinder. Different diameter holes can be produced by changing the

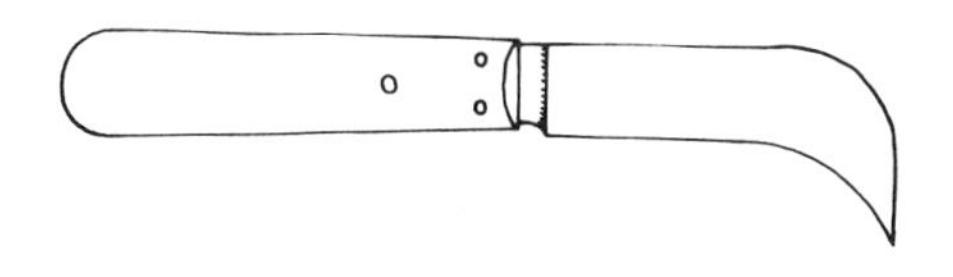

Fig. 2.1 Lead knife used for cutting sheet lead

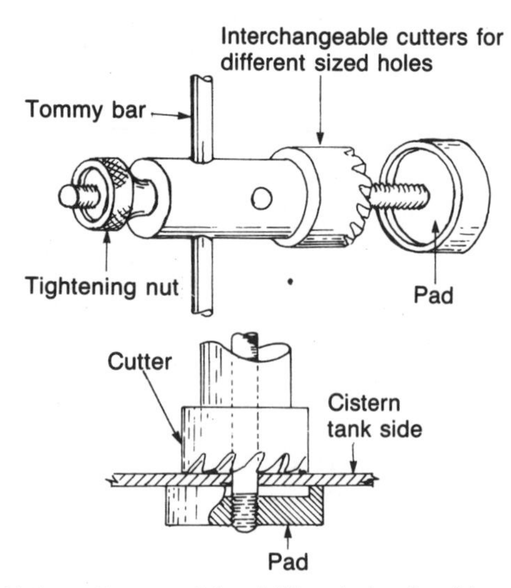

(a) Hole cutter used for drilling holes in sides of tanks or cisterns

Tightening nut is slowly rotated in a clockwise direction as the tommy bar is turned, forcing the cutting teeth into the metal thus cutting the hole

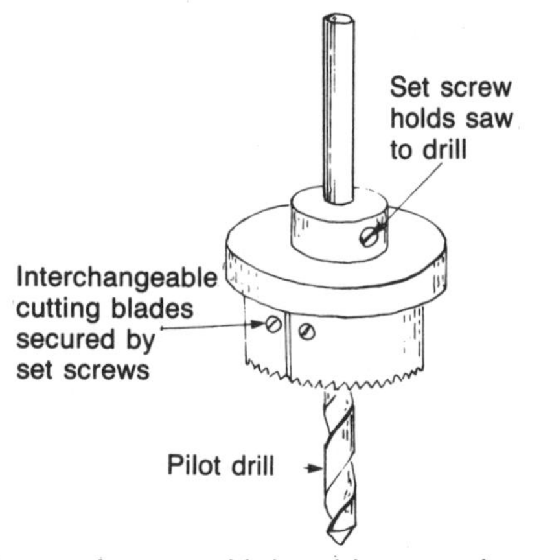

(b) Hole saw for use with hand brace or low speed electric drill

Fig. 2.2 Tools for cutting holes in metal

cutter and pad or by using another size of hole saw.

A variety of tools are available for cutting pipe, the choice being determined by the material to be cut. The *hacksaw* is the most common tool, the best pattern having a tubular frame. Three factors are important when buying hacksaw blades: the nominal length, the blade steel and the number of teeth per 25 mm. These factors are explained in Table 2.1 and Fig. 2.3.

Table 2.1 Hacksaw blades.

Blade steel	
Low steel tungsten	*High speed steel*
General purpose work	Maximum blade life and speed of cutting. For hard material.

Both types available as follows:

Flexible – hard on cutting edge only; unbreakable in use; for unskilled operators.

All hard – hardened (brittle) over whole blade: rigid; gives accurate cutting for skilled operator.

Teeth per 25 mm		
Metal thickness being cut	*Hard material*	*Soft material*
Up to 3 mm	32	32
3–6 mm	32	24
6–12 mm	24	18
12–24 mm	18	14

Arrow on blade indicates direction of cut. Care must be taken to ensure blade teeth are pointing forward, towards front of hacksaw. In use pressure is exerted on forward stroke but not on backward stroke. (See also Table 2.1.)

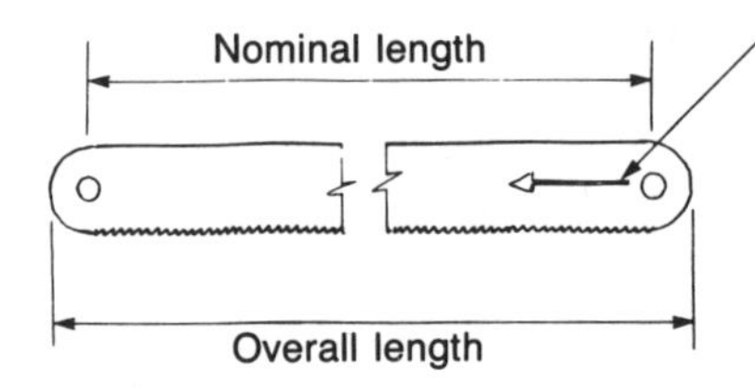

Fig. 2.3 Hack-saw blades

When hand cutting, the blade teeth point forward and at least three consecutive teeth must be in contact with the section being cut (this avoids broken teeth). Sparing application of lubricant will prolong the life of the blade.

It should be noted that the teeth on all saw blades are 'set' so that the actual cut is wider than the saw blade. This ensures the saw blade does not bind in the cut and gives adequate clearance round the blade to allow the cuttings to fall clear. Saws for cutting timber have larger teeth than those used for cutting metal and can be reset using a saw set.

Unless the reader has had some instruction on saw sharpening and setting, this job is best left to a specialist, or make friends with a good carpenter.

Pad saws

Pad saws (see Fig. 2.4) are sometimes called 'keyhole' saws and were indeed used for that purpose in the days when keys were very large. Nowadays these saws are mainly used for cutting large circular or irregular shapes in timber boards and sheet material such as plywood, plasterboard or hardboard. The blades can be obtained separately from the handles and range from fine to coarse cutting. Use a fine toothed blade if cutting thin wooden sheets, as this will prevent persistent bending or snapping of the blade.

Handles are made in both wood and metal. Those made of wood have the advantage that the blade length can be varied but if this type is used it is important to ensure the blade is firmly secured by the set screws. Should the blade work loose it may slip and penetrate the wrist. Those with a metal handle lack the flexibility of blade length but are safer in use and are made to permit broken hack saw blades to be fitted. These are very useful to a plumber where a pipe must be cut in a position where it is impossible to use even a junior hack saw.

Pipe cutters are of three basic types: *wheel* roller (Fig. 2.5(b) & (c)) and link.
A wide range of multi-wheel and roller cutters are available for use on the various pipe materials (with the exception of cast iron for which the link cutter only is used). Reference to a manufacturer's catalogue is advised where there is doubt as to the best cutter to be used for particular pipe materials.

The three or four wheel cutter is of most use where a full turning circle cannot be achieved. These need only be rotated through 120°–130° and this is a real advantage where space is limited. The formation of both external and internal burrs produced by these cutters is a disadvantage as the

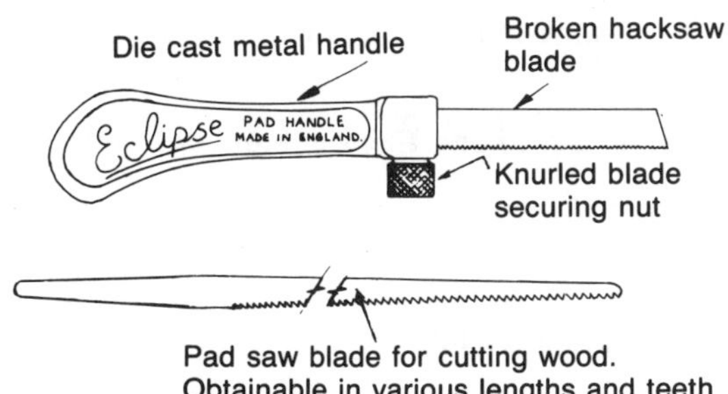

Fig. 2.4 Pad saw

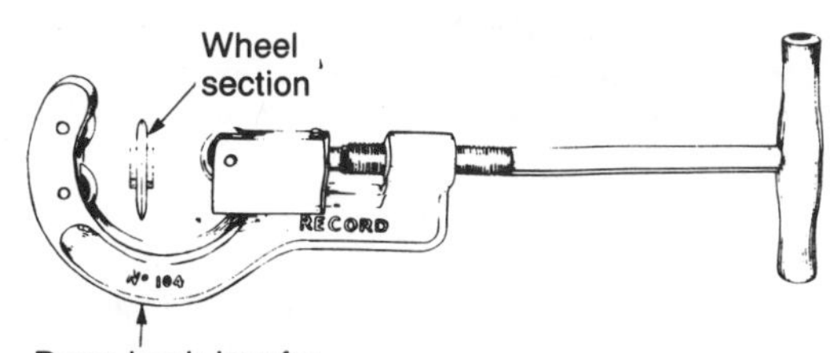

(a) Three-wheel pipe cutter (for steel pipes)

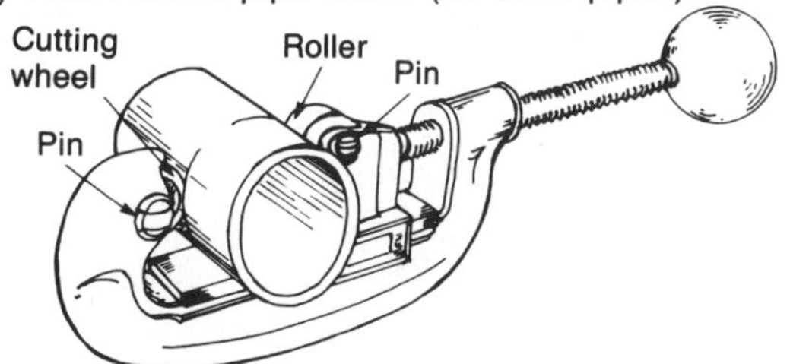

(b) General purpose single wheel roller cutter for copper tubes

The rollers steady the tool and enable a thin fast, cutting wheel to be used. The ends of a cut should meet when cutter is rotated round the pipe if the pins in the roller or cutting wheel become worn, two or three cuts will be made on the pipe – this being known as 'tracking'

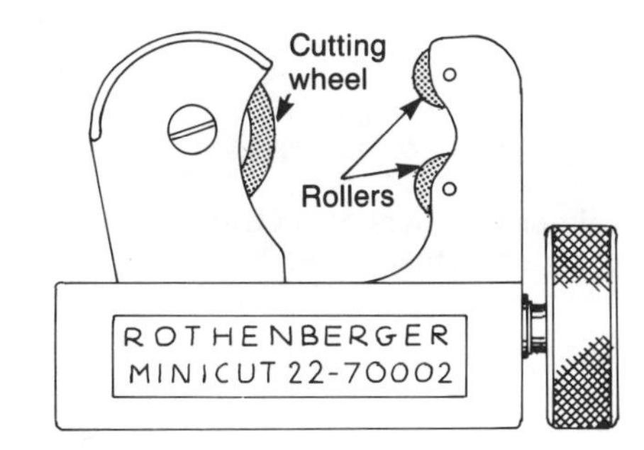

(c) Mini tube cutter

These small cutters are very useful in situations where those of a larger size are difficult to operate, i.e. under floors or in corners. The type shown will cut diameters of from 8 mm to 22 mm copper tubes.

Fig. 2.5 Pipe cutters for copper and steel tubes

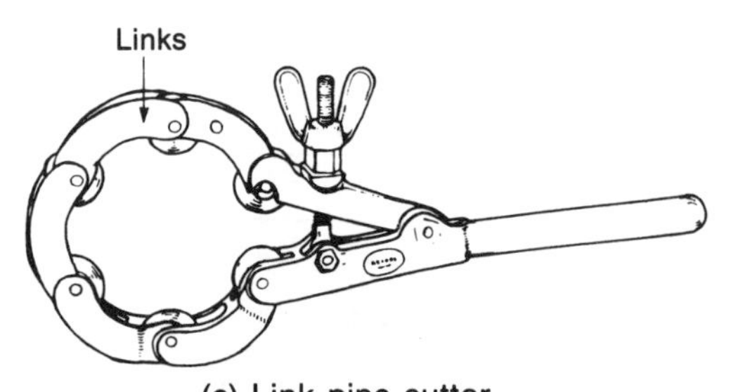

(c) Link pipe cutter

Used for cutting cast iron pipes. Adjustable in that links may be removed for smaller pipe

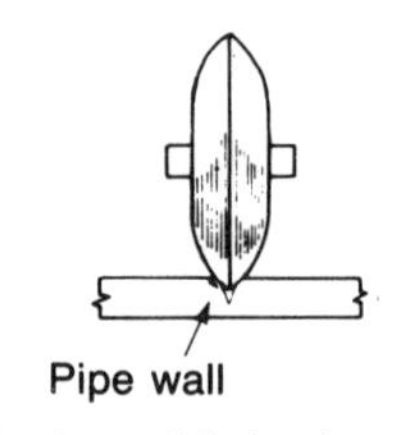

(d) Action of link pipe cutter

The hard outer skin of the casting is penetrated by the wheels and parted by splitting due to the wedging action of the wheels

Fig. 2.5 *Continued*

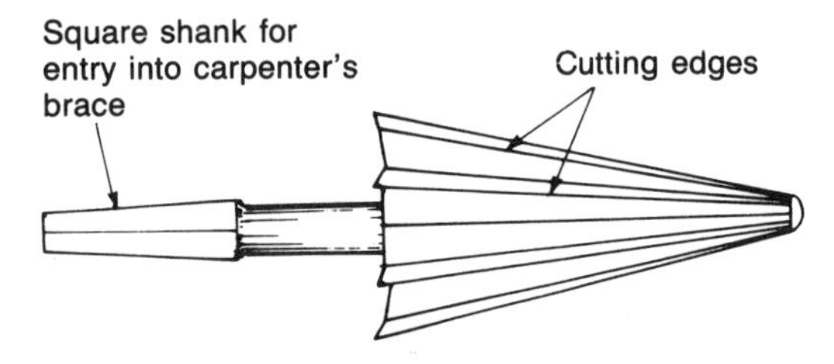

(a) Pipe deburring reamer

Reamers are made in a variety of sizes, each size being used for three or four pipe sizes, to remove burrs which have formed on the inside of the pipe when cutting

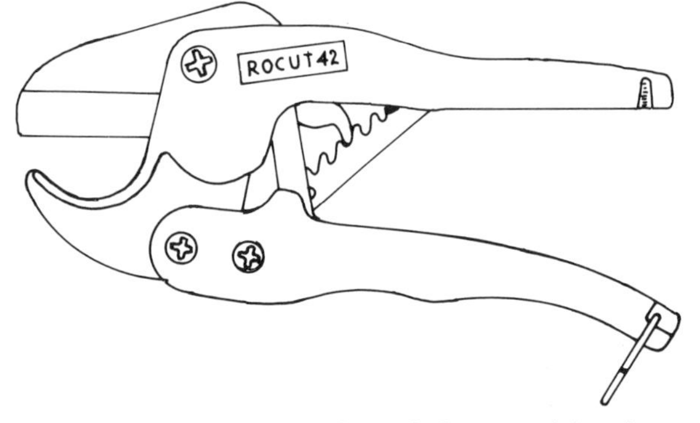

(b) Plastic pipe cutters for polythene, polybutylene and cross-linked polythene

Fig. 2.6

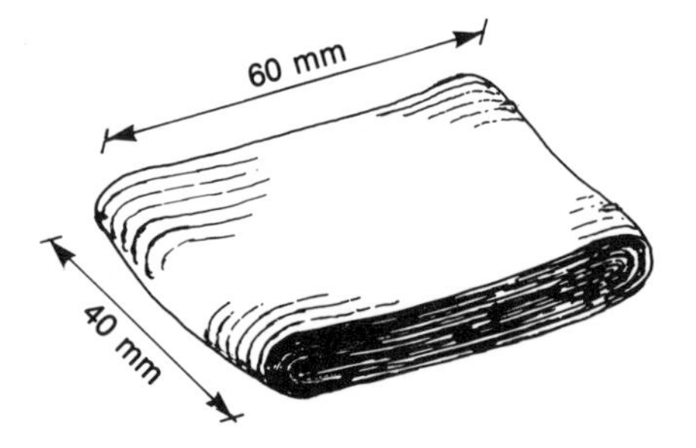

Fig. 2.7 Wiping cloth. Suitable for patching and wiping solder dots on lead sheet. Sizes shown are approximate.

burrs have to be removed, but the latter can be avoided if the cut is completed using a hacksaw.

As roller cutters require to be turned through 360°, they have a limited use. The rollers prevent an external burr being formed, and also position the cutter at 90° to the pipe preventing the 'spiral cut' that is common with the multi-wheel type.

Link cutters are used mainly for cast iron drain, discharge and flue pipes. Cutting does not actually occur with this brittle pipe material, the break being achieved by the even pressure exerted around the pipe by the knife edges (see Fig. 2.5(d)).

Burrs form inside a pipe when cutters are used, and as these restrict the bore they must be removed using either a pipe *burring reamer* or a *round/half round file*. The reamer illustrated in Fig. 2.6 is for use in a carpenter's brace but other patterns can be obtained.

Plastic pipe cutters

Figure 2.6(b) shows the cutters that are available for making a clean cut on polythene, polypropylene and polybutylene pipes, avoiding the woolly fibres which have to be removed when such pipes are cut with a saw.

Tools for laying and fixing sheet roof materials

Wiping cloths

These are made from a hard-wearing heat-resistant fabric known as 'moleskin', a member of the 'fustian' group of fabrics. Close inspection of this material will reveal a grain which should run along the length of the cloth. A cloth having dimensions as shown in Fig. 2.7 is a good general-purpose tool suitable for wiped solder patches or solder dots on sheet lead fixings. Wiping cloths may still be obtained from specialist plumbing tool stockists, or offcuts are sometimes available from manufacturers of clothing made from moleskin.

Bevel

This is a simple tool similar to a square but having a movable blade to enable angles other than those of 90° to be set out. Typical examples of its use are for setting out step flashing and pipe bending. A 600 mm steel folding rule is a good substitute and is probably more useful to a plumber than bevels obtained from tool stores, which are mainly used by carpenters.

In the absence of a folding rule a bevel can be simply made out of two convenient lengths of steel strips. The ends are rounded off and secured with a rivet or nut and bolt as shown in Fig. 2.8.

To prevent damaging roof-covering materials, many tools are made from a non-metallic substance. Plumber's wood tools were traditionally manufactured from boxwood, beechwood, hornbeam or lignum vitae, but due to the high cost of these timbers, manufacturers now produce some of them from high density polythene.

The more common tools are as follows.

Lead dresser (Fig. 2.9) This is used for flattening sheet lead. Its face must be kept smooth to avoid damage to the lead.

Chase wedge (Fig. 2.10) This is used for chasing in angles, upstands and drips, and is used in conjunction with a drip plate. A *drip plate* is shown in Fig. 2.11 and is made from approximately 1 mm thick steel sheet (100 mm × 150 mm in size). It is positioned between the overcloak and undercloak when forming details such as lead drips, to allow the sheets to slide freely over each other. (See also Fig. 11.22(a) for another use of the drip plate.)

Mallets (Fig. 2.12) These are made either of boxwood, lignum vitae or hard rubber and are used mainly on hard metals such as copper and aluminium sheet.

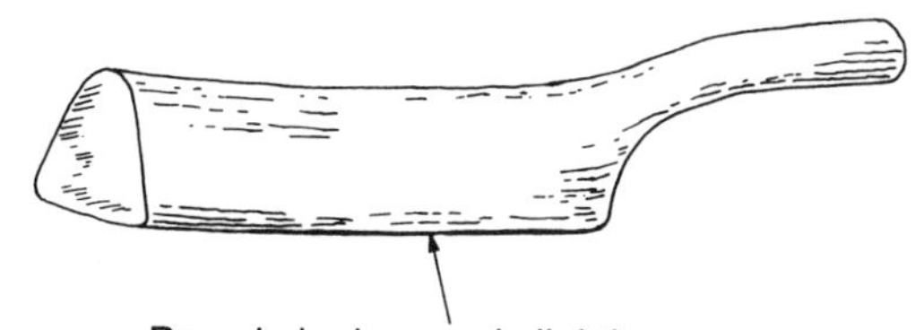

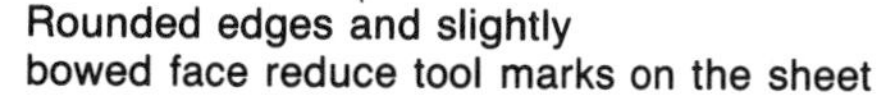

Fig. 2.9 Dresser

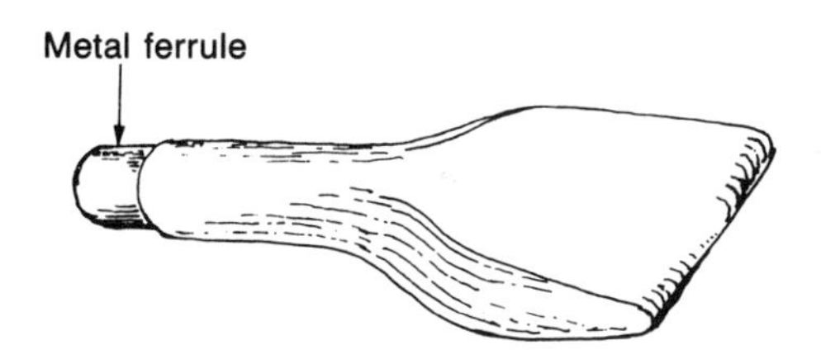

Fig. 2.10 Chase wedge. Used to move sheet lead into a corner and for straightening corners after bossing is completed. In latter case great care is needed to avoid stretching the lead which would cause cracking.

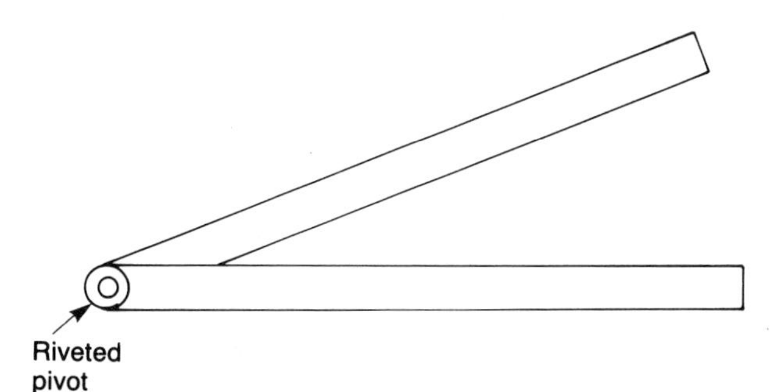

Fig. 2.8 Bevel. A useful tool for measuring angles. The illustration shows a bevel made from two pieces of bright low carbon steel, 20 mm wide × 3 mm thick and approximately 300 mm in length.

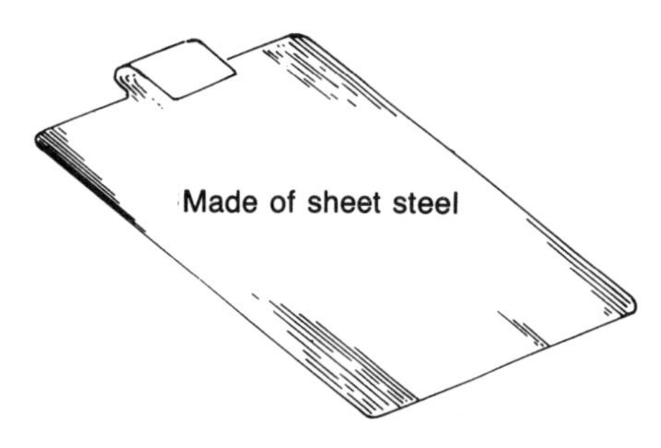

Fig. 2.11 Drip plate. The drip plate is laid between two sheets of lead when the top sheet is to be pushed into a corner by a chase wedge.

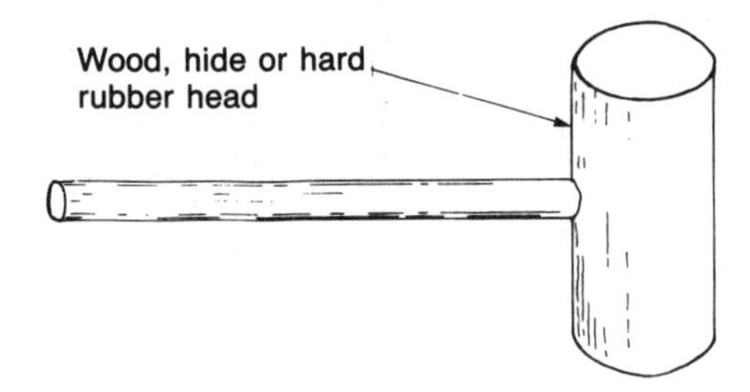

Fig. 2.12 Flat face tinsmith's mallet for hard metalwork

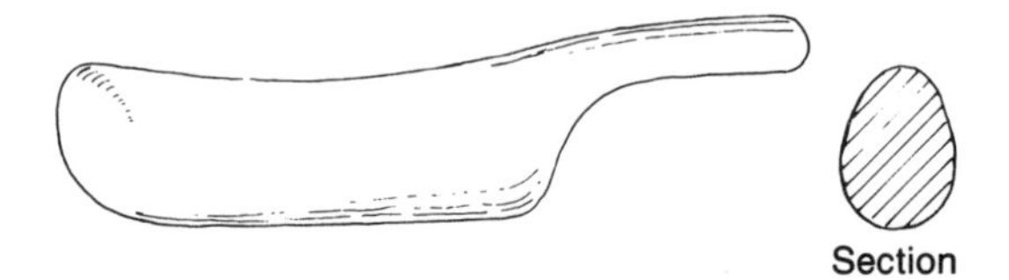

Fig. 2.13 Bossing stick. Used for bossing corners and rolls in sheet lead.

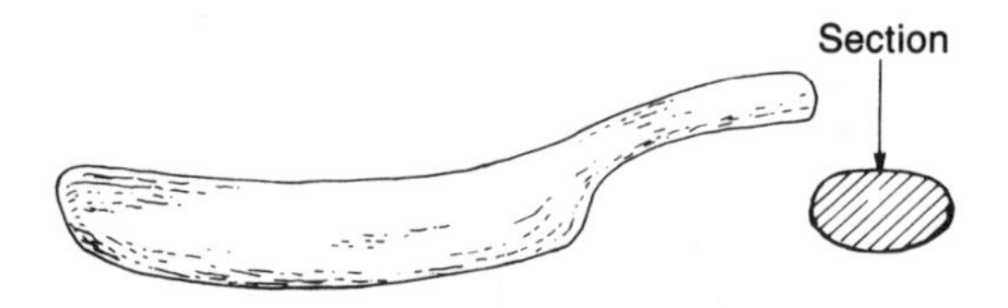

Fig. 2.14 Bending dresser. This is made of boxwood and used when bending large-diameter lead pipes.

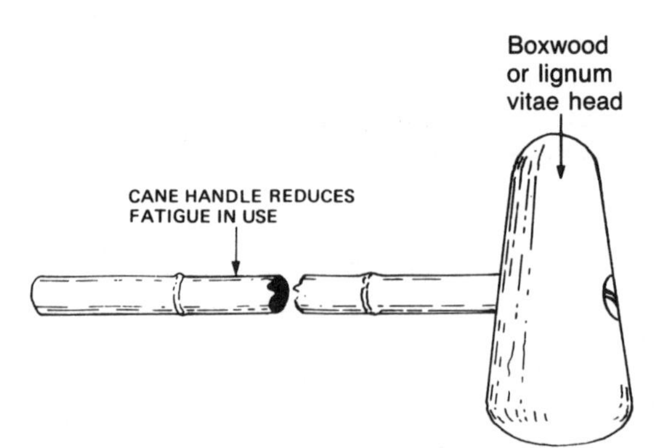

Fig. 2.15 Bossing mallet. An all-purpose bossing tool for sheet lead. Some plumbers also find them useful for copper and aluminium sheet work. Mallets used with these latter two materials lose their smooth surfaces and should not be used on sheet lead.

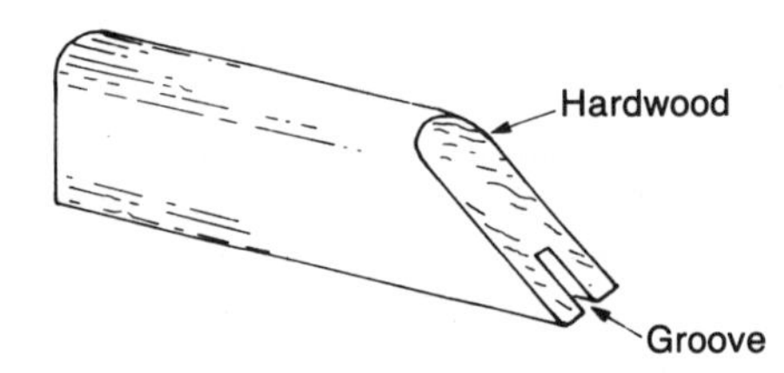

Fig. 2.16 Step turner. Used for folding or turning the edge of sheet lead for flashing.

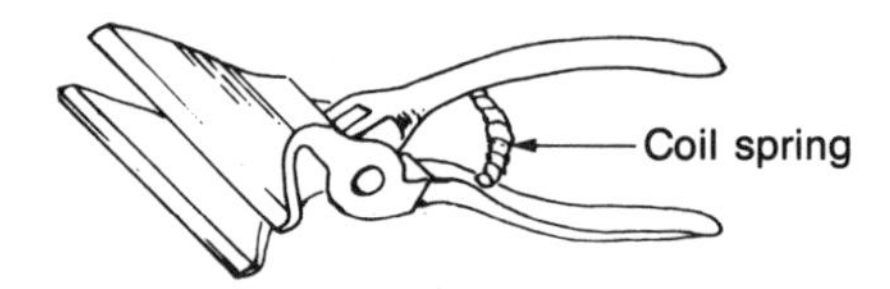

Fig. 2.17 Seaming pliers for closing welts and seams on hard metalwork

Bossing stick (Fig. 2.13) This is a tool for sheet lead work and can be used with advantage where it is difficult to work with a bossing mallet, as for example on roll ends and overcloaks. Sometimes used for bossing external corners in conjunction with a bossing mallet.

Bending dresser (Fig. 2.14) The name of this tool is derived from the period when large diameter lead pipes were bent by hand. They are similar in appearance to the bossing sticks but have a broader face. They are mainly used on lead sheet for bossing chimney aprons and back gutters or internal corners where their broad rounding face avoids excessive tool marks on the lead.

Bossing mallet (Fig. 2.15) Also a tool for sheet lead work, this is used in forming various roofwork details such as bossing corners, roll ends and drips. To reduce fatigue in use, the handle is made flexible by using malacca cane.

Step turner (Fig. 2.16) This is used to form the 'turn-in' on step flashings. These can be bought or made from a piece of hardwood.

Seaming pliers (Fig. 2.17) These are used to ensure seams and folds in hard metal roofwork are locked tightly together.

Copper bit or soldering iron (Fig. 2.18) Copper soldering bits have a wooden handle and are

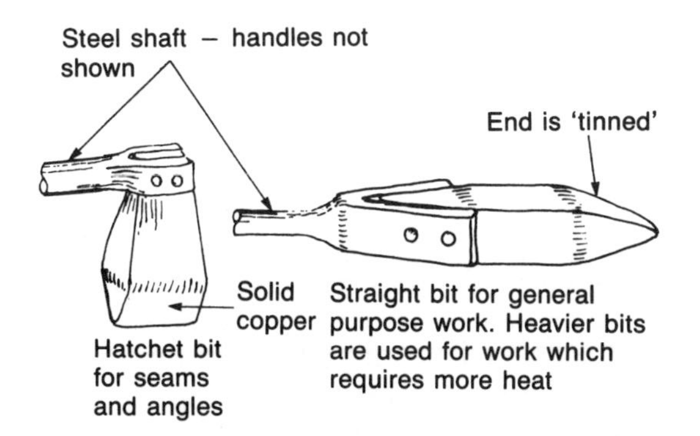

Fig. 2.18 Copper bits or soldering irons

usually specified by the mass of the copper head. Soldering irons are used for soft soldering of certain roofwork details, particularly zinc sheet. Hatchet-type bits are useful when soldering angles.

Many tools required to work roofing materials are purpose made and cannot be purchased. They can, however, be easily fabricated in the workshop. Uses of these special items, together with the applications of the common tools described, are detailed in Chapter 11.

Rasp (Fig. 2.19) Similar in shape to a file but has much coarser teeth enabling it to be used for soft materials such as lead or shaping wood. Although originally a tool principally used for lead pipe jointing it is still a very useful addition to the jobbing plumber's tool kit.

The shavehook (Fig. 2.20) This is used to clean the surface of lead sheet prior to soldering or welding. Various shapes of blade are available – that shown is heart shaped.

Pipework-jointing tools

Sanitation and water services require the use of a wide variety of pipework materials and, in order to identify suitable tools, it is necessary to consider briefly the method of joining each pipe material. More detailed information on jointing procedures can be found in Chapter 4 which deals with pipework-working processes.

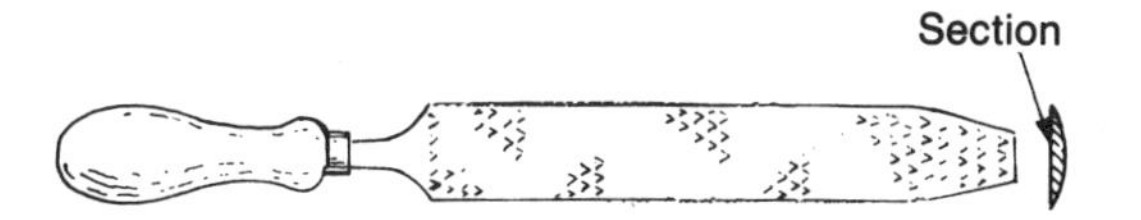

Fig. 2.19 Rasp. Used for shaping soft materials such as lead or wood and has large teeth which will not clog easily. The teeth are cleaned from time to time using card wire.

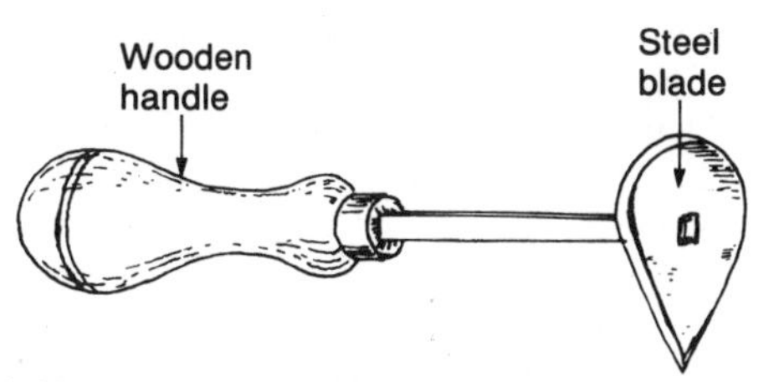

Fig. 2.20 Shavehook. Used for removing the lead oxide surface film on sheet lead and lead pipe prior to soldering or lead burning.

Steel pipe threading tools

Chaser die stocks The tools used for cutting threads on steel pipes are known as stocks and dies, the stocks being the body and handle of the tool, the die being the actual thread cutter. The pattern shown in Fig. 2.21(a) has interchangeable sets of dies (chasers), allowing threads to be cut on a range of different diameter pipes. It should be noted that chasers which are interchangeable in a common stock are numbered 1 to 4. It is important that these numbers correspond to those marked on the stock when they are changed as failure to do this will result in a defective or stripped thread. Dies can be obtained to cut parallel or taper British Standard Pipe (BSP) threads.

Die stocks that cut a thread on only one size of pipe are also available (Fig. 2.21(b)) separate die heads being required for every pipe diameter. They are very useful for threading pipes in position, and by reversing the dies, close-to-wall threading can be achieved.

Receder dies Chaser dies are drawn on to the pipe by means of the thread which is being cut. This action places great strain on the pipe and can, in certain circumstances, result in the pipe being crushed. To obviate this, receder die stocks as illustrated in Fig. 2.21(c) are designed with a screw feed that is firmly clamped on to the pipe.

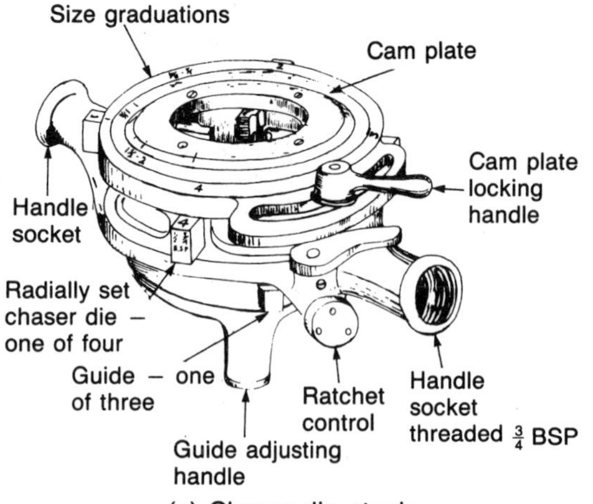

(a) Chaser die stock

Chaser die stocks are adjustable and can be used for threading pipes of different sizes.

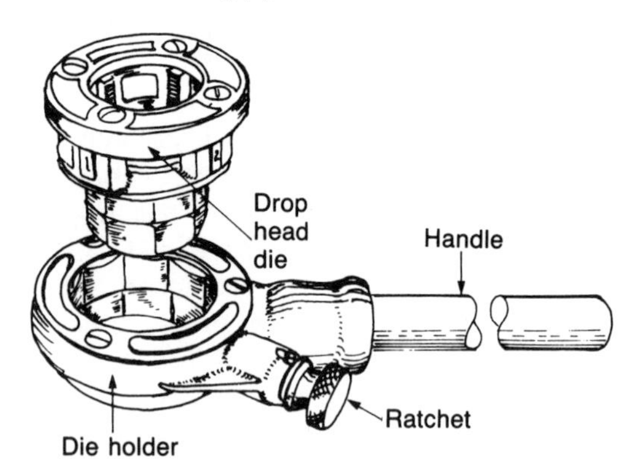

(b) Ratchet drop head die stock

Interchangeable heads save time when threading pipes of different diameters in that the dies for different sizes of pipe will fit into the same die holder.

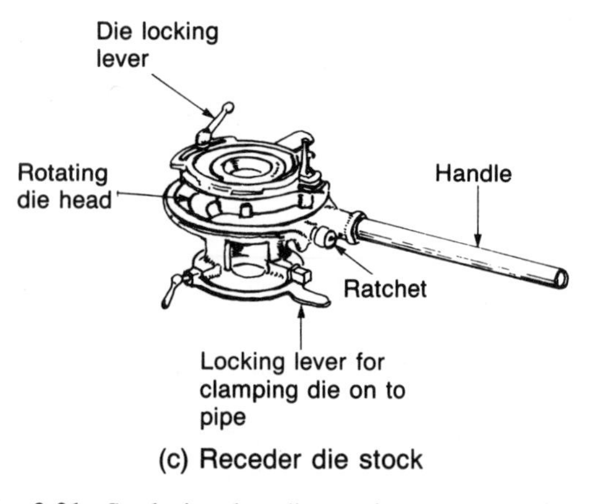

(c) Receder die stock

Fig. 2.21 Steel pipe threading tools

When the die is rotated it is pulled on to the pipe along the screw feed in a similar way to a nut pulling itself on to a bolt. This operating principle removes the strain on the pipe which is encountered with chaser dies. The threads on receder dies are parallel, but taper threads can be cut by adapting the stock and causing the chasers to move outwards from the pipe (to recede) as threading is in progress (see Fig. 2.21(c)).

Other patterns of stocks and dies are available, such as solid or block, circular split and two piece. These are, however, used mainly for bolt threading, although the solid die is still in use for light-weight tube such as electrical conduit.

Hexagon dienut and rethreading files The tools shown in Figs. 2.22 and 2.23 are used for repairing damaged or rusty threads – they do not replace stocks and dies.

British Standard Pipe (BSP) threads are always used for pipework in the plumbing industry, but for general threading a variety of thread forms are available, ranging from coarse to fine, the principal threads being British Standard Fine (BSF), British Standard Whitworth (BSW), British Association (BA) and Metric (ISO).

Copper tube jointing tools

Certain basic tools are required for jointing copper pipes such as tube cutters, reamers and smooth cut files, depending on the particular requirements of the type of fitting being used.

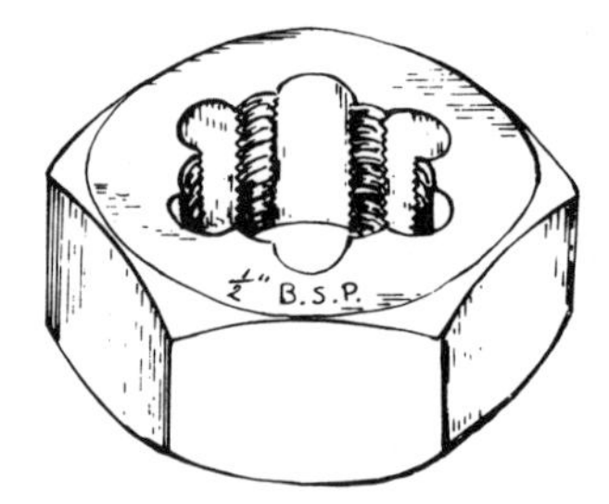

Fig. 2.22 Hexagon die nut. The die nut is used to clean used or badly cut threads. It is made of very hard tool steel.

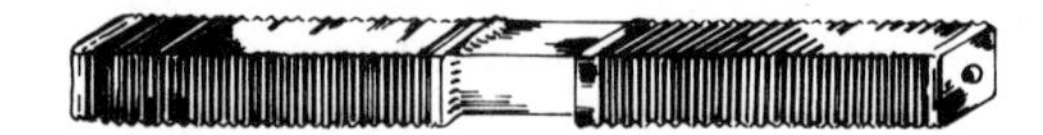

Fig. 2.23 Rethreading file

Swaging tools

Many special tools are available for capillary jointing but are outside the scope of this book. A common tool widely used, however, is the *socket former* (Fig. 2.24(a)) which provides an economic method of making soldered or brazed capillary joints. Figures 2.24(b) and (c) show swaging tools for manipulative compression fittings.

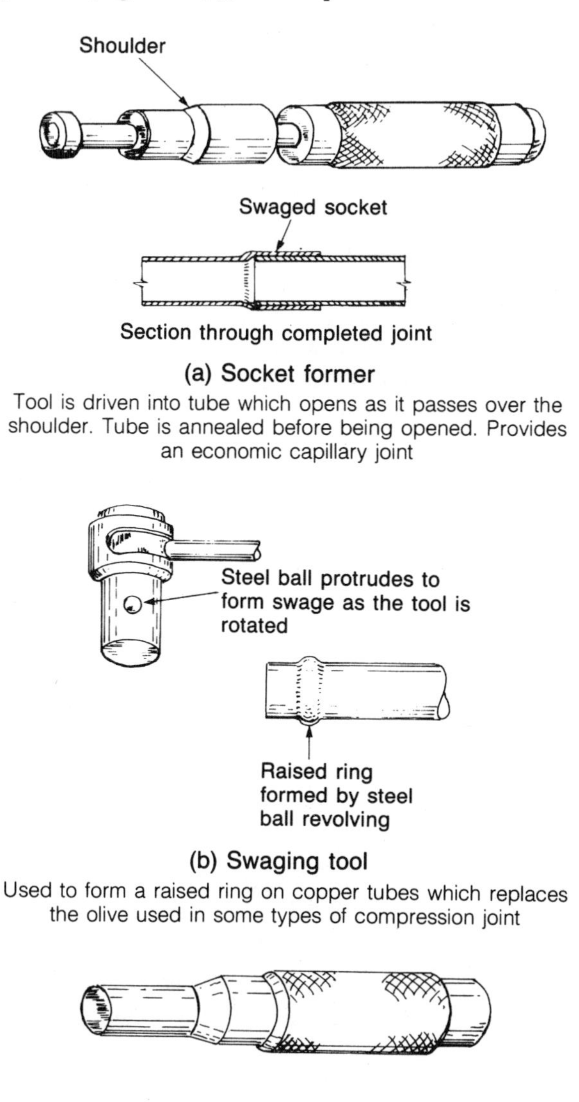

(a) Socket former

Tool is driven into tube which opens as it passes over the shoulder. Tube is annealed before being opened. Provides an economic capillary joint

(b) Swaging tool

Used to form a raised ring on copper tubes which replaces the olive used in some types of compression joint

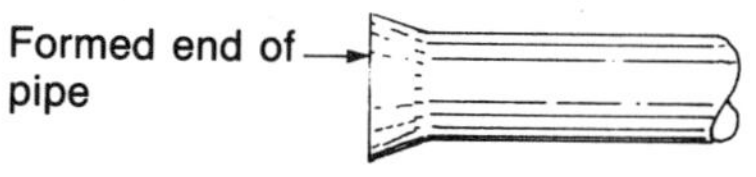

(c) Drift

Used for forming a bell mouth when preparing a manipulative compression joint or bronze welded joint on copper tubes

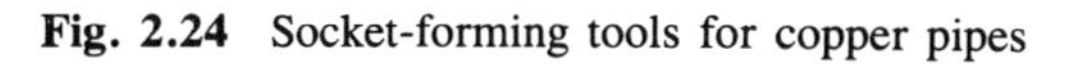

Fig. 2.24 Socket-forming tools for copper pipes

The bent bolt (Fig. 2.25) This is used in conjunction with a hammer to open up branch holes in copper pipe. This technique is illustrated in Book 2, Chapter 2.

For soft-soldered joints, paraffin or propane blowlamps are required. When using low temperature brazing alloys, the smaller pipes can be jointed using a blowlamp, but larger diameters will require the use of oxyacetylene heating equipment.

Non-manipulative compression fittings, as the name implies, do not require any shaping or forming of the tube end other than cutting square. Open ended or adjustable spanners are needed to tighten the coupling nuts of the fitting.

Jointing tools for cast iron pipes

No special tools are needed for jointing cast iron flue pipes or externally fixed rainwater pipes. However, drain, discharge and internally fitted rainwater pipes must have watertight joints, and special tools are needed.

While in many cases flexible joints for cast iron pipe are mandatory, in situations where rigid joints are permissible, e.g. above ground, lead run caulked joints may be used.

The usual procedure for making watertight joints is to pack yarn into the sockets and then to fill them with either molten lead or lead wool. *Yarning* and *caulking irons* (see Figs 2.26(a) and (b)) are used to consolidate the yarn and the lead in the sockets. When jointing pipes in the horizontal position using molten lead, a *squirrel*

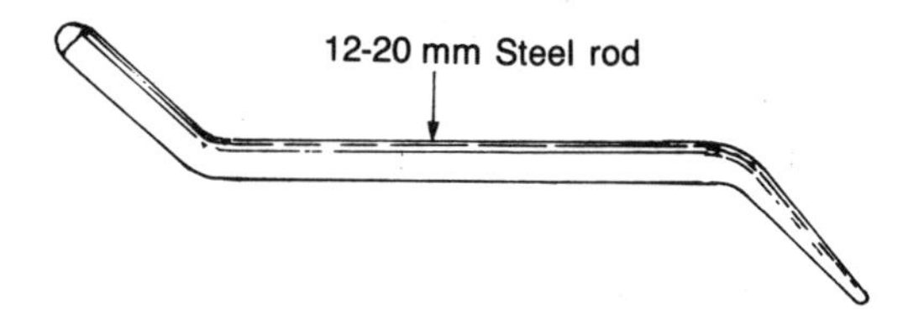

Fig. 2.25 Bent bolt. The bent bolt has many uses for the plumber including its use when opening holes for branch joints in copper pipes.

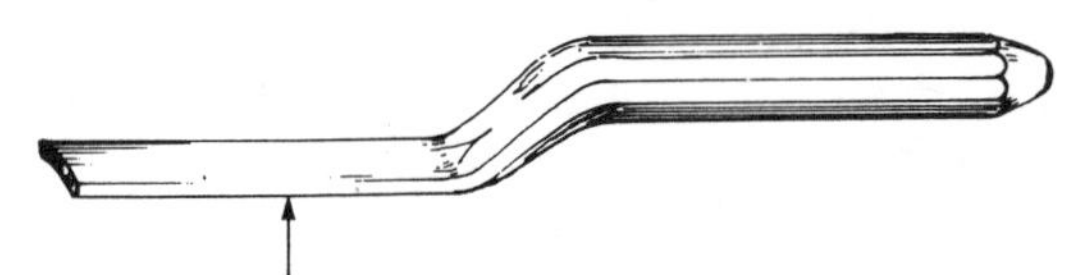

Long thin blade

(a) Yarning iron

Used for consolidating the yarn in the bottom of a spigot and socket joint which is to be caulked

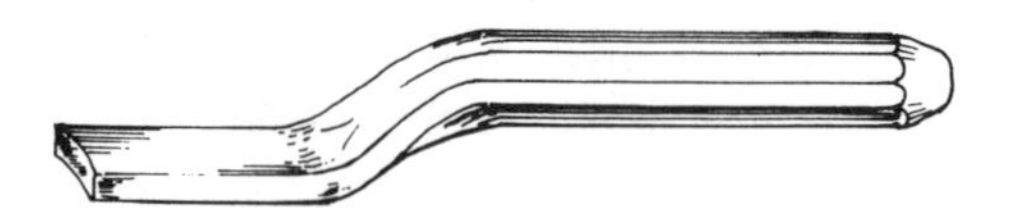

Short thick blade

(b) Caulking iron

Used for caulking lead into spigot and socket joints on cast iron pipes

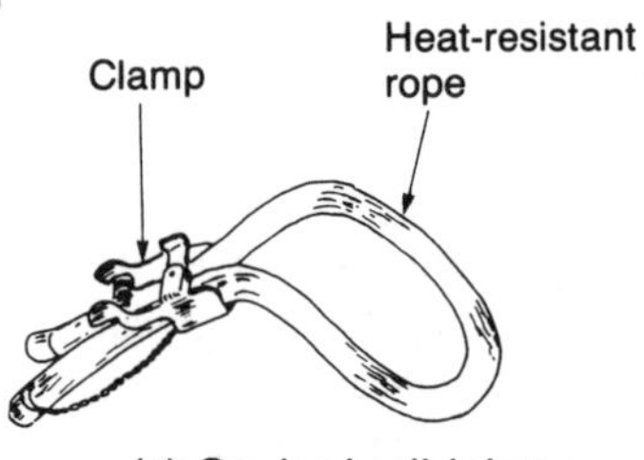

(c) Squirrel tail jointer

This part fits over a joint in a cast iron pipe which is to be lead caulked and controls the molten lead, preventing it from escaping

Fig. 2.26 Jointing tools for cast iron pipes

tail pipe jointer (see Fig. 2.26(c)) is needed to retain the lead in the socket until it solidifies.

Although other methods of jointing cast iron pipes are now being increasingly used, lead caulked joints are still used in certain circumstances. It is important to note that the heat resistant rope shown in the illustration is made of asbestos and as yet no suitable alternative has been found. Until another material is available or such joints are no longer necessary, it is suggested that to prevent loose asbestos fibres becoming inhaled, the flexible rope is treated periodically with a light smear of tallow to bind the fibres. As an additional precaution a suitable face mask should be worn when handling this tool. It may be of interest to note that before these tools became popular, clay was used to retain the lead in the joint. Assuming that suitable clay is available and the installer is prepared to spend the extra time making the joint, this is, at present, the only alternative.

Pipe-bending tools

The various bending *machines* that are available for copper and steel pipes are described in Chapter 4.

Bending springs

Small diameter copper tube can be easily bent by the use of a bending spring (Fig. 2.27) made from square section polished steel wire. They are positioned inside the pipe before bending is commenced and provide internal support to prevent collapse of the pipe wall. Springs should be lightly oiled before use as they are then more easily removed after the bend has been made.

Tools for fitting and fixing pipework

In this section various general tools will be considered which enable plumbing work to be fixed and fitted to the building fabric.

Wrenches and spanners

There are three basic wrenches, illustrated in Fig. 2.28, used for rotating pipes when jointing, e.g. when tightening screwed joints.

Straight pipe wrench (Fig. 2.28(a)) These are commonly referred to as 'Stilsons' because the original pattern was produced by an American company called Stilson. A range of patterns and sizes (150–1,500 mm long) are available, the length to be used being dependent on the pipe diameter or the leverage required. Never extend the handle with pipe as this can cause the tool to be strained and damaged.

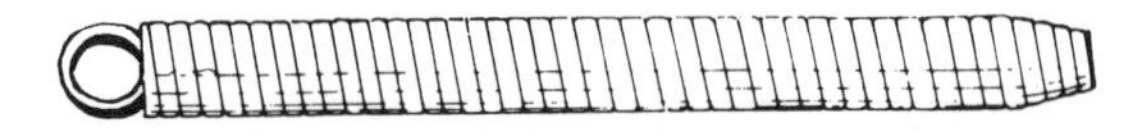

Fig. 2.27 Bending spring. The spring is placed inside the pipe at a point where a bend is to be made to maintain the true bore of the pipe during the bending process.

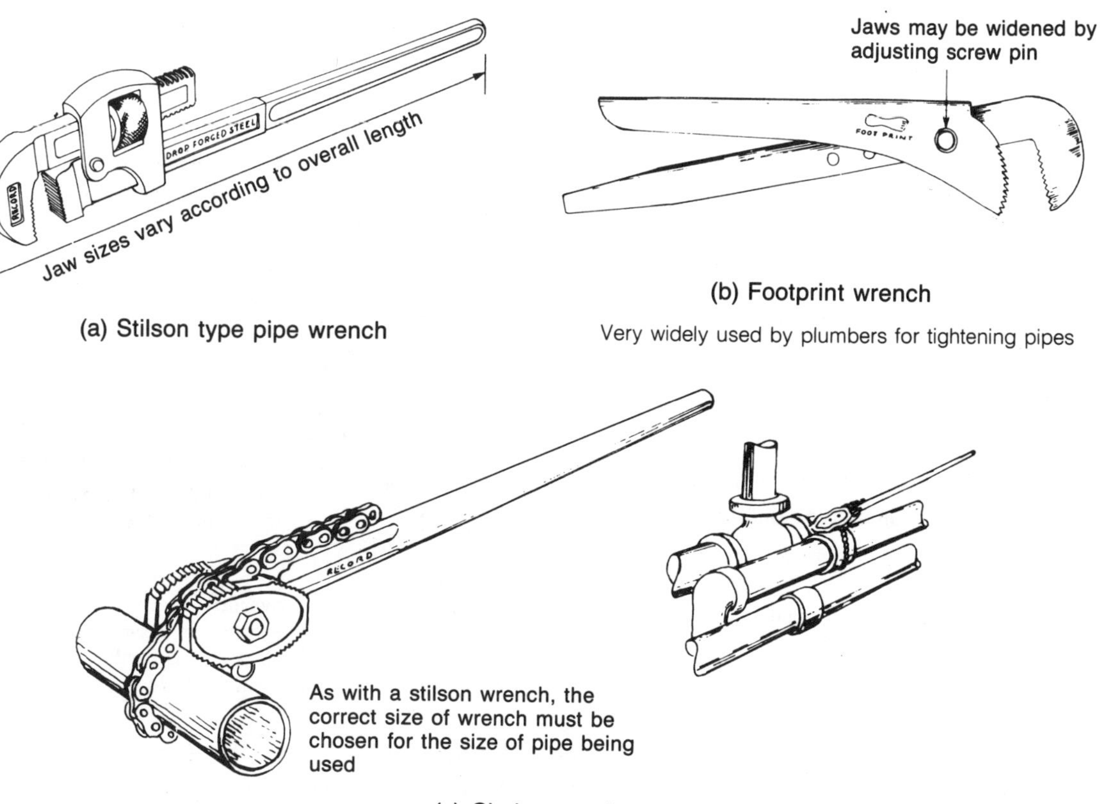

Fig. 2.28 Basic wrenches

Footprint pipe grip or wrench (Fig. 2.28(b)) 'Footprints' is the common name given to this type of wrench because the trade mark of the manufacturer is a footprint. They can be obtained in lengths from 150 mm to 350 mm to suit pipe size and leverage.

Chain wrench (Fig. 2.28(c)) These are obtainable in lengths of 500 mm to 1,283 mm, and for nominal pipe diameters of from 3 mm to 200 mm.

The hardened teeth of the tools described above make them suitable only for steel pipe, and any application to soft materials such as copper and brass must be avoided since the hardened steel will damage the soft metal. Use of these tools must also be limited to circular sections such as pipe, and they must never be used on square-faced coupling nuts. Teeth marks left on pipe and fittings must be removed by file, as the sharp edges can be a hazard.

Spanners The correct tool to use on square, hexagonal or octagonal nuts and couplings is a spanner. Three types are in common use in the plumbing industry: open-ended, ring and adjustable, and these are illustrated in Fig. 2.29.

Ring spanners should be used where possible as they completely encircle the nut and prevent slipping. Obviously they cannot be used for tightening the nuts on compression type fittings as the copper tube would not permit access of the spanner to the nut. They are very useful, however, for carrying out maintenance work on boilers and other mechanical appliances.

The *open-ended* pattern is used where it is only

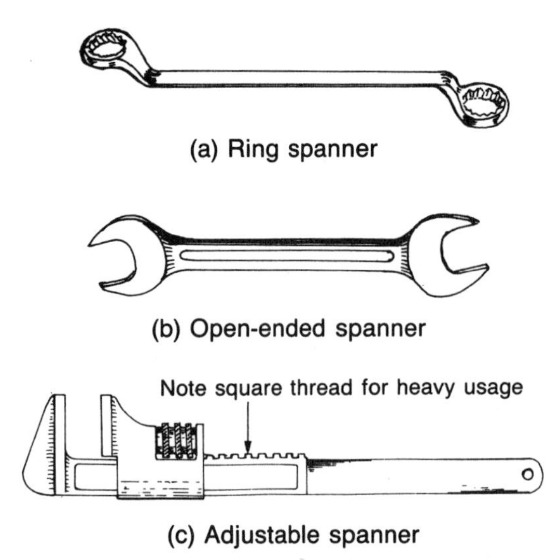

Fig. 2.29 Common types of spanner

possible to apply the tool from the side, e.g. when tightening the coupling nut of a pipe fitting.

An *adjustable* spanner can be used instead of the open-ended pattern, and depending on its size, can be adjusted to suit a range of nuts.

Basin wrenches (Fig. 2.30) These are for use in tightening nuts where an ordinary spanner will not reach, such as the nuts on lavatory basin taps, bath taps and taps fitted to sink units. These wrenches are available in a range of sizes to fit bath and lavatory basin wastes and tap back nuts.

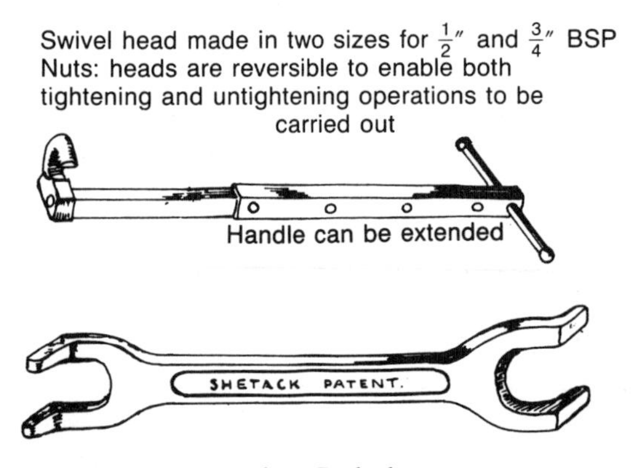

Fig. 2.30 Basin wrenches. Both these spanners are useful when hot and cold water connections are made to baths, basins and sink units.

Waste fitting holder (home made) A useful tool that can easily be made by a plumber for screwing up bath overflows and preventing waste fittings turning while the backnut is being rotated, is shown in Fig. 2.31. Generally a short 25 mm nominal bore steel pipe is the most suitable material to use.

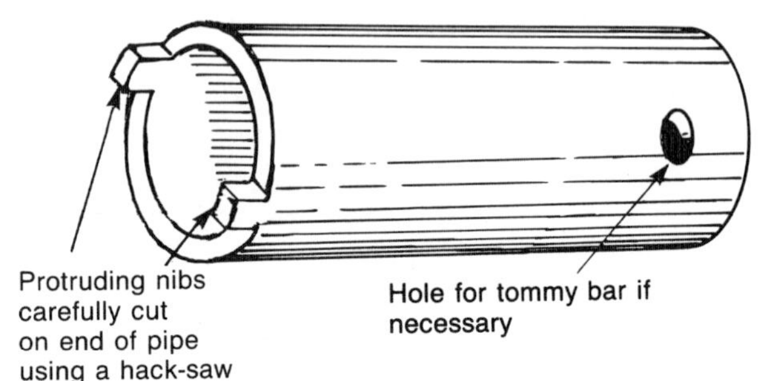

Fig. 2.31 Waste and overflow spanner. This is a useful tool for holding a waste fitting in a sanitary appliance while tightening the back nut; it is easily made from a short length of 25 mm nominal bore steel pipe.

Leverage (Fig. 2.32) The length of any wrench or spanner affects the force being applied to the pipe. This can be expressed by the following simple formula:

$$\text{Leverage} = \text{Length} \times \text{Effort}$$

This means that the total leverage applied by using a wrench or spanner is the result of multiplying the length of the wrench and the effort

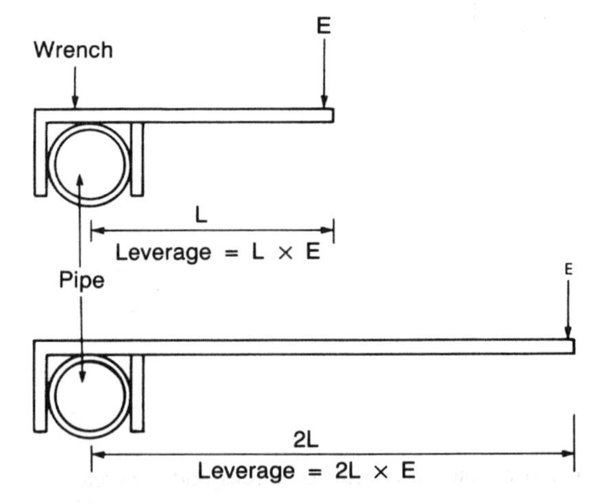

Fig. 2.32 Leverage

used by the operator. From the following diagrams it can be seen that if the effort E remains constant but the length of the lever is doubled, then the force being applied to the pipe will double. A longer lever, therefore, enables either (a) less effort to be used, or (b) more work to be done.

It is important to remember and apply this principle of leverage carefully, and the appropriate size of wrench should be used for the size of pipe being used. A wrench which is too small would not be capable of exerting the leverage required, whereas a wrench which is too large may damage the pipe. Each size of wrench is designed for a certain maximum leverage, and if the leverage is increased by extending the wrench with a piece of tube, then the wrench itself may be permanently damaged. Also by using a longer tool than is needed, the increased force produced could crush a pipe or shear a bolt.

Drills

Twist drills (see Fig. 2.33) are used for making holes in conjunction with hand, breast or power drills. Two main types of steel, *high-speed steel* and *high-carbon steel*, are used in twist drill manufacture.

High-speed steel (HSS) contains under 1 per cent carbon to which is added alloying agents such as tungsten, chromium, vanadium, cobalt and molybdenum. These drills are able to cut even at a dull red heat without loss of hardness or rapid blunting of their cutting edge.

High-carbon steel contains 1–2 per cent carbon to which is added alloying agents such as tungsten or chromium. If used at high speeds, i.e. in a

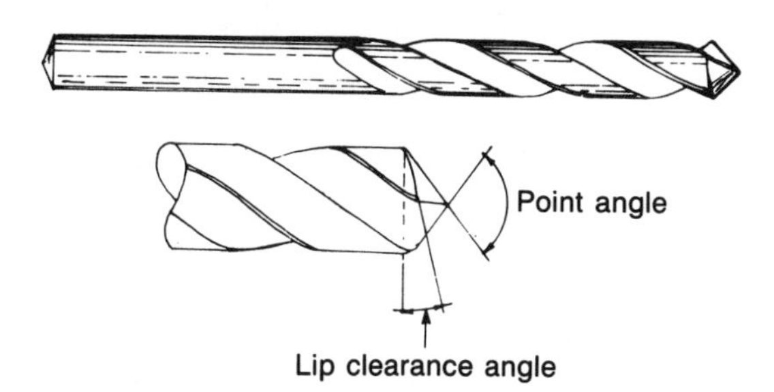

Fig. 2.33 Twist drill. This is used for drilling holes in almost any material except for bricks and concrete, where a masonry drill is needed.

power drill, and the operating temperatures exeed 250 °C, the drill becomes blunt due to loss of hardness.

Correct cutting lubricants must be used to prolong drill life and improve the quality of drilling, the following lubricants being recommended with differing materials:

Low-carbon steel	Soluble oil
Copper and brass	Dry, paraffin or soluble oil
Aluminium	Paraffin or soluble oil
Cast iron	None required
Plastics	None required

Twist drills for general use have a point angle of 118° and a lip clearance angle of 10°–12°. Other materials require different angles for effective cutting, e.g. for plastic, point and lip clearance angles are 75° and 20° respectively.

The *masonry drill* is similar to the twist drill but has a tungsten carbide tip which is very hard and wear-resistant. These drills are suitable for brick, tile, stone and concrete, and can only be resharpened on a special grinding wheel.

Hammers

A hammer consists of a steel head and a wooden shaft, the face of the head and pein being hardened and tempered. The wood used for the shaft must be well seasoned, straight grained, and is usually ash or hickory. The head must be securely fixed to the shaft by means of wedges.

Figure 2.34(a)–(c) shows the different patterns of pein available. The plumber uses all of these, although the straight pein is traditionally referred to as a 'plumber's hammer' as it is useful for forming hard sheet metal.

For cutting brickwork or concrete the use of a 'club' or 'lump' hammer is recommended (see Fig. 2.34(d)) which has a heavy head and two striking faces. The short handle reduces fatigue in use yet enables a heavy blow to be delivered.

Hammers are classified by the shape of the head and the mass of steel used.

Chisels

A selection of 'cold' or 'hard' chisels (see Fig. 2.35(a)) are required by the craftsman to cut holes in brick and concrete. These are specified by

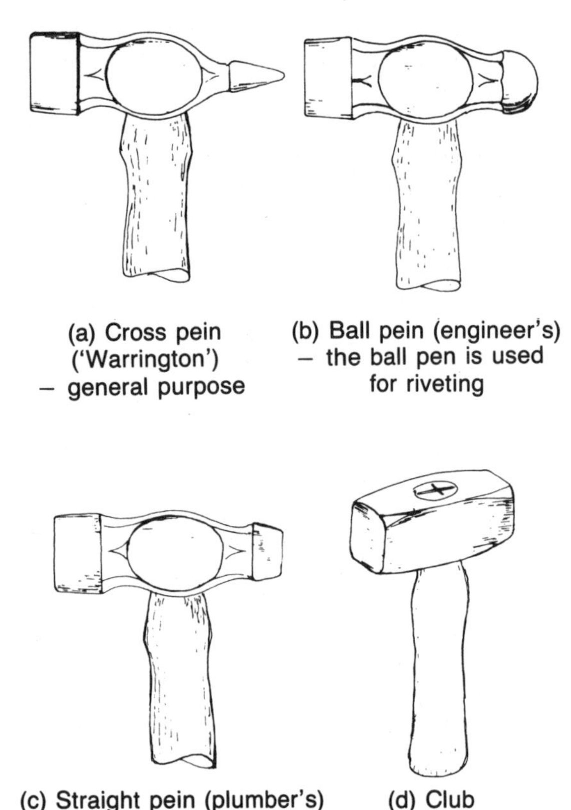

Fig. 2.34 Hammers. Hammer shafts are usually made of hickory wood which endures the impact of hammering without breaking.

length and diameter, a useful size being 15–20 mm diameter and 250–350 mm long.

Pointed chisel (Fig. 2.35(b)) This type of chisel is also of use when concrete or hard brickwork has to be removed. The point, however, tends to split rather than cut. These are also specified by diameter and length.

Plugging chisel (Fig. 2.35(c)) This type of chisel is used for removing the mortar joint between bricks in order to insert the edges of flashings or to position wooden plugs for fixings.

Bolsters These belong to the same family as cold chisels, the difference being in the shape of the blade which is wider and thinner. These tools are not generally used by the plumber, the exception

(a) Cold or hard chisel

A general-purpose cutting tool for use on most materials.

(b) Brick point

Useful for cutting away hard bricks or concrete.

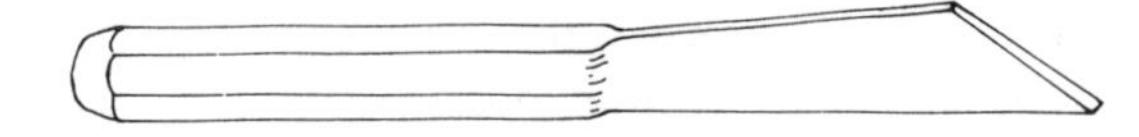

(c) Plugging chisel or joint raker

Used to remove the jointing media between bricks so that a wood-fixing plug or lead wedge may be inserted.

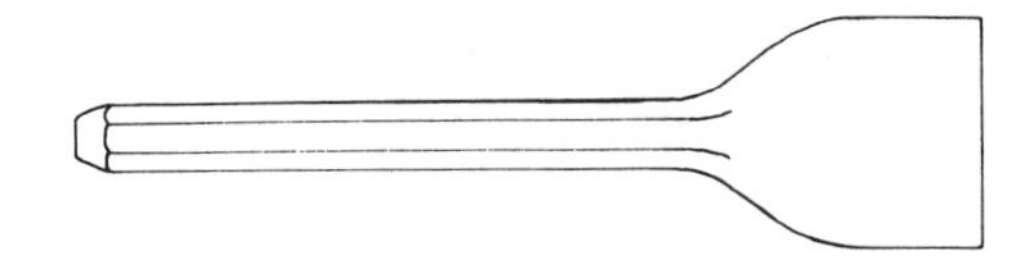

(d) Bolster or floorboard chisel

Can be used to cut the tongues of floorboard to facilitate lifting.

Fig. 2.35 Types of chisel

being the 'electrician's' or 'floorboard' chisel (see Fig. 2.35(d)) which is of the same shape as a bolster with a thinner blade. This is required for cutting through the joint between tongued and grooved flooring in order that a saw can be inserted to cut away the tongue thus enabling the board to be lifted.

Chisels, bolsters and points are all manufactured from high quality steel often alloyed with a small percentage of chromium or tungsten to increase their toughness.

Screwdrivers

It is essential to select the correct screwdriver from the many patterns available (see Fig. 2.36). For slotted screws they are identified by length and width or diameter of the blade, the point size being used only to specify those for recessed-headed screws. The blade of the screwdriver is of cast or alloy steel with a wood or plastic handle which may be electrically insulated. When using

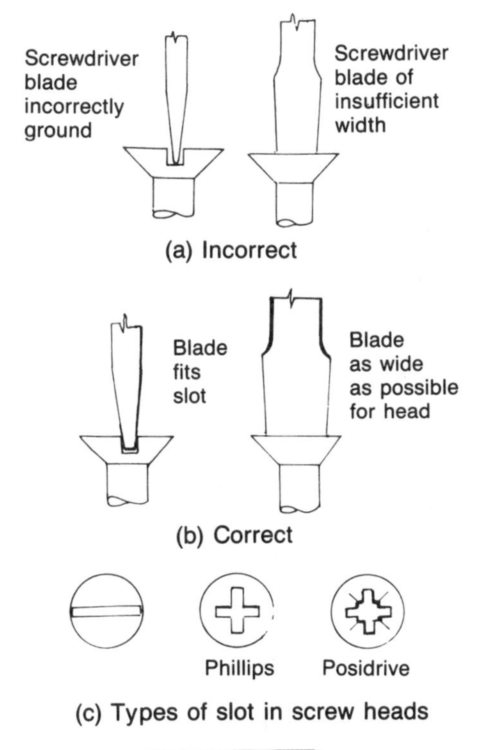

(d) Phillips type screwdriver

Fig. 2.36 Screwdrivers. A screwdriver of the correct size should always be used, i.e. the end of the screwdriver should fit snugly into the slot at the end of the screw.

slotted screws the blade should fit the width and length of slot.

Files

These tools are manufactured from high carbon steel and are classified according to shape of section, length, and spacing of the teeth as shown in Table 2.2 and Fig. 2.37.

Table 2.2 Classification of files.

File type	*Teeth per 25 mm*
Rough	20
Bastard	20–25
Second cut	30–40
Smooth	50–60

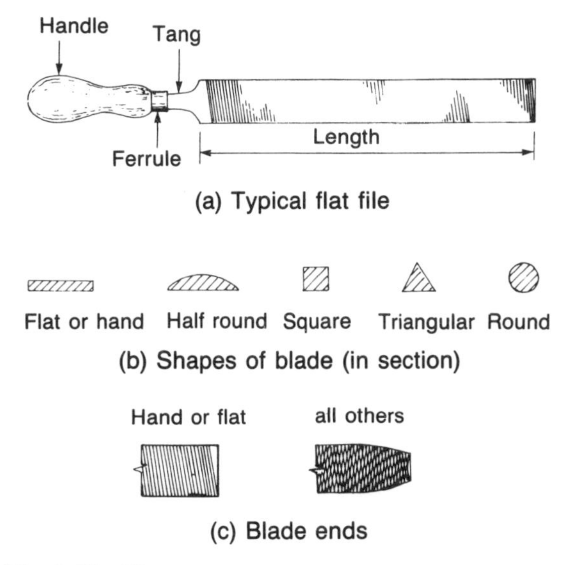

Fig. 2.37 Files

The choice of file type depends on the operation being undertaken, i.e. the amount of metal that has to be removed. For general filing the most popular lengths are 250 and 304 mm. It is essential for safe use that a good quality handle is fitted to the file tang. Apart from the different shapes, files are also made in various cuts for use on different materials (see Table 2.2).

Thread taps

A thread tap as illustrated in Fig. 2.38(a) cuts an internal (female) thread, either right or left hand, being made to produce the different types of thread listed previously in the section dealing with external (male) threads.

To form an internal thread three taps are available as illustrated in Fig. 2.38(b) a taper tap, (c) a second, and (d) a plug or bottoming tap, plus the tap wrench (e) which is required to hold the tap. To cut an internal or female thread the following procedure should be adopted.

A hole is drilled using the correct tapping drill size, the diameter of the drill being equal to the minor diameter of the bolt that is to be fitted into the tapped hole. An alternative site method is to use the largest drill which will pass through the nut or die that fits the bolt (see Fig. 2.39). The taper tap is then inserted, the taper allowing the

(a) Typical tap for cutting female threads

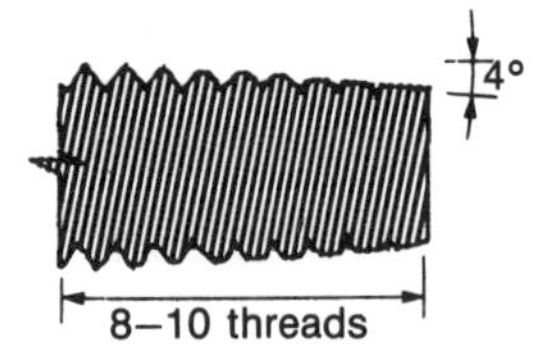

(b) The taper tap is first screwed through the hole and is followed by the second tap

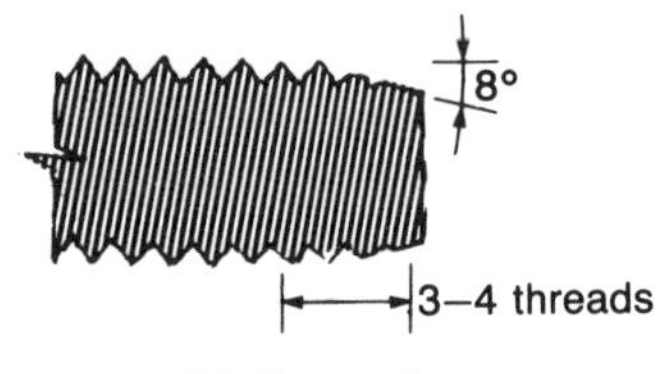

(c) Second tap

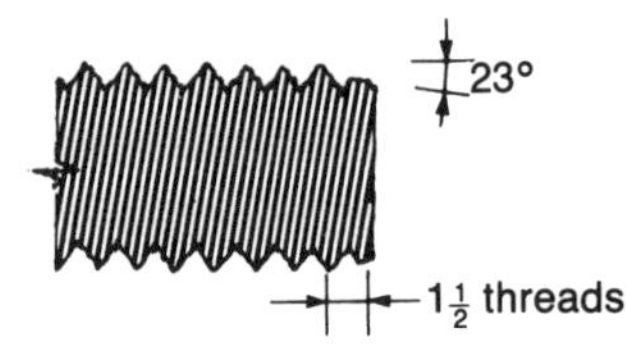

(d) The plug, or bottoming tap, provides the completed thread

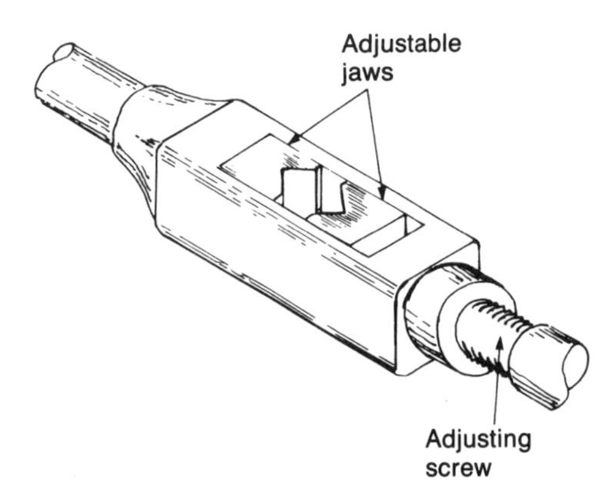

(e) Tap wrench for holding taps

Fig. 2.38 Thread taps. When a threaded hole is required in a thick sheet material, first a hole is drilled and then the taps are screwed into the hole to form a thread. Taps are also used for cleaning old or damaged female threads. To provide a good thread, taps (b), (c) and (d) are used.

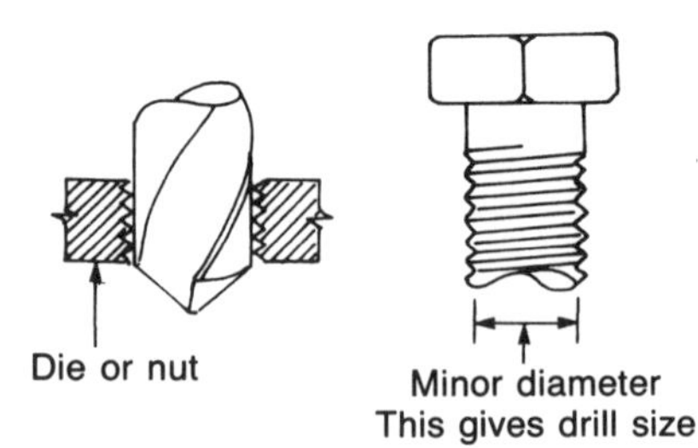

Fig. 2.39 Approximate method of determining a suitable tapping drill

tap to centre correctly. It is essential to keep the tap at 90° to the surface or it will break. Rotation of the tap wrench commences the thread which is then followed by the use of a second tap, and finally by the bottoming taps. A lubricant must be used throughout the whole operation, except when tapping cast iron, brass or plastic.

When tapping a blind hole, i.e. where the hole does not pass completely through the metal, it is necessary to drill a deeper hole than the required thread length, to allow swarf to collect and to prevent the tap binding.

Split circular dies

The dies illustrated in Fig. 2.40 are made specially for cutting male threads on solid round rods. The diameter of the die varies to that of the thread and it is important that the die stock is of suitable size for the die. Although for a 'one off' job both die holders and dies can be obtained separately, they are usually sold in sets covering a range of sizes and thread forms.

Electrical tools

Screwdrivers Many modern plumbing appliances require an electrical supply and it is not unusual for the plumber to make the electrical connections to the appliance from a socket outlet. The following tools will be a useful addition to the plumber who has to make electrical connections.

Smaller screwdrivers than those normally used will be required so that damage to the small setscrews used in electrical appliances is avoided. These screws mainly range from 2, 4, 6, 8 BA threads and it is worth noting that an 8 BA set screw has an approximate diameter of only 1.5 mm.

Two other useful tools will also be necessary.

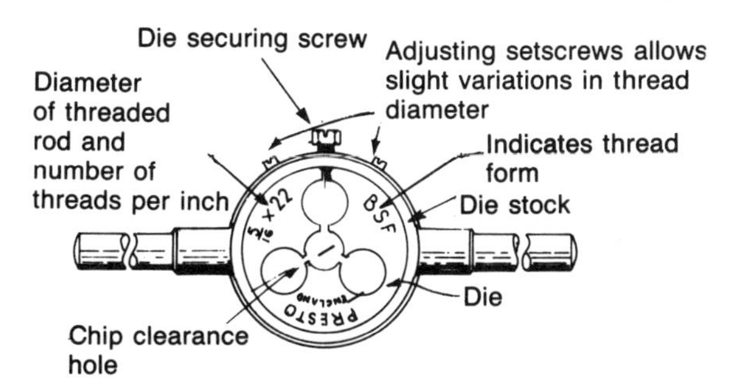

Fig. 2.40 Illustration of split circular die

Cable stripping pliers Figure 2.41 shows one type of these pliers which are adjustable to allow for various diameters of cable. It is important that prior to use the jaws are correctly set to avoid damage to the cable. The cable to be stripped is then placed in the jaws, the pliers tightened and rotated through 90°, the insulation may then be simply pulled off.

Non-adjustable pliers are also available which have a series of cutters in the jaws, which can accommodate various cable diameters, the main

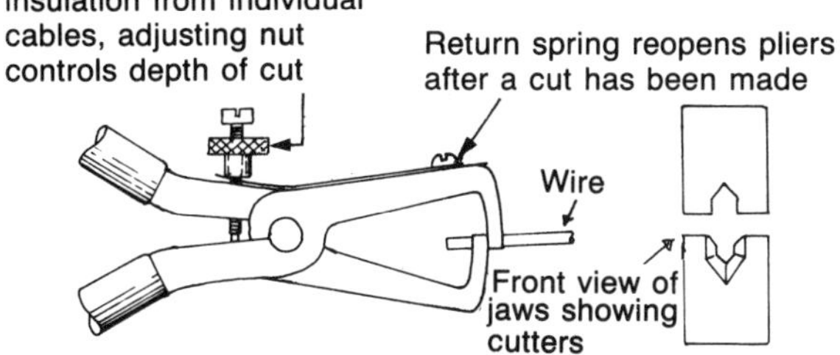

Fig. 2.41 Wire-stripping pliers. To avoid damage to the conductor it is suggested that the correct setting of the cutting jaws is checked on a piece of scrap cable prior to use.

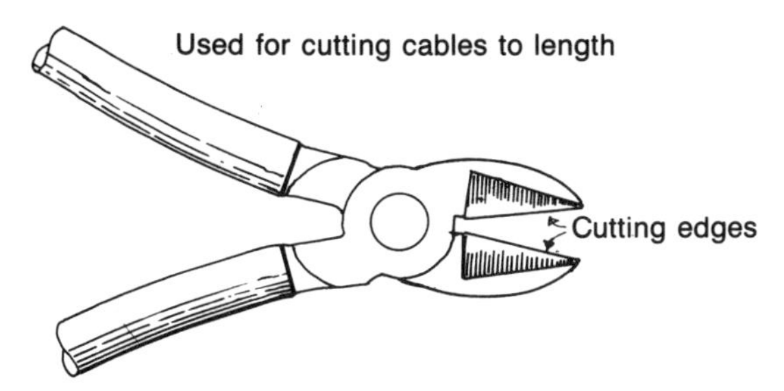

Fig. 2.42 Side-cutting pliers

disadvantage of these being a direct pull cannot be obtained when the insulation is drawn off.

Side-cutting pliers These are used for cutting cables to the required lengths. A typical example is shown in Fig. 2.42.

Care and maintenance of tools

To ensure tools have a long life and remain effective in use, it is essential that they are cared for and correctly maintained. It is bad practice to wait until a tool is not suitable for use before maintenance is carried out. Routine maintenance, and this includes cleaning, must be undertaken regularly.

It will be useful to consider the more important aspects of care and maintenance which apply to common hand tools.

The working parts of all tools must be lubricated, steel tools lightly coated with oil to prevent rusting and the edges of cutting tools maintained in a sharp condition.

Wood tools

(a) Apply raw linseed oil to prevent drying and cracking.
(b) Maintain a smooth working surface by treating with sandpaper.
(c) Ensure mallet handles are securely wedged and not broken.

Cold or hard chisels

(a) The 'mushrooming' on these and other cutting tools must be removed at the earliest stage of formation by grinding.
(b) Sharp cutting edges must be maintained by grinding.
(c) Repeated grinding will leave the cutting end too 'stubby' or 'thick'. When this stage is reached the chisel should be 'drawn out' to its correct shape as follows.
   (i) Heat the cutting end of chisel to bright red and by using hammer blows 'draw out' the end to the required shape as shown in Fig. 2.43 using a new chisel as a pattern.

   Hammering must only be carried out

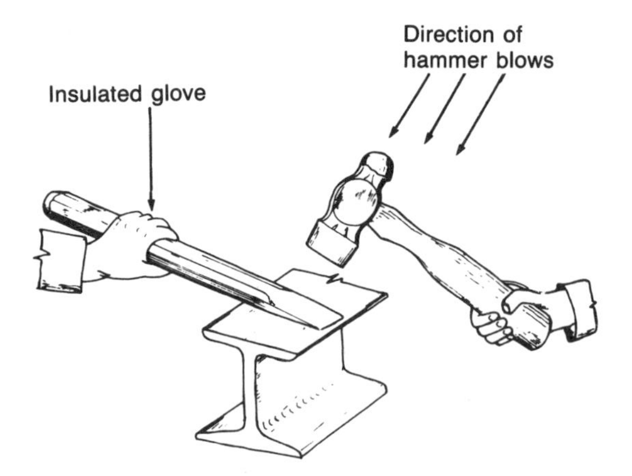

Fig. 2.43 Drawing out a cold chisel

while the chisel is at the red heat or cracking will occur.

(ii) When the correct shape is obtained, the chisel end must be heated again to bright red and allowed to cool slowly. The blade is then in a soft condition (annealed) and the final shaping can be carried out using a file or grinding wheel.

(iii) After final shaping of the blade it is again heated to a cherry red and cooled quickly (quenched) in oil or water depending on the type of steel, i.e. water or oil hardening. This process leaves the steel in a hard brittle condition (hardened).

(iv) To obtain the correct hardness, the chisel must be tempered. The blade is cleaned to bare metal and heat is applied to a point roughly 50 mm from the cutting edge. Coloured oxide films will appear where the heat is applied, these colours indicating the temperature of the steel (see Table 2.3) and as the heat is conducted so colour changes occur.

(v) When the tip colour is purple (270 °C) the chisel is quenched in water. The eventual hardness of the chisel depends upon the temperature indicated by the colour at which it is quenched. At certain specific temperatures it is known that the iron/carbon relationship will provide steel of a certain hardness, the colour purple indicating a suitable hardness for chisels used in general plumbing work. Rapid quenching then 'freezes' or 'sets' the steel at the desired state of hardness.

Table 2.3 Oxide colour and corresponding temperature.

Oxide colour	*Temperature (°C)*
Pale straw	230
Dark straw	240
Yellow brown	250
Purple/brown	260
Purple	270
Deep purple	280
Deep blue	300

Dies

(a) Worn or damaged chasers must be discarded or imperfect threads will be cut.
(b) Chasers are manufactured as a set of four and they must never be interchanged with those of another 'set'.
(c) When placing chasers in the die stock it is essential they are inserted into their correct number slot.

Files

(a) Clean file teeth using card wire.
(b) Store files away from other tools.
(c) Wrap files individually in rag when not in use.
(d) Do not use split handles.

Hammers

(a) Ensure the shaft is of the correct size for the head and secured by wedges.
(b) If the shaft dries out the head will become loose – soak the hammer head in water from time to time. Never use a hammer with a loose head.
(c) Replace damaged shafts.

Pipe cutters

(a) Replace blunt cutter wheels making sure the correct wheel is used – wheels are available for plastic, cast-iron, copper and stainless steel.
(b) Replace worn rollers.
(c) Roller and wheel pins (spindles) must be removed when worn or spiral cutting will occur.

Pipe grips and wrenches

(a) Ensure the teeth are not 'clogged' with jointing paste or other matter or the tool will slip when in use.

(b) When teeth become worn, the defective part or the whole tool must be replaced.

Screwdrivers

Never grind the end of a blade for slotted screws to a knife edge. The blade end must be the same thickness as the slot in the screw head.

Spanners

Open-ended spanners which are 'splayed' (strained open) must be discarded as they are dangerous. Ring or open-ended spanners which are worn must also be replaced for safety reasons.

Twist drills

(a) Maintain correct point angle and sharp cutting edge.

(b) If accurate drilling is required it is inadvisable to sharpen by hand – use a grinding jig or have the drill sharpened by an expert.

(c) Approximate angles and edges can be achieved by grinding, a special wheel being needed for tungsten-tipped drills.

Further reading

Information relating to tools may be obtained from the following sources:

Clarkson Osborne International, PO Box 37, Penistone Road, Sheffield, S6 3AH.

Monument Tools, 22–24 Balham Hill, London, SW12 9EB.

Record Tools Ltd, Parkway Works, Sheffield, S9 3BL.

Ridge Tools, Arden Press Way, Pixmore Avenue, Letchworth, Herts, SG6 1LH.

Rothenberger (UK) Ltd, 14–18 Tenter Road, Moulton Park Industrial Estate, Northampton, NN3 1PS.

S.K.S. Dormer Ltd (drills and threading equipment), Summerhill Street, Birmingham, B1 2PD.

Self-testing questions

1. State the fault in pipe cutters which leads to tracking or spiral cutting.
2. List the information necessary when selecting hand saw blades.
3. State the maintenance requirements for wood tools.
4. List the tools that would be required to form and install the lead flashings round a chimney stack.
5. Select a suitable tool for removing the head work of a tap.
6. State the type of wrench that can be used for the installation of LCS pipework systems.
7. Describe two methods of selecting a tapping drill size.
8. Select a suitable tool for cutting a male thread on a small circular bar.
9. Explain the terms hardening and tempering when related to tool steels.
10. Name the tool shown in Fig. 2.44 and describe its use.

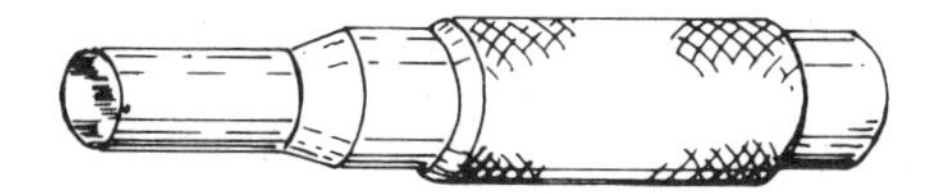

Fig. 2.44 Question 10

11. Describe why it is necessary to use a drip plate and roll overcloaks in sheet lead.
12. Why is it necessary to use a lubricant when cutting threads on LCS pipes.

3 Materials and fittings for gas, water services and sanitary pipework

After reading this chapter the reader should be able to:

1. Name the material used for pipes, fittings and components.
2. Understand the SI metric system of calculating and expressing force (pressure).
3. Describe the constituents of the main alloys used in the manufacture of plumbing materials.
4. Recognise and describe the types and grades of steel and copper pipework and principles of jointing.
5. Select materials, fittings and components for differing types of work.
6. Describe the effects of corrosion on metals and identify methods of protection.
7. Identify the characteristics and limitations of pipework used for sanitation.
8. State the advantages and disadvantages of various pipework materials.
9. Describe simple site tests to identify the main thermoplastic materials used for sanitation.
10. Demonstrate knowledge of the physical and chemical properties of thermoplastics used for pipework services.

Introduction

The installation of pipework systems for the supply of hot and cold water – and to a lesser extent gas – forms the major part of the work of the plumber. This means that it is essential for the craftsman to have a thorough understanding of the performance of pipework materials in current use.

The earliest recorded use of metallic pipes for water supplies is about 2750 BC when lead and copper piping was used by the Egyptians. Lead piping was also used by the Greeks and Romans, and apart from terracotta (clay), wood (bored trunks of oak, elder and elm) and the introduction of steel, it was the principal pipe material used by plumbers until after the Second World War. Since then there has been a gradual decline in the use of lead pipe due mainly to its cost and weight, the high levels of skill required in its use and the danger it has been found to pose to health. It was, however, still in use for repair and maintenance work until the 1986 Water Bylaws prohibited the use of lead or lead alloys for potable water supplies.

The principal materials used nowadays for plumbing and gas services are steel, copper, lead, brass and plastics. To appreciate the qualities of each material, certain factors such as pressure must be understood in addition to those physical and working properties given later in this chapter.

Pressure in pipes and fittings

It is necessary for a plumber to have a basic knowledge of the effects of pressure or force on the pipes and fittings he installs. The internal pressure to which a pipe or vessel is subjected will affect the strength of the material specified, i.e. a higher pressure will require the use of a thicker or stronger material.

The basic unit of pressure or force in the SI system is the *newton*. Its name is derived from the seventeenth-century scientist Isaac Newton, who discovered from his experiments that the centre of the earth exerts a positive pull on all forms of matter. When an object is said to be heavy, two things are really being considered:

(a) its weight in relation to other objects
(b) its mass in relation to the pull of gravity.

It is the latter that is considered in relation to water pressure.

As confusion often arises due to the close relationship between force and pressure the following explanation may help. Pressure is force, expressed in newtons (N), distributed over an area (in metres squared) thus the unit of pressure is expressed in newtons per m^2. This unit is also known as the pascal (Pa) and is accepted as the official SI unit. Thus pressure expressed in newtons per m^2 (N/m^2) has numerically the same value. The reader should note that the examples in this book will be expressed in N/m^2.

Calculating intensity of pressure

The pressure exerted by a column of water may be given in terms of metres head (i.e. how high it is) but this height does not indicate the force or pressure to which a pipe or vessel will be subjected, and for this reason it is converted to newtons. To find the pressure created by a given mass of water the following formula is used:

$$\text{Pressure} = \frac{\text{Mass of water (kg)}}{\text{Area (m}^2\text{)}}$$

One kilogram is the force exerted by a mass of one kilogram of water and can be expressed as 9.81 newtons thus:

$$1 \text{ kg force} = 9.81 \text{ newtons}$$

So the formula becomes:

$$\text{Pressure} = \frac{\text{Force (N)}}{\text{Area}}$$

and the answer will be in newtons per square metre, or N/m^2.

If a force of 10 kg is involved it will be equal to 98.1 N as illustrated in Fig. 3.1(a) and (b).

A cistern holding one cubic metre of water will have a mass of 1,000 kg. The pressure on its base of one square metre will be 1,000 kg, which when converted to newtons becomes:

$$1{,}000 \times 9.81 = 9{,}810 \text{ N/m}^2 \text{ (newtons per square metre)}$$

The reader should note that the number 9.81 is often approximated to 10 in industry for ease of calculations. Consider now the force exerted by a column of water 5 metres high having a base of 1 square metre.

$$5 \times 1{,}000 \times 9.81 = 49{,}050 \text{ N/m}^2$$

The newton is a very small unit and its use leads to excessively large numbers which become difficult to manipulate and can lead to error. For this reason it is usual to convert newtons to kilonewtons. A kilonewton (kN) is 1,000 newtons and, as indicated previously, 1 cubic metre of water has a mass of 1,000 kg so the figures in the example can be changed as follows:

$$49{,}050 \text{ N/m}^2 \div 1{,}000 = 49.05 \text{ kN/m}^2$$

This is a more convenient number to use for water pressure and is usually adopted for plumbing calculations.

The foregoing relates to the intensity of pressure over an area of 1 square metre, due to, in this instance, the height of a column of water. As the area may be more or less than 1 square metre this term is often omitted, using the term newtons or kilonewtons only.

Total pressure

This relates to the pressure exerted over a given area subjected to a column of water of a given height. Supposing a vessel which had a height of 1 metre and a base measuring 3 metres by 2 metres, then it would hold 6 cubic metres of water. Over the whole area of the base there would be a force of

$$6 \times 9.81 \text{ kN} = 58.85 \text{ kN}$$

i.e. a *total force* of 58.85 kN yet the intensity of pressure would still be 9.81 kN/m^2, as shown in Fig. 3.1(d). If the vessel was 3 metres in height but the base was still 3 metres by 2 metres then

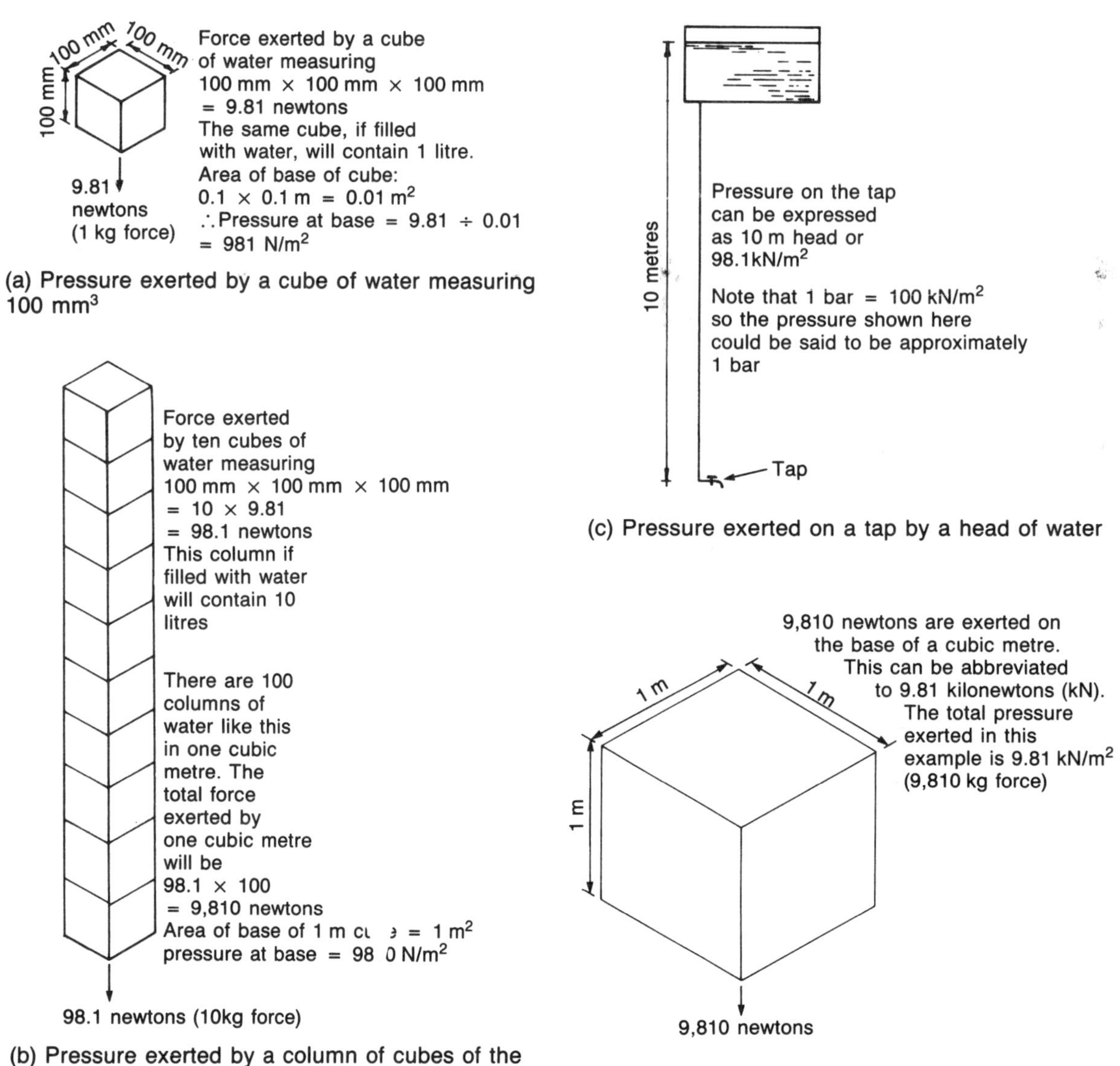

(a) Pressure exerted by a cube of water measuring 100 mm³

(b) Pressure exerted by a column of cubes of the same measurement

(c) Pressure exerted on a tap by a head of water

(d) Pressure exerted by a cubic metre of water

Note that one cubic metre of water contains 1,000 litres and has a mass of 1,000 kg

Fig. 3.1 Water pressure

the intensity of pressure would be $3 \times 9.81\ \text{kN/m}^2 = 29.43\ \text{kN/m}^2$, and total force would be area of base $\times\ 29.45 = 6 \times 29.43 = 176.58$ kN.

Taking a common example in plumbing, that of the pressure on a tap washer, assume a tap is fed from a tank as shown in Fig. 3.1(c). The 10 metres head of water would give an intensity of pressure of $9.81 \times 10 = 98.1\ \text{kN/m}^2$, i.e. 98.1 kN on each square metre. Assume the tap washer had an area of $0.0012\ \text{m}^2$, then the total force on the tap would be $98.1 \times 0.0012 = 0.11772$ kN. In this instance it would be more sensible to revert to newtons, i.e. multiply 0.11772 by 1,000 which gives 117.72 N acting on the tap washer.

In some cases involving the use of very high

pressures, for example compressed oxygen or mains water supply, a unit of pressure called the *bar* is often used. Permissible working pressures on pipes are also given in bars. While the bar is not a unit officially approved by the mathematicians responsible for the evolution of the SI metric system, it is a very convenient unit commercially, 1 bar being equal to 100 kN/m^2. To give a practical example of its use, the maximum working pressure to which 15 mm Table X copper tube should be subjected is 58 bars. Given in kN/m^2 the same pressure would be 5,800 kN/m^2.

Pipework materials

There is no perfect pipework material that is suitable for all applications, each must be considered in relation to such factors as type of water, pressure, cost, bending and jointing methods, corrosion resistance, compatibility with other materials, expansion and appearance.

Alloys

Alloy is a term which is commonly used when describing metals, and it is important to understand the meaning of the term before discussing pipework materials and fittings.

Alloys can be produced either by mixing different metals or by mixing metals with non-metallic elements, such as carbon or phosphorus, to form a new solid having different characteristics to the parent metal. The reason for the formation of these alloys is that in many instances a metal in its pure form is unsatisfactory for a particular function. When it is blended with certain other materials it acquires additional properties such as hardness, toughness and corrosion resistance. Alloys commonly used by plumbers are shown in Table 3.1.

Table 3.1 Composition of alloys.

Alloy	*Main elements*
Brass	Copper and zinc
Bronze	Copper and tin
Steel	Iron and carbon
Soft solder	Lead and tin

Low carbon steel

One of the alloys which has a wide use is commonly known as *mild steel* but the more correct term *low carbon steel* (LCS) will be used. From Table 3.1 it can be seen that steel is an alloy of iron and carbon The typical carbon content for various steels is indicated in Table 3.2.

Table 3.2 Percentage of carbon in tube steel.

Type of steel	*% Carbon*
High carbon	0.50–1.40
Medium carbon	0.25–0.50
Low carbon	0.15–0.25

High carbon and medium carbon steels are generally used for tools and specialist engineering purposes. LCS is the usual material for steel tubes used by plumbers. It is manufactured to BS 1387 in three grades of weight: light, medium and heavy. The outside diameter of a tube of each grade is similar, the difference being the tube wall thickness, as can be seen from the examples given in Table 3.3 which also specifies the painted 50 mm wide colour band used to identify each grade.

Only medium or heavy tubes are used for gas and water services, the choice being dependent on the internal operating pressure to which it is subjected. Heavy tube is normally required for water or heating services while medium quality is specified for supplies fed from a storage cistern within the building. Tubes to BS 1387 can be obtained either painted black or coated with zinc (galvanised).

Table 3.3 Wall thickness and diameters of LCS tubes.

Nominal bore	*Outside diameter (average) in mm*		*Wall thickness in mm*		
	L	*M&H*	*L*	*M*	*H*
25 mm	33.5	33.7	2.65	3.25	4.05
50 mm	59.9	60.2	2.9	3.65	4.5
100 mm	113.4	114	3.65	4.5	5.4
Colour band			Brown	Blue	Red

Galvanised tube must always be used for domestic water supplies as the zinc coating prevents corrosion and the process of oxidation. Chapter 10 explains fully how when rusting takes place the oxide (rust) flakes off ferrous material thus exposing new metal to be affected. Water contains oxygen which causes oxidation, the rust discolours the water and eventually the process leads to the pipe being 'eaten away'.

Medium and heavy grade tubes are obtainable in 6 m lengths of 6 mm to 150 mm nominal bore – the bore is known as *nominal* as the *actual* internal diameter is dependent on the tube wall thickness (see Table 3.3). It is very important when storing construction materials on site that floors are not overloaded or they will collapse. This means the quantity of tube stored in one position must be carefully controlled and it is necessary to calculate the mass involved. The average mass of screwed and socketed tube of more common sizes is given in Table 3.4.

Table 3.4 Mass/kg of LCS tubes.

Nominal bore (mm)	*kg per metre run*	
	Medium	*Heavy*
15	1.23	1.46
20	1.59	1.91
25	2.46	2.99
32	3.17	3.87
40	3.65	4.47
50	5.17	6.24

Joints and fittings Steel tube for water and gas services is usually joined by means of screwed joints, the threads conforming to BS 21. Black tube (i.e. not galvanised) can also be joined by welding which is a process beyond the scope of this book. Galvanised tube, however, must not be fusion welded as the heat required would remove the zinc coating and leave the steel unprotected against corrosion attack. A comprehensive range of pipe fittings is available both for screwed and welded joints, the latter type having no threads but the outer edge bevelled to provide the necessary joint preparation. Some of the more common screwed fittings are illustrated in Fig. 3.2.

Pipe fittings are manufactured to four British Standards. Fittings made from white cast iron and having parallel threads on the female end with taper threads on the male end are covered by BS 1256. Fittings made from white cast iron and having taper threads on both male and female ends are covered by BS 143. As cast iron is brittle, fittings to these standards are heat treated to give some ductility, i.e. the treatment for BS 1256 decomposes the hard carbides at the surface of the fitting into ductile iron and soft graphite, while that used for BS 143 anneals the surface by burning out a large proportion of the carbon. After heat treatment the fittings are referred to as malleable cast iron or malleable iron.

Steel is used to make fittings to BS 1740, these having parallel female threads and tapered male threads. Although more expensive than those made from cast iron, these fittings are able to withstand higher internal pressures. Another group of fittings often encountered are known as *tubulars* and are formed from steel tube to BS 1387.

Malleable cast iron pipe fittings are usually provided with a beaded end, while steel fittings have a plain end. A beaded end (see Fig. 3.2(f)) provides a reinforcement on malleable fittings which strengthens the outer end and prevents splitting when it engages with a taper pipe thread.

Compression joints for LCS tubes These are made of malleable iron, but instead of threads, they have special locking rings and seals which are tightened on to the pipe. They are suitable for use with gas, water and compressed air. The same type of joint using different locking rings may also be used for polythene pipes. This feature enables joints between steel and polythene to be made quickly and easily. They are extensively used by the gas and water utility companies for underground services due to their flexibility for both new and repair work. Figure 3.2(i) shows a straight coupling with one end having a locking ring for polythene pipe, the other threaded for steel. A full range of tees, elbows, reducers, etc., is manufactured for steel pipelines above ground for heavy duty work, i.e. industrial and commercial work.

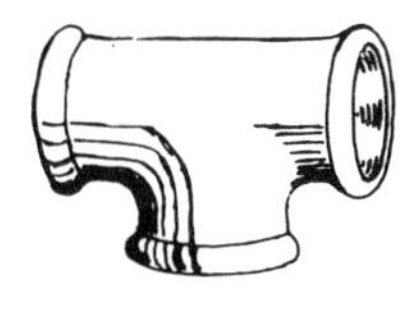

(a) Tee

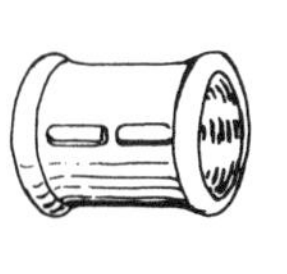

(b) Socket

(c) Hexagon nipple

(d) Sweep tee

(e) Male and female ended elbow

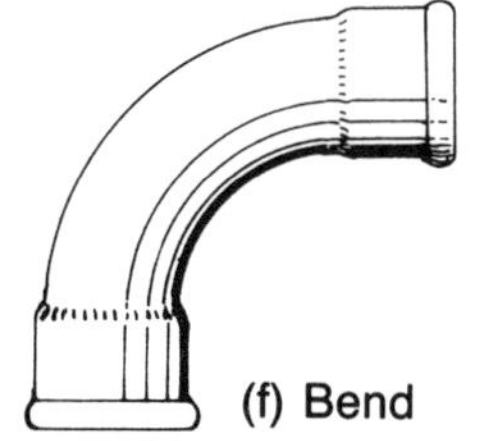

(f) Bend

Malleable iron fittings can be identified by the strengthening beads round each socket. These beads prevent the sockets splitting as the pipe is tightened home

(g) Union connector

Ground in, flat or domical seatings are available with this type of connector

Parallel threads

Tapering threads

(h) Threads

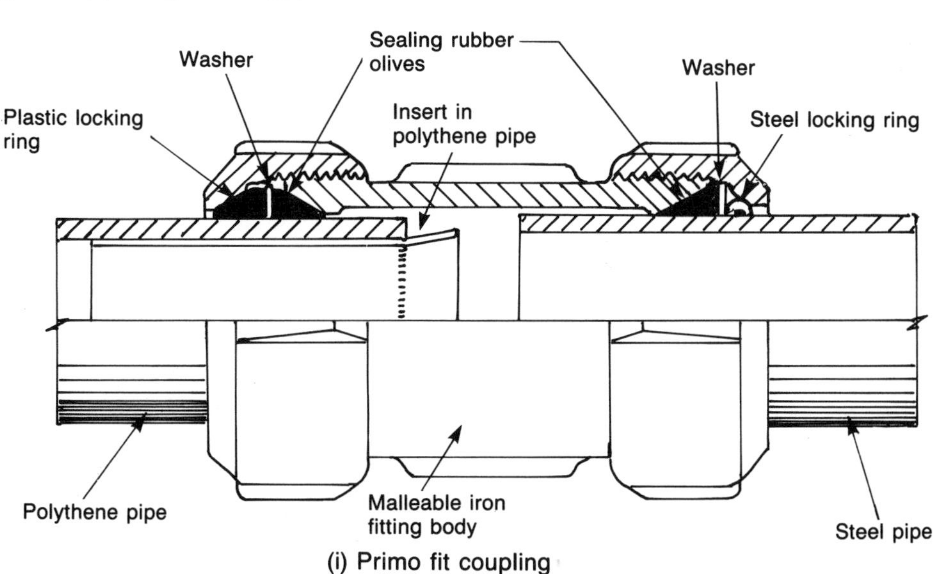

(i) Primo fit coupling

This fitting is designed to enable steel pipes to be jointed without threading. By using different olives and locking rings they can also be used for polythene services. They save time and are especially useful for emergency repairs in public buildings. They are not suitable for steel pipes that are severely corroded externally.

Fig. 3.2 Typical fittings and threads used for steel pipework (BS 21)

Stainless steel

This is the most recently developed metallic pipe material for domestic water and gas services. The alloy used is an austenitic type, made to BS 4127 and having a chemical composition of 18 per cent chromium, 10 per cent nickel, 1.25 per cent manganese, 0.6 per cent silicon, a maximum carbon content of 0.08 per cent, the remainder being iron with small amounts of sulphur and phosphorus. The chromium and nickel provides the tube with its shiny appearance and its resistance to corrosion, protection being due to a microscopic film of chromium oxide which rapidly forms on the tube surface and prevents further oxidation.

This tube is available in nominal bores of 6 mm to 35 mm but only one grade (light gauge) with an average wall thickness of 0.7 mm, the outside diameter and mass being similar to copper tube to BS 2871 (see *Table X* of Table 3.5). Tubes of 15 mm to 35 mm diameter are supplied in 6 m lengths, with smaller diameters in 3 m lengths, or longer to special order.

Joints and fittings for copper tubes Standard copper and gunmetal fittings to BS 864, Part 2, are generally suitable for jointing this tube. The British Water Association has approved three categories: compression types of either manipulative (type B) or non-manipulative (type A), and capillary (see Figs 3.3 and 3.4), although special fluxes are required for soldered joints on stainless steel tubes. It is not generally used for gas work but where it is, the use of compression fittings is recommended, bearing in mind they must always be accessible.

Complete systems of discharge pipework are made of this material, and properly installed will have a service life in excess of fifty years. Push-fit joints with synthetic rubber rings, similar to those used for plastic pipes, form the main method of jointing for this purpose. It is especially useful in commercial kitchens where its low expansion rate is an advantage. It is very strong, and having a high polish it is easily cleaned, which is a very desirable characteristic when used in food preparation processes. Due to the special chamfer on spigot ends, if the pipe has to be cut, the end must be re-chamfered with a special tool which is obtainable for purchase or hire from the material supplier.

Table 3.5 Comparison of 15 mm diameter copper tube to BS 2871. (EN 1057: 1996)

Specification	*Mass (kg/m)*	*Temper annealed*	*Wall thickness (mm)*
Table X (R250)	0.2796	Half hard	0.7
Table Y (R220)	0.3923	Fully annealed	1.0
Table Z (R290)	0.2031	Hard	0.5

* NOTE: The text in brackets in Table 3.5 relates to the latest British and European Standards for copper tubes. It should be noted howover that references to copper tubes throughout this chapter relate to BS2871, e.g. Tables X, Y, Z.

Stainless steel is also commonly used for sparge pipes and other integral tubes used with sanitary appliances. Like chromium-plated copper it can be polished, but is much stronger and offers more resistance to the vandalism of appliances in public buildings.

Lead

Where connections from existing lead services to other materials have to be made, the best and most economic method is to use one of the special adaptors which are available, two of which are illustrated in Fig. 4.1.

Lead has also been superseded for sanitary and rainwater pipes, mainly because of its high cost and the special skills necessary for its installation. Its use is now limited to pipework in old historic buildings, where it is installed and maintained by plumbers specialising in this work.

Copper

Copper tube has been used in many public buildings since as early at 1900 although in those days connection was by screwed and socketed joints which required a pipe wall thick enough to thread. The first British Standard was issued in 1936 to standardise the various sizes of tube on the basis of their outside diameter. When a method of jointing copper tube without threading was introduced, the tube walls were made much thinner and were identified by the designation of 'light gauge tubes'. Wall thickness of light gauge tube has been reduced twice since 1936 in order to use less copper and thus lower the costs. A typical example of this can be seen in 15 mm tube which in 1936 was 1.2 mm thick, reduced in 1954 to 1 mm, and subsequently to 0.7 mm in 1964.

Copper tube for use in the construction industry is manufactured to BS 2871. Part 1 covers tube to three specifications, these being denoted as *Tables X, Y* and *Z*. (Parts 2 and 3 of the specification relate to tube for general purposes and are not relevant to building practice.) *Table X* specifies the requirements for light gauge tube of half-hard temper. *Table Z* refers to tubes of hard temper, the increased hardness allowing for significant reductions in wall thickness and weight but making the tube unsuitable for bending. Tubes complying with these two tables are not intended to be buried underground. *Table Y* covers tube of either half-hard or fully annealed temper, that is tube to be laid underground for the conveyance of gas or water. The basic differences between these tubes is their temper and wall thickness, the latter of course affecting the mass. A comparison of 15 mm diameter tube made to the three different specifications is given in Table 3.5.

Tube to *Table Y* specification, often called soft copper, is supplied in sizes from 6 mm to 54 mm in 20 metre coils and from 6 mm to 108 mm in up to 6 metre straight lengths. The other two grades are only available in 6 metre lengths in sizes from 6 mm to 159 mm. The outside diameter is similar for each type of pipe but there are differences in the internal bore due to the variations in wall thickness.

Joints and fittings Jointing methods for three types of copper tubes are shown in Table 3.6, the most commonly used being capillary or compression fittings.

The design and manufacture of capillary and compression fittings of copper and copper alloy (brass and gunmetal) are covered by BS 864:Part 2. Capillary fittings are of two main types, the difference being in the method by which solder is introduced into the joint, these being (a) at the end of the fitting, or (b) from a reservoir of solder in the form of a ring incorporated inside the fitting during manufacture. These fittings are illustrated in Fig. 3.3.

It should be noted that to conform with the water regulations only lead-free solders may now be used on water services. Several alloys are available including tin silver, tin antimony and tin copper. Tin silver has very good 'wetting' or tinning characteristics but is expensive, tin copper solders are more common due to lower costs. Higher soldering temperatures than those for lead tin alloys are required, but in practice are negligible. Joints made with lead-free solder are stronger, especially at high temperatures. Traditional fluxes used for tin lead alloys can also be used for tin silver and tin copper solders but special fluxes have been developed to better withstand the higher soldering temperature necessary.

Table 3.6 Jointing of copper tubes.

Jointing method	*Type of tube* X	Y	Z
Capillary fittings	✓	✓	✓
Compression fittings			
Type A	✓	—	✓
Type B	✓	✓	—
Bronze welding (not normally employed or recommended for water services)	✓	✓	—

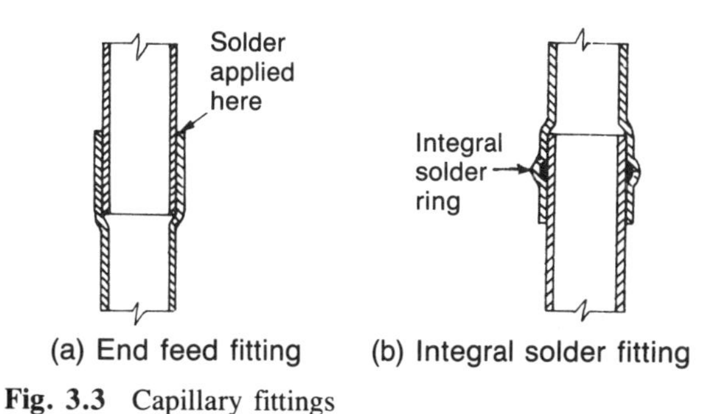

Fig. 3.3 Capillary fittings

Compression fittings are of two basic types, manipulative and non-manipulative, these names merely indicating the preparation of the tube end. The non-manipulative pattern requires no special preparation but for the manipulative type the tube end has to be formed – manipulated – using special tools. An example of each type is given in Fig. 3.4.

A comprehensive range of both capillary and compression fittings such as elbows, bends, tees and unions plus the necessary adaptors for joining metric to imperial sizes is available. Generally all fittings have a catalogue number which should be quoted with the size of fitting required. If no catalogue is available it is necessary to indicate the branch and end sizes of tees in the following order:

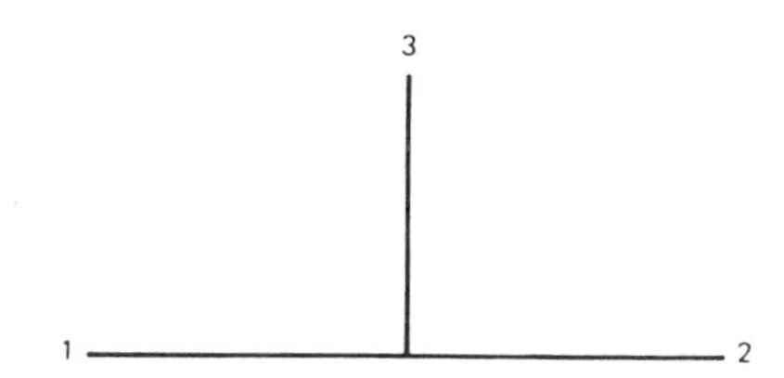

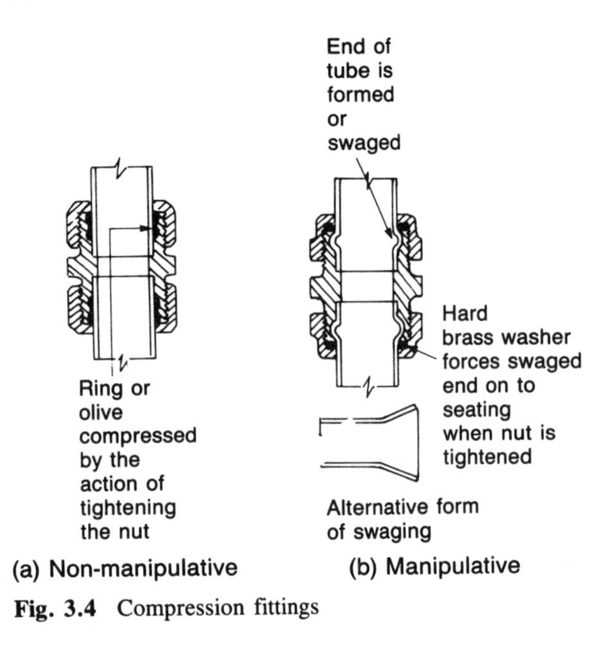

Fig. 3.4 Compression fittings

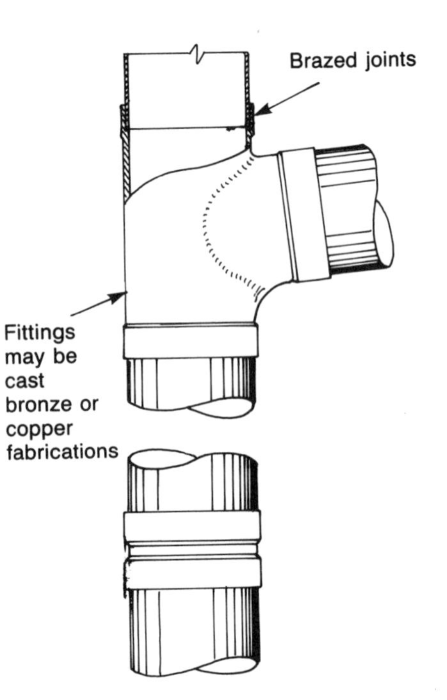

Fig. 3.5 Typical brazing fittings for large-diameter (54 mm upwards) copper tubes. These fittings are sometimes used in public and commercial buildings where a long maintenance-free life is required. The type shown are for sanitary pipework, but similar fittings are available for use with large-diameter hot and cold water services.

Therefore a 25 mm tee with the end reduced to 22 mm with a 15 mm branch would be ordered thus: 1 Tee, 25 × 22 × 15. The same information and method of ordering applies to tees for steel pipes.

Copper tubes to BS 2871 are often used for sanitary pipework in public and commercial buildings where a long working life is essential. It is an expensive material, but its resistance to corrosion and its smooth internal bore make it very suitable for discharge pipework. The rigidity and comparative light weight of copper pipes reduce the number of fixings necessary and also enable it to be fabricated off the site and fitted as a complete unit. Individual sections of large units are then fitted on site by straight couplings.

A wide choice of methods for jointing copper tubes is available. The capillary and compression joints described above are obtainable for pipes of up to 54 mm diameter. Compression fittings are not normally used as a means of making large-diameter joints due to their high cost, but they are very useful in cases where it is necessary to dismantle sections of pipework periodically for cleaning purposes, a typical example of this being discharge pipes fitted to urinals which frequently become obstructed. A special form of solder capillary fitting is made for water services having pipe diameters in excess of 54 mm. For sanitary pipework, joints can be fabricated by hand or by special forming tools using bronze welding or brazing to make the joints. An alternative is the use of large capillary fittings made of copper or cast bronze with brazed joints shown in Fig. 3.5.

Cast iron

For many years cast iron has been used as a material for both underground and overground sanitary pipework. Its use for sanitary pipework is confined to main discharge and ventilating pipes, the branch pipes being of copper or synthetic plastic materials. Pipes of less than 50 mm diameter are not normally produced, as cast iron is insufficiently flexible for direct connection to sanitary fittings other than WCs.

Cast iron is an alloy of carbon and iron.

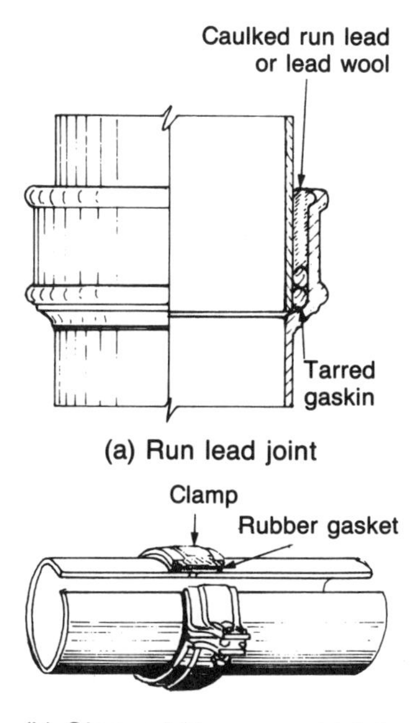

This method of jointing cast iron pipe employs the use of plain ended pipes and fittings which are jointed with a rubber gasket secured by a cast iron clamp bolted in position

Fig. 3.6 Jointing cast iron pipes

Although it is brittle in comparison with other iron alloys, it is capable of taking some very hard knocks and will stand up to exceptionally hard wear. Its chief drawback is its weight, being a very heavy material to handle.

The standard length for overground use is 1.8 m, but shorter lengths can also be obtained, stock sizes being 1.2 m, 900 mm and 600 mm. Sanitary pipework is fabricated on site using the appropriate lengths of pipe and the choice of a wide variety of fittings.

Casting Cast iron pipes are made by sand or centrifugal casting. Sand cast pipes are cast with the socket in the downward position to give the metal greater density, enabling it to take the imposed stresses when it is caulked. Pipes made in this way are identified by the mould marks on the back and front of the casting, and a slight bow is often perceptible throughout its length due to shrinkage of the metal on cooling. Centrifugally cast iron, sometimes called 'spun iron', pipes are made in a revolving mould, which as it turns throws the molten metal outward by centrifugal force, thus giving the casting as a whole greater density and a more even thickness to the pipe wall. Spun pipes are recognised by their more even texture and the absence of fixing ears on the sockets. As a protection against corrosion, to all cast iron pipes a coating of bituminous solution is applied by the manufacturer.

When an iron casting has been made, internal stresses are set up in the metallic structure as it cools. These stresses are relieved by a method called normalising during which the casting is re-heated to approximately 880 °C and allowed to cool slowly. During this process, the grain structure of the casting adjusts to a stress-free condition.

While it is very rare, a split or crack can occur in castings, due to either a manufacturing defect or by careless handling. It is good practice to test castings before they are fixed, by holding them clear of the ground and delivering a light blow with a hammer. A clear ringing note will indicate that the casting is sound.

Fixing Cast iron pipes are fixed to the structure of a building by ears cast on the socket, or where no ears are provided, a holderbat bracket may be built into or screwed to the wall. Special pipe nails are available for fixing, but the materials of which modern buildings are constructed are sometimes so hard that stout screws and plugs are used for fixing. The pipe should be fixed in such a way that it is spaced off the wall to permit the back to be painted. Special spacers can be obtained for this purpose but unless they are specified, short pieces of steel pipe can be used as a substitute.

Jointing The traditional method of making joints in cast iron pipe is yarning and caulking with run lead (see Fig. 3.6(a)). Yarn or gaskin used in making these joints is a soft stranded hemp rope which may be used in its natural condition or may be impregnated with tar. (The tarred variety is best, the tar acting as preservative.)

Lead-caulked joints are no longer permitted underground as they lack the flexibility required by the building regulations. They are, however, permitted for above ground drainage and sanitary pipework systems and have some advantages over flexible joints in this situation.

The object of using yarn is two-fold, it centralises the spigot in the socket and it prevents the molten lead from running inside the pipe. The yarn should occupy no more than one-third of the socket when it is consolidated – usually two turns are sufficient. Molten lead is poured into the socket and caulked down using the caulking tools described in Chapter 2, Fig. 2.26(a) and (b). For joints in the horizontal position a squirrel tail is used (see Chapter 2, Fig. 2.26(c)). A check must be made to ascertain that the pipe socket is free from water or moisture before the lead is poured, as failure to do this may result in a serious accident. The hot lead would cause the water to boil violently and result in what is best described as an explosion of molten metal. A little tallow or similar grease in the socket prior to pouring will help to avoid any danger, but goggles should always be worn when carrying out this operation.

Finely shredded lead called lead wool is sometimes used as an alternative to molten lead to avoid the dangers described, this being fed into the socket and caulked home in the same way as poured lead.

It is now mandatory that all underground joints on drains are flexible, and in many cases such joints have been adapted for above-ground use thus avoiding all the problems associated with run lead. One such joint is made by using plain ended pipes jointed by synthetic rubber sleeves which are held in place by a bolted clamp (see Fig. 3.6(b)). One obvious advantage of this method of jointing is where new connections need to be made to existing stacks, when a complete section can be removed to facilitate the additional junctions.

Synthetic plastics

General characteristics The plastics industry has made a big impact over the last 3 decades in the supply of materials for the mechanical services industry. The advantages of these plastic materials are their light weight and the simple methods used in the jointing process.

The term 'plastics' is a very loose one, and can be applied to almost any material that can be manipulated. Synthetic plastics is a better term to use but even this can be confusing as only a few of these materials are widely used in plumbing. It is important for the plumber to be able to identify the type of plastic he is using, as the jointing technique suitable for one type may be unsuitable for another. For this reason it is essential that the plumber is aware of a few simple tests and observations which will enable him to correctly identify a particular material. These simple tests will be described as each of the principal materials are discussed.

Plastic materials used in plumbing have many physical characteristics in common. Those used for pipes are invariably thermoplastics, all of which have a high resistance to corrosion and acid attack, a useful attribute especially where chemical or laboratory wastes are concerned. They have a low specific heat, a term which implies that they do not absorb the same heat quantity as metals. They are also poor conductors of heat.

One of the biggest drawbacks in the use of synthetic plastics is their high rate of expansion. Manufacturers make allowance for this in the design of their products, and providing the instructions for use of the material are followed, no problems from this source will be encountered during its service life.

Plastics are not such stable materials as metals and are all, to some degree, affected by the ultraviolet rays of sunlight. The long-term effect of these rays is to cause embrittlement, usually referred to as 'degradation'. Most plastics in their natural state are clear and colourless, and to offset the effects of degradation, manufacturers add colouring agents.

There are two main categories of synthetic plastics:

(a) thermosetting
(b) thermoplastics

Thermosetting plastics are generally used for mouldings, i.e. some types of flushing cistern shells and electrical equipment are typical examples of thermosetting plastics. Thermoplastics are those which can be resoftened when heated.

Most pipework materials used by the plumber fall into this category.

Most plastics are unaffected by strong acids or alkalis but they may be softened and permeated by hydrocarbons. This is not surprising as the basic elements used in the manufacture of these materials are in the hydrocarbon family, i.e. materials composed chiefly of the atoms of hydrogen and carbon. Petrol, oil, paint strippers and cellulose thinners are all hydrogen and carbon compounds which will soften and degrade thermoplastic materials.

The reader should note there are certain exceptions to this, for example fuel oils may be stored in vessels made of polythene. This is because there are many grades of polythene, these grades varying as to the basic material or 'polymer' of which a particular grade is made. Because of this vessels made of plastic should be clearly labelled with instructions for their use. Generally however, unless there is no doubt about their suitability, components made of synthetic plastics should not be used with materials derived from hydrocarbons.

In a practical context cooking fats and oils are basically hydrocarbons, but while a domestic sink discharge pipe made of plastic will remain serviceable for many years, it would be an unsuitable material to use in a large commercial kitchen where vast amounts of hot water and grease would quickly lead to its degradation.

All synthetic plastics used for sanitary pipework are marketed as complete systems and it is not advisable to use the components produced by one manufacturer with those of another, even if the basic material is the same.

Traditionally, synthetic plastics have not been used for hot water supplies due to their relatively low melting point. Recent developments have, however, enabled some types of plastics, namely cross-linked polythene (xLPE) and polybutylene (PB) to be used successfully for this purpose.

The method of making joints on plastic materials varies as to its type. Table 3.7 is not exhaustive but gives a good general guide to the types of joint suitable for the most common plastics used by the plumbing industry.

Plastics commonly used for water services and sanitary pipework

Polythene This is possibly the most commonly used plastic in the polyolefin group of plastics. Polythene pipes up to 150 mm nominal bore are made to BS 3284 and are classified as 'high', 'medium' or 'low' density. High-density polythene is capable of withstanding slightly higher temperatures and is more rigid than medium- and low-density types, which are comparatively flexible and have a slightly lower melting point. It is available in straight lengths and is mainly used for chemical discharge pipework and cold storage cisterns.

The use and installation of polythene pipelines for water supply should conform to BS 6730 (for above ground use) and BS 6572 (for underground use). Both these standards cover pipes of up to 20 mm to 63 mm OD. A standard is in preparation for pipes of larger diameters.

Pipes to both these standards are suitable for pressures up to 12 bar. Polythene pipes for use above ground are black in colour and for

Table 3.7 Jointing methods for synthetic plastic pipes in plumbing services.

Types of plastic	*Mechanical joints**	*Solvent welding*	*Fusion welding*	*Push fit 'O' ring (discharge pipes/overflows only)*
Polythene	✓	×	✓	×
Polypropylene	✓	×	✓	✓
Polbutylene	✓	×	×	×
Polyvinyl chloride	✓	✓	×	✓
Acrylonitrile butadiene styrene	✓	✓	×	✓
Cross-linked polythene	✓	×	×	×

*Note that mechanical joints may be combinations of 'O' ring seals, grab rings, rubber or metal olives

underground use blue. Polythene in its natural state is colourless but due to the ultra-violet rays of the sun it degrades and hardens causing it to crack. To prevent this a black colouring agent is added.

The colour blue was chosen for plastic underground water services to fulfil a commitment made by all public utility companies to enable their services to be readily identified, as several fatalities have occurred as a result of the mistaken identity of electrical cables. It is worth noting the different colours used for underground polythene services. They are as follows:

Water	Blue
Gas	Yellow
Electricity	Black and Red
Telecommunications	Grey

In comparison with some other plastics, polythene has low mechanical strength, a tendency to 'creep', i.e. expand the tube bore under pressure, and is very light, having a density of 900 kg/m^3. Medium- and low-density polythene for water supplies is usually marketed in coils of 24, 50 and 100 m in length.

Identification of polythene and polypropylene
Polythene and polypropylene are both members of the polyolefin group of plastics. They have a waxy appearance and touch, and when ignited, burn with a steady flame very similar to that of a wax candle. Because most polyolefins used in plumbing are less dense than water, they can also be identified very easily as they will float, unlike many other synthetic plastics.

On freezing, water expands approximately 1/10th of its volume and this pressure can cause water-filled vessels or pipes to burst, but the elastic property peculiar to polythene makes it superior in its resistance to damage by frost. Polythene is unsuitable for jointing using solvent welding or adhesives. The most common jointing methods used are fusion welding, or one of the many special fittings developed by manufacturers for this type of pipe. Non-manipulative compression fittings as used for copper tubes (but one size larger than the nominal bore of the polythene pipe) can be used with an appropriate liner.

Polypropylene This material is in the same family group as polythene. It is a tough, lightweight plastic of high rigidity and surface hardness. Its main characteristic is its ability to withstand high temperatures, and in this respect it is superior to PVC, ABS or polythene. Articles made of polypropylene can be used with boiling water for short periods of time. For this reason most manufacturers of plastic waste systems produce traps made of this material.

Some also produce a range of pipes made of polypropylene for discharge systems which are jointed by means of 'O' rings. Having similar characteristics to high-density polythene, it is also used extensively for the manufacture of cold water storage and feed cisterns. When used for pressure pipework, fusion-welded and mechanical joints are the principal method of jointing.

Polybutylene Being another member of the polyolefin family of plastics, it has similar characteristics to those previously mentioned, and pipework systems made of this material can be used for both hot and cold water supplies. Its melting temperature is 125 °C, so that, with some reservations, it can be connected directly to hot water equipment such as boilers and hot store vessels. As with polythene, if the pipe is frozen it will expand without bursting and revert to its original shape at temperatures above freezing point. It is a poor conductor of heat, as are most plastics, and this does have the advantage that hot water in deadlegs will not cool as quickly as that in metal pipes. Hot water and heating pipes should, however, be insulated to the same standards as those of copper and steel pipelines. Based on simulated tests, this material is claimed to have a life expectancy of 50 years at pressures between 12 bar at 20 °C and 6 bar at 90 °C when installed in domestic systems.

Should failure of a thermostat produce temperatures for short-term periods of up to 100 °C, the material will not fail, but its long-term life may be shortened.

Because pipes made of materials in the polyolefin family of plastics permit the permeation of oxygen through their walls, polybutylene is produced both with and without a protective

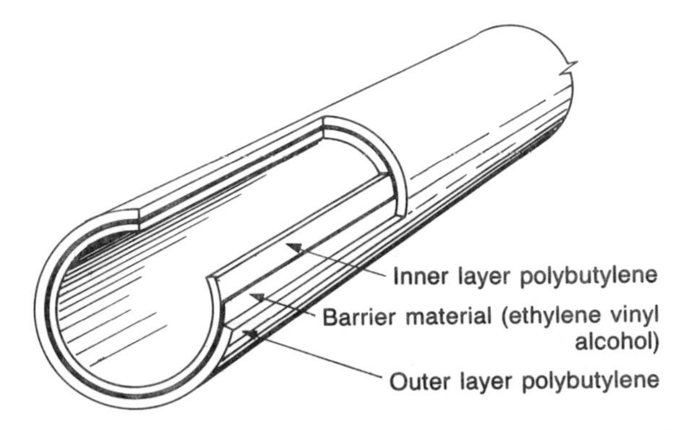

Fig. 3.7 Section of laminated barrier pipe. This type of pipe prevents the ingress of oxygen through the pipe walls. The main use of barrier pipes is in connection with heating systems where no inhibitor is used.

barrier. Barrier pipes should be used in heating systems to prevent corrosion where, for various reasons, the installer does not wish to use an inhibitor. Barrier pipe contains within its walls a material called ethylene vinyl alcohol (OVHO) which effectively prevents the ingress of gases shown in Fig. 3.7. As with other polyolefins, polybutylene can be identified in a similar way to that of polythene and polypropylene. The British Standard for this material is BS 7291:Parts 1 and 2. BS 5955:Part 8 relates to the installation of thermoplastic pipes for hot- and cold-water services.

Cross-linked polythene This material is another type of plastic that has been developed for both hot and cold water services. Its main characteristics are very similar to those of polybutylene. Tubes are obtainable in both coils of up to 100 m and straight lengths of 2 m or 3 m in nominal sizes of 15 mm and 22 mm. This material is designed to last for 50 years, but this is dependent on its working temperature. The fittings used with both this material and polybutylene may also be used on copper tubes. They are easily dismantled by pressing in a collet which forms part of the fitting.

Polyvinyl chloride Possibly the most common of the synthetic plastics used for sanitary pipework is polyvinyl chloride (PVC) to BS 4514. This Standard requires that the material should not soften below 70 °C for fittings and 81 °C for pipes, and it should be capable of discharging water of higher temperatures for short periods of time. PVC, unlike many other plastic materials, will not burn easily, a fact which can be used for its identification. If a doubt exists, cut a thin strip from a piece of scrap pipe or gutter and hold over a lighted match. It will ignite initially and give off a heavy vapour, which excludes the oxygen to the flame causing it to be smothered. For this reason PVC is said to be self-extinguishing, and while it is not the only plastic having this property, it could reasonably be assumed on site that it is indeed PVC. It is obtainable either plasticised (softened) or unplasticised, the former being flexible and not used for building services. There are two types of unplasticised PVC, both of them being specified for water and chemicals above ground where there is a risk of impact damage. The requirements of standard PVC are covered by BS 3515 and BS 3506 deals with 'high impact' strength pipe. The impact strength (ability to withstand a blow) of both materials is greatly reduced as temperatures approach 0 °C and great care must be taken when handling in cold weather or cracking will occur. For continuous operation the maximum temperature is around 60–65 °C and if this is exceeded the mechanical properties are lowered in a similar manner to polythene.

It is a light material (density 1,360 kg/m^3) that can be handled easily, a 6 m length of 75 mm diameter pipe weighing only 8.5 kg. Its major application in the plumbing industry is for sanitary pipework and rainwater goods. The maximum length of pipes for sanitation is usually listed as 4 metres with diameters from 32 to 150 mm nominal bore. The usual method of jointing PVC pipe up to 50 mm is the use of solvent cement, the only exceptions normally being the connection between the discharge pipe and trap, using either a rubber olive and nut or an 'O' ring joint. Both of these allow for expansion. Connections to the main stack may be made using solvent welding, a rubber 'O' ring or olive, the latter two being used where provision for expansion is necessary.

Acrylinitrile butadiene styrene (ABS) ABS is a rigid material having a much higher impact strength than PVC at low temperatures. Its maximum operating temperature is 80 °C and it can be used for pressures up to 15 bar or 150 m head. Although a British Standard is not available for this material, ABS is made to conform to the pressure rating classes used for PVC and to the dimensions specified in BS 3506.

Although it is not suitable for hot water services, it does have the advantages over PVC in that it can withstand higher water temperatures for a longer period of time. It also retains its impact strength at subzero temperatures which give it more resistance to mechanical damage. For these reasons many manufacturers produce a range of pipes and fittings for discharge pipework, although this material is more expensive than PVC. It is also marketed as a pressure pipe for cold water services but is not in general use for this purpose in the plumbing industry.

The usual method of jointing ABS is by solvent welding. It has a rather more matt exterior finish than PVC which has a polished or shiny appearance. If it is ignited it burns with a bright white flame which gives off particles of carbon, very similar in fact to that of a welding blowpipe when it is burning only acetylene.

Synthetic rubber There are many types of synthetic rubber, some of which contain a proportion of natural rubber. The most common synthetic rubber used in plumbing is called Neoprene which is usually used for the manufacture of 'O' ring expansion joints. It is more stable against oxidation than natural rubber and is only affected slightly by oils.

Another of its main features is its resistance to sunlight, an important factor when pipes are situated externally. Unlike most types of rubber, it is flame resistant and self-extinguishing.

Polystyrene This material has many applications in industry for moulded components but its principal use in the mechanical services industry is for insulation where it is known as expanded polystyrene. It is obtainable in both sheet and sectional form for pipe insulation, is usually white in colour and has the appearance of a coarse rigid foam. It is very light and brittle and must be handled with extreme care. During manufacture it is made in large blocks which are cut to size by an electrically heated wire. It is best cut on site with a very sharp knife.

Acrylic This is one of the acrylic resin group of plastics which is obtainable in clear sheet form or mouldings where it is extensively used for sanitary appliances. It can be cut with a fine-tooth saw and drilled in the usual way, providing sharp drill bits are used. It is not so resistant to scratching as components made of metal, and extra care must be taken when it is handled during installation.

Nylon This is a very strong and tough thermoplastic material with very good heat-resistant qualities. It is one of the more expensive plastics to produce and this is probably why it is not produced in tubular form for plumbing services. It is extensively used, however, for small components in taps, and valves, the seatings in float-operated valves being typical. It is also used for gearing and bearings in shower mixers and water meters, where due to its self-lubricating properties, little maintenance is necessary.

Jointing materials Joints made on plumbing components, e.g. threads, require some material to make good the small gaps between the mating surfaces. Even machined joints such as those on radiator unions are more reliable if a suitable jointing material is employed.

For many years the mechanical services industries used proprietary paste and hemp for this purpose, but because the oil used in their manufacture was generally vegetable based, it was found they supported the growth of harmful bacteria in pipelines. It was also found that joints on gas services using similar materials could be permeated by natural gas. For these reasons it is important that the plumber employs the correct jointing media for the type of work to be carried out.

Polytetra-fluoroethylene (PTFE) This material is supplied in the form of white tape and is suitable for water services, including potable water and gas. It is used to seal threads on metal and plastic materials, is clean to use and is non-toxic. Unlike pastes it is not affected by oxidation, enabling any joint to be easily taken apart. The only reservation to its use is that the pipe threads must be in good condition, e.g. no stripped threads, and the male and female ends must be free from distortion.

Pastes These are best described as thick pastes and have traditionally been used with hemp to fill the gaps between threads. In fact if the threads are well formed, hemp should not be necessary. It was never good practice to use it on steam services, as at high temperatures it could burn out.

For potable water special pastes have been developed based on silicone compounds and filler that will not support bacteria. It can be used where necessary with a special plastic fibre which swells when wet and is used as a substitute for hemp. Most pastes that are suitable for potable water are also suitable for gas heating and low-pressure steam services. Traditional paste-jointing compounds are still available but they must not be used for potable water services. The inclusion of graphite to a compound increases its lubricative qualities and enables joints to be broken more easily. Manganese-based compounds are especially suitable for flange joints, as they set more quickly and the joints are harder.

Special jointing compounds are necessary for joints on oil pipelines; oil-based pastes are not suitable as fuel oils will soften and permeate through the joints. Suitable compounds include finely ground ferrous oxide (rust) which is bound with alcohol and phenolic resins. Due to their alcoholic content they are flammable and give off a harmful vapour. For this reason they should be used with care and in a ventilated environment.

Soldering

Fluxes The object of using a flux is to prevent oxidation of the joint during the soldering process. Oxygen in the air will oxidize and tarnish metals, and soldered joints cannot be made satisfactory under such conditions. One of the major prerequisites of soldering is the work must be thoroughly clean.

Fluxes traditionally used by the plumber are dealt with more fully in Chapter 10, where they are related more specifically to the requirements of sheet weatherings using soft solder alloys of tin and lead. The use of tin, copper, and in some cases tin—silver solders, can cause problems due to their high melting temperatures in connection with some grease-based fluxes. What actually appears to happen is the grease burns at these high temperatures, which tends to blacken the copper pipes and fittings being soldered. It must be said that in many cases the fault is sometimes due to the flame being too large, and in the case of butane or propane blowlamps the pressure may be too high. If the equipment is provided with a regulator, try reducing the pressure, which will result in a softer flame giving a more even heat, and will, in many cases, solve the problem. Flux manufacturers are aware of this and many are now producing fluxes which are more compatible with the higher temperatures required for soft solder alloys used for hot and cold water supplies.

Grease-based fluxes, not being soluble in water, are not recommended for pipe jointing whatever type of solder is used, as they have been found to be a common cause of corrosion. Paste fluxes which are water soluble are now readily available, and any residue may be washed out with water when the work is commissioned.

Fluxes in general may be divided into three main groups:

(a) Strong oxide-dissolving fluxes made with organic salts, i.e. zinc, ammonium, and stannous chlorides which are sometimes combined with various acids. Most fluxes used for capillary joints are of this type and a close look at the ingredients should make the reader realise how important it is to remove any flux residues after a joint is soldered.
(b) Less active fluxes made of stearic and lactic acids and helagons, e.g. various hydrochlorides.
(c) Non-aggressive resin usually based on undissolved natural pine resin. These fluxes are more suitable for use with soldering irons rather than a flame and are referred to more fully in Chapter 9.

Flow characteristics of pipe materials

An important consideration when designing pipework installations is the *frictional resistance* caused by the pipewalls. This resistance can be appreciated by comparing the speed of someone running in a race in the normal manner with the speed of the same person running a race while in contact with a wall. Obviously the result of rubbing on the wall will cause a considerable reduction in speed. Pipe walls have the same effect on water flow, i.e. the water is 'rubbing' on the pipe, and this is further influenced by the rough or smooth nature of the material. The effect of this is known as 'frictional resistance' or frictional loss. Where the influence of smooth and rough pipe surfaces is compared, the flow of water from a 25 mm diameter pipe under a head of 10 m would be approximately 22 litres/minute for copper, lead and plastic pipes having smooth internal surfaces, and 18 litres/minute for galvanised low carbon steel which has a rougher inner surface.

The effects of friction and subsequent loss of head or pressure is also influenced by the number and types of fittings in the pipe run. This resistance is measured in terms of equivalent pipe length. To give one example, the frictional loss caused by water passing through a 28 mm elbow will be the same as a length of the same diameter pipe one metre long. Where possible machine-made or large radius bends should be used as their resistance to the flow of water is negligible.

Resistance of pipes to corrosion

Corrosion is dealt with in detail in Chapter 9, but in brief it may be described as the gradual eating away of a material by chemical action.

The nature of the water to be conveyed, the chemical nature of the ground in which it is to be laid, or the environment above ground, will influence the choice of a suitable pipe material to minimise the effects of corrosion.

Steel

Black steel (i.e. not galvanised) is not acceptable for hot and cold water supply as it is liable to corrosion both internally and externally. Water supply pipes must be galvanised but gas supplies can be conveyed in black steel pipes although a non-corrosive external covering is required for buried pipes. Most new gas service pipes, however, are now made of polythene, identifiable by its bright yellow colour and for use in the building, copper tubes have superseded steel.

Copper

Almost all waters are 'cuprosolvent', i.e. liable to dissolve copper, this property being due to the same conditions that cause plumbosolvency. Only rarely does cuprosolvency reach a level where it is harmful to health but if the copper content of water is in excess of 1.5 mg/3 cm^3 an unsightly green strain is deposited on sanitary fitments. When used underground in certain soils, corrosion can occur and it may be necessary to wrap with a protective tape, such as a self-adhesive plastic type or that made of cotton, impregnated with petroleum jelly and an inert silicious filter (Denso tape is a well-known trade name for this type of material). An alternative is to use plastic-coated pipe to BS 2871:Part 1, *Table Y*. Plastic covered tube to BS 2871:Part 1, *Tables X* and *Z*, is also available when copper tubes are to be installed in plaster, brickwork or concrete.

Stainless steel

This metal is unaffected by corrosion with the exception of water containing high chloride concentrations which affect the protective oxide film. This can be prevented by using metal having a higher chromium content such as austenitic 18/8 steel.

Plastic pipe

All plastic pipes are unaffected by the normal corrosive effects of water and soil conditions. Where brass compression fittings are used underground for jointing, it will be necessary to wrap them for protection against corrosion.

Brass pipe fittings

The action of certain water and soil conditions (usually in soft-water areas) can cause dezincification of duplex brass which is a zinc-rich alloy. This is a particular form of electrolytic corrosion which is fully described in Chapter 9 where it can be seen that zinc and copper are widely separated in the electrochemical series. An aggressive water or soil provides an electrolyte which accelerates the destruction of the anodic metal (zinc) leaving the copper porous and brittle – the alloy has dezincified, i.e. has had its zinc content removed.

To overcome this problem there are several alternatives. Gunmetal, an alloy of approximately 85 per cent copper, 5 per cent tin, 5 per cent lead and 5 per cent zinc may be used. This is an expensive alloy and is normally only used for heavy duty fittings. Compression fittings made of brass but specially treated, and referred to as dezincification resistant, are a cheaper alternative.

The most commonly used fittings are, however, made of copper which is unaffected by dezincification.

Pipework fixings

Care must be taken to ensure pipes are adequately supported according to their strength and rigidity. Details of the horizontal and vertical spacing of the fixings and a selection of the various types of fixings available are dealt with in Chapters 5 and 6 relating to cold and hot water supplies.

Water storage vessels

Cisterns, tanks and cylinders are all containers used for the storage of water. Suitable definitions of these terms are as follows.

Cistern An open top circular or rectangular cold water storage vessel with a loose lid or cover.

Tank A closed rectangular vessel used for the circulation and storage of hot water.

Cylinder A closed cylindrical vessel with parallel sides and domed ends. Used for the circulation and storage of hot water.

All vessels used for water storage must be constructed of a material that will not impart taste or odour and be able to resist any corrosive action by the water.

Steel water storage vessels

Cisterns Cisterns of galvanised steel are manufactured to BS 417 and are available in capacities ranging from 18 litres up to 3,364 litres. Capacities of vessels can be referred to as *actual* or *nominal*, the latter being based on calculations using extreme overall dimensions, while the actual capacity takes the depth to be that of the water below the water line. They are made in Grades A and B which relate to the thickness of the steel sheet used, and are of either riveted or welded construction. Loose covers are supplied in the same material but of a lighter gauge. Greater capacities are catered for either by coupling two or more of these vessels together or by using sectional steel cisterns, which can be built into any required size from 1 metre square sections. The sections have flanged edges which are bolted together and are erected on site. These cisterns, which are made to BS 1564, must be protected by galvanising, metal spraying or by bonding rubber or plastic coatings to the plates and fittings.

Tanks Tanks of galvanised steel are manufactured to BS 417 and can be obtained in various capacities from 95 litres up to 155 litres. Metal thickness is related to maximum working head pressure, Grade A (3.2 mm) and Grade B (2.5 mm) being suitable for heads of 4.5 m and 3.0 m respectively, the maximum working head pressure being the vertical distance between the water level in the cold water storage cistern and the bottom of the tank or cylinder.

Plastic storage vessels

Plastic is the most recent category of material to be utilised for the production of storage vessels. The type of plastic could be glass fibre, polythene or polypropylene. The British Standard dealing with polythene and polypropylene cisterns is BS 4213. Actual capacities of these cisterns range from 18 litres to 455 litres and they are comparatively light in weight. These vessels are not affected by electrolytic corrosion and are able to resist the effects of freezing due to their elastic properties. Some can also be flattened so that passage can be gained through a confined space and being black in colour are resistant to algae growth. Polythene cisterns are usually circular but rectangular patterns can be obtained made of polypropylene, this material being slightly more rigid.

Cylinders

Cylinders of galvanised steel are manufactured to BS 417 in a range of capacities from 73 litres up to 441 litres. The metal thickness is related to maximum working head pressure, Grade A (3.2 mm) Grade B (2.5 mm) and Grade C (2.0 mm) being suitable for heads of 30 m, 18 m and 9 m respectively. These cylinders are of the direct pattern but there is also an indirect type made to BS 1565. The nominal water content of the indirect pattern may be from 136 litres up to 455 litres. Further details of the two types of cylinder are given in Chapter 6.

The only vessels commercially manufactured from sheet copper are hot water storage cylinders. The direct pattern is made to BS 699, while both single and double feed patterns of indirect cylinders are covered by BS 1566. Cylinders are specified in accordance with the maximum working head of the system, three grades being available for heads of 25, 15 and 10 metres: Grades 1 to 3 respectively. Capacities of cylinders to BS 699 are from 74 litres to 450 litres, although *preferred sizes* having capacities of 116, 120, 144 and 166 are those normally stocked for installation in new dwellings. In addition to these four sizes, 98 and 200 litre capacities are also available for indirect cylinders to BS 1566.

Specially adapted cylindrical storage vessels for unvented hot water systems are available made of steel, stainless steel or copper. They are in effect a self contained unit incorporating the pressure vessel and the associated controls. Those made of steel are protected internally by a plastic coating bonded on to the steel which ensures a long service life.

Corrosion

The problems of electrolytic corrosion have previously been mentioned and it is only necessary at this stage to suggest that caution must be exercised if pipework, components and fittings of different metals are to be used in the same installation. Special care must be taken to avoid using galvanised low carbon steel with copper. This would result in a particular form of electrolytic corrosion known as galvanic action which results in the destruction of the zinc coating leaving unprotected steel. The effects of this form of corrosion are greatly increased in hot water systems where soft/acid water is encountered. Reference to the electromotive series in Chapter 9 will indicate that mixing of some metals, e.g. copper and lead, is not likely to create problems in plumbing systems.

Storage calculations

Calculations in respect of water storage are undertaken by the plumbing designer who takes into consideration the legal requirements involved, the number of consumers, type of building, rate and regularity of supply, and the consequence of exhausting the storage supply. Some reference is made to these factors in Chapter 6, but the subject of design is outside the scope of this book.

The plumber is, however, required to undertake calculations which involve the following:

(a) specifiying the dimensions of a vessel to hold a given quantity of water
(b) calculating the capacity of a storage vessel from given dimensions
(c) deriving the mass of a storage vessel either empty or when filled

To solve these problems certain factors must be understood. One cubic metre (1 m^3) of water contains 1,000 litres. One litre of water has a mass of 1 kg, therefore 1 m^3 of water has a mass of 1,000 kg. Diameter (D) is the distance across the centre of a circle (or cylinder), and the radius (r) is half this distance, while the circumference is the distance around the extreme edge of a circle.

Pi (π) is equal to 3.142 and is the number of times the diameter of any circle can be divided into the circumference.

The following formulae are used in calculating the size and capacity of a vessel.

Area of a rectangle Multiply length by breadth.
Circumference of a circle Multiply π by the diameter.
Area of a circle Multiply π by the radius squared (πr^2).
Area of a cylinder side Multiply circumference by height.
Volume of rectangular vessel (i.e. a tank) Multiply area of base by depth of vessel (usually in metres).
Volume (m^3) of cylindrical vessel Multiply area of end by length.
Mass of material in vessel Multiply total volume (m^3) by mass/unit volume (i.e. water has a mass of 1,000 kg/m^3).
Capacity of vessel in litres Multiply volume (m^3) by 1,000.
Mass (kg) of water in vessel Multiply capacity (litres) by 1 kg (i.e. 1 litre of water has a mass of 1 kg).

Examples

1. A galvanised steel storage cistern measures 1.5 m by 2.0 m and is 1.0 m deep. Calculate the capacity of the vessel in litres when filled to the top edge.

   Volume = Area of base × depth
   = 1.5 m × 2.0 m × 1.0 m
   = 3.0 m^3
   Capacity = Volume × 1,000
   = 3.0 × 1,000
   = 3,000 litres

   (Note that this would have a mass of 3,000 kg.)

2. A copper hot water cylinder measures 1.8 m high and is 0.6 m in diameter. Calculate the mass of water in the cylinder.

   Volume = Area of end × height
   = πr^2 × height (radius is half of diameter, i.e. 0.300 m)
   = 3.142 × 0.3 × 0.3 × 1.8
   = 0.51 m^3
   Mass = Volume × 1,000 kg
   = 0.51 × 1,000
   = 510 kg

   (Note that the capacity of the cylinder is 510 litres since 1 litre of water has a mass of 1 kg.)

3. A galvanised low carbon steel storage cistern has a nominal capacity of 3,000 litres and base measurement of 1.0 m by 2.0 m. What is the depth of the vessel?

   Capacity = Volume × 1,000
   3,000 = Volume × 1,000
   ∴ Volume = 3,000 ÷ 1,000
   = 3 m^3
   Volume = Area of base × depth
   3 m^3 = (2 × 1.0) × depth
   3 m^3 2 × depth
   Depth 3 ÷ 2
   = 1.5 m

Sizes and capacity can also be obtained by reference to the British Standards for each type of vessel.

Further reading

Much useful information can be obtained from the following sources.

BS 1990:Part 1 Specification for spigot and socket systems of cast iron pipes for soil waste and ventilating pipes.
BS 1990 Specification for socketless systems of cast iron pipes.
BS 4514:1969 'Unplasticised PVC soil and ventilation pipe fittings and accessories'.

Copper fittings and tubes
Conex Sanbra Ltd, Whitehall Road, Tipton, West Midlands, DYA 7JU.
Peglers Ltd, PO Box 182, St Catherines Avenue, Doncaster, South Yorkshire, DN4 8DN.
Wednesbury Tubes, Oxford Street, Bilston, West Midlands, WV14 7DS.
Yorkshire Imperial Metals, PO Box 166, Leeds, LS1 1RD.

Steel pipes and fittings
George Fischer Ltd, Paradise Way, Coventry, CV2 2ST.

Plastic pipes for water services

George Fischer Ltd, Paradise Way, Coventry, CV2 2ST.

Hepworth Building Products, Hazelhead, Stocksbridge, Sheffield, S30 5HG.

John Guest Ltd, Horton Road, West Drayton, Middlesex, UB7 8JL.

Plastic materials for sanitary pipework

Hunter Building Products, London, SE28.

Caradon Terrain Ltd, Aylesford, Kent, ME20 7PJ.

Marley Plumbing, Dickley Lane, Lenham, Maidstone, Kent, ME17 2DE.

Plumbing solders and fluxes

CFS Products, Rowberry House, Copse Cross Street, Ross-on-Wye, HR9 5PN.

Frys Metals, Tandem House, Harlow Way, Beddington Farm Road, Croydon, Surrey, CR9 4BT.

Jointing pastes

Fernox Manufacturing Co. Ltd, Britannica Works, Clavering, Essex, CB11 4QZ.

Cast iron drainage

Windsor Ltd, 763 Harrow Road, London, NW10 5N7.

Glnwed Drainage Pipes and Fittings, Sinclair Works, PO Box 3, Ketley, Telford, Shropshire

Self-testing questions

1. A head of water is given as 35 m. Explain this in terms of kN/m^2.
2. Explain the term 'alloy' in relation to metals and give three examples used in the plumbing industry.
3. State the colour used to identify low carbon steel pipes made to BS 1387 and relate the colours to the wall thickness.
4. Explain the term 'galvanising' and how it is used to protect steel against corrosion.
5. Explain why steel has a high resistance to corrosion.
6. Explain why the maximum working pressure is important when selecting a hot storage vessel.
7. A cold water storage cistern measures 2.0 m long × 1.5 m wide × 0.8 m deep. Calculate its actual capacity in litres when the water level is 0.075 m below the top edge.
8. Describe the difference between manipulative and non-manipulative compression fittings used for copper tubes. State which of these can be used with (a) table 'W' and (b) table Z tubes.
9. Explain the term 'dezincification' of fittings used for water supply.
10. Describe two methods of producing cast iron pipes for discharge pipework and underground drainage.
11. The following are all methods of making joints on synthetic plastic pipes:
    (a) synthetic rubber 'O' rings
    (b) fusion welding
    (c) solvent welding
    (d) mechanical joints employing grab rings or rubber seals

    Which of these methods may be used for (a) polybutylene and cross-linked polythene water services and (b) PVC discharge pipework? Which of these jointing methods permit pipework to expand and contract?
12. (a) State the purpose of a flux when making soldered joints.
    (b) Specify the type of flux recommended for pipework installation.

4 Pipework jointing and bending processes

After completing this chapter the reader should be able to:

1. Understand the importance of cleaning and removing the oxide film on metals prior to soldering.
2. Select suitable methods of jointing various pipework materials.
3. Describe the various methods of jointing low carbon steel pipes.
4. Describe the correct methods of jointing different types of plastic pipework.
5. Describe the basic principles of making bends in various pipework materials.
6. State the working principles of pipe-bending machines and their requirements for maintenance.
7. Describe how to adjust the working parts of pipe-bending machines to produce acceptable bends.

Connections to existing lead pipe

The use of lead pipe for water services in post-war housing has been rare, mainly due to its high costs and the fact that other materials, especially copper tubes where they are exposed, are neater in appearance and easier to keep clean. Quite apart from the foregoing, because lead is a poisonous substance and soft acidic waters are capable of dissolving it, public health authorities have been averse to its use not only for water pipes but also in paint and leaded petrol. The model Water Bylaws 1986 expressly forbid the use of lead pipes not only for new work but for repairing existing services. To replace all lead water services would be too costly but as their service life comes to an end they must be replaced with copper or a suitable plastic material. Irrespective of any regulations, all water authorities treat water to ensure that lead will not be taken into solution and it is extremely unlikely that consuming water supplied through lead pipes is likely to result in any fatalities.

Lead to copper mechanical joints

Where repair work becomes necessary and the consumer cannot afford to replace all the lead pipes in the system, the most economic method is to use one of the mechanical joints shown in Fig. 4.1(a) and (b) that are available for making connections to the new copper or plastic pipe used to replace the defective section. The ends of the fitting for copper or polythene pipes are made with compression, or push fit joints. Connections to existing lead pipes present greater difficulty as lead, being soft, is not a suitable material on which to make a compression joint; it would simply pull out when subjected to longitudinal stress and any lateral movement would almost certainly result in a leaking joint.

These problems are prevented in the two fittings illustrated. Fig. 4.1(a) shows a clamping arrangement which is tightened when the compression nut has been screwed home. The ridges inside the clamp bite into the lead when the hexagon screw is tightened, preferably with a suitable spanner rather than a screwdriver. This type of fitting is excellent for underground use.

Figure 4.1(b) shows a similar fitting having a different method of positively securing the lead pipe. As the nut on the lead end of the fitting is tightened, the conical-shaped locking ring is forced flat causing the inside edge to bite into the lead pipe thus preventing any lateral withdrawal of the pipe.

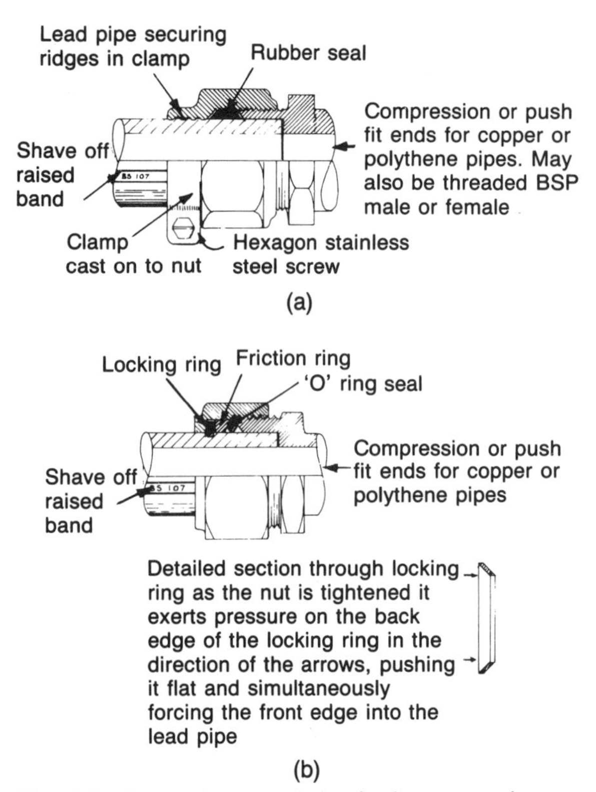

Fig. 4.1 Connections to existing lead water services

In all cases where lead pipe is jointed using these fittings, a few simple points must be observed:

(1) The lead pipe must be as round and straight as possible, the seating area being free from deep dents or score marks.
(2) The pipe must be cut square and any burrs removed. Entry into the fitting will be assisted if the outer edge is slightly chamfered.
(3) The raised bar showing the BS code of the pipe should be shaved off to ensure it is a sliding fit into the socket; it must not be loose or slack.
(4) Fittings having a locking ring have a positive stop when the nut is fully tightened. This eliminates overtightening and ensures the locking ring secures the lead.

Taft joints

When lead was the principal material used for pipework, taft joints were considered only for temporary connections. They did not conform with the previous requirements of most water authorities as being a wiped joint in the true sense of the term.

The only occasion where the use of a taft joint would be permissible is for the repair of lead discharge pipework, but even here it would be more economic in most cases to replace the pipe completely using copper or plastic. Figure 4.2 shows a typical taft joint. The socket is 'tafted' out with a turn pin and mallet enabling it to be thoroughly cleaned. The spigot end is chamfered with a rasp to ensure a good fit in the socket prior to cleaning. The joint is then fluxed and wiped as shown using plumber's wiping solder and a small wiping cloth.

Jointing copper tube

Copper tubes have thinner pipe walls than most comparable materials. It is light, corrosion resistant and has a much smaller outside diameter than most other tube materials of a similar nominal bore. Except for those of fully annealed temper, copper tubes are quite rigid and do not require continuous support. This rigidity also permits them to be spaced off any surface to which they are fixed enabling the surface to be cleaned and painted.

The three main types of fittings used for making joints in copper tubes have been described in Chapter 3 but the methods of jointing, using these fittings, are described here.

Capillary solder joints

Probably the capillary solder fittings are the most popular due to their reliability, neatness of

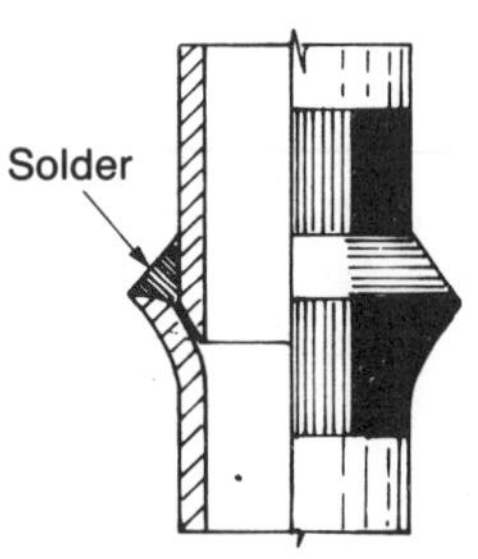

Fig. 4.2 Taft joint

appearance and low cost in comparison with most fittings of the compression type.

The tolerance between the tube and fitting is vitally important, and if either the tube or fitting is distorted, capillary action will not take place and result in a leaking joint. Reference should be made to Chapter 10, Fig. 10.2, which explains the basic principles of capillarity. When working with fully annealed tube, which is usually supplied in coils, the tube is slightly flattened and before a joint is made the ends of the tube should be restored to their correct bore using a mandrel which may be obtained from the tube manufacturers.

To make a capillary solder joint, both the inside of the fitting and the tube end should be cleaned with a fine or medium grade steel wool. Care must be taken when handling this material as if it is pulled from the skein carelessly it may result in a cut finger or hand. Small circular wire brushes are obtainable for cleaning the sockets of fittings and these save time and ensure that the socket is clean to its maximum depth. A thin coating of flux is then applied to the mating surfaces before they are pushed home. If the fitting has an integral solder ring, no solder need be applied, the joint being heated until the solder appears at the mouth of the fitting. In the case of end feed fittings, solder is applied by touching each end of the socket until the solder is seen to melt and enter the fitting. With both types of joint, a complete ring of solder should be seen round the mouth of the fitting as proof of a sound joint. When the joint has been completed, any surplus solder should be wiped off with a dry cloth before it solidifies. Do not forget to remove any surplus flux when the joint is cooled.

Straight joints in copper tubes can be formed by making capillary sockets on plain ends of tube by using a socket-forming tool as described in Chapter 2. The soldered joint is made by using similar techniques as those used for end feed fittings.

A jointing material marketed as solder paste is available for making capillary joints of the end feed type. It consists of finely ground solder mixed with a liquid flux forming a paste which can be painted on the tube ends and inside the socket of the fitting. It is effective but expensive, and the flux is very corrosive. Any residue must be carefully washed off after the joint has been made.

Unlike lead tin solder alloys, joints made with tin copper solders are difficult to remake after the joint has been soldered. It is essential that the pipe and fitting is correctly located before the joint is made, as it will be almost impossible for alterations to be made later.

Compression joints

Reference should be made to the illustrations shown in Chapter 3 and Table 3.6 which indicate the type of fitting which can be used on the three main types of copper tube used in the industry.

Manipulative and non-manipulative joints The manipulative type of compression joint will resist high internal pressures due to the swaged end formed on the tube. Under service conditions the tube will split before the fitting pulls off. When making these joints, do not forget to put the nut on the tube before the swaged end is formed.

Non-manipulative joints, as their name implies, require no swaging on the tube end. They rely instead on the compressive effect of the nut on the olive for their soundness.

Although most fitting manufacturers do not recommend any form of jointing paste to be used on these joints, this may be found necessary in practice, especially with fittings with a hard brass olive.

Two very important points of preparation must always be remembered whatever joint is used. The tube must always be cut square and any burr formed internally by cutters or externally by a hacksaw must be removed by a file or reamer.

Dezincification of fittings The water supply in some parts of the country causes breakdown in the structure of brasses which have a high zinc content. The name given to this form of corrosion is dezincification. All fittings used in such areas should be made of a zinc-free metal such as copper or bronze neither of which are affected by dezincification. Bronze is expensive and copper, while suitable for capillary fittings, would not be satisfactory for compression-type fittings. Brass compression fittings that have been specially heat treated are also available and it is claimed that these are unaffected by dezincification. They are slightly more expensive than ordinary fittings but not so costly as a comparable fitting made of bronze and are stamped DR (Dezincification Resistant). The reader should refer to Chapter 3 under the heading of corrosion for further information on this subject.

Branch joints

When branches are inserted into existing copper pipework, it is sometimes difficult to spring the pipe apart a sufficient distance to admit the new fitting. 'Slip' straight joints and tees of the capillary type are available for this situation and have no internal stop, allowing them to be slipped along the pipe. When using one of these fittings it is necesary to make sure it is centralised over the cut pipes before making the joint in the usual way (see Fig. 4.3).

Where it is necessary to cut into existing supplies and make new joints it is essential to remove any traces of water that may be laying in the pipes. This water will prevent the solder from reaching melting point and thus result in a defective joint. In such circumstances a compression joint may be used but if for some reason this is not desirable there are two ways of overcoming the problem:

(a) Attaching a hose over the cut end of the pipe and blowing the water out through a tap on the downstream of the cut.
(b) Where there is no such convenient tap, water can be withdrawn from the pipe as shown in Fig. 4.4.

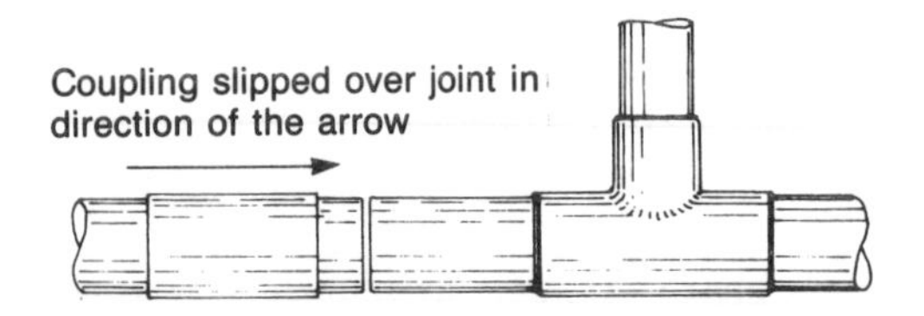

Fig. 4.3 'Slip' straight coupling for copper tubes. Coupling has no stop which permits it to slide over the pipe. Ensure centre of fitting is over meeting point of pipes.

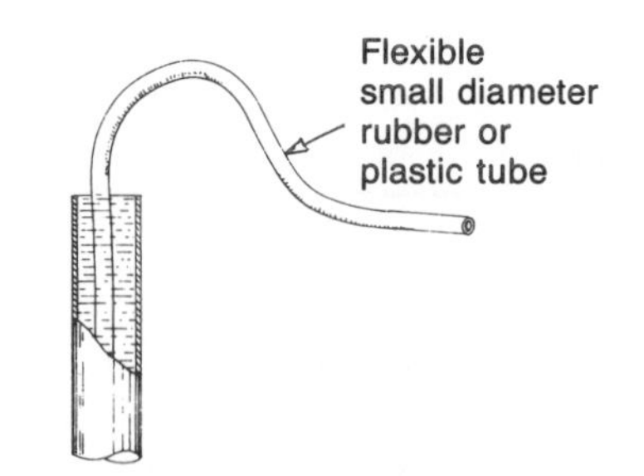

Fig. 4.4 Removing water by siphonage from a pipe prior to soldering

Jointing steel pipe

The smaller sizes of steel tubes used in plumbing for hot and cold water services are almost invariably joined by cutting and screwing using BSP (British Standard Pipe Threads) to BS 21. Pipes are usually threaded on site by hand dies or power machines, and when using the latter the appropriate safety regulations must be strictly observed. A thread cut on the outside of a pipe is called a male thread while the internal thread in a fitting is referred to as female.

Threaded joints

The cutting head on threading machines and some types of hand dies is adjustable. Each set of dies is designed to cut two or more thread sizes, and being adjustable they allow for a little wear to take place. Care must be taken to ensure that the marks on the face of the die are correctly set so they will remove the correct amount of metal from the pipe. Failure to do this will result in the thread being too loose or too tight in the fitting. When the size of the dies are changed, the numbers 1 to 4 stamped on them must correspond to the same number on the die stock to avoid damaged or stripped threads. Some stocks employ block dies, one die being required for every tube diameter. These are not adjustable and when they become worn must be replaced.

As a comparatively large amount of metal is removed when a thread is cut, to avoid the dies becoming overheated or threads stripped, cutting oil or paste must be freely applied during the threading operation.

The normal type of dies are designed to cut a tapering thread which when screwed into the socket of the fitting causes it to expand slightly. This ensures that both thread surfaces are in close contact, making a watertight joint. A well-formed thread cut with correctly adjusted dies should enable a parallel threaded fitting to be screwed on the pipe by hand for four threads. When it has been fully tightened, only two or three threads should be exposed.

Care should be taken to avoid damage to the fittings with grips and wrenches during the tightening operation. With elbows or tees a piece of pipe may be screwed into the socket at 90° to the joint being made, the fitting being tightened home using this short length of pipe. In this way the excessive use of grips can be avoided (see Fig. 4.5).

Prior to making these joints the male thread is smeared with a proprietary jointing paste before being tightened home. Some plumbers use hemp with the jointing paste but this should not be necessary if good threads are cut and new fittings are used. If hemp is used any 'whiskers' remaining visible on the joint after tightening should be removed with an old hacksaw blade. Hemp should not be used for joints on potable water pipes.

A cleaner method of making joints is the use of polytetrafluoroethylene (PTFE) tape, which although suitable for normal joints on gas, water and steam pipes, should not be used for making joints where the thread is slightly distorted. A typical instance where this occurs is the welded connection fitted into sheet steel radiators. The welding operation distorts the thread slightly and often the only effective method of making a watertight joint in such a situation is by using jointing paste and hemp. PTFE has a high resistance to chemicals and a working temperature range of −80 °C to 250 °C. It has a very low coefficient of friction which enables the joints to be tightened up easily.

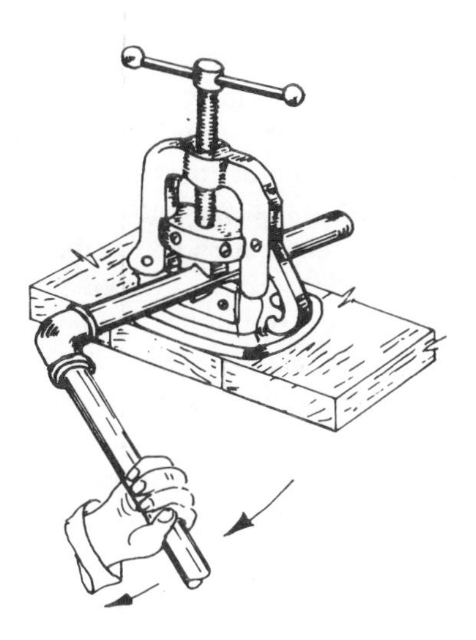

Fig. 4.5 Tightening steel fitting where possible without the use of pipe grips. Short length of pipe used as a lever.

Low carbon steel pipes are cut with a hacksaw or wheel cutters. When a hacksaw is used the work should be rigidly clamped in a pipe vice to avoid breaking the saw blade. The application of a little cutting oil or paste will lengthen the blade life and reduce the manual effort required to cut the pipe. Wheel cutters, due to their compressive action, leave a burr in the bore of the pipe and unless this is removed it will obstruct the flow of water or gas. All cut ends should be deburred with a tapered reamer, most threading machines having one fitted as standard. If hand methods are used a suitable reamer may be obtained for fitting into the chuck of a carpenter's brace (see Chapter 2, Fig. 2.6).

Flange joints

Joints on large steel pipes and in industrial installations where pipework may have to be periodically renewed are often joined with flanges which are more easily removed than screwed joints (see Fig. 4.6(a)). The flanges are fitted to the pipe with BSP threads, or may be fusion welded provided the pipe is not galvanised. The material used as a gasket between the flanges will depend upon the fluid the pipe is carrying. Insertion rubber is suitable for many applications and consists of a sandwich of one or more layers of loosely woven linen between heat-resistant rubber. It is also possible to obtain synthetic rubber sheets from which suitable gaskets may be cut, this material having no need for reinforcement linen. Purpose-made corrugated brass rings are another alternative (Fig. 4.6(b)). The corrugations give slightly on tightening and make a sound joint.

When flanges are bolted up, the correct

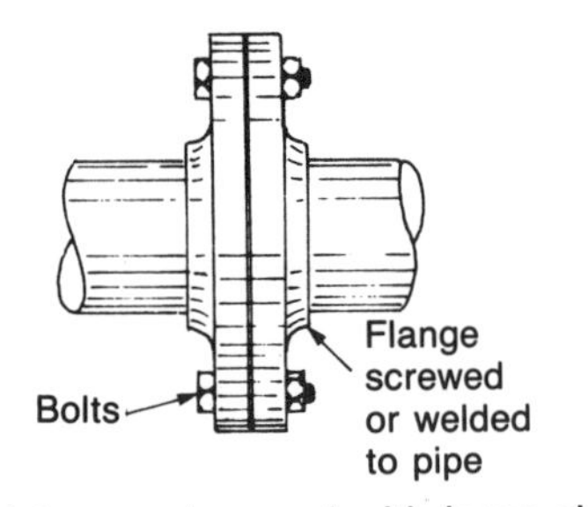

(a) Flange joint can be used with large size steel, copper, and plastic pipe installations

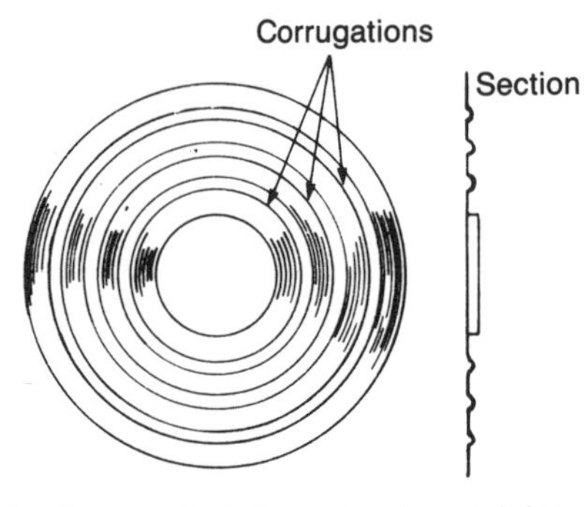

(b) Corrugated rings for flange joints

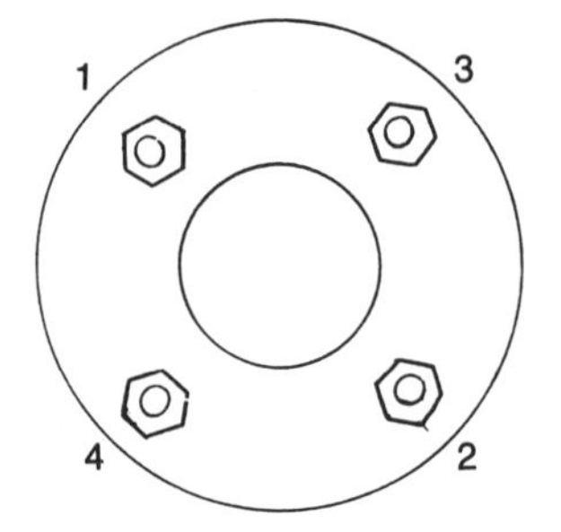

(c) Order of tightening nuts to avoid stressing the flange

Fig. 4.6 Flange joints

sequence of tightening the nuts is important to ensure an even pressure on mating faces. Figure 4.6(c) shows the order of tightening a small flange with four bolts. Assuming the nuts are all hand tight and the flange faces and the pipes are correctly aligned, the nuts should be turned approximatley one revolution at a time through the sequence shown. This will avoid undue stress at any one point in the joint.

Jointing synthetic plastics

The most common of these materials in use for underground water services is medium density polythene, methods of jointing being fusion welding or compression/mechanical joints made of gunmetal, high density polythene or polypropylene. Fusion welding, requiring specialist equipment, is mainly used on gas and water main work and, in some cases, chemical discharge pipework.

British Standards list two grades of medium density polythene tubes coloured blue for underground use and black for installation above ground. The liners are inserted into the end of the tube, their purpose being to support the tube against the crushing effect of tightening the cone. Typical compression fittings for adapting polythene to a male iron thread are shown in Fig. 4.7(a) and (b).

To make a joint on (a) the nut is placed over the end of the pipe and the liner inserted, a flange on the liner preventing it being pushed too far into the bore. The olive or cone is then pushed over the end of the pipe before it is inserted into the socket of the fitting. The nut should be tightened as far as possible by hand and completed by one and a half turns with a spanner. Do not over tighten as this will cause the edges of the cone to cut into the polythene. When the joint is tight it should not be possible to twist the pipe within the fitting.

Figure 4.7(b) shows a push fit connection. Unlike simple 'O' ring joints of the type used for discharge pipes, which would not be capable of withstanding high internal pressures, this joint is designed to tighten as the water pressure increases. Reference to the illustration shows that due to the tapering shape of the body any increase in pressure tends to force the 'grip' and 'O' ring more tightly into the body. To make the joint, the polythene tube should be bevelled (a special tool is available for this purpose) and a mark made on the pipe to indicate the length of pipe to be inserted into the fitting. Do not make the mark with a sharp instrument. The liner is pressed into the bore of the pipe, the pipe end then being inserted into the socket with a slight twist. When the pipe is fully home the original mark on the pipe should correspond to the end of the fitting. As further

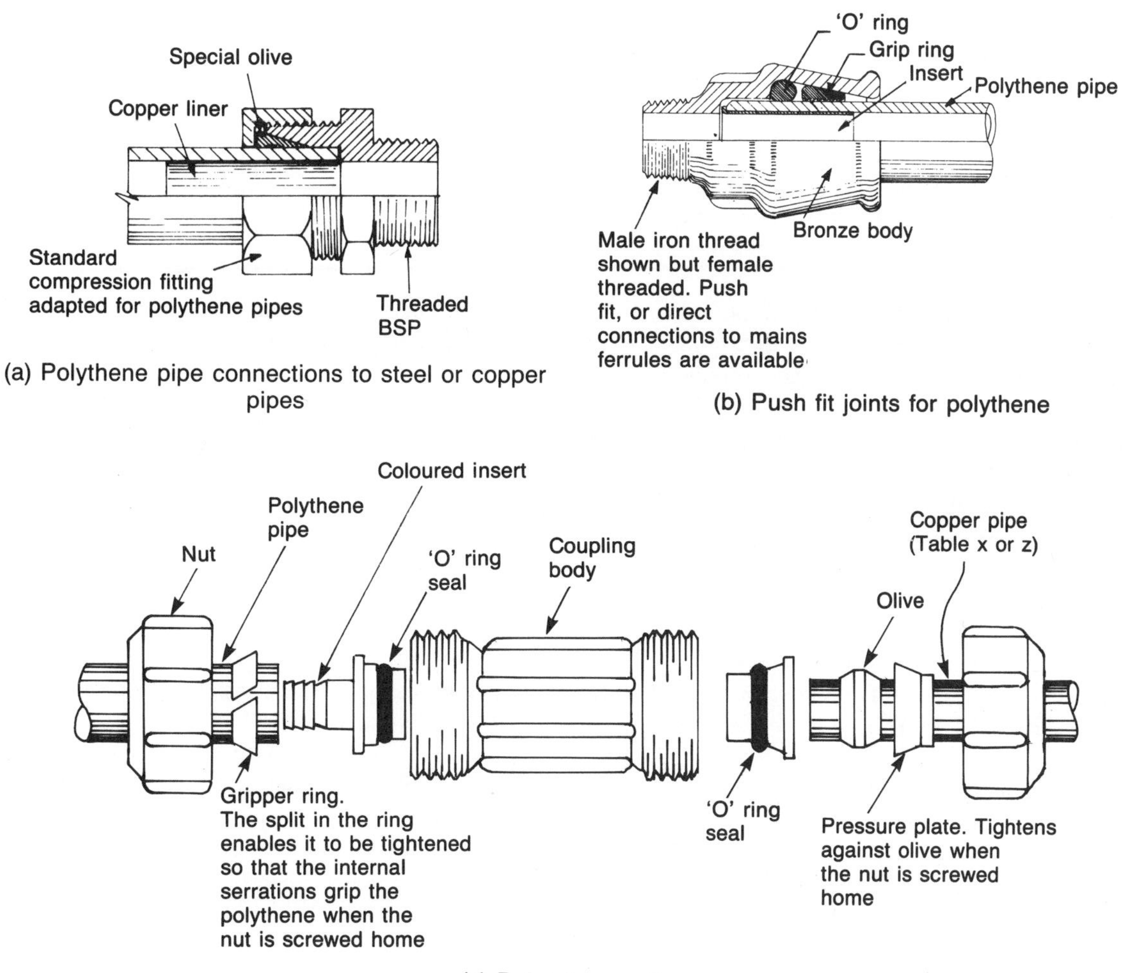

(a) Polythene pipe connections to steel or copper pipes

(b) Push fit joints for polythene

(c) Polygrip coupling

A standard body can be used for most connections between a wide range of differing pipes, e.g. lead, copper and various grades of polythene. Adaptors for this material are colour-coded to ensure the correct insert is selected. The illustration shows the adaptors necessary to make a joint between Table X copper tube and blue medium-density polythene.

Fig. 4.7 Pipe connections

proof of a sound joint, two points of resistance will be felt as the pipe is pushed through the grip and 'O' rings.

Should it be necessary to dismantle the joint, special extractor tools must be used. They are inexpensive and available from the fitting supplier. The fitting illustrated has a bronze body but those for connecting plastic to plastic are made of high density polythene, a slightly tougher material than medium density.

The connection shown in Fig. 4.7(a) is a copper compression joint which is adapted for connection to polythene pipes. An alternative is the use of fittings made entirely of plastic, usually high density polythene or polypropylene. The body can be used with a variety of inserts to accommodate connections to different pipe materials. The type shown in Fig. 4.7(c) shows the fitting with inserts for copper and medium density blue underground water service pipe.

Plastic pipes for hot and cold services

The two main materials that have been developed for this purpose are polybutylene and cross-linked polythene, the latter being used extensively by specialist firms for underfloor heating. Both types of pipe have the same external diameter as copper tube and can be adapted for direct connection to this material. The above-mentioned plastics have similar characteristics relating to pressure resistance and maximum working temperatures. Until recently the maximum diameter available was 22 mm but one company is currently producing fittings and tube of 28 mm OD.

The basic principle of joints used for these materials is best described as mechanical. They are push fit and employ what is called a grab ring which is designed to embed into the pipe when the joint is made. On no account should the fingers be pushed into the joint – the grab ring will ensure it is difficult and painful to remove them!

The water seal in all joints of this type is an 'O' ring made of heat-resistant synthetic rubber. When cutting pipes on any plastic in the polyolefin group, a tool of the type shown in Fig. 2.6 should be used as it produces a clean square cut.

Prior to making a joint using these materials a support sleeve must be inserted into the end of the pipe to give a greater degree of rigidity at the joint.

Figure 4.8(a) shows a joint marketed by Hepworth Building Products using the Hep 'O' system. It is made of polybutylene and is marked throughout its length as shown, which enables the installer to check visually that a satisfactory joint has been made. Figure 4.8(b) shows a joint produced by John Guest Ltd, which is similar in that a rubber seal is used to make a watertight joint, and the grap ring principle is employed to prevent the withdrawal of the pipe under pressure. These fittings are designed for use with cross-linked polythene but both types may also be used as a substitute to traditional fittings on copper tube.

Connection to other pipework material

In cases where polybutylene or cross-linked polythene tube has to be joined to steel pipes, a range of adaptors, having male or female BSP threaded ends are marketed. Where it is necessary to connect them to medium density polythene, e.g. the entry of a water service into a building, stopcocks are available having the inlet end adapted for medium density polythene pipe, the outlet end being suitable for direct connection to polybutylene or cross-linked polythene. Because these materials have a much better resistance to high temperatures than other plastics they can, in some cases, be connected directly to boilers. Figure 4.9 illustrates the recommendations of Hepworth Building Products.

Although both these materials are relatively rigid, any bends made will tend to spring back to their original shape because, like polythene, they have a plastic memory. Any bends made will be of a large radius and they must be well secured with suitable pipe clips. Hepworth Ltd produce a special former illustrated in Fig. 4.10 which not only provides a good fixing for bends but enables a smaller radius to be used.

Both types of pipe have sufficient rigidity for surface mounting, although more fixings must be employed than for metal pipes. However, because plastic pipework of any material lacks the mechanical strength of metal, it should not be used in circumstances where it could be subjected to mechanical damage. Long exposed surface-mounted runs of pipe are not recommended, as while they may be fitted in neat straight runs, an increase in temperature will cause them to expand and undulate (look wavy) which is unsightly.

Fusion-welded joints in plastic pipe

The most satisfactory way of jointing polythene for water and gas mains is by means of fusion welding. Special tools containing an electrical heating element are used for this purpose and a pictorial illustration of a hand-held tool for smaller pipe sizes is shown in Fig. 4.11(a). By changing the bushes and spigot heads a variety of pipe sizes can be jointed. All these tools are designed for use with an electrical intake control box. This enables accurate assessment of the temperature and the period of joint heating time to be made automatically; an indicator light shows when the

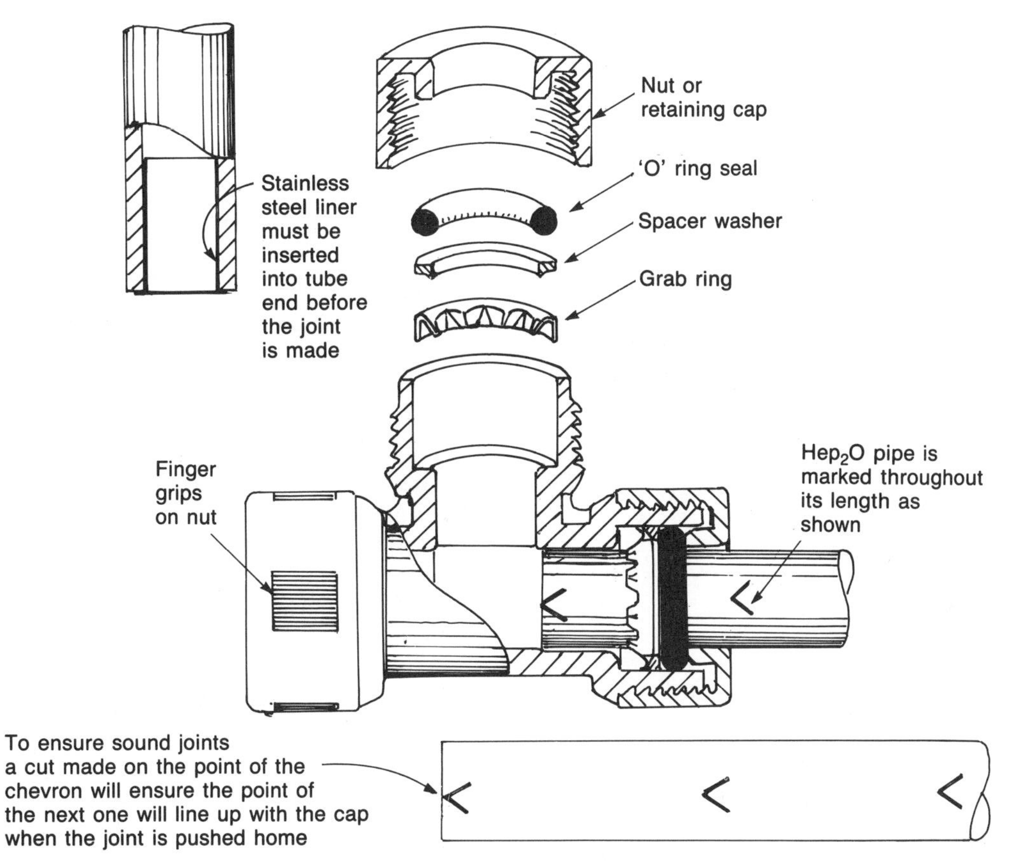

(a) Hepworth Hep$_2$O demountable fitting for polybutylene pipes; also available with fixed retaining caps

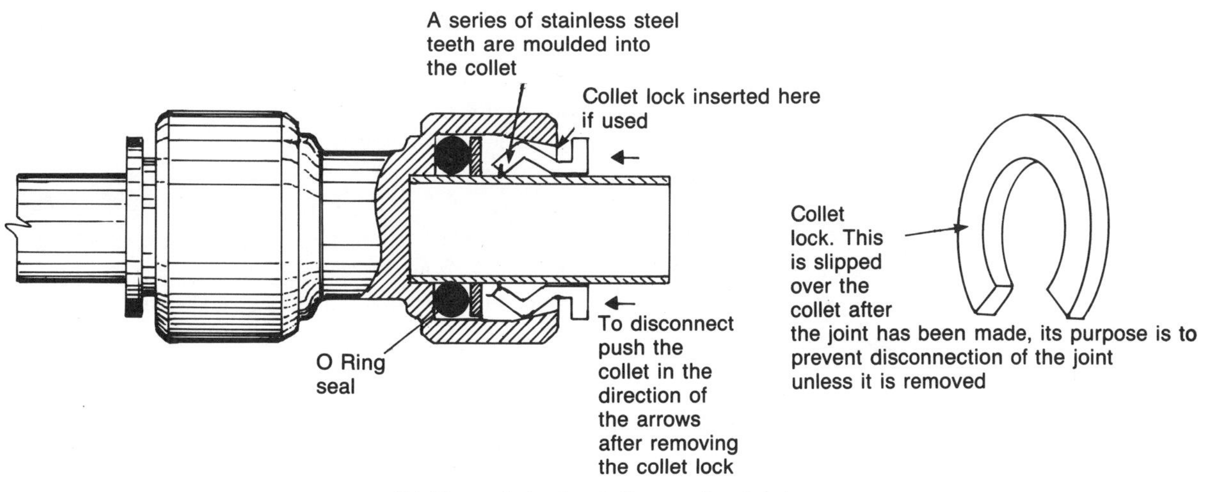

(b) Cross-linked polythene pipe joint

Collect covers (not shown) are available in a variety of colours for identification purposes if required. When the joint is subjected to internal pressure the fitting body tends to be pushed off the pipe, but, due to its internal taper, the stainless steel teeth in the collet are forced more firmly into the pipe thus preventing its withdrawal.

Fig. 4.8 Joints

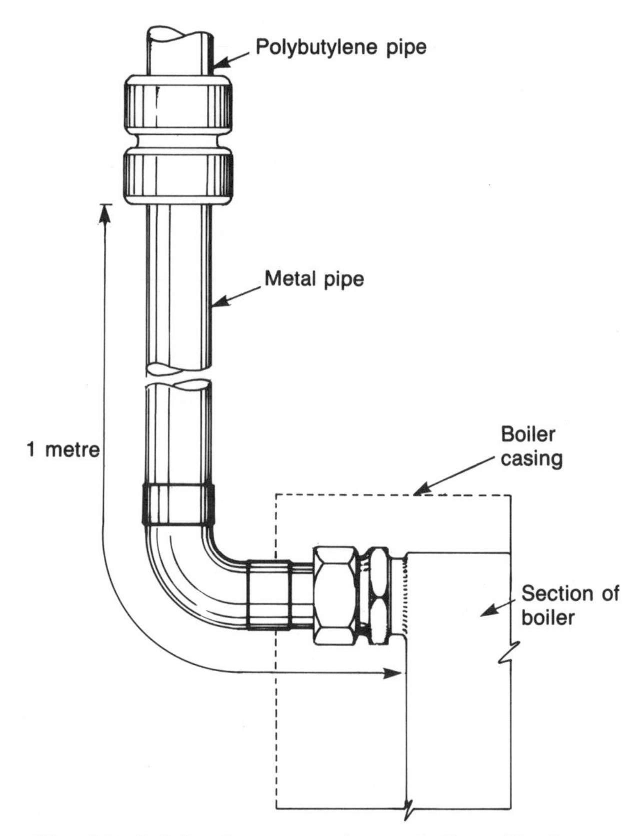

Fig. 4.9 Polybutylene connections to boilers. Metal pipework must be used if the connections are inside the boiler casings, if the heat source is within 350 mm of the polybutylene pipe or if the boiler has no positive thermostatic control, e.g. solid fuel boilers.

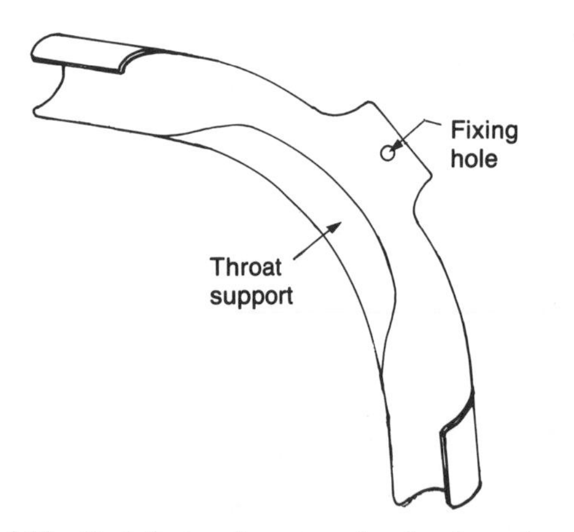

Fig. 4.10 Steel former for supporting bends made with Hep 'O' polybutylene pipe

shows when the correct joint temperature is achieved. Obviously a larger joint will require a longer period of time to achieve fusion temperature and this can be varied by the timer on the control box.

The voltage on which these tools operate varies, but for on site work it should not exceed 110 volts. In the event of there being no mains supply a portable generator can be used. To make a satisfactory joint the tube end should be cut square with a suitable cutter and any burrs or wooliness adjacent to the cut removed with a sharp knife. The surface film caused by the ultra-violet rays of light must then be removed as it has a detrimental effect on fusion jointing processes. It may be scraped off with a sharp knife but a more accurate method is to use a cutting tool as shown in Fig. 4.11(b).

After first ensuring the spigot and bush are of the correct size for the pipe being jointed, a check should be made to ensure they are quite clean and dry. The socket of the fitting and pipe end must then be positioned on the spigot and bush and the machine switched on. The temperature and heating recommendations must be followed exactly. When the indicator light shows, the fitting and pipe are removed from the tool and the pipe pushed firmly into the socket of the fitting. All fusion joints must be allowed to cool completely before a pressure test is conducted, and a general rule is to wait at least an hour after the final joint as been made. A diagrammatic illustration showing how these joints are made can be seen in Fig. 4.11(c).

Another method of making fusion-welded joints is shown in Fig. 4.11(d). A special fitting is used which has an electrical heating element moulded integrally in each socket. The joint is made by connecting a low voltage electrical supply to the heating element made integrally in the socket. This heats and fuses the mating surfaces of the socket and the pipe simultaneously. These joints are equally successful on both high and low density polythene and are also used for polypropylene, a plastic used in industry where a more rigid material than polythene is required.

A method of fusion welding using a hot inert gas such as nitrogen or hot air is sometimes employed to make joints on thermoplastics. It has

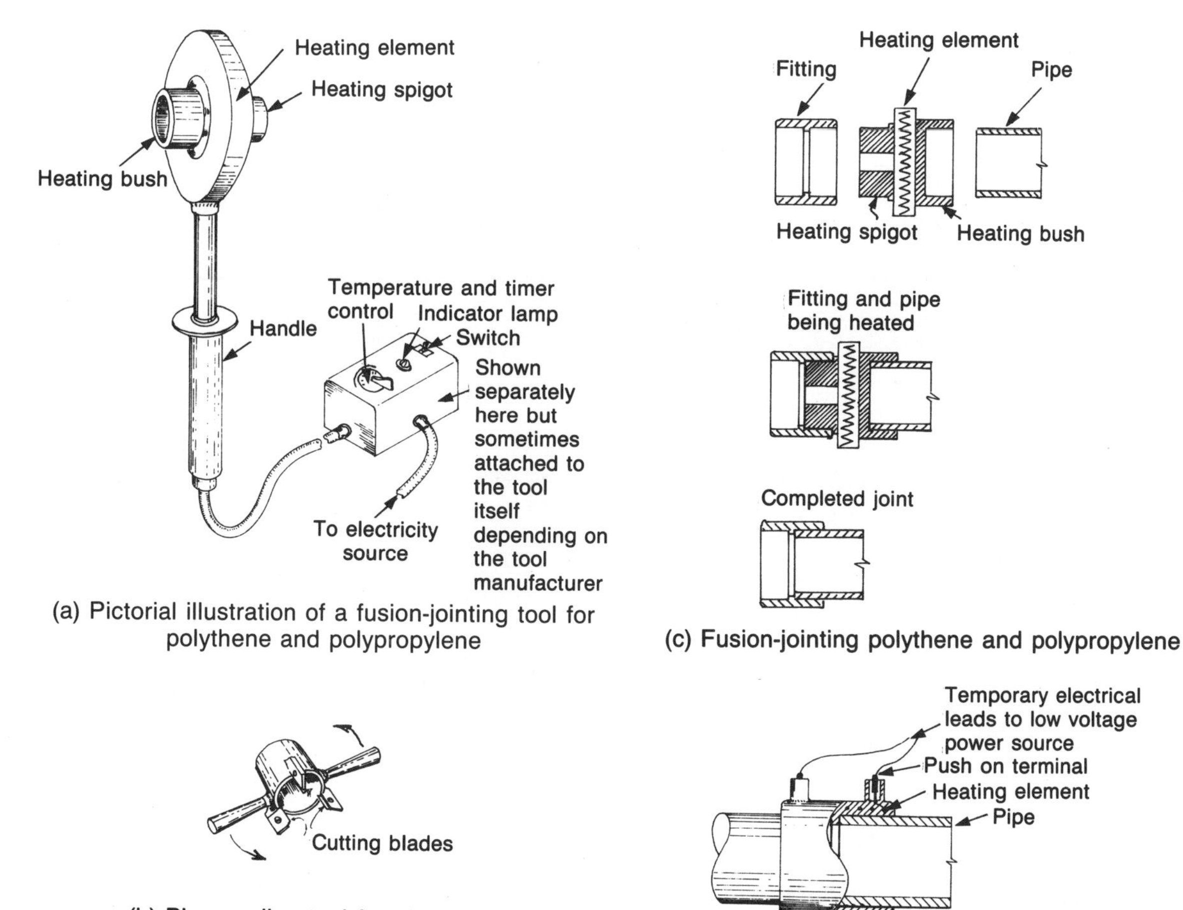

To remove the surface film this tool is rotated round the end of the pipe

(d) Fusion-welded socket with integral heating element for polythene or polypropylene

Fig. 4.11 Fusion-jointing equipment

some industrial uses but is not generally used in the plumbing industry due to the special equipment required and the readily available alternatives which are more suitable for site work.

Jointing plastics used for discharge pipework systems

Solvent welding is the usual method of jointing PVC and ABS pipes for both water and discharge pipes. A typical example is shown in Fig. 4.12. To make a joint, both the internal surface of the socket and the pipe end should be degreased with cleaning fluid. The mating ends are then lightly brushed with solvent welding cement and pushed home. This cement is not an adhesive; its action is to permeate the surfaces to be joined and soften them, thus permitting fusion of the surfaces. On exposure to the atmosphere the liquid content of the cement evaporates causing the joint to harden. Any surplus cement must be quickly removed as soon as the joint has been made or it will solidify and be almost impossible to remove without damage to the joint.

This method of jointing is best described as a cold welding process. Although the cement sets in a very short period of time, 12 to 24 hours should elapse before the joint is subjected to full working pressures. When joints of this type are used for

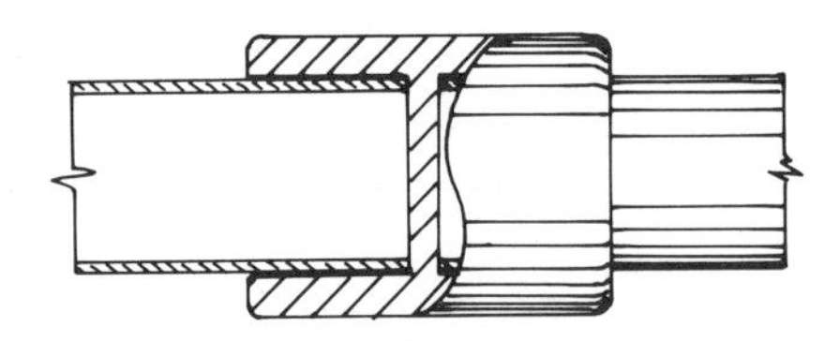

Fig. 4.12 Solvent welded joint. To make a successful joint the fitting and pipe should be degreased with fine grade steel wool or proprietary degreasing fluid prior to applying solvent cement to the end of the pipe. As the pipe is pushed home a slight twist will ensure its even distribution over the mating surfaces. Clean off any surplus cement from the edge of the socket with a clean cloth before it sets.

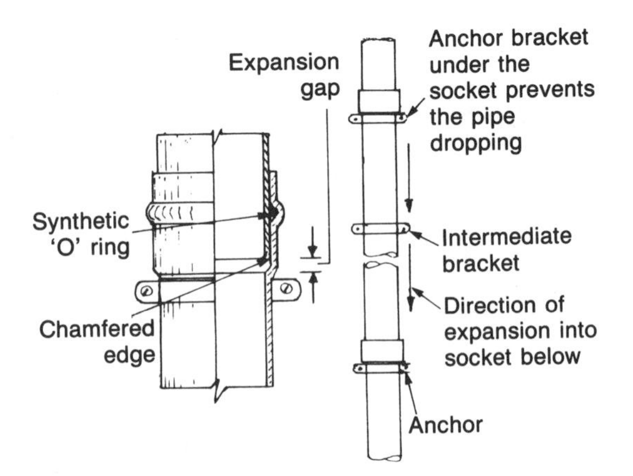

Fig. 4.13 'O' ring joint. This method of joining PVC pipes provides for freedom of movement to allow for the effects of expansion.

pipes of a large diameter a slower setting cement is used to enable the larger surfaces to be covered before the cement solidifies.

Welded joints obviously make no provision for expansion within the sockets, but as such short lengths of pipe are normally used on a well-designed system, any expansion is usually taken up by bends and compression joints at trap connections. Most systems, however, include a special expansion coupling if excessively long branches are necessary.

'O' ring joints The predominant type of joint used for long runs such as the main discharge pipe is the 'O' ring which, by leaving a gap of approximately 12 mm between the spigot and the base of the socket (see Fig. 4.13) provides for any expansion when hot water is being discharged, or for any variation in air temperature which may occur. The socket of each length is secured with a bracket and acts as an anchor, so that the expansion is accommodated within the joint. The pipe must be capable of moving freely through any intermediate brackets.

The method of making 'O' ring joints is similar whatever type of plastic material is used. When it is necessary, pipes should be cut with a fine tooth saw, a hacksaw being most suitable for this purpose. The spigot of the pipe to be inserted should be chamfered and marked with a pencil to determine the length of insert into the socket allowing for the expansion gap. When cut it should then be lubricated with soapy water or a proprietary lubricant prior to being pushed into the depth of insert indicated by the predetermined mark. Always make sure that the 'O' ring is not distorted and is secure in its groove before making the joint. PVC and polypropylene are suitable materials for this method of jointing.

Adjustable bends To provide greater flexibility with their material many manufacturers produce adjustable bends (see Fig. 4.14(a)). These enable a standard bend to be cut and fitted with a special adaptor, thus allowing bends to be fabricated to any angle.

Figure 4.14(b) shows a flexible joint for 100 mm nominal bore discharge pipes. It is made of PVC and is designed to allow flexibility between the angle of a WC outgo and the branch discharge pipe into which it is fitted. They are especially useful when connecting to a range of 'p' trap WCs where, due to the fall on the 'float' or main branch, the angle of each WC branch pipe will vary slightly.

Jointing to cast iron and clayware Plastic discharge pipes are joined to cast iron or clayware drains by means of cast iron or PVC connecting pieces, which provide for a normal run lead or cement joint to be made while allowing for the expansion of the plastic pipe (see Fig. 4.15).

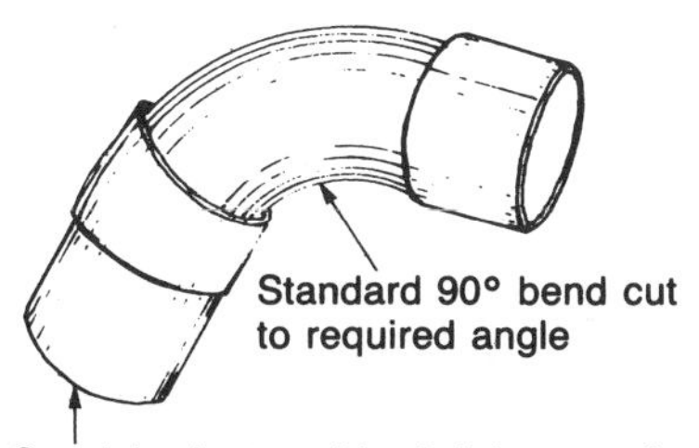

Fig. 4.14 Adjustable bends

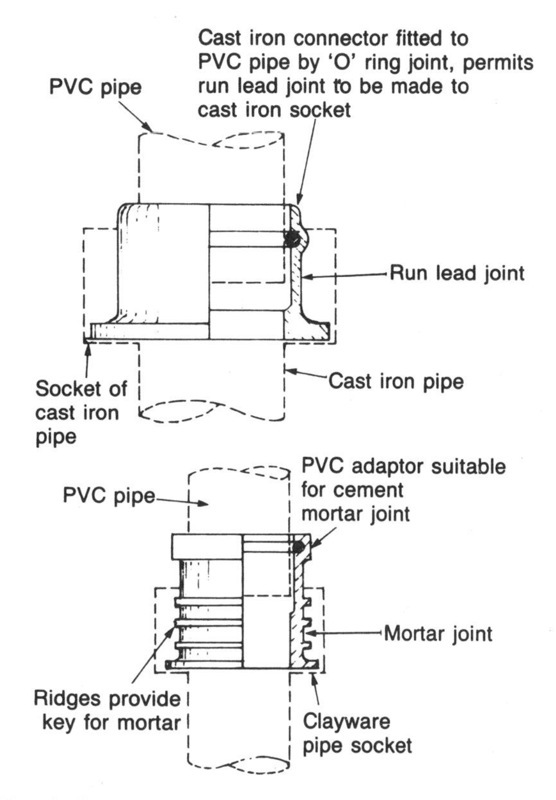

Fig. 4.15 Adaptors for PVC discharge pipes to pipes of other materials

Pipe bending

One of the most important skills a plumber should possess is the ability to bend pipes of various materials quickly and accurately. This is complicated by the variety of materials the plumber handles and the differing techniques used to make bends whether by hand or in purpose-made machines.

Bending by hand

It will be obvious that any attempt to bend a thin-walled pipe will result in it being crushed or kinked before it has been pulled through many degrees unless certain methods are used. To enable such pipes to be bent, various methods of supporting or filling the inside of the pipe are employed to ensure against collapse during bending. Of all the materials a plumber uses, only the heavier grades of polythene have pipe walls of sufficient thickness to enable them to be bent without internal support. One other possible exception is the method of hot bending steel pipes, but even these pipes tend to flatten slightly when the bend is made, and on the occasions when this method is used, the sides are squeezed inward by pressure from a vice while still at red heat, causing the section of the bend to assume a truly circular shape throughout its length. With the increased use of hydraulic bending machines this method of bending is little used except in the cases where a radius, different from that produced by the machine former, is required and a screwed or weldable fitting is unsuitable.

Bending springs for copper tube
Spring bending is a useful method of bending small diameter pipes, especially if the radius of a machine made bend is unsuitable for any reason.

The spring, which should be a running fit inside the pipe bore, is located centrally at the point where the bend is to be made. The most common method of pulling the bend is across the knee, taking care to shift position slightly to avoid too sharp a bend, which may result in causing the throat to wrinkle. The physical strength required to make a bend using this method does to some extent limit its use to the smaller sizes of the tube. Copper tubes of 22 mm diameter or more should be annealed prior to bending, to reduce the physical effort needed and to avoid rippling on the throat of the bend. It is not usual to bend copper pipes of larger diameter than 28 mm by hand as the large radius which would have to be used to prevent rippling would be unacceptable.

It should be appreciated that when a bend is pulled, the throat and back of the bend tightens on the spring, and to enable it to be withdrawn without damage, it is usual to over pull the bend by a few degrees and then open it to the desired angle as shown in Fig. 4.16 to reduce the gripping effect of the bend. If the spring still cannot be withdrawn easily, a slight clockwise turn with a tommy bar will tighten the coils, and slightly reduce its diameter facilitating its removal. Prior to bending, the spring should be oiled to facilitate its withdrawal. Springs can be used to make bends in the middle of long lengths of pipe by fastening a stout wire extension to the loop at the end of the spring, fencing wire being very suitable for this purpose. String should not be used as it will not permit the coils of the spring to be tightened

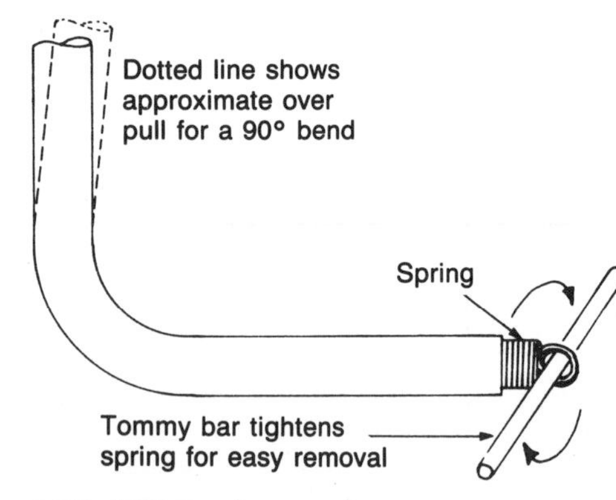

Fig. 4.16 Withdrawing a spring

should this be necessary to enable it to be withdrawn.

Bends having a radius of less than 6 × diameter are not recommended on copper tubes due to the possible difficulty of removing the spring.

Other materials such as sand or a low melting point alloy containing lead, tin and bismuth were at one time used in the plumbing industry for bending copper pipes of large diameter to a relatively small radius. The thinner wall thickness now used has made this very difficult, and where such bends are required they are usually made by companies who have the special bending machines necessary for the work.

Setting out

Hand made bends of all materials should be 'set out' to give the same accuracy of position and radius as those made by a bending machine. The main point to understand about all bends is the apparent 'gain of material' when the bend is formed. A study of Fig. 4.17(a) will show the reasons for this. The distance from 'a' to 'c' through 'b' along the dotted line is in effect the measure length of the bend, but when it is actually pulled, its path follows the arc a–c which is a shorter distance than the measured length.

To make allowances for this – and to ensure the bend is pulled in the right position in relation to a fixed point – the length of pipe actually occupied by the bend in relation to the measured distance can be calculated in the following way. The first step is to decide on the centre-line radius of the bend. In the absence of any other specification it is assumed to be four times the nominal diameter of the pipe, this being written as 4D. To find the length of pipe occupied by a 90° bend the following formula may be used:

$$\frac{\text{Radius} \times 2 \times 3.142}{4}$$

Example 1 Assuming a 25 mm diameter pipe is to be bent to a radius of 4D, find the length of pipe which will be occupied by the bend.

$$\text{Radius of bend} = 4 \times 25$$
$$= 100\text{ mm}$$

$$\therefore \text{ Length of bend} = \frac{\text{Radius} \times 2 \times 3.142}{4}$$
$$= \frac{100 \times 2 \times 3.142}{4}$$
$$= 157.1\text{ mm}$$

The length of pipe required to form this bend will be 157.1 mm. This method is very accurate but is rather a long way round to ascertain a simple length. In practice an approximation is made by assuming the length of pipe forming the bend to be one and a half times the radius, assuming a 4D radius, of course.

Example 2 Assume the same size of pipe bent to the same radius as Example 1, but use the following formula: length of bend equals one and a half times the radius:

$$\text{Radius} = 4 \times 25$$
$$= 100\text{ mm}$$
$$\text{Length of bend} = 100 \times 1.5$$
$$= 150\text{ mm}$$

When the answers to Examples 1 and 2 are compared it will be seen that the approximate method is incorrect by only 7.1 mm and is sufficiently accurate for normal site calculations.

To make the bend, mark off the required length from a fixed measuring point (probably the end of the pipe) to the centre line of the bend as shown in Fig. 4.17(b). Divide the calculated length of pipe which will form the bend by three, which in this example gives three equal parts of 50 mm. From the original centre line, mark 50 mm forward and 100 mm back as shown in Fig. 4.17(c). This shows the relationship of these measurements to the centre line and the fixed point or end of the pipe. When the bend is pulled it must be confined to the areas indicated by the shading, which will ensure its centre will be at the correct distance from the fixed point as shown in Fig. 4.17(d). This method can be used for making bends of less than 90°, but it is necessary to understand that a bend of 135° will occupy only half the length of pipe required for a right angled bend. By using this technique, bends can be made as accurately by hand as with a bending machine, with the added advantage that, unlike machine bends, which have

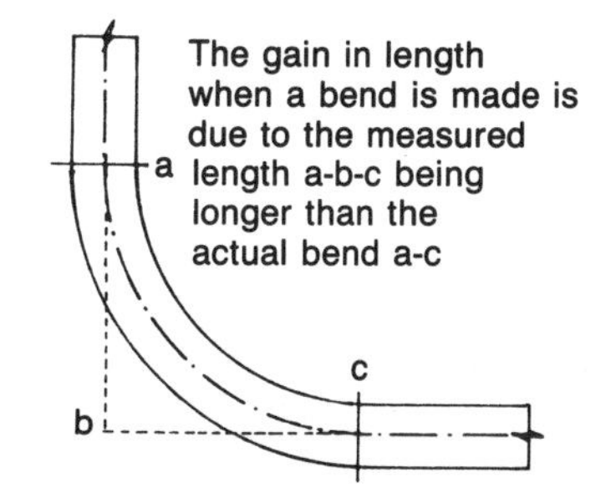

(a) Apparent gain in length made when a bend is formed.

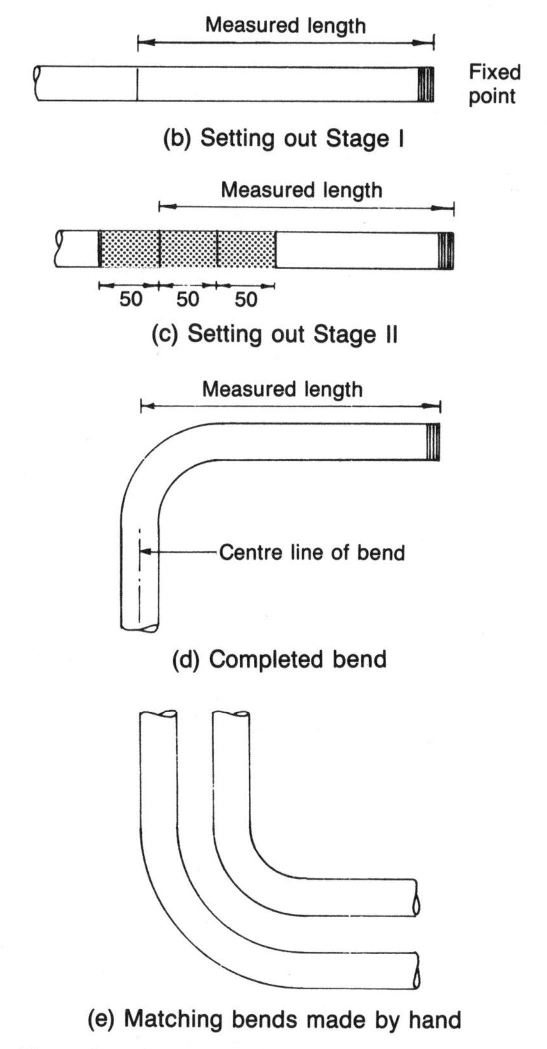

(b) Setting out Stage I

(c) Setting out Stage II

(d) Completed bend

(e) Matching bends made by hand

The radius of each bend is different, so the bends are spaced evenly apart. If one bend is made by machine the other must be made by hand so that it follows the curve

Fig. 4.17 Making bends

a fixed radius, the radius can be altered to make following or matching bends if necessary (see Fig. 4.17(e)).

Bending by machine

Light gauge copper tubes

Copper tube is probably the most widely used material for both hot and cold water supply and domestic heating and it is essential that a plumber can bend it quickly and accurately. While a bending spring is useful for small repair jobs or perhaps making a bend on a pipe that is already fixed, mechanical means of bending are essential on large installations if they are to be economic in time and materials.

There are many differing types of machines available, from the hand operated machines used for bending small diameter tubes to those of large powered type which can bend pipes of up to 150 mm in diameter. A selection of the smaller benders used for site work are illustrated in Fig. 4.18.

Figure 4.18(a) shows a simple tool for bending soft copper tubes of 8 and 10 mm diameter, this being simply a former around which the tube is bent by hand. Due to the comparatively large radius bends that are made on these small-diameter pipes no support is required at the back of the bend.

The small hand bender shown in Fig. 4.18(b) is a very useful tool both to the jobbing plumber and the heating fitter for bending 15 mm and 22 mm tubes. The machines are very light and can easily be carried in a tool bag. They are most useful when working in occupied premises as they take up little space and can easily be carried from room to room.

Figure 4.18(c) shows a larger machine capable

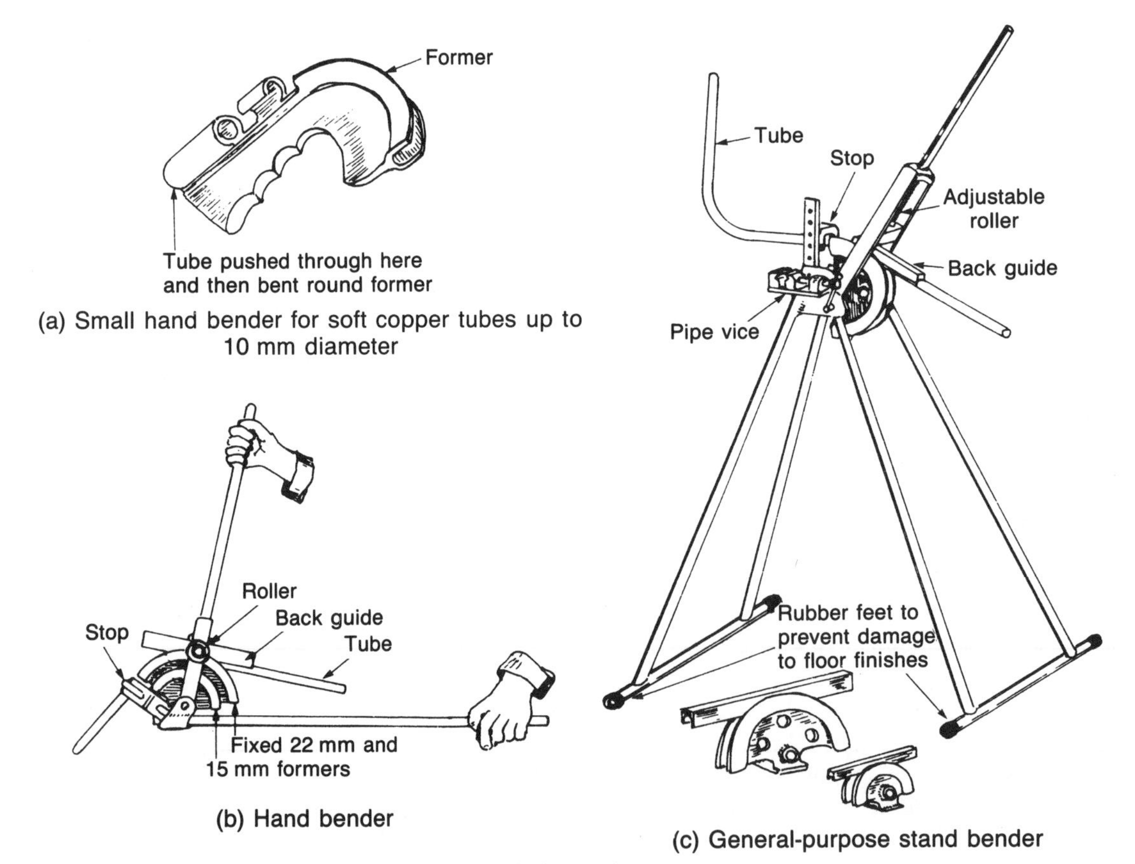

Fig. 4.18 Small pipe-bending machines for general site work

of bending up to 35 mm pipes and is quite suitable for general site work. Similar types are made for bench fitting.

For bending pipes of 42 mm to 54 mm special machines are made which are in effect geared down to reduce the manual effort required to make bends of this size. The type shown in Fig. 4.19 employs a screw thread arrangement for rotating the tube former, giving it a similar effort-weight effect to that of the simple lever. This machine can also be equipped for bending mild steel tubes to BS 1387 of up to 25 mm nominal diameter.

Bends on tubes of larger diameters than 54 mm are made on special equipment called mandrel machines which not only support the outside of the tube but also the inside by means of a travelling mandrel. Bends of small radius can be produced on these machines, a typical example being the fabrication of lead and copper tube traps.

Light gauge stainless steel pipes

Light gauge stainless steel pipes may be bent in all the machines mentioned with the exception of the hand bender which has a roller supported on one side only which would not sustain the stresses imposed by the tougher and more rigid stainless steel pipe. It is generally recommended that the

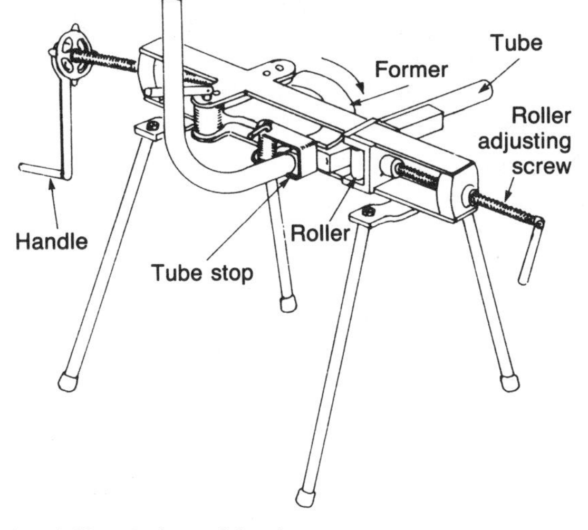

Fig. 4.19 Horizontal bender

capacity of benders should be reduced by one size when bending this tube, i.e. a bender with a capacity of up to 28 mm copper tube should not be used for bending stainless steel in excess of 22 mm diameter. Steel back guides should always be used with stainless steel.

Components of bending machines

Figure 4.20(a) shows the two principal components of a typical rotary bending machine of the type used in the plumbing industry. The former supports the throat of the pipe while the back guide rotates around the bend as pressure is applied by the roller. Figure 4.20(b) shows a sectional view of how these components support the tube during the bending process and prevent distortion taking place, provided the machine is correctly adjusted. Incorrect adjustment of the roller can have two effects, as illustrated in Fig. 4.20(c). Excessive throating will be the result of the roller 'A' being too tight, compressing the throat of the bend into the former. Slight throating of the tube is acceptable but too much is unsightly and will also obstruct the flow. The wrong adjustment of the roller 'B' will produce ripples on the throat of the bend due to the lack of support at this point. Some machines have a guide fixed to the handle which, when lined up to the tube, indicates the correct roller position. The roller, is, of course, adjustable.

There are other machines which have no provision for roller adjustment, and while they produce good bends when new, they have no provision for wear of the working parts. To obtain good bends with these older machines, especially when the larger sizes of tube are used, may necessitate the insertion of a thin piece of steel between the roller and back guide to avoid rippling the bend.

Maintenance

Very little maintenance is required to keep tube-bending machines in good working order. They should be cleaned regularly and lightly oiled with thin machine oil. The parts most prone to damage are the knife edges of the back guide. Any deviation from a sharp straight edge will transmit dents and cuts to the tube during bending. Light

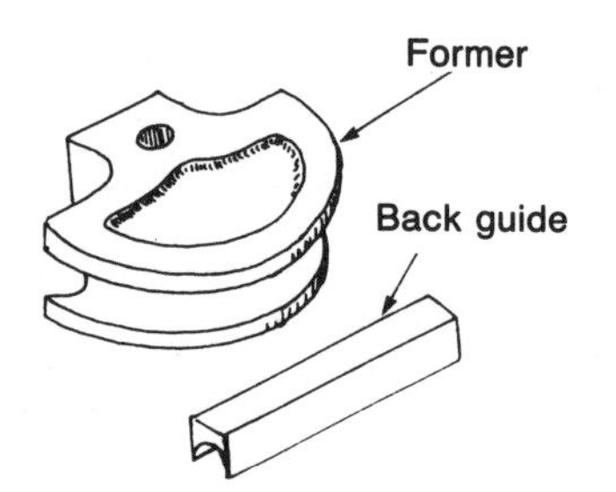

(a) The essential components of a rotary bender

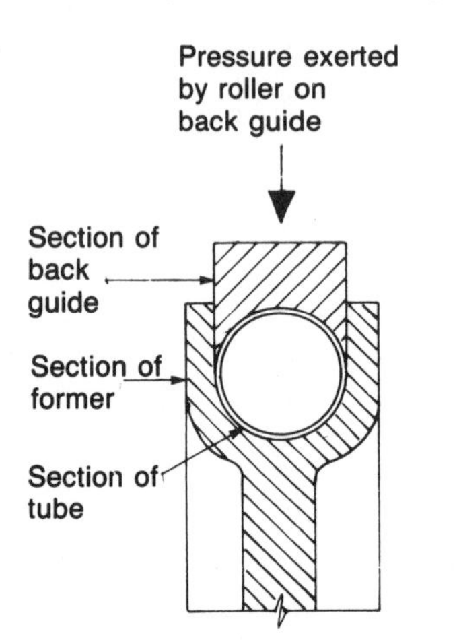

(b) How the tube is supported during bending

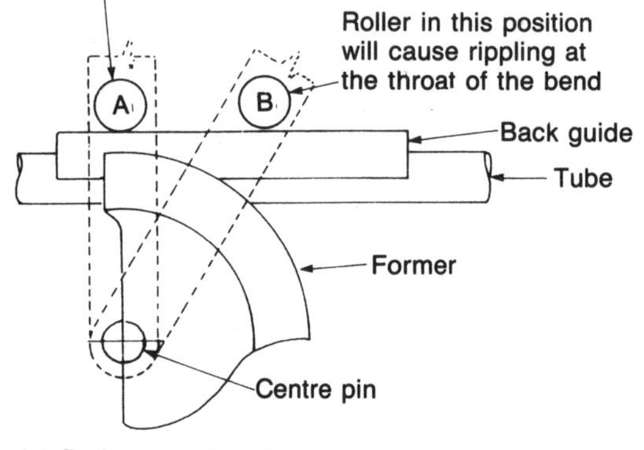

(c) Defects to bends caused by incorrect roller position

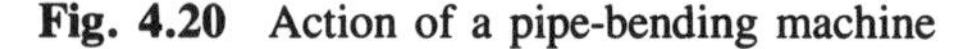

Fig. 4.20 Action of a pipe-bending machine

distortion can be erased with a fine file but in cases of severe damage, they are best replaced. Care should be taken to avoid dropping the guide when the handle of the machine is raised after a bend has been made. When not in use protect the guide by wrapping in a clean soft cloth.

Machine-bending processes

There are two basic operations to be fully understood and mastered in order to use a bending machine effectively, one being the ability to form 90° or square bends, the other being to bend offsets. Both of these operations must be carried out accurately to given measurements.

90° or square bends

The measurements of square bends given on a drawing are usually taken from the centre line of the pipe, as illustrated in Fig. 4.21(a). It is often more convenient on site to express the measurement back to back, as shown in Fig. 4.21(c). Those taken from the inside to back as at Fig. 4.21(b) will be the same as (a) and are only given in this way to enable the tube to be squared across from the inside or the outside of the former.

The example in Fig. 4.22(a) shows a bend made from a fixed point to the back of the bend. A pencil mark is made from the end of the tube, in this case 300 mm, the end of the tube being referred to as the fixed point (see Fig. 4.22(b)). The tube is set up in the machine with the mark squared off from the outside of the former. With the back guide in position and the roller correctly adjusted the bend is ready to be pulled (see Fig. 4.22(c)). If the measurement has been quoted to

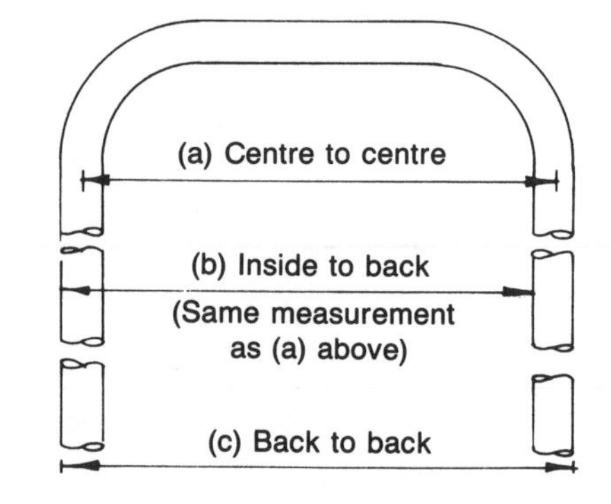

Fig. 4.21 Methods of taking measurements for tube bending

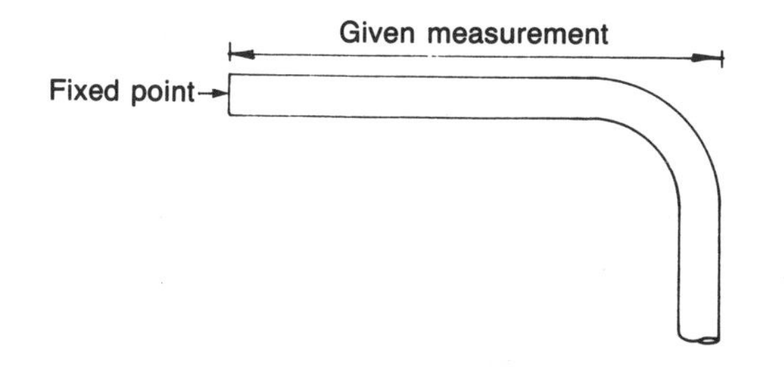

(a) The given measurement here is from a fixed point to the back of the bend

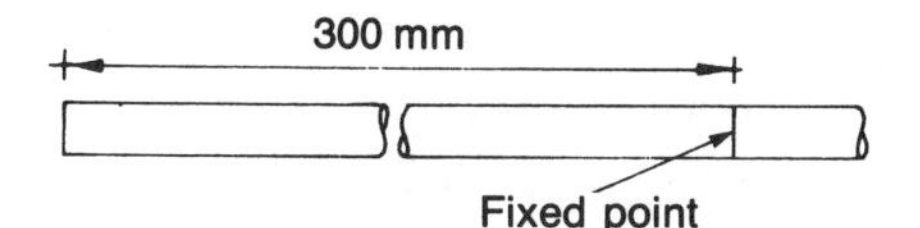

(b) Marking tube for bending

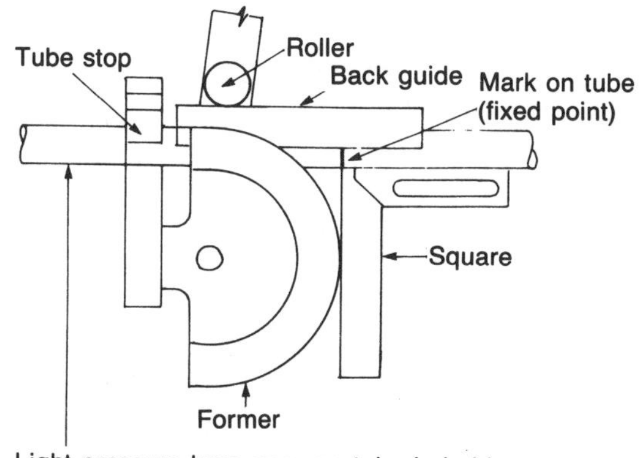

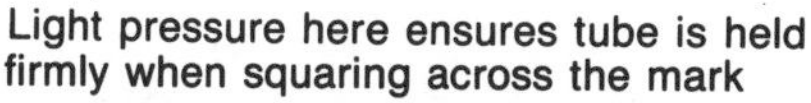

(c) Squaring across the mark to the outside of the former

Fig. 4.22 Bend made from a fixed point

the centre of the bend, then half a tube diameter must be added to the original measurement. To give an example of this, assume the tube diameter to be 22 mm, and the given measurement 300 mm from the fixed point to the centre of the bend, then this measurement would be 311 mm from the fixed point to the back of the bend.

To make a return bend or to bend the pipe again in another place, the technique is the same, the only difference being that the first bend has now become the fixed point. Figure 4.23 shows the set up for making a return bend from the centre-line measurements but actually measuring from the inside of one bend to the back of the other bend (this is equivalent to measuring centre to centre – see Fig. 4.21(a)).

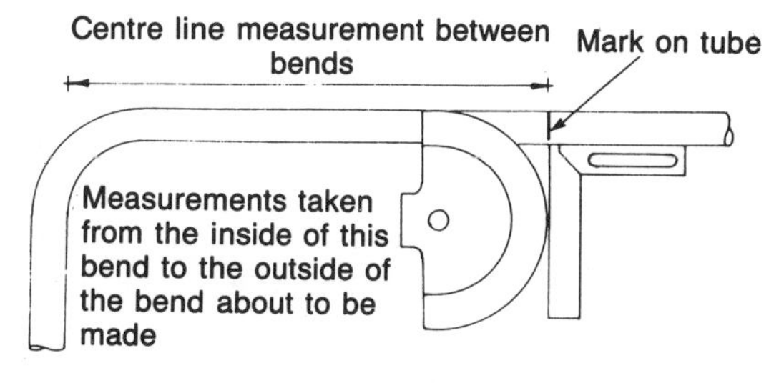

Fig. 4.23 Setting out return bends

Offsets

There are various methods for measuring offsets. The method shown here will give accurate results for small offsets and is most suitable for use with bench benders. The first set is made to the desired angle as shown in Fig. 4.24(a). In many cases the angle is not critical and can be judged by eye, but if a fixed angle is essential, it should be taken from the actual job using a bevel. When making very small offsets take care not to make the first set with too large an angle, or when the second one is pulled it will be found that they run together, often kinking the tube. Having made the first set the tube is reversed and pushed through the machine, a straight edge being set against the tube former, parallel to the tube, as illustrated in Fig. 4.24(b). The required offset measurement is taken from the inside of the tube to the inside edge of the straight edge. Note the relationship of the second set shown as a dotted line parallel to the straight edge. It will be seen that the measurements are actually taken from the inside to the back of the tube, this being the equivalent of a centre-line measurement. When the tube has been adjusted against the stop of the machine, it is advisable to mark the tube at the edge of the former, as shown at ‘A’ in Fig. 4.24(b). This permits the tube to be correctly repositioned in the machine if it is accidentally moved after the measurements have been taken.

While the foregoing is not impossible with stand

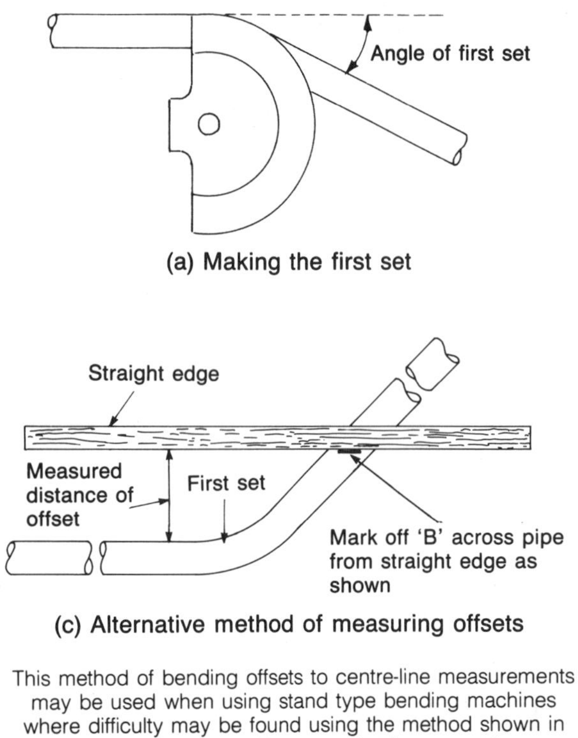

This method of bending offsets to centre-line measurements may be used when using stand type bending machines where difficulty may be found using the method shown in (b). It must be stressed however that it may not be as accurate. It is important to note that the channel of the former must be as deep as the outside diameter of the pipe and is not suitable for some types of hand benders. It is assumed that the first set has been formed to the required angle.

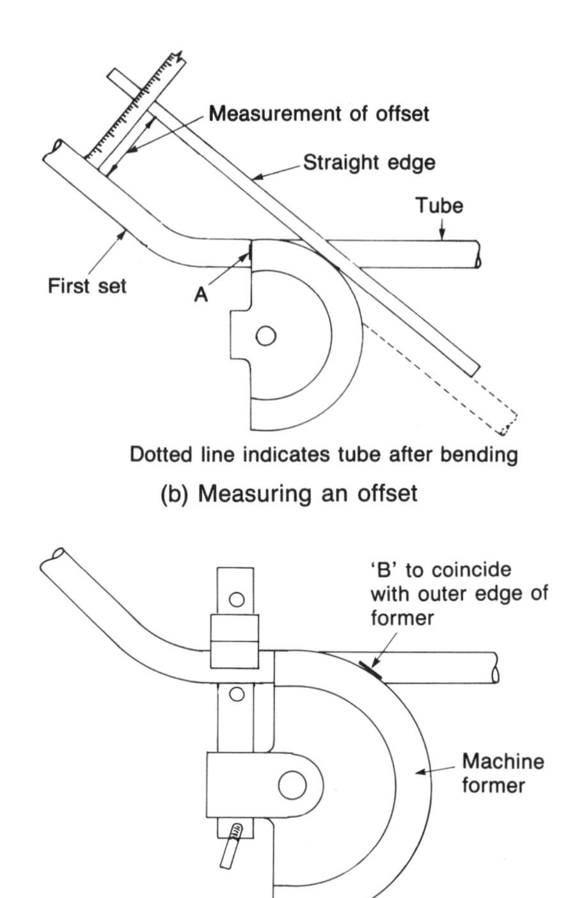

Fig. 4.24 Making an offset

type machines, it is not quite so easy to use a straight edge. The method of measuring offsets shown in Fig. 4.24(c) and (d) will produce offsets to a reasonable degree of accuracy. It must be stressed however that if an accuracy of ± 2 mm is required then the method using a straight edge must be employed.

Passover bends

It is often necessary to make bends to clear obstructions across a run of pipe. Bends of this type are called 'passovers' and fall into two categories. The type shown in Fig. 4.25(a) occurs when a branch is made in a service pipe and has to pass over another service pipe above. The measurements in this case are taken in the same way as for an ordinary offset, the only difference being that the second bend is over pulled until the desired angle is obtained to fit the tee.

The other type of passover is sometimes called a 'crank set' because of its similarity to the crank on an engine, this being shown in Fig. 4.25(b). A two or three millimetre gap should be left between the throat of the passover bend and the obstruction, which avoids any possibility of action between dissimilar metals or chafing between the two pipes. The required clearance is determined after the centre set has been made, the measurement being taken as shown in Fig. 4.26(a). Marks 'A' and 'B' are then made across the tube with a straight edge as shown in the illustration before replacing the pipe in the machine. When

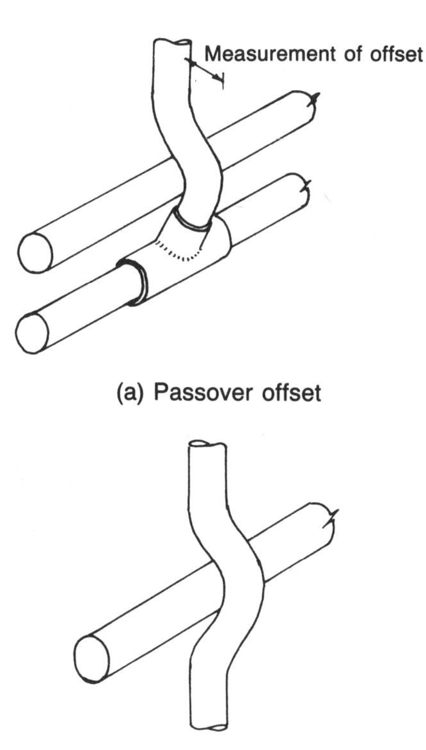

(a) Passover offset

(b) 'Crank' passover bend

Used to clear an obstacle such as another pipe

Fig. 4.25 Types of passover

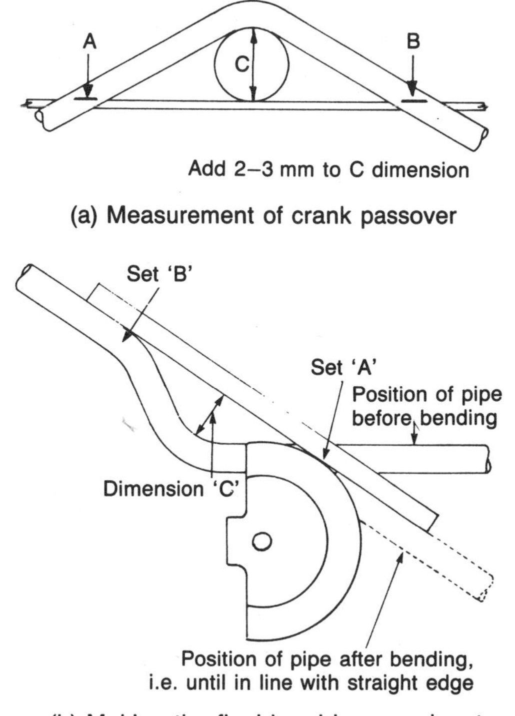

(a) Measurement of crank passover

(b) Making the final bend in a crank set

Fig. 4.26 Making a crank passover bend

mark 'A' is at a tangent to the former the pipe is in the correct position and should be pulled to the angle indicated by the dotted line. The pipe is then reversed in the machine and mark B lined up to the former. A straight edge should be used at this stage to check both the clearance of the passover and the alignment of marks A and B. The last set can then be pulled to complete the bend.

Location of sets

It is sometimes necessary to locate the exact position of a single set in relation to that of another, a typical example of that being when a pipe has to be bent around the angles of a bay window.

Some plumbers make a mark on the former to locate the measured mark shown in the illustration. This doesn't always give accurate results as marks on the former will vary as to the degree of the angle to which the tube is to be bent. For absolute accuracy when using this method a simple drawing is required indicating the degree of the angle. This method is illustrated in Fig. 4.27(a)–(c). In the absence of a bevel a 600 mm steel folding rule can be used to determine the angle.

Steel pipe bending

Steel pipes to BS 1387, due to their thicker walls, do not need to be fully supported during the bending process, and a method known as press bending is used. This involves a hydraulic bending machine of the type shown in Fig. 4.28.

Hydraulic pipe-bending machines

Some idea of hydraulic power can be seen on most building sites where bulldozers and trenching machines are commonplace. The power source of all hydraulic machinery is the ram which in many ways is similar to a pump. Liquids, for all practical purposes, are incompressible and as such are capable of exerting a similar force to that of a

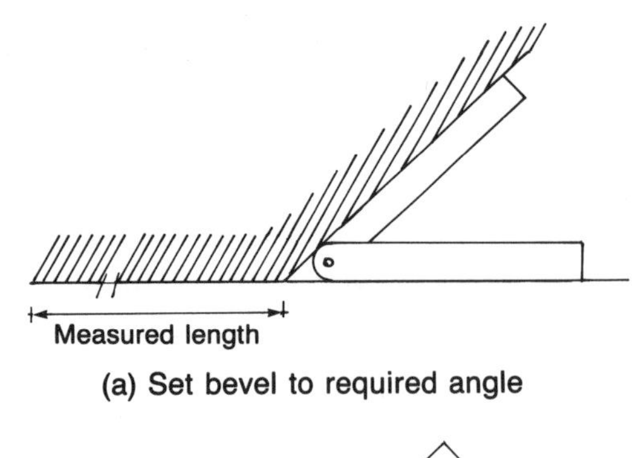

(a) Set bevel to required angle

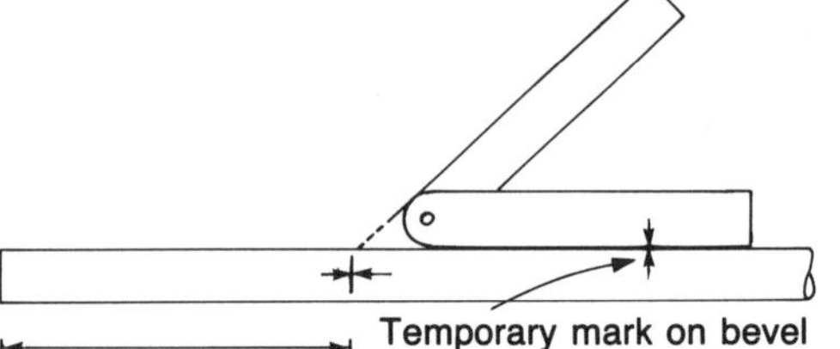
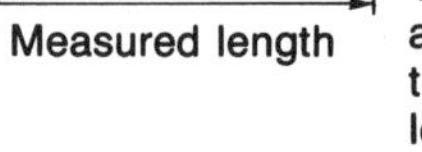

(b) Line up to top edge of bevel on marked pipe

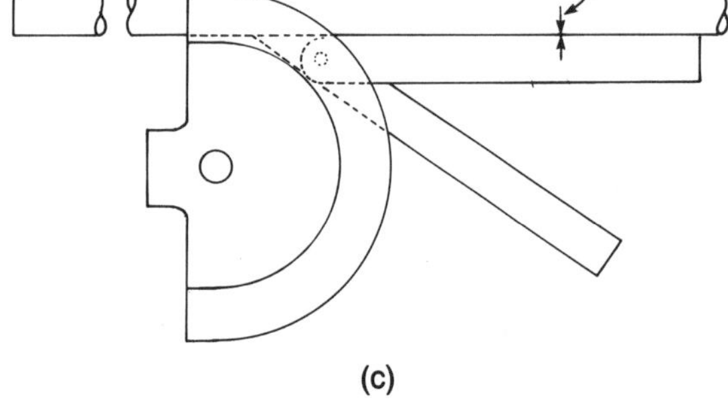

(c)

This method can be adapted as an alternative to that shown in Fig. 4.24(a) and (b) for measuring offsets on tripod bending machines.

Fig. 4.27 Accurate location of sets in copper tubes

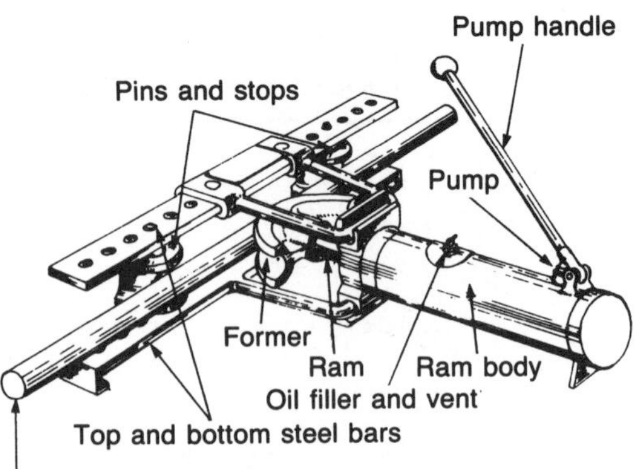

Fig. 4.28 Hydraulic press bender for steel pipes. The action of moving the pump handle backwards and forwards is to pressurise the oil which, acting against the piston, moves the ram and former against the pipe which is to be bent.

steel bar when they are subjected to pressure. Figure 4.29(a) shows the basic principle of a hydraulic ram which can be explained simply as follows. The fluid is pumped into the cylinder of the ram and exerts pressure against the piston which is forced outward to work many differing types of equipment. Figure 4.29(b) illustrates the pump details of the ram.

In the case of a hydraulic pipe-bending machine, a former of the appropriate size for the pipe to be bent is selected and fitted on the end of the ram. When the pump is operated the ram moves forward pressing the tube against the stops, which are secured by pins through two steel bars. These bars have a series of holes throughout their length which permit the stops to be moved to accommodate various pipe sizes. The appropriate holes are clearly marked indicating for which pipe size they are suitable. The position of the stops must always be checked before a bend is made to ensure that the pins are in the correct holes. Failure to do this could cause very serious damage to the machine such as broken formers, a bent ram or damaged seals, all of which are expensive to repair.

Some machines require the bleed screw (which is usually the filler cap) to be in the open position during bending to allow the machine to 'breathe' during operation. When it is necessary to top up with fluid, always use the correct grade of hydraulic oil as recommended by the maker as the wrong oil could degrade the synthetic rubber seals rendering them useless. This information is usually marked on the machine, near the filler as a rule.

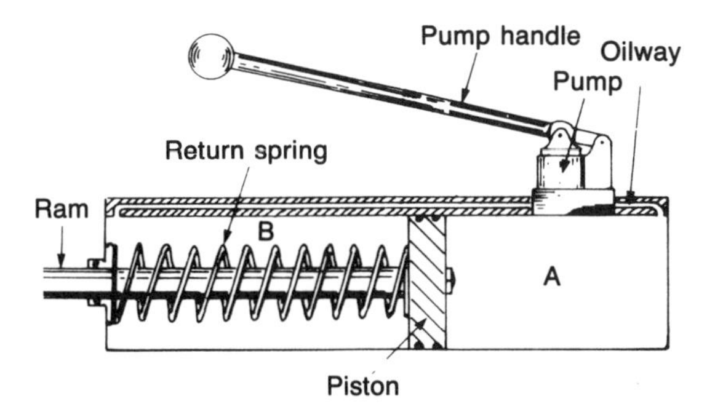

(a) Principle

When the pump is operated oil is pumped from B and A, forcing the ram outward and tightening the spring. By opening a bypass valve on the pump (see (b)) the pressure exerted by the spring pushes the piston back, simultaneously allowing the oil to be displaced from A to B.

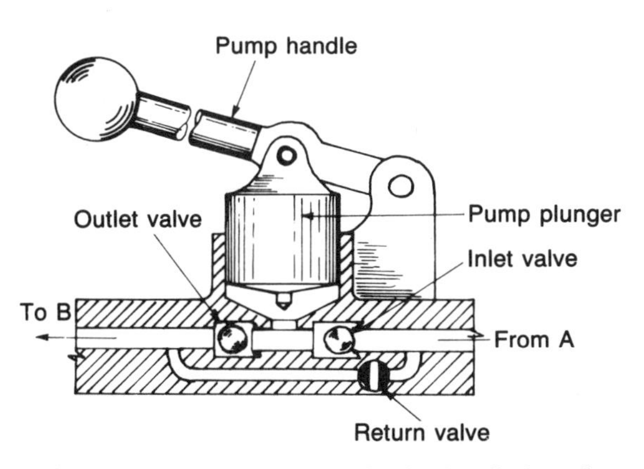

(b) Detail of ram pump used for hydraulic bending machines

Fig. 4.29 Basic principles of a hydraulic pump

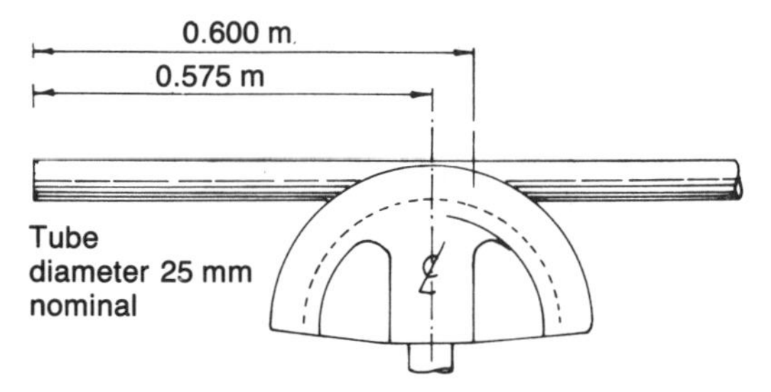

(a) Location of pipe in former for 90° bend

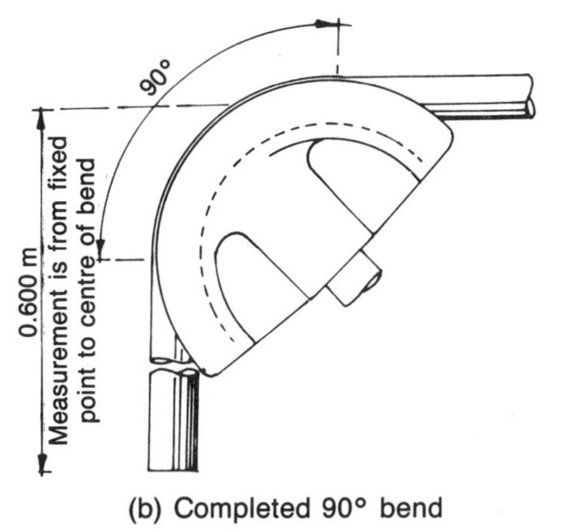

(b) Completed 90° bend

A 90° set square (workshop size) can be laid over the bend to check the accuracy of bending

Fig. 4.30 Using a hydraulic press bender

Do not overfill or the excess fluid will be pumped out through the bleed screw when the pump is operated.

90° bends

The method or system of using press benders to make accurate bends is very simple. To make a 90° bend from a fixed point, e.g. the threaded end of a tube, mark off the required measurement of the centre line which in Fig. 4.30(a) is 600 mm. Bearing in mind the gain in length when bends are made, one pipe diameter (in this example 25 mm) is deducted from the measurement, which leaves 575 mm. This should be maked off from the end of the tube. The pipe should be placed in the bender with this mark corresponding to the centre line of the former as shown. Having checked the stops are in the correct position, the bend can now be made. Due to the fact that even low carbon steel is slightly springy, the bend should be over pulled by about 5° to allow for springback. When the bend is complete it should be checked for measurement as shown in Fig. 4.30(b).

It will sometimes be found difficult to remove the former from the pipe when the bend is completed, especially when the former is new but this need not be a problem if the pipe is held as shown in Fig. 4.31 and the pipe end is tapped on the wooden block until the former becomes disengaged. Small formers may be held by hand to prevent them dropping on to the ground, larger ones should be protected by allowing them to fall on soft material. (Care must be taken to avoid the former dropping on the operative's feet when it falls, this being a typical example of why shoes

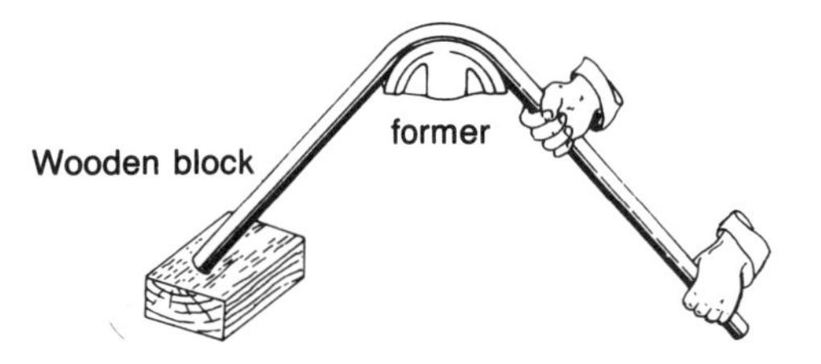

Fig. 4.31 Removing a former from the pipe. Former must not be allowed to fall on hard ground or it may be damaged.

with steel toe caps are necessary.) Some machine manufacturers recommend greasing the tube where the bend is made. While this is messy, it does aid the bending operation and enables the former to be removed more easily.

Offsets

To make offsets in steel pipes, the marking out method is very similar to that used for copper tubes. The required measurement for the first set is marked off on the pipe and placed in the machine with no deduction, as shown in Fig. 4.32(a). The measurement indicated is 0.450 m from the end of the pipe to the centre of the set. After the set has been pulled and removed from the machine, a straight edge is placed against the back of the tube (Fig. 4.32(b)), and the measurement of the offset marked from it to point 'A' on the tube. This mark should coincide with the centre of the former when the pipe is replaced in the machine. After the second set has been made it is important to ensure that both ends of the pipe are parallel which is done by checking the measurements from a straight edge laid across the back of one of the bends (see Fig. 4.32(c)). The remarks concerning the setting out of bends, sets and passovers in light gauge copper tube also apply to steel tube. Once the methods governing the accurate bending of 90° bends and offsets have been mastered, this knowledge and skill can be applied to most other applications.

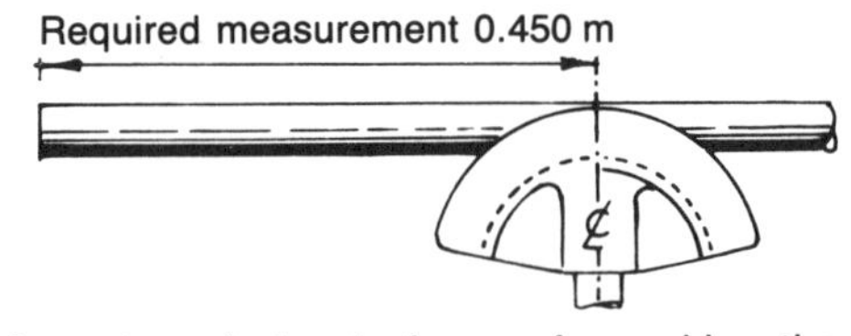

(a) Location of pipe in former for making the first bend

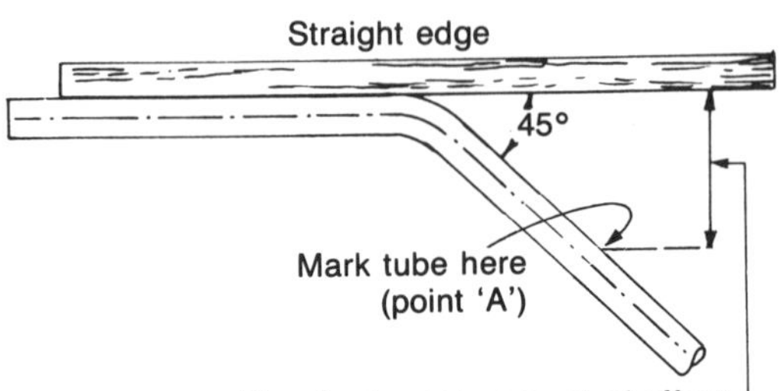

(b) Setting out the mark for the second set

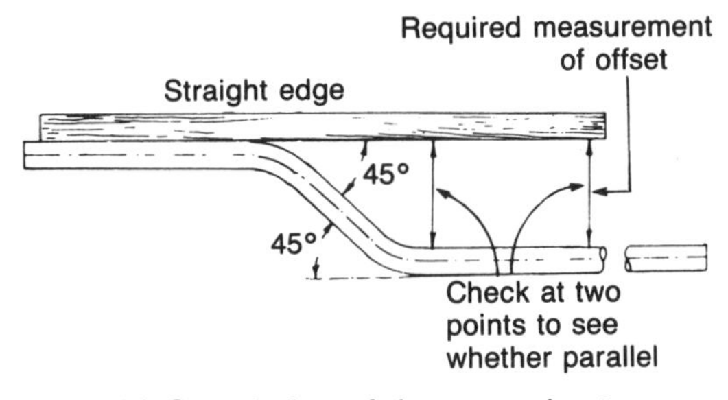

(c) Completion of the second set

The angle of 45° which is shown is only approximate, but both angles must be exactly the same

Fig. 4.32 Making offsets in steel pipe

Further reading

Much useful information can be obtained from the following sources:

Copper tube bending

Record Tools Ltd, Parkway Works, Sheffield, S9 3BL.

Hilmore Ltd, Caston Way, Stevenage, Herts, SG1 2DQ.

Steel pipe bending

Tubela Engineering Company Ltd, 2–6 Fowler Road, Hainault, Essex, 1GB 3UP.

Plastic pipes and fitting

Hepworth Building Products, Hazelhead, Stocksbridge, Sheffield, S30 5HG (polybutylene pipe and fittings).

John Guest Ltd, Horton Road, West Drayton, Middlesex, UB7 8JL (cross-linked polythene pipes and fittings).

Philmac Pty Ltd, Diplocks Way, Hailsham, East Sussex, BN27 3JF (plastic–metal; couplings and adaptors).

Self-testing questions

1. State the reason for 'rounding up' the ends of coiled soft copper tube prior to making capillary joints.
2. Explain why the correct sequence of tightening the nuts should be employed when flange joints are made.
3. State the reason for slightly overbending steel pipes when a hydraulic bending machine is used.
4. Describe why it is necessary to deburr steel and copper pipes after they have been cut with a wheel cutter.
5. Which type of compression fitting must be used with Table Y copper pipe?
6. State the cause of rippling the throat of a copper tube bend made with a bending machine.
7. Explain why it is important to position the pins in the correct holes of the bars on a press bender.
8. State the effect of using the wrong type of oil in hydraulic-bending equipment.
9. Using the approximate method, calculate the length of 40 mm pipe occupied by a 90° bend made to a radius of four times its diameter.
10. Name two methods of jointing PVC pipes.
11. Explain why a lubricant is necessary when cutting threads on steel pipes.

5 The cold water supply

After reading this chapter the reader should be able to:

1. Identify the types of rainwater least likely to present problems in its subsequent purification.
2. State the physical properties of water.
3. Identify the causes of hardness in the water supply and its effect on pipework systems.
4. Describe the working principles of taps and valves used for domestic hot and cold water supplies.
5. Describe the methods used to repair and maintain taps and valves in good working order.
6. Recognise different types of float-operated valves and understand their working principles.
7. Describe the methods used to protect plumbing installations from frost damage.
8. State the need for correct positioning of tappings in storage cisterns.
9. List the main points requiring consideration in relation to the positioning of water storage cisterns in a roof space.
10. Describe the methods of securing pipes of various materials to the building structure.
11. Describe methods of making a fixing to various types of building components.

The supply of water

An adequate supply of pure water on tap is one of the prerequisites of modern living, a fact sometimes not fully appreciated by the average person. Local water authorities are required by law to provide a pure and wholesome supply of water, often referred to as 'potable' water, a term implying that it is fit for drinking and culinary purposes.

Before water is distributed throughout the water authorities' distribution systems to individual premises, it must be collected and treated to rid it of any harmful water-borne bacteria and suspended organic and non-organic solids. Water authorities, through their bylaws and inspectorate, constantly guard against pollution of main water supplies and every effort is made to ensure the supply of pure water to their consumers.

The 1986 Water Bylaws include many new requirements relating to the prevention of pollution, the reduction of flushing cistern capacities, unvented hot water systems and much stricter control over the supply of water to bidets having inlets lower than the flood level of the appliance. The main purpose of these bylaws may be summed up as follows:

(a) prevention of contamination
(b) prevention of waste
(c) prevention of misuse
(d) prevention of undue consumption

Because much of the work of a plumber is concerned with the supply of water and the fitting of appliances which are connected to mains supplies, it is necessary that he should have a good knowledge and understanding of the methods used to implement these four main requirements and also be able to interpret the bylaws relating to water supply.

Sources of water supply

Rainwater

All water is derived from rainfall, which after use in various ways is returned to lakes, rivers or the seas, from where it evaporates to form water vapour and eventually falls again as rain. The term applied to this continuous succession of events is the 'Water Cycle'.

Falling rainwater is quite pure, but as water is a natural solvent it is unlikely to be pure when it reaches the water storage areas such as the reservoirs and lakes. Rainwater falling through air which has been polluted by industrial gases, notably sulphurous compounds or heavy carbon dioxide concentrations, will absorb some of these gases and become a very weak acid, a quality which will increase its natural powers of dissolving other substances. This acidic rainwater as it runs through the earth's crust or as it flows over various rocks will dissolve and absorb further substances and, instead of being the pure water that it was in its original state, its nature is changed and it becomes soft or hard (i.e. acidic or alkaline) water, terms which will be explained later in this chapter. The sources of supply from which water can be taken may be broadly classified under four main headings:

(a) wells
(b) springs
(c) upland surface water
(d) lakes and rivers

Wells

Before the introduction of the mains water supplies provided by local water authorities, the main form of supply to each individual dwelling or small community was a well or pit dug into the ground, the water being extracted by a pump of some description. These wells were, and indeed still are, classified as being *deep* or *shallow*. Some water authorities still maintain deep wells and their associated pumping equipment to supplement their supplies taken from other sources. Others retain them for standby use in cases of drought.

The terms deep and shallow do not necessarily refer to the physical depth of the well, but whether or not its supply is derived from above or below the first impervious stratum of the earth. The earth is made up of differing layers of material, some of which permit the percolation of water and are called pervious, and some of which are more dense, usually a form of rock, through which water cannot pass and are referred to as impervious. Some water does find its way under this impervious strata due to what the geologist calls 'faults' or 'slips' and outcrops of the impervious strata.

Shallow wells are those which are dug into the earth above the first impervious stratum until a water-bearing vein is tapped. The water yielded by shallow wells is usually derived from the immediate locality and in bygone days was often in close contact with cesspits and middens, there being no waterborne systems of drainage. As a result the water was often heavily contaminated and where such wells are still in existence, water taken from them should be regarded with the utmost suspicion.

A *deep well* is one which penetrates the first impervious stratum. Figures 5.1(a) and (b) indicate the difference in the source of water provided by deep and shallow wells. Water found under impervious strata will usually have travelled long distances underground and as such will have been subjected to natural purification, making it pure and wholesome. Such a well, if sunk into the ground in an area where the first impervious stratum is relatively near the surface, may not be as deep as a well classified as shallow.

Artesian wells (see Fig. 5.1(c)) are formed as a result of the character of the surrounding strata and its water table. The term water table relates to the natural level of water under the earth. In the case of most wells this is below the outlet, which necessitates mechanical means of raising it to the surface. In the case of artesian wells, however, the water table is above the mouth of the well which forces water to the surface and may even form a natural reservoir or lake. If the water from such a well can be piped it can prove to be an effective source of supply for small communities. It is unlikely to dry up during prolonged periods of low rainfall due to the wide area from which it is supplied.

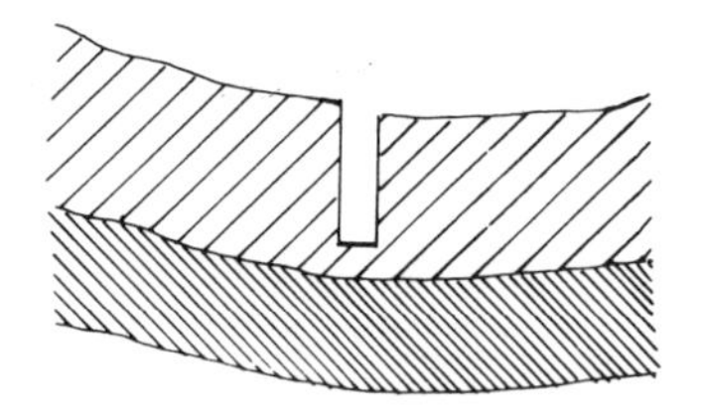

(a) Shallow well

Water collected from immediate locality is often polluted by animal grazing and human habitation. A shallow well does not penetrate an impervious stratum

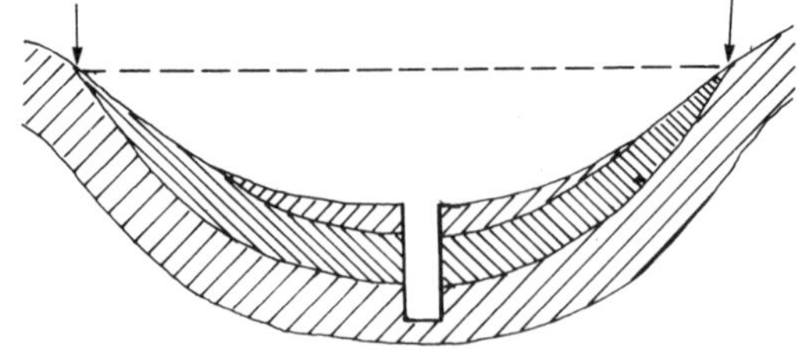

(c) Artesian well

Water will be held underneath the impervious stratum to the level shown by the dotted line thus there is a head of water available to force the water out of the well under pressure

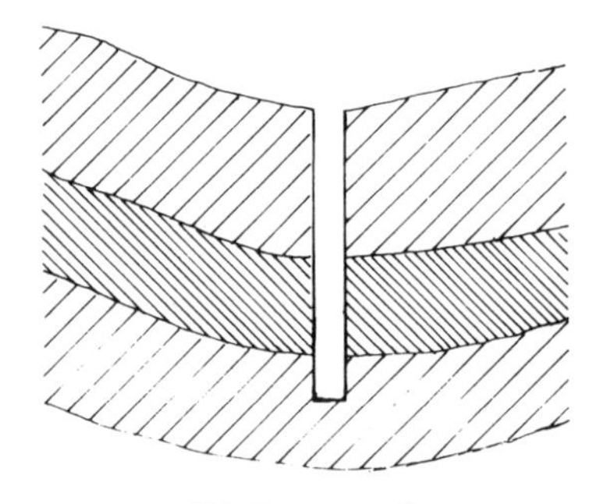

(b) Deep well

Generally pure wholesome water is yielded which has been purified by natural means. A deep well has penetrated an impervious stratum

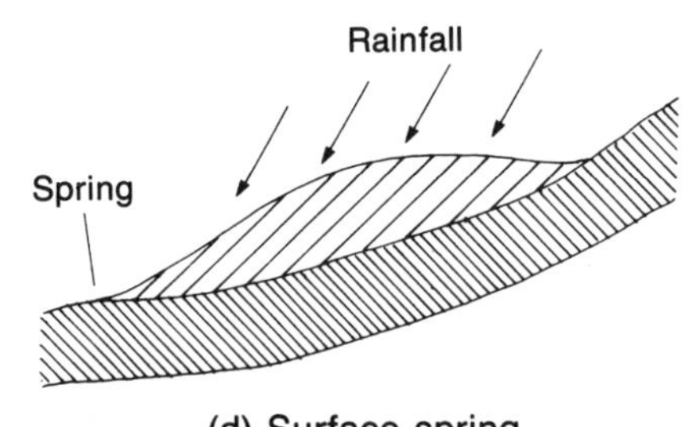

(d) Surface spring

Water yield is of doubtful quality

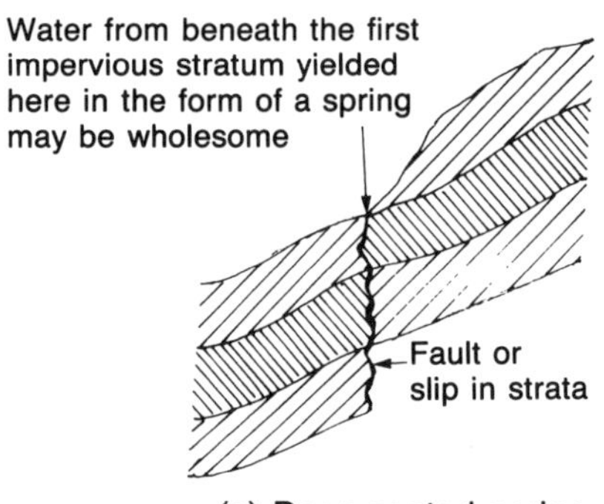

(e) Deep-seated spring

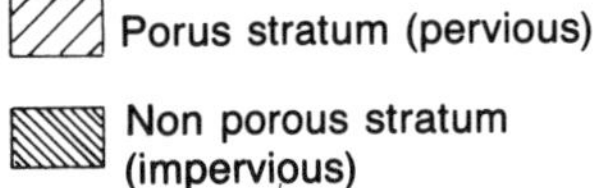

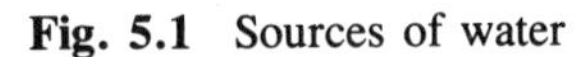

Fig. 5.1 Sources of water

Springs

Water obtained from this source, like well water, depends for its purity largely upon the distance it has travelled underground and whether or not it may be classified as a deep-seated spring. This term, as with well water, relates to its origin and whether it is derived above or below the first impervious stratum. While spring water is seldom capable of supplying more than a small number of isolated properties, a combination of several in the same area often form the source of rivers. For the different types of springs see Fig. 5.1(d) and (e).

Upland surface water

Several large water undertakings in this country depend almost entirely upon this source of supply.

This source is mostly found in the north and north western parts of the country where the isolated ranges of hills and mountains are favourable to the damming of streams to form impounding reservoirs. Lakes and natural reservoirs are also classified under this heading. Water from such a source is usually of good quality being comparatively free from human and animal contamination and suspended solids. As the majority of the water collected and stored in these reservoirs is derived from the 'run off' of mountainous surfaces and has not percolated through the earth in great quantities, it tends to be soft, and if it has had any contact with moss or peat, it is likely to be acidic to the detriment of metals used in plumbing systems – as will be explained later.

River water

Water taken from rivers provides the main source of supply where other forms of catchment are not available. The purity of river water is generally very suspect, especially if taken from the lower reaches. Rivers are liable to pollution by drainage from manured fields, effluents from sewage works, waste water from industrial processes and solid waste from factories and waste tips. The classification of water abstracted from rivers will, therefore, depend largely upon its source and the characteristics of the tributaries which flow into it. Despite its disadvantages, many parts of the country rely solely on this source of supply.

Characteristics and physical properties of water

Pure water is clear, tasteless and colourless. It is a compound of two gases, two parts hydrogen to one part oxygen. The chemical equation for water is given as H_2O.

The maximum density of water occurs at a temperature of 4 °C. This is to say the molecules in a given quantity of water are as small as possible and occupy the smallest volume possible at this temperature. Expansion of the molecules occurs if the water is heated or cooled above or below this temperature. Expansion takes place rapidly when the temperature drops to 0 °C and the water solidifies to become ice. An interesting fact emerges here: during the period of time that a change of state, i.e. liquid to solid, is taking place, heat is given off by the water while no visible drop of temperature is recorded by a thermometer. This heat is called 'latent' or hidden heat. The thermometer will not record any change of temperature until all the water has changed to ice. A total of 335 kJ must be given off by 1 kg of water at 0 °C before it changes to 1 kg of ice at 0 °C.

A similar occurrence takes place when water is heated to boiling point, i.e. 100 °C. At this temperature a change of state takes place in that the water begins to change from a liquid to steam which is a gas. Although heat is continually applied during this change, a thermometer will record only 100 °C until all the water has been converted to steam, and then the temperature will begin to rise, providing the steam is under pressure. It requires 2,258 kJ to convert 1 kg of water to 1 kg of steam. It will be seen that 1 kg of steam at 100 °C has a greater amount of energy stored in it than water at 100 °C. This being so, when the steam is converted back to water, it will release its heat energy. This fact is useful when steam heating of water is considered. A good example of this can be seen in cafés and restaurants where steam is injected into a cup of hot water so it will boil within seconds for tea-making purposes.

Yet another fact should be noted when the temperature of water is raised or lowered above or below 4 °C. In both cases an increase in volume takes place. When water turns to ice it expands by approximately 1/10, which is quite considerable. This causes pipes to burst when the water inside freezes and expands. Figure 5.2 shows how two ice plugs form in a pipe which is exposed to

Fig. 5.2 Frost bursts. Pressure by plugs of ice cause expansion thus weakening the pipe wall, finally causing it to split or burst. Note: water will not run from the burst pipe until it thaws.

draughts causing the water between them to be compressed, first stretching the pipe walls and finally causing them to split. The damage caused by leaks when the temperature rises and the pipe thaws is often very serious and every effort should be made to position water pipes where they are least likely to become frozen, or when this is impossible, to provide the best possible insulation. Where pipes are fitted in very exposed positions, i.e. stand pipes, farm buildings, etc., polythene pipes can be used with advantage as the plastic material, due to its inherent elasticity, will stretch as ice forms in the pipe and return to its original size on thawing.

Water, like most other fluids is a poor conductor of heat and a simple experiment is often used to illustrate this point (see Fig. 5.3). The experiment shows how the water in the top of the test tube can be made to boil while the ice at the bottom remains unmelted.

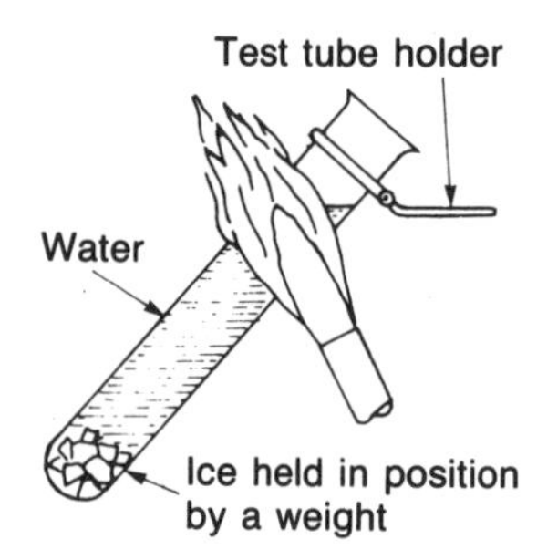

Fig. 5.3 Experiment to show water is a poor conductor of heat. Water at top end of test tube is boiling while ice remains unmelted at bottom of test tube.

Relative density

The density of pure water at 4 °C is taken as the standard with which to compare the density of other materials. The term density relates to the mass per unit volume of a substance. For example, a thermolite building block, having a sponge-like structure which contains large quantities of air, will have less mass and a lower density than an engineering brick.

When comparisons are made between the densities of materials, it is important to ensure that the unit volume of each substance is the same. As an example, if the relative density of a piece of lead pipe is to be found, its volume must first be ascertained. The simplest way to do this is to immerse the pipe in a container of water carefully filled to the invert of the overflow prior to immersing the length of pipe, as shown in Fig. 5.4. The amount of water displaced will be exactly the same volume as the piece of pipe. If the water and the pipe are weighed, it will be found that the lead weighs 11.4 times the weight of the water, and the relative density of the lead is, therefore, said to be 11.4. Materials less dense than water, for example, polythene, which has a relative density of approximately 0.9, will float.

Volume

Volume is the word used to describe the space taken up by an object and two basic units of volume in the SI (Système International) metric system are the cubic metre and the cubic centimetre. The latter is very small and for purposes of calculations a rather unmanageable unit when dealing with the relatively large quantities of water met in plumbing calculations. If one can imagine a small cube, roughly the size of a small cube of sugar, measuring 10 mm on all sides, it will be realised just how small it is. Although the cubic metre is rather a large unit and many of the tanks and cisterns used in plumbing would not hold such a quantity of water, it is the most convenient unit to use to express volume. Volumes of less than 1 are expressed as a fraction of 1 m^3.

An alternative, although technically not an approved SI unit, is the litre, this being the unit

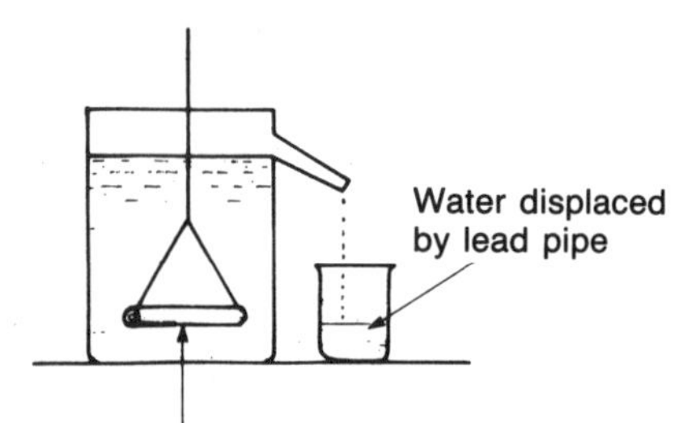

Fig. 5.4 Calculating the volume of an irregular shape

commonly used for plumbing and heating calculations. A litre of water at 4 °C will occupy a cube with sides of 100 mm square, and its mass will be equal to 1 kilogram. It is useful to remember that 1 m^3 contains 1,000 litres. This is very convenient, as if the contents of a cistern are calculated in litres, not only is the volume obtained but also the weight of the water it contains. This has already been dealt with in Chapter 3 but is worth repeating here for clarity. Assume that the contents and the mass of water contained in a rectangular cistern are to be calculated. The cistern measures 0.450 m wide, 0.500 m high and 0.600 m long. Using the formula for finding volume:

$$\begin{aligned} \text{Volume in cubic metres } (m^3) \\ &= \text{Width} \times \text{Height} \times \text{Length} \\ &= 0.600 \times 0.450 \times 0.500 \\ &= 0.135\ m^3 \end{aligned}$$

As there are 1,000 litres in a cubic metre and its mass is 1,000 kg, by multiplying the volume in $m^3 \times 1{,}000$, i.e. $0.135 \times 1{,}000 = 135$, it will be found that the capacity of the cistern is 135 litres and the mass of water 135 kg. (Furthermore, note how easy it is to multiply a number by 10, 100 or 1,000: the decimal point is simply moved one, two or three places to the right.)

Soft water

Water dissolves many of the metals and minerals with which it comes into contact, especially soft acidic waters which, if used with lead pipes for domestic supplies, can cause lead poisoning. Such water is said to be plumbosolvent.

While it is true to say that lead poisoning is not common since lead service pipes are not used in most modern plumbing systems, many older properties are plumbed with this type of material. Fortunately, the water authorities are aware of the danger and in districts where water is plumbosolvent steps are taken to minimise the problem. Unfortunately as the cost of replacing all existing lead service pipes would be extremely high, they are simply replaced by alternative materials at the end of their service life.

Soft water also dissolves and takes into solution small quantities of copper and iron. As only small amounts are dissolved there is very little danger to health but copper-bearing water does tend to leave a bluish stain on sanitary fittings, while dissolved iron often discolours the water, especially in hot water systems. This can readily be seen if the sediment in the base of the hot water storage vessel is disturbed by a brisk circulation.

Hard water

Much of the rain which falls percolates into the earth where it dissolves many of the minerals with which it comes into contact. Water that falls on and percolates through chalk and limestone is said to be 'hard'.

Unlike soft water which lathers easily, hard water makes the formation of lather very difficult. The reaction of the stearates in the soap and the lime content of the water produces an objectionable scum on its surface, and on the sides of sanitary fittings.

It should be made clear that there are two types of hardness called 'temporary' and 'permanent'. Both forms are undesirable as they present difficulties in laundering and in industries where steam plant is necessary for heating or manufacturing processes. Another disadvantage of hard water is the extra cost of special soap and detergents that its use entails.

Permanent hardness occurs due to the natural solvency of pure water which enables it to dissolve the sulphates of limestone. This type of water, however, is not such a problem in plumbing schemes as temporary hard water which is responsible for the formation of a hard scale which accumulates on the inside of boilers and circulatory pipes. This scale has the effect of insulating the water contained in the boiler from the source of heat, causing waste of fuel and, in the case of mild steel boilers, being responsible for overheating of the boiler plates with the resultant possibility of buckling and distortion.

Temporary hard water occurs when the water, containing carbon dioxide gas absorbed as it fell as rain, dissolves the chalk and calcium carbonates which are very common in the earth's strata in many parts of the country. These carbonates are insoluble in pure water but, due to the carbon dioxide in the water collected from the atmosphere, the carbonates are converted into soluble bicarbonates. When such water is heated to approximately 65–70 °C the gases it contains, including the carbon dioxide, are given off. The water is then no longer able to retain the calcium in solution and it is precipitated in the form of scale or fur. The examination of the inside of a kettle used in a hard water area will illustrate very clearly how serious this scale build up can be. For this reason it is not advisable that water known to have a high temporary hardness content should be heated to high temperatures, 65 °C being the maximum recommended.

Although it would be too expensive to remove the hardness content of water completely, most water authorities in the hard water districts do their best to reduce it to manageable proportions. Various methods are used for softening the water depending on the nature of the water concerned. The usual method used by water undertakings to remove some of the hardness in the water is the soda lime process. Extra lime is added to the water which amalgamates with that already held in solution. This causes the total lime content to reach saturation point and when this happens, both the lime which was added and some of that which was already in the water is precipitated. This has the effect of reducing the temporary hardness content. The addition of soda reduces the permanent hardness content of the water by acting on the calcium sulphates to convert them into insoluble carbonates which fall out of the solution as solid particles. Both these processes are carried out in sedimentation tanks, the water being kept as still as possible during the process.

The treatment of the water by the local water authorities for removing hardness is not sufficient for all purposes and in some cases water is softened to a greater degree by private individuals using water softeners. This process is described fully in Vol. 2 of this series.

Water is classified as being hard or soft by the number of parts of calcium it contains per million parts of water. As water may be both temporarily and permanently hard, both types are grouped together for this purpose to give the 'total' hardness content, as shown in Table 5.1.

Table 5.1 Classification of water hardness.

Classification	*Parts per million*
Soft	0–50
Moderately soft	50–100
Slightly hard	100–150
Moderately hard	150–200
Hard	200–300
Very hard	300+

Water treatment

It is not intended to deal with this subject in great depth, but the plumber should have a general knowledge of the main processes of water purification.

Sedimentation

When water is stored and allowed to remain undisturbed in a lake or reservoir, solid impurities suspended in it such as grit, mud and decaying vegetable matter sink to the bottom, this being known as *primary sedimentation*. Storage also has the effect of reducing the bacteriological content of the water, as lack of suitable food, competition with harmless organisms and low temperatures all have the effect of preventing the multiplication of harmful or *pathogenic* bacteria. Local authorities, therefore, take advantage of this fact and storage is often used as a form of primary purification. When it is required, water is pumped from the storage reservoir, through a coarse metal strainer and allowed to flow into sedimentation tanks, where further controlled sedimentation takes place before the water is filtered.

Filtration

Slow sand filtration is an old established method of water purification and is still used by many authorities. It consists basically of allowing the water to flow over a graded sand bed on top of which large colonies of minute vegetable growth called algae have formed naturally. It is largely the action of the algae and filtration that purifies the water.

Due to the slow action and the necessity for constantly replacing the top 50–75 mm of the sand bed, *rapid sand* and *pressure filters* are gradually replacing those of the slow sand type. Rapid sand filters are very similar to slow sand types but operate under a greater head of water. There is also a difference in the depth of the sand and the grades used.

Pressure filters are usually housed in a steel container and are capable of dealing with 4,500 litres of water per hour. Filters worked at this rate very quickly become clogged which necessitates frequent back-washing. This process is, however, entirely mechanical and does not involve the manual labour necessary to clean slow sand filters.

Sterilisation

The final treatment of water to ensure its purity before it enters the main is its dosing with chlorine or a mixture of chlorine and ammonia. These are both sterilising agents which rapidly kill off any harmful bacteria which may have escaped filtration.

Special treatment

This is carried out by water undertakings in areas where the water is either very hard or very soft. In many cases, especially in the case of the former, it would be too costly to produce a perfect water, but most local authorities undertake to produce water of an acceptable quality. Figure 5.5 shows a diagrammatic illustration of the main processes of water treatment.

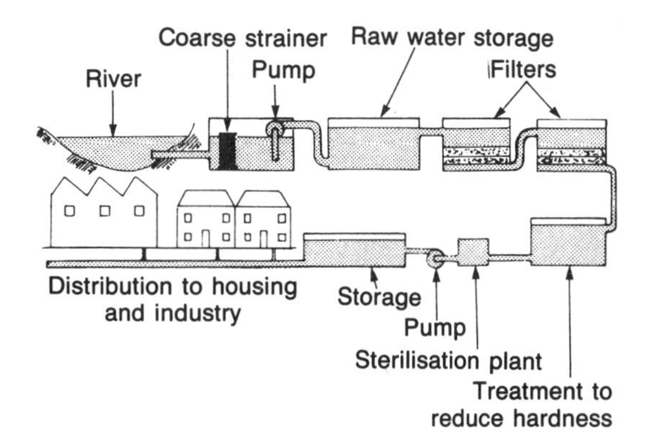

Fig. 5.5 Water treatment flow chart

The cold water supply to the consumer – taps and valves

Taps and fittings used for both hot and cold water supplies should conform to BS 1010:Parts 1 and 2. They are usually made of brass pressings or castings and in many cases are chromium plated to improve the appearance and facilitate easy cleaning. More recently a thermosetting plastic called acetal has become increasingly popular for the manufacture of taps. Marketed under various trade names, it is an engineering plastic which is hard wearing and capable of being machined in a similar way to metal. It has a lower impact strength than metal, is more easily damaged and is not so resistant to abrasion as brass. It is mainly these factors which have at present prevented it from becoming a serious competitor to metals. It should be borne in mind, however, that taps made of this material have been following traditional lines. Like most plastics, acetal is a very versatile material and in the future taps and valves of completely new design may be available, these new designs utilising to the full the characteristics of this and other modern materials.

Polypropylene is another plastic material that has been used successfully for the manufacture of float-operated valves, especially the diaphragm type which, due to its design, lends itself to the limitations of this particular material.

Screw-down taps

Prior to 1986 only screw-down taps were permitted on water supplies taken directly from the mains, as these are designed in such a way that they cannot be closed quickly, thus avoiding water hammer. This is a term used to describe noise, usually a banging sound or noisy vibrations throughout the system, which occurs when the flow of high pressure water is suddenly arrested. Not only is this noise disconcerting but water hammer can be damaging to pipework, especially where it has been weakened due to exposure to frost.

Since the 1986 Water Bylaws, however, quick closing, usually quarter-turn taps, are permitted, but the reader should be aware that by installing them they may cause water hammer. To avoid this, especially in areas where water is supplied by the water authorities at very high pressure, it may be necessary to provide pressure-reducing valves or water hammer arresters.

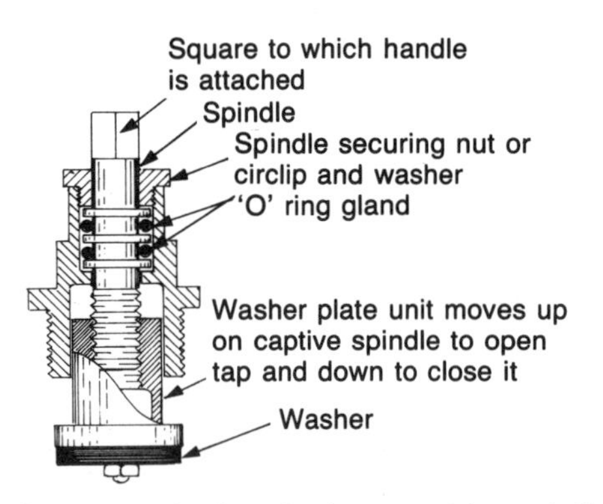

Fig. 5.6 Headwork of tap having non-rising spindle. This type of headwork is used by many manufacturers of mixer taps for modern bathroom equipment. To enable the 'O' rings to be replaced the spindle securing nut or circlip is removed. The spindle is rotated thus unscrewing it from the washer plate unit when it can be lifted out of the headwork.

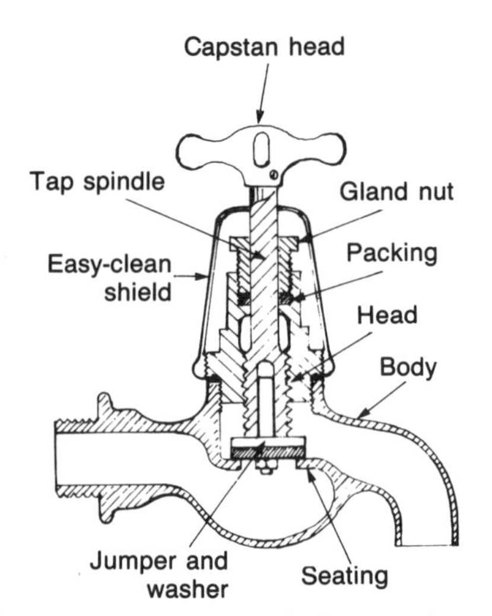

Fig. 5.7 Bib tap. The action of all screw down taps or valves is the same. When the spindle is turned it engages on the threads in the head allowing the gradual opening or closing of the jumper.

Screw-down taps are made in a variety of patterns depending upon the fitting they supply. However, the headwork of all screw-down taps and valves is basically the same, incorporating a threaded spindle which has to be turned through several revolutions to effect its closure. Many modern taps are made having non-rising spindles. This is an advantage as it reduces wear on the spindle as it passes through the gland. It also enables manufacturers more freedom in the aesthetic design of taps (see Fig. 5.6).

Bib taps

The bib tap is a typical example of a screw-down tap, being designed to be fitted into bosses screwed to the wall over such appliances as butlers' or cleaners' sinks. They are also used on standpipes when the bib is threaded which enables a hose union to be fitted. When bib taps are provided over a sink, always be sure that they are fitted high enough to enable a bucket or bowl to be comfortably placed beneath. A sectional elevation of a bib tap is shown in Fig. 5.7. The shroud which covers the headwork is called an easy-clean shield and it serves to prevent the build up of dirt and corrosion in the head. The spindle, gland and jumper shown in the illustration of the bib tap are common to all types of screw-down valves. When the capstan head is turned, the threads on the spindle engage the threads inside the head, causing the spindle to move upwards or downwards, thus opening or closing the water supply.

Until the revision of CP 1010 all cold water taps were fitted with loose jumpers. Hot water taps have always been provided with fixed jumpers as the pressure on most hot water systems is low and sometimes insufficient to lift the jumper. By pinning it to the spindle, this problem is overcome. As so many taps fitted in modern plumbing systems are also fed from a low pressure supply, i.e. the storage cistern, it has been found necessary to provide fixed jumpers in those taps used for cold supplies for the same reason. Figure 5.8 shows a detail of the usual method of fixing jumpers.

Pillar taps

These are also of the screw-down pattern, but unlike bib taps, are designed for fitting directly into the sanitary appliance and secured to it by a back nut. Chapter 7 deals with the methods used for fixing these taps to ensure a watertight joint. The pillar tap illustrated at Fig. 5.9 is of modern design, the easy-clean shield being made in such a way that it also serves to turn the tap on or off.

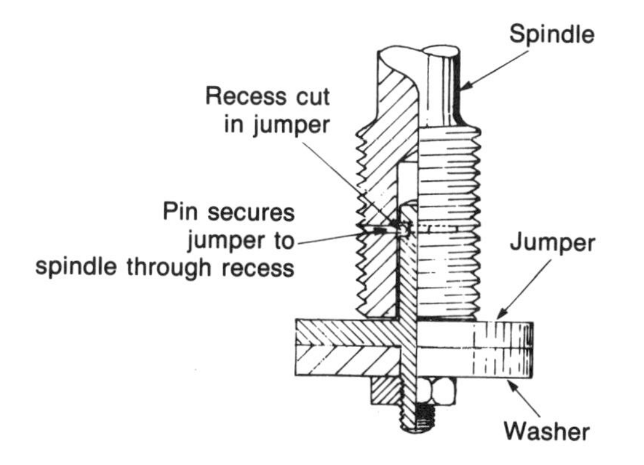

Fig. 5.8 Fixed jumper in screw-down tap

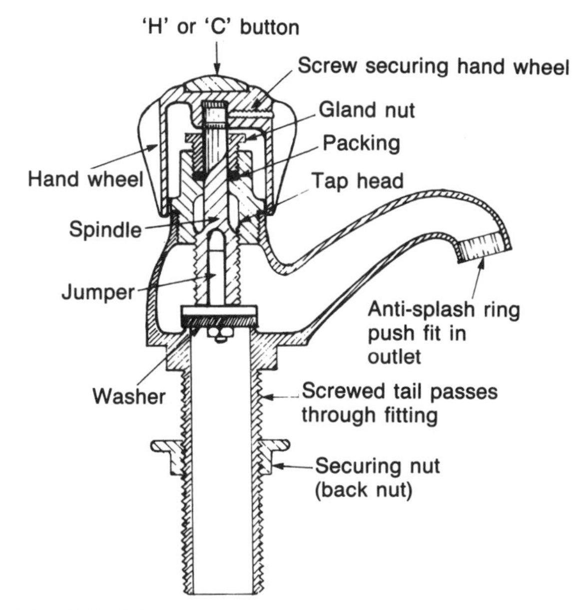

Fig. 5.9 Pillar tap

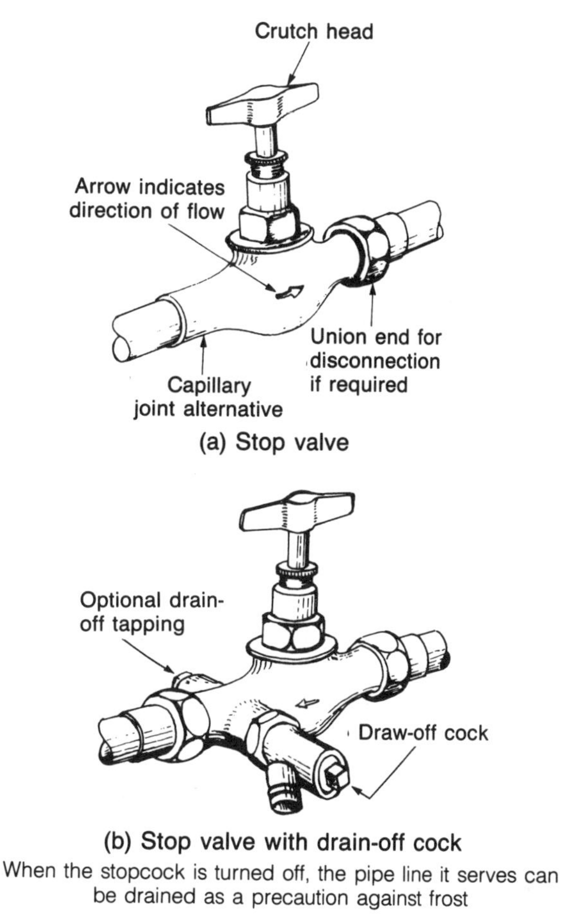

Fig. 5.10 Stop valves

Many of these modern taps employ acrylic plastic for the manufacture of this type of head. Pillar taps of a similar design but with extended bodies are used with sink units. The upward sweep of the outlet enables sufficient clearance under the tap for filling kettles and culinary equipment.

Stop valves

Yet another member of the family of screw-down valves is the stop valve (see Fig. 5.10(a)). These are fitted in a straight run of pipe, their purpose being to control the supply of water into a building, or control one fitting or a group of fittings independently of others in the same building. They are available with union ends as shown suitable for copper, or they can be obtained with threaded ends conforming to BS 21 which enables connection to low carbon steel pipes. Stop valves are also made with capillary and compression ends for copper tubes made to BS 2871. Most copper fitting manufacturers also provide adaptors so the compression ended stop valves can be used with various types of plastic pipework.

The stop valve illustrated at Fig. 5.10(b) incorporates a drain-off cock, and is made in such a way that the drain-off cock can be fitted and be accessible irrespective of the position of the stop cock. Such valves are intended for use as the main

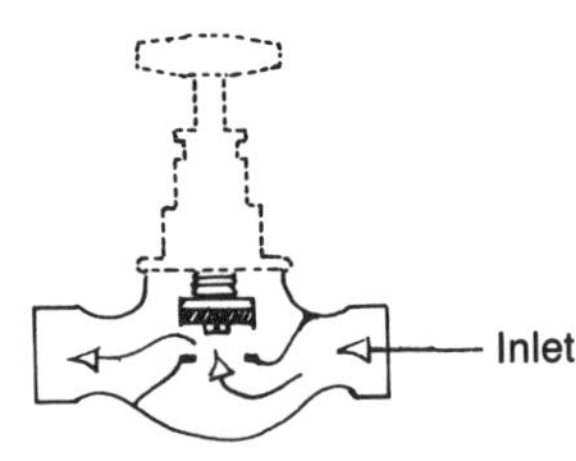

Fig. 5.11 Fitting stop cocks. Care must be taken to fit the stop cock in the correct direction of flow.

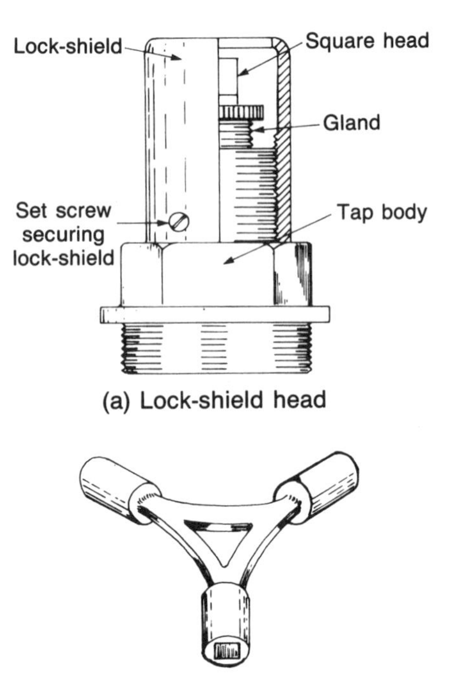

(a) Lock-shield head

(b) Isle of Man key

Ends are of different sizes – for use with lock-shield valves and taps.

Fig. 5.12 Lock-shields

control of the water supply on entry into a building, the drain off enabling all the water in the pipework to be emptied as a precaution when the building is left without heat in periods of cold weather.

Stop cocks for use below ground must comply with BS 5433. They are made of corrosion resistant material such as gunmetal or bronze, have a loose jumper and are of heavier construction than those complying to BS 1010.

Stop valves are stamped with an arrow indicating the direction of the water flow. Figure 5.11 shows the correct installation of a stop valve and the flow of water through it.

Lock shields

These are fitted to taps not intended for public use to prevent interference from unauthorised persons. They are provided on stop valves and bib taps (used by cleaners, etc.) in schools, hospitals and public toilets. Figure 5.12(a) shows the head of a tap fitted with a lock shield. The spindle terminates as a square below the top edge of the lock shield which is screwed over the head of the tap and secured with a set screw. These taps are provided with a loose key which should be handed over to a responsible person for safe keeping and use when the tap is installed.

A three-headed key (Fig. 5.12(b)), sometimes called an Isle of Man key, is one of the tools that should be part of a plumber's tool kit. Each of the three ends are recessed with squares of differing sizes for operating lock shield valves if the original is not available when required.

'Supa' taps

There are several patent types of tap available which, although of the screw-down type, do not conform to a standard pattern. One of the best known of these is the 'Supa' tap (see Fig. 5.13) which features a device allowing it to be rewashered without turning off the main supply of water. It embodies a valve in the head which closes when the valve body is unscrewed beyond a certain point. This point can only be reached after the gland nut, which for normal use restricts the movement of the body, has been unscrewed. The body can then be removed to give access to the built in anti-splash device which also houses the jumper. When rewashering is necessary, the anti-splash device can be removed from the body by a light tap; the jumper can then easily be removed. Although washers are obtainable for these taps it is advisable to replace the jumper as a complete unit as they are very lightly constructed and in some types of water completely disintegrate as a result of corrosion when an attempt is made to remove the washer. It should be noted that although these

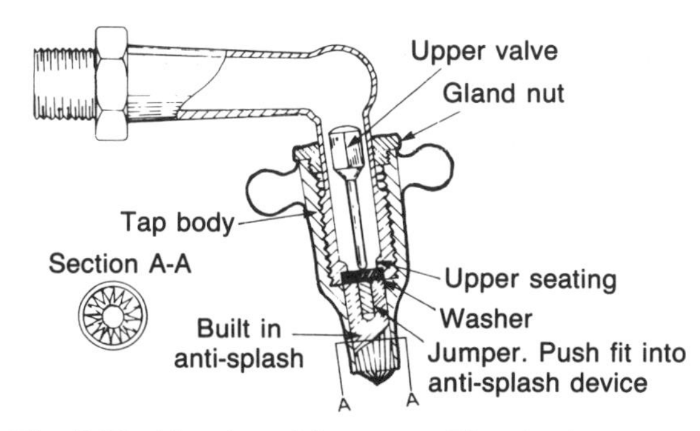

Fig. 5.13 'Supa' tap bib pattern. The gland nut prevents the body from being completely removed during normal use. When it is unscrewed it allows the body to be completely removed to gain access to the jumper. The upper valve is then able to seat effectively shutting off the water while the washer is changed.

taps are no longer made there are many in existence and spares are still available from some builders merchants.

Rewashering and maintenance of taps

Generally speaking, the most common maintenance required to screw-down taps and valves is renewal of the washer. Soft rubber washers should not be used as these can cause water hammer, a loud knocking noise in the pipes.

The gland may also occasionally require repacking and PTFE tape will be found, in most cases suitable for this purpose, but if on inspection the spindle itself is worn, the tap should be replaced. To gain access to the gland on those taps fitted with an easy clean shield, the crutch or capstan head must be removed after first removing the set screw which secures it to the spindle. (Put the plug in the sink outlet first or the set screw is sure to go down it!)

If the head is still tight it is best removed from the spindle by using light taps on a copper drift with a hammer. A short piece of scrap copper tube is ideal for this purpose and, being soft, will not mark the chromium plating. When the gland has been repacked, lightly file the square on the spindle to allow the head to be replaced without resorting to force and secure by replacing the set screw.

The seatings of taps sometimes become pitted and, although the washer may have been renewed, a constant drip is still evident. It is possible to buy reseating tools which enable new seatings to be cut in old taps, but a defective seating in a tap is generally a sign of age and it will almost invariably be found that the spindles on such taps are badly worn. The most economic procedure in such circumstances is to renew the tap.

The gland seal on taps having non-rising spindles are usually of the 'O' ring type. To replace these the top of the operating handle or wheel must be removed by unscrewing the screw which secures it to the spindle. This will expose the tap headwork which must be unscrewed from the tap body. Great care is required here to avoid both damage to the sanitary appliance and the tap. Having removed the head the spindle may be taken out by pressing out the circlip housed in a groove on the spindle. This will expose the 'O' ring seals which should be replaced carefully by new ones. A little silicone grease will assist their replacement. During this operation the washer should be inspected for wear and replaced if necessary, and the hexagonal guides, to which the washer plate is attached, should be checked to ensure there is no corrosion product present. It is a common failure with these taps that a build up of corrosion will prevent the tap from functioning correctly.

One of the problems met by the plumber is the fact that taps not complying to BS 1010 have non-standard seal washers and ceramic discs, and unless the manufacturer is known, it can be difficult finding the right replacements.

Before taps are fitted to sanitary fittings they should be 'broken', that is to say the head should be removed from the body and a suitable oil or grease applied to the thread. This ensures that when rewashering is necessary at a future date, the head can be removed without resorting to excessive force. Many cracked basins are the result of attempts to remove the tap head to rewasher the taps. One method of supporting a pillar tap against the force applied by a spanner is to wrap a piece of cloth round the spout of the tap and carefully place a piece of pipe of suitable size over it to act as a lever against the pressure applied by the spanner (see Fig. 5.14). To be on the safe side, however, especially when

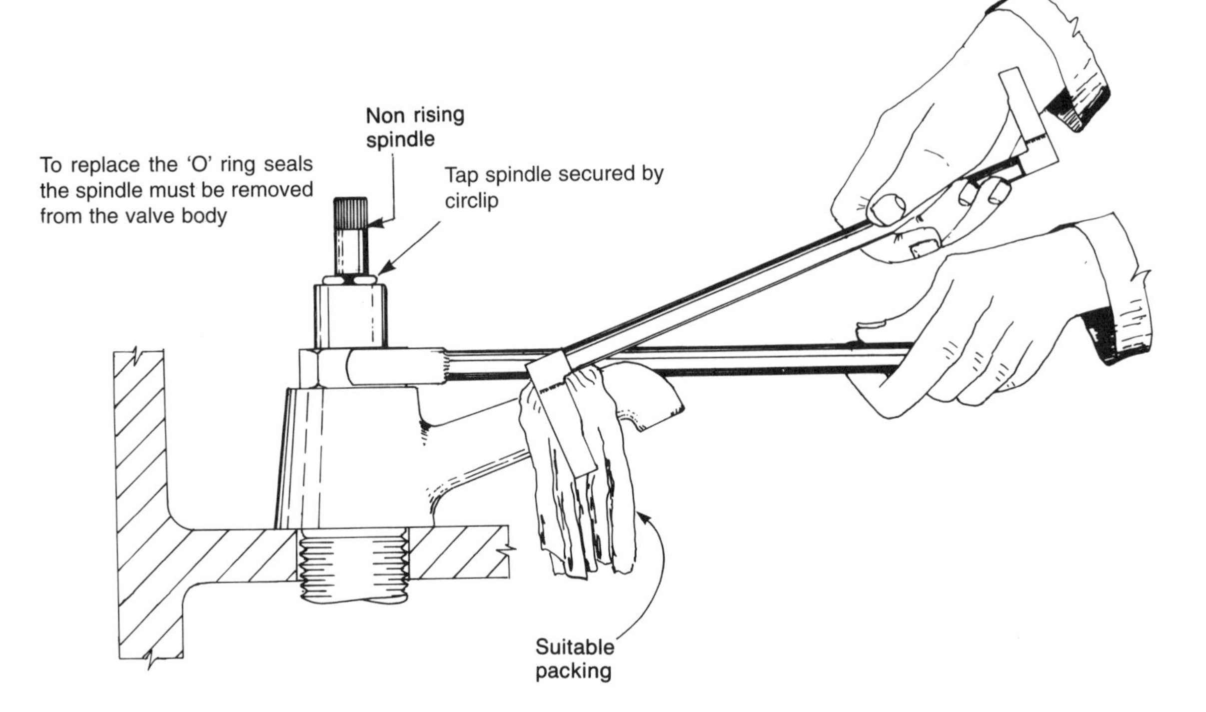

Fig. 5.14 Removing a tap head for maintenance. The spout is held with a basin (shetack type) spanner to prevent the body twisting in the sanitary fitting. Note the packing across the spout to avoid damage by the spanner to the surface plating. An alternative to the spanner shown is a suitable length of copper pipe held over the spout. Packing will still be necessary.

considering wash basins, taps are best completely removed and held in a vice, care being taken to avoid damaging the chromium plating.

Ceramic disc taps

Taps of this type are quite different in operation to the screw-down pattern which were designed in Victorian times to replace the plug cock commonly in use at that period. Figure 5.15 illustrates in diagrammatic form the working principle of a simple disc valve. Unlike screw-down valves they have no seating or washer. These are replaced by two ceramic discs, one static, being fixed to the valve body so it cannot rotate; the other can be turned through 90°, usually by a lever.

The discs are provided with waterways which line up when the tap is turned on. Like most modern taps they have non-rising spindles which eliminate leakage from glands and it is claimed that discs, unlike rubber washers, will last the lifetime of the tap.

The principle of the ceramic disc is also used for mixing taps on wash basins. They are more complex than the simple illustration shown and should it be necessary to dismantle the tap, it is advisable to refer to the makers instructions.

Non-concussive taps

These are special taps which are designed to save water in public buildings and operate only when downward pressure is exerted on the head. The original types were quite simple, the taps being opened by compressing a spring, closure being affected automatically by the spring when pressure was released on the head. It was found, however, that they were very noisy in operation, the sudden closure causing water hammer (see Book 2, Chapter 4), especially in areas of high pressure. Modern non-concussive taps are rather more complicated, closure being effected gradually by means of a hydraulic valve inside the tap. In all cases where repair and maintenance of these valves

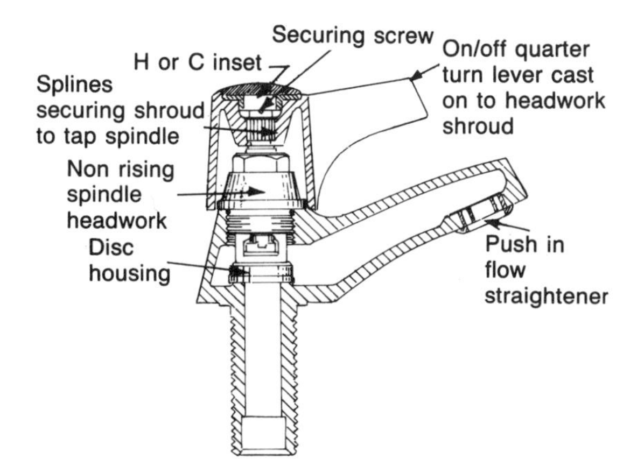

(a) Ceramic disc tap – pillar type

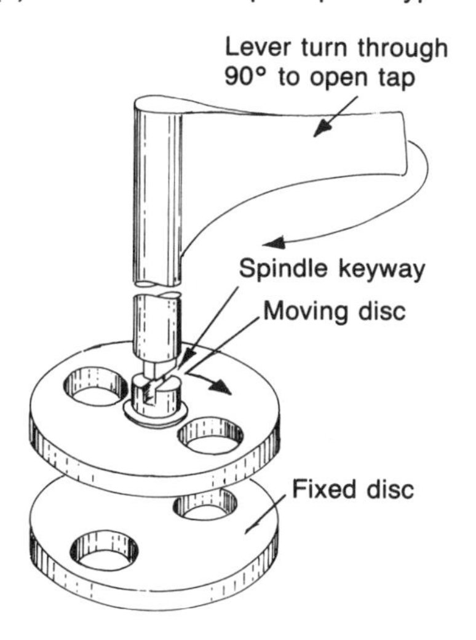

(b) Detail illustrating the operating principles of a simple disc tap

Note that although the discs are shown apart here, in practice the faces of each are close together in the housing. When the tap is in the off position no water can pass. By rotating the lever through 90° the holes in the discs line up allowing water to flow to the tap outlet.

Fig. 5.15 Ceramic disc tap

becomes necessary, it is important to refer to the manufacturers instructions, as wrongly adjusted, they are not non-concussive and will give rise to water hammer.

Service valves

One of the requirements of the 1986 Water Bylaws is the ability to shut off for servicing purposes,

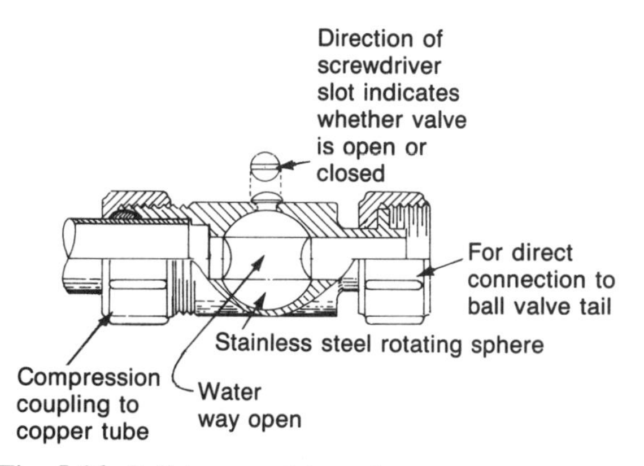

Fig. 5.16 Ball type servicing valve

plumbing fixtures such as flushing cisterns and storage cisterns, one of the advantages being that maintenance work can be carried out without shutting down the entire system. Stop cocks to BS 1010 and gate valves to BS 5154 can be used, but a much neater and cheaper alternative is the use of a ball type servicing valve (not to be confused with float operated valves) which is similar to a plug cock but has a circular seating instead of one that tapers. They are commonly used for servicing valves on central heating pumps and as a means of isolating washing machines when they are not in use. They are made with a variety of connections and may be straight or angle pattern. The ball valve shown in Fig. 5.16 has a compression joint for copper tubes on one end, the other being suitable for direct connection to a float-operated ball valve tail.

Mixer taps

It is often desirable to use a mixed or blended supply of hot and cold water. Typical applications are where the water is blended in a special mixer valve, as in the case of a shower fitting or a bath or sink when hot and cold water is delivered through a common spout. (Shower mixers are generally more complicated than those used on baths and sinks and are dealt with more fully in Volume 2 of this series.)

Figure 5.17(a) shows in diagrammatic form a mixer tap where the water mixes in the tap body. It will be seen that if this mixer is used with unequal pressure, the water having the higher pressure (usually the cold water) would flow back into the hot water service subjecting the hot store vessel to excessive strain and possibly causing it to overflow. Despite the foregoing, the new Water Bylaws permit their use with unequal pressures, and providing a single check valve is fitted on the cold inlet and the mixer has an inbuilt type 'A' air gap, they may be connected directly to the mains supply. Such mixers may now be used for sinks, wash basins and baths having both equal and unequal supplies providing appropriate precautions are taken against contamination. They can be used with advantage where the hot water is provided from an unvented system.

To meet the demand for taps capable of providing a mixed supply at differing pressures without the disadvantages previously mentioned, the 'Biflow' mixer has been developed (see Fig. 5.17(b)). It will be seen that by using this type of mixer, both hot and cold supplies are quite

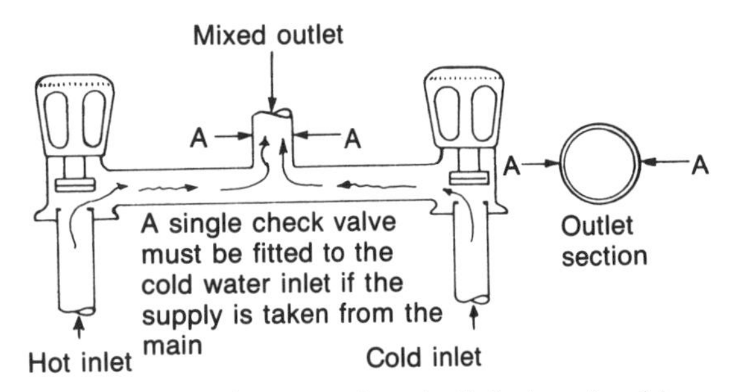

(a) Mixer tap for use when both hot and cold supplies are at equal pressures

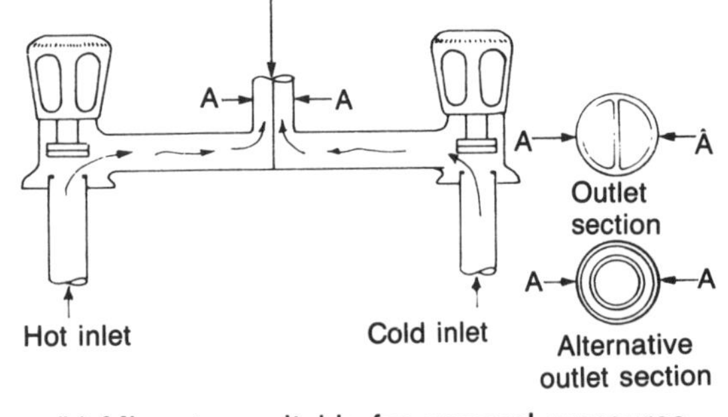

(b) Mixer tap suitable for unequal pressures

Fig. 5.17 Mixer taps

separate until they are discharged from the outlet. The outlets are usually made to swivel through 180° so they can be used with a double sink. The one weakness in their design is the joint between the swivel of the outlet and the body of the tap which is made watertight by a neoprene 'O' ring. These seem to wear very quickly and require frequent replacement which necessitates the removal of the swivel spout. Always make sure the instructions for servicing these taps are left with the client as different manufacturers recommend various methods for removal of the swivel.

Gate valves

Gate valves to BS 5154 are used to control the low pressure cold water supply from a cistern to the hot store vessel or cold down services. They are ideal for the purpose, as unlike screw-down valves, they have a full way water passage and offer no resistance to the flow of water. Figure 5.18 shows a typical gate valve which, like stop cocks, can be obtained with female threaded ends for low carbon steel pipes or for direct connection to copper tubes.

Gate valves employ no washer, having a metal to metal seating. A circular wedge-shaped gate closes into a matching wedge-shaped seating when the valve is closed. They are always fitted with a wheel top and non-rising spindle. Some plumbers and fitters avoid installing these valves in an upright position, preferring a horizontal or even upside down position. This prevents any sludge or solids in the water from settling in the base of the seating and preventing the gate from closing.

Plug taps

Sometimes referred to as 'quarter turn' taps these are probably the oldest type of tap in use (see Fig. 5.19). Until the 1986 Water Bylaws they were not permitted on mains cold water services, but since this date they have applications as service valves in some circumstances and are used almost exclusively by the gas industry for domestic appliances, main cocks on gas services and control taps on gas boilers.

The plug is operated by turning the tapered valve which is tightened into a corresponding

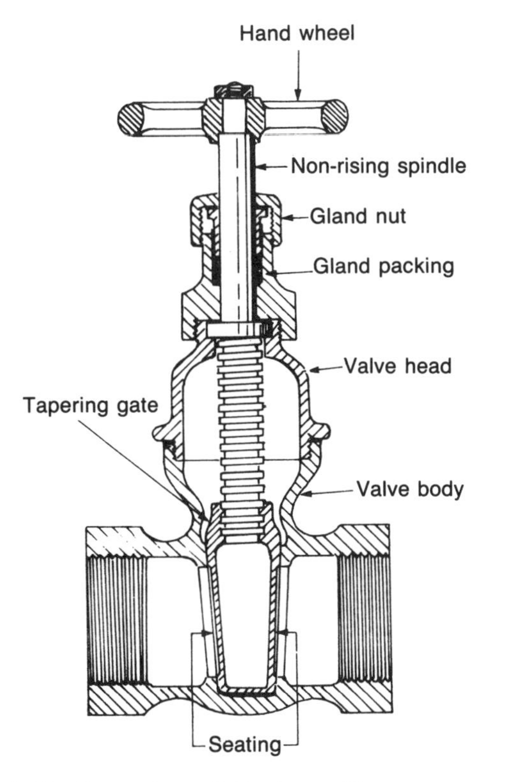

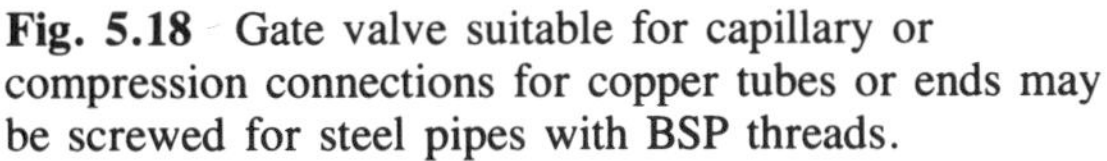
Fig. 5.18 Gate valve suitable for capillary or compression connections for copper tubes or ends may be screwed for steel pipes with BSP threads.

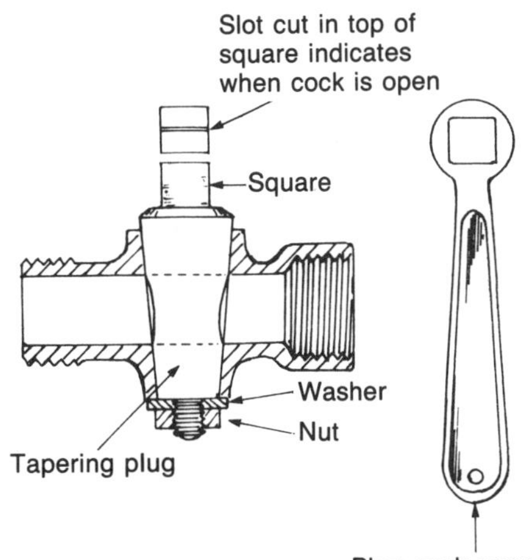

Fig. 5.19 Plug cock. Turns on or off by a quarter turn. Used mainly on gas supplies and appliances.

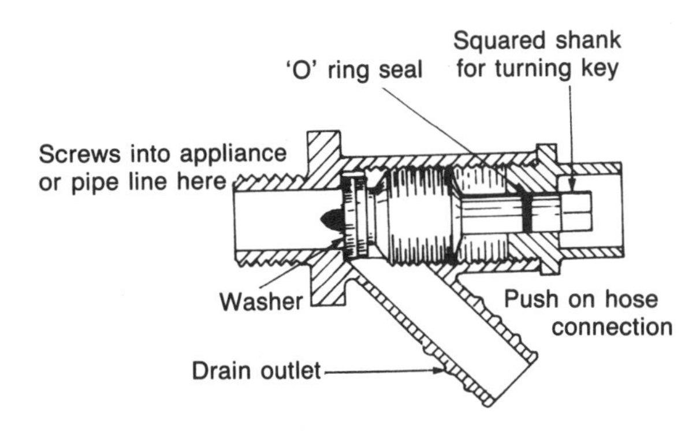

Fig. 5.20 Drain-off cock

circular tapering seating, a lock nut securing the valve into the body of the tap. Plug cocks do not require a great deal of maintenance; if they become difficult to turn, the valve body can be removed by undoing the lock nut and applying a light smear of grease. Plug taps may be provided with a square head for which a suitable spanner is provided, and so that the position of the valve can be determined, i.e. whether it is in the open or closed position, a slot is cut into the square. When the slot is in line with the pipe run the valve is open, when it is at 90° to the pipe run, it will be in the closed position. Some of the smaller valves are provided with an elliptical-shaped finger plate instead of a spanner.

Drain-off cock

As the name implies, the chief use of these cocks (see Fig. 5.20) is to drain down hot or cold water services for maintenance purposes or to empty a system to prevent it becoming frozen. As these cocks are not in constant use it is often found that the washer has deteriorated, especially those used on hot water systems. It is good practice to replace the washer when the system is drained down as nothing is more annoying and time consuming than to find the cock leaks when the system has been refilled, thus involving another drain-down to replace the washer.

Float-operated ball valves

These valves are best described as taps which close as the water level raises the float and automatically shut off the water at a predetermined level. They were commonly called ball valves until the ball service type valve illustrated in Fig. 5.16

became popular. To avoid confusion they should be described as float-operated valves. There are three main types: Croydon, Portsmouth and Diaphragm. The Croydon type, which is the oldest, is no longer available in smaller sizes. Its main distinguishing feature was that the ram, unlike that of the Portsmouth type, operated in a vertical plane and had an unfortunate habit of jamming in the open position, especially when it became worn.

As with the Croydon type, Portsmouth valves in the smaller sizes have been superseded by the diaphragm type mainly for two reasons. Adjustment of the water level when using this valve was made by bending the float arm, which is not permitted on half-inch valves by the water bylaws. The main reason, however, is due to the fact that, having a bottom outlet, the air gap requirements specified by the bylaws are difficult to achieve. In larger installations requiring float-operated valves supplied by pipes of 25 mm nominal bore or more. Portsmouth valves, usually of the equilibrium pattern are still in use but special arrangements are made to accommodate these valves to ensure their use meets the requirements of the water bylaws. Reference to air gaps and larger cold water installations is made in Book 2 of this series.

Diaphragm float-operated valves Diaphragm float-operated valves (see Fig. 5.21(c)) should comply with BS 1212:Part 2 (brass body) and Part 3 (plastic body). They were used almost exclusively in flushing cisterns and due to the requirements of the water bylaws of 1986 they are becoming increasingly used in cold water and feed cisterns.

Unlike Portsmouth and Croydon ball valves the diaphragm type will not permit backsiphonage or backflow to take place even when completely submerged. Diaphragm ball valves, have the advantage of fewer working parts than those previously described, and like the BS 1212 Portsmouth type, the seatings are interchangeable. These ball valves are made of brass or high-density polythene or polypropylene, two plastic materials which are unaffected by corrosion such as dezincification.

The water level can be finely adjusted by a screw, a further advantage over the Croydon and Portsmouth types where adjustment was made by bending the arm – a difficult procedure, especially when the valve is fitted in a narrow flushing cistern. Another advantage is the position of the outlet which can be situated on top of the valve thus reducing the risk of water pollution. The outlet can be moved through 180° to facilitate right or left hand entry into the cistern.

Equilibrium valves Unlike those previously described which rely solely on the force exerted by the float to overcome the pressure exerted by the water, these valves utilise the pressure of the water, to help close the valve. It will be seen in Fig. 5.21(b) there is a waterway through the centre of the valve, one end of which carries a washer in the normal way, the other a cup washer, very similar in appearance to the type used in cycle pumps. Water passing through the hole in the valve exerts the same pressure on the cup washer (which tends to push the valve towards the seating) as the pressure of the incoming water tends to push it off. As these pressures are equal they cancel each other out, hence the name equilibrium valve. The float on this type of ball valve has only to lift the arm whereas in the case of ordinary ball valves, the effort provided by the float not only has to overcome the weight of the arm but also the pressure of the incoming water.

Equilibrium valves are used with advantage in areas where very high mains pressures exist and where persistent water hammer may be encountered. All valves of over 50 mm nominal diameter are of the equilibrium type to reduce the size of the float that would otherwise be required to close the ram in an ordinary valve of this size.

Valve floats were at one time made almost exclusively of spun copper in two halves and usually soldered together. Acidic waters acting on the solder often cause it to disintegrate, resulting in the float becoming waterlogged and useless. While copper is still usual for the construction of large floats, polypropylene is almost exclusively used for those of small size. They are cheaper than those made of copper and generally speaking give less trouble. Floats made of expanded urea formaldehyde, a thermosetting plastic, are also available, these having a cellular structure enclosing small air pockets. Even if the surface of these floats becomes chipped or damaged in any way, they will still function satisfactorily.

One of the chief objections to float-operated valves of all types is the noise they make because of the high velocity of the water flowing through a small orifice and the noise of the water falling into the cistern. Little can be done to prevent the former other than to feed the appliance from a storage cistern, it being good practice to supply flushing cisterns in this way, using a full way orifice in the valve if in an ordinary house. While it has been common in the past to fit silencing pipes to float-operated valve outlets, it is now only permissible to do this if the supply is not connected directly to the main. Only silencing pipes made of thin polythene tube should be used as they will collapse and flatten if a negative pressure occurs as the use of silencing pipes made of metal can cause water to siphon back from the cistern into the water main which constitutes a risk of water pollution.

Float-operated valves are designated as high pressure, medium pressure or low pressure and are marked HP, MP, LP. These terms relate to the size of the orifice inside the valve. Due to the considerable force imposed by a high-pressure water supply on the cross-sectional area of the orifice there is a limit to the pressure against which a valve will close in relation to the orifice diameter. An increase in the size of the float would provide a greater opposing force to close the valve or a longer arm would provide more leverage, but neither of these two alternatives are a practical proposition, due to the limited space into which they are often fitted, i.e. flushing cisterns. The pressures against which half-inch valves should close are given in Table 5.2.

Table 5.2 Float-operated valve closing pressures.

Valve type	*Diameter of orifice*	*Max. pressure (kN/m^2)*
High pressure	3 mm	13 80
Medium pressure	4–5 mm	7 00
Low pressure	6 mm	2 75

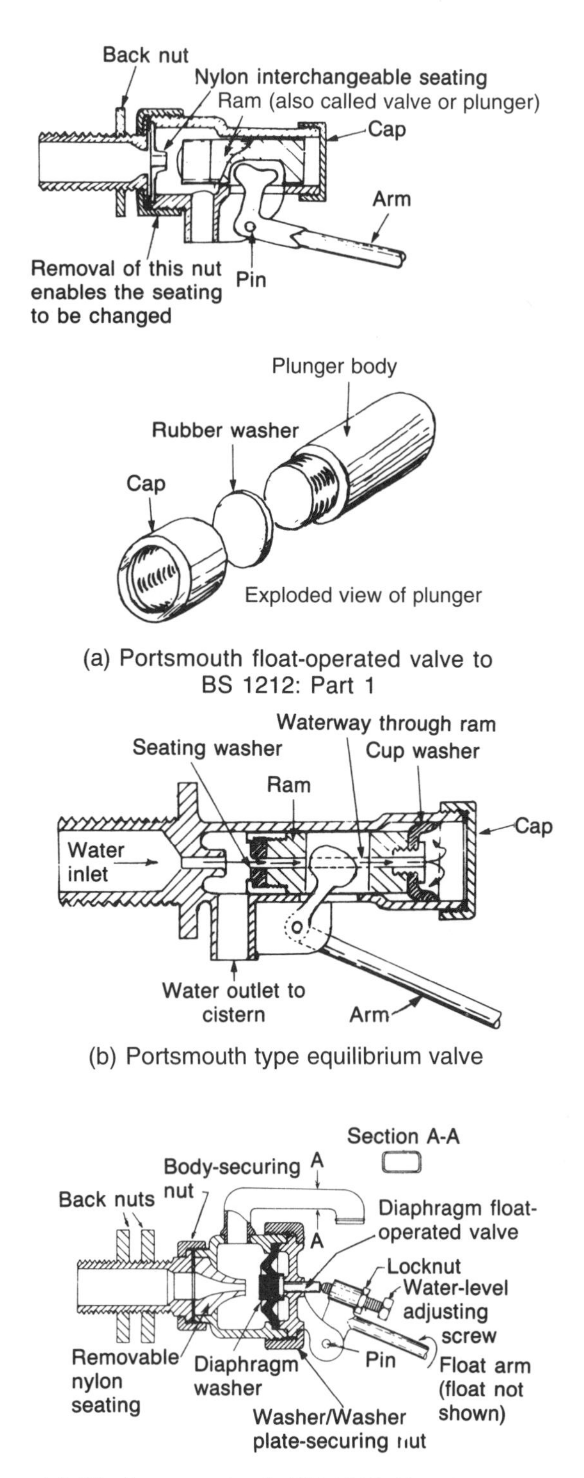

(a) Portsmouth float-operated valve to BS 1212: Part 1

(b) Portsmouth type equilibrium valve

(c) Diaphragm valve to BS 1212: Part 2 or 3

Fig. 5.21 Float-operated valves

Portsmouth valves The Portsmouth float-operated valve has a valve which moves in a horizontal plane, one of the best types of this pattern being that which complies with BS 1212:Part 1 (see Fig. 5.21(a)). It has a removable seating, usually made of a tough nylon which has two advantages, one being it is easily interchangeable so that a high-pressure valve can be converted for medium or low pressure simply by changing the seating. The other advantage is that nylon is unaffected by cavitation, which is a form of erosion which causes the pitting of brass seatings. Cavitation is best described as the effect of air bubbles in high pressure supplies bursting as they pass through the restriction of the orifice. When this occurs with brass seatings a small quantity of metal is removed which, over a period of time, causes pits or cavities on the surface of the seating making it impossible for the rubber washer to seat properly and shut off the water. Portsmouth valves employ rubber washers which require occasional replacement by dismantling the piston. It is often found difficult to remove the cap which holds the washer in position on the valve body due to the seepage of water corroding the thread. A light tap round the circumference of the cap will enable it to be removed more easily. When the washer has been replaced, all the many parts should be cleaned to remove any corrosion or fur that could cause malfunctioning of the valve when it is replaced.

Water supplies to consumers

When a new supply of water is required, an application must be made to the local water authority who will provide the necessary forms to be completed by the building owner. When approval has been obtained and the fees paid, the supply will be laid to a point approximately 150 mm outside the consumer's boundary, terminating with a square-headed stop cock. All the work is generally done by the water authority's own employees but in some cases it can be done by plumbing contractors, and for this reason it is necessary for the plumber to have a working knowledge of the materials and methods used for this type of work.

High-pressure water mains are made of cast iron, PVC (coloured grey) and polyethylene (coloured blue). These two latter materials are not strong enough to accept a BSP thread and special components are used to permit tapping into these mains. To make connections to existing asbestos mains a gunmetal clamp or saddle is used (see Fig. 5.22(a)). These may be bolted as shown or fixed by drive on dovetail wedges. Tappings to PVC and polyethylene mains is achieved in a similar manner, saddles of PVC being jointed to the main by solvent cement, and those of polyethylene by fusion welding using electrically heated elements. The ferrule itself with these two materials is moulded on to the saddle as a complete unit. The use of plastic materials makes the actual operation of tapping much easier as the ferrule body is made as a cutter. By screwing it down to open a supply of water to the service pipe, it simultaneously cuts a hole in the plastic pipe thus avoiding the necessity of the special machine required for tapping cast iron (see Fig. 5.22(b)).

When tapping into cast iron mains a special machine (see Fig. 5.23(a)) is used which enables a new connection to be made without turning off the water supply and inconveniencing other consumers. The machine is constructed to be used in conjunction with the ferrule body shown in Fig. 5.23(b) which must be fitted into spindle 'A' of the machine after first screwing down the plug valve on to its seating. The other spindle 'B' carries a combined drill and tap (see Fig. 5.23(c)). The machine is then clamped on to the main over a leather washer and secured by an adjustable chain. The leather washer ensures a watertight joint between the main and the machine during the tapping operation. Pressure is imposed on the spindle 'B' by the screw which forces the combined drill and tap into the main, its rotation being achieved by a ratchet lever fitted to the square on the top of the spindle. When the tapping has been completed the centre of the machine is rotated through 180° so the spindle 'A' lines up with the hole. By rotating the ratchet lever the body of the ferrule can then be screwed securely into the main and at this point the machine may be removed.

Figure 5.24 illustrates a half section of the

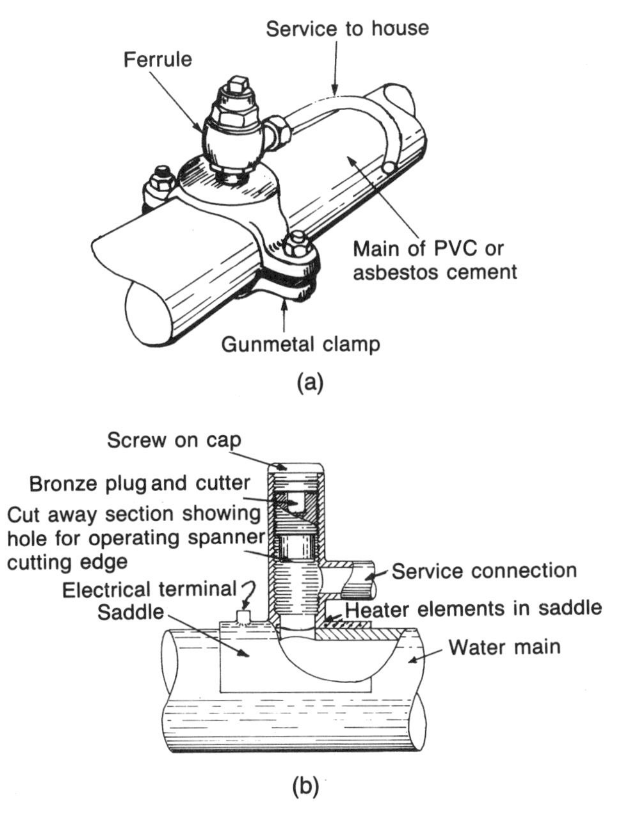

The joint is made between the main and saddle by fusion welding while the saddle is clamped on to the pipe. To cut a hole in the main the bronze plug is screwed down to its full extent with a hexagon spanner, forcing the cutting edge of the plug through the main; the disc cut from the main remaining in the plug. Having pierced the main the plug is screwed back as shown in the fully open position.

Fig. 5.22 Bolt-on ferrule connection for asbestos mains

complete ferrule, the shaded portion is the part that has been screwed into the main. The upper part of the ferrule is sometimes called the 'banjo' due to its shape and is made to swivel on the body of the ferrule, being secured to it by the upper body and made watertight by two rubber washers. It should be fitted so the outlet is in line with the main (see Fig. 5.25) which ensures that a gooseneck bend is made at the connection thus allowing for subsidence or shrinkage of the soil surrounding it. For the same reason service pipes should be 'snaked' from side to side of the pipe trench as shown in Fig. 5.26.

This is true in the case of services made of

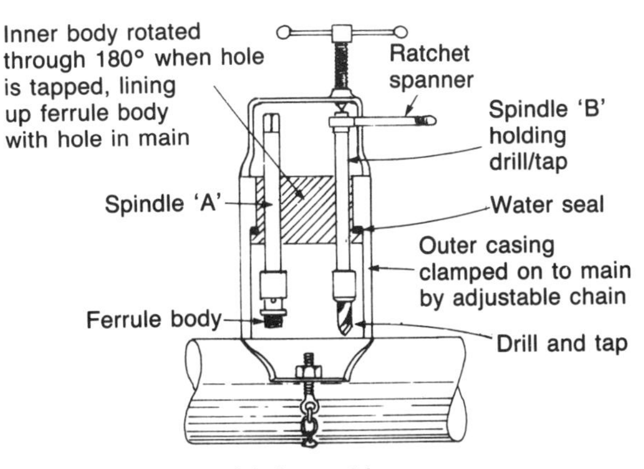

(a) Assembly

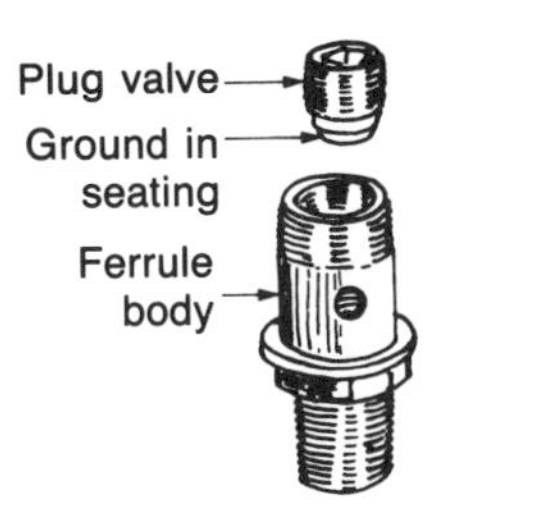

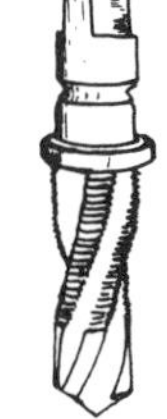

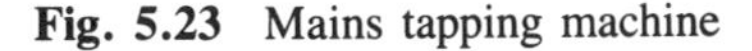
(b) Detail of ferrule body (c) Detail of combined drill and tap (used in water and gas main tapping machines)

Fig. 5.23 Mains tapping machine

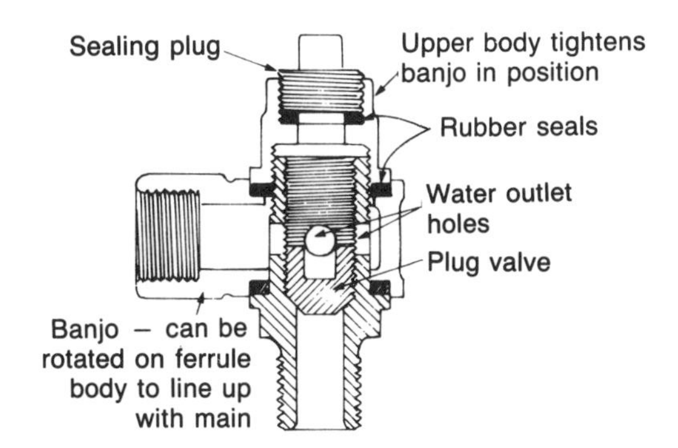

Fig. 5.24 Main ferrule details

metal. Some water authorities do not consider any provision for subsidence necessary for polythene service pipes due to their inherent flexibility.

When the connection is completed, the water supply is turned on at the ferrule by unscrewing

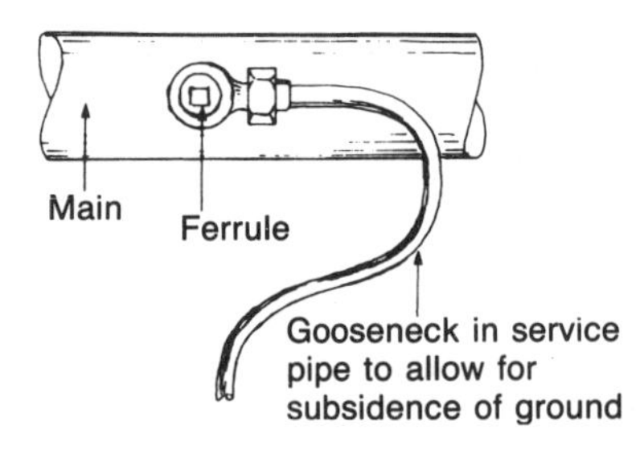

Fig. 5.25 Connection of service pipe to water main

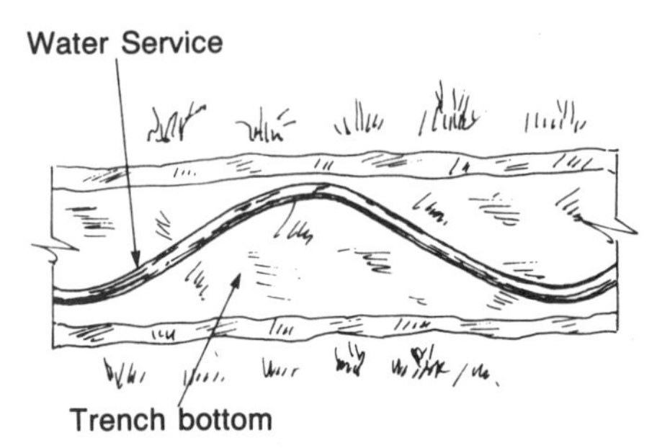

Fig. 5.26 Running service pipes in trenches. Note that subsidence provision in the case of polythene services is at the discretion of the supplier.

the plug valve with a special spanner which fits into its square recess to the top of the thread. To prevent any seepage of water past the valve and to prevent the ingress of earth into the ferrule, a plug is fitted loosely into the top of the upper body.

The maximum size of a tapping in a main will depend upon its diameter and the diameter of the required service. It will be obvious that a large tapping into a relatively small main would be difficult to make watertight; it could unduly weaken the main and excessive draw off might well cause starvation of water to other consumers at a higher level. For these reasons tappings exceeding 25 mm diameter are rarely used. In cases where large quantities of water are likely to be required, larger storage cisterns are usually recommended.

Protection of service pipes

Some soils are very acidic and metal service pipes should be adequately protected against corrosion. Pipelines can be completely sheathed in PVC or they can be wrapped with an anti-corrosive bandage before laying. Any covering removed for jointing must be suitably wrapped when the joint has been made. An alternative is to duct the service completely which has the added advantage of allowing the service pipe to be withdrawn for repair or renewal without a great deal of excavation, an important advantage when the service passes under a main road. It is important that the joints on these ducts, which are usually made of earthenware pipes, are made watertight. Any service pipe passing through or under the footings of buildings must be ducted similarly.

All mains and service pipes should be at least 750 mm beneath the surface of the ground as a precaution against damage by frost. The maximum depth of cover should not exceed 1.350 m as pipes buried at greater depths would not be readily accessible.

A diagrammatic illustration of a service pipe, its connection to the main and entry into the building is shown in Fig. 5.27. The square-headed stop cock fitted just outside the boundary line of the building is usually the limit of the water authority's responsibility for the service. The only exception to this is where the main is laid under a public footpath and a stop cock ferrule (see Fig. 5.28) is used serving the combined purpose of connecting both the service to the main and as a stop cock.

Access to the stop cock may be obtained via an earthenware pipe 150 mm in diameter placed over the stop cock, carefully supported on two bricks placed alongside the service. The top of the pipe is fitted with a cast iron cover which should terminate at the same level as the pavement (see Fig. 5.29(a)).

An alternative to the foregoing is a complete unit which is moulded in plastic materials and shown in Fig. 5.29(b). Its height is adjustable and the base is formed to offer full support to the stop cock.

Water meters

The metering of water supplies in commercial premises has been usual for a long period of time and is gradually being introduced for domestic properties. Large water meters for commercial premises are installed in a brick-built pit with a cast iron cover, but complete units for domestic

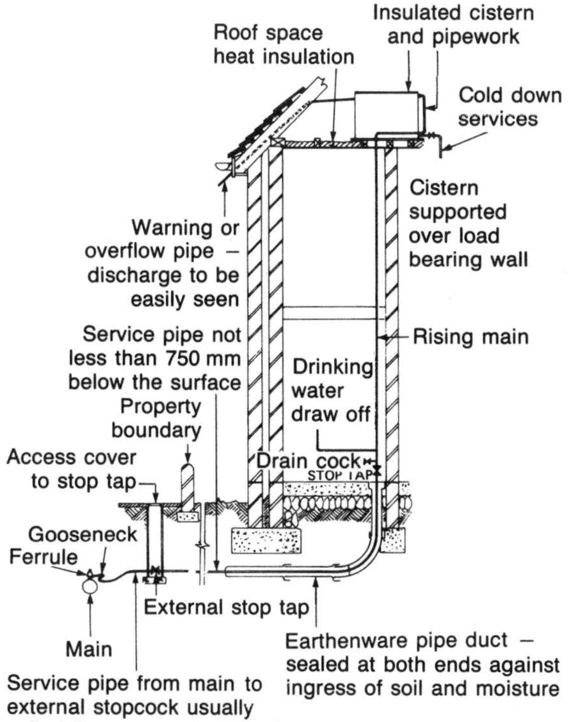

Fig. 5.27 Connection of water service pipe to water main and entry into building

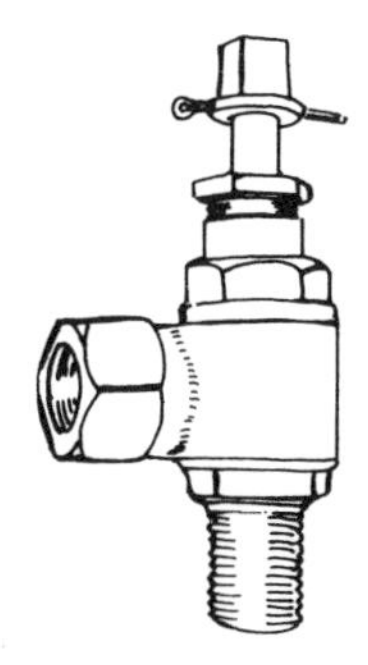

Fig. 5.28 Stop cock ferrule. It should be noted that the square head shown here and in Fig. 5.29(a) is being superseded by crutch-headed stop cocks as shown in Fig. 5.29(b).

properties having small meters have been developed. They not only incorporate the meter but also the service stop cock. Where a meter is fitted into a service having metal pipes, i.e. copper, a permanent earthing bond must be fitted

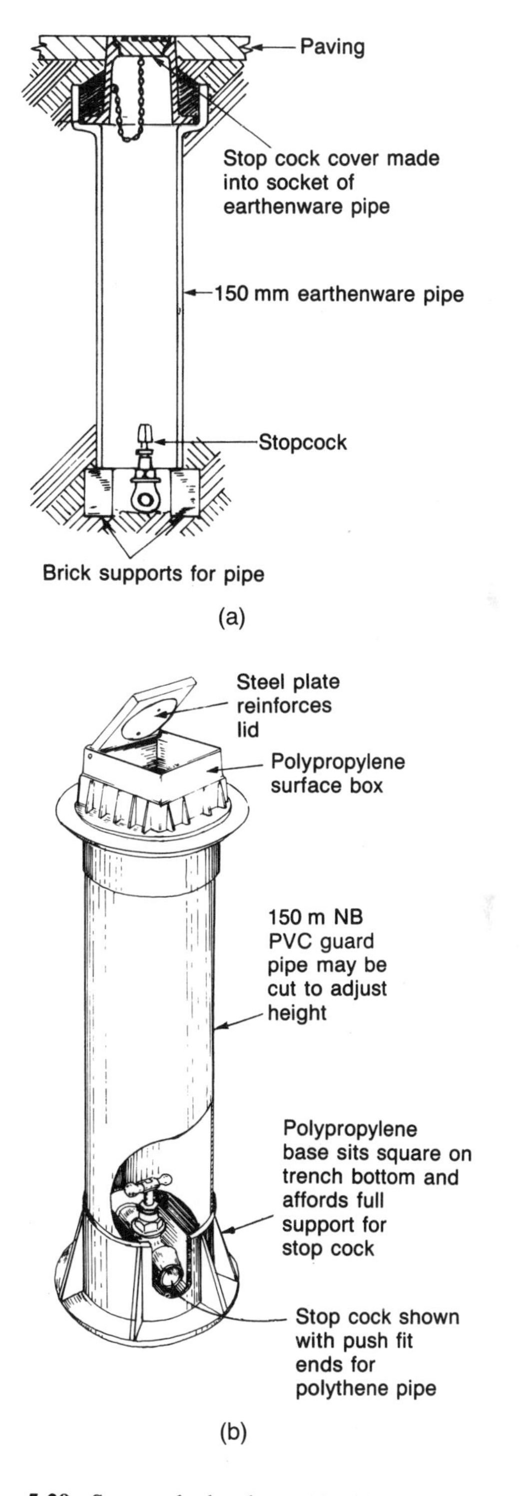

Fig. 5.29 Stop cock chambers: (a) older method of access to outside stop tap; (b) all plastic chamber

across the inlet and outlet of the meter, so that in the event of removal of the meter earthing continuity is maintained.

CP 6700 lists the following recommendations for the installation of meters for both outside and inside the premises as illustrated in Figs 5.30(a)–(d). In all cases the position of the meter should conform to the measurements shown. This ensures that it can be easily read and serviced when necessary. The special unit shown in Fig. 5.30(d) is made so that it can be fitted into the bodywork of an existing stop cock conforming to BS 1010 and BS 5433. When the existing stop cock is conveniently situated, its use can save a lot of time and expense.

Prior to the installation of a meter, a careful check should be made to ensure the underground service pipe is sound. Any possible source of water wastage such as defective taps and valves should be serviced or replaced.

Entry of the service pipe into the building

A consumer's stop cock is fitted where the service pipe enters the building at as low a level as possible with a drain-off tap immediately above it. This enables all the water in the cold water pipes to be drained off if necessary in frosty weather or when repairs are needed. Any low points on the service which cannot be drained by the draw-off tap should have a separate drain tap. Pipes in unheated buildings should be well insulated and turned off and drained during long periods of frost when they are not in use. Pipes should not be fixed to external walls, to avoid freezing, but if this is unavoidable, they should be insulated and spaced off with suitable clips for copper or steel pipes.

Pipe runs

Pipe runs in roof spaces should be as short as possible, well insulated and never fixed within two metres of the eaves to avoid cold draughts, nor should pipes be run near ventilators and outside doors as these are normally very draughty positions. The storage cistern should be sited over a load bearing wall to provide adequate support for the volume of water it contains. A situation above or near the airing cupboard is an added precaution to prevent frost damage. The storage vessel should be well lagged with non-flammable materials such

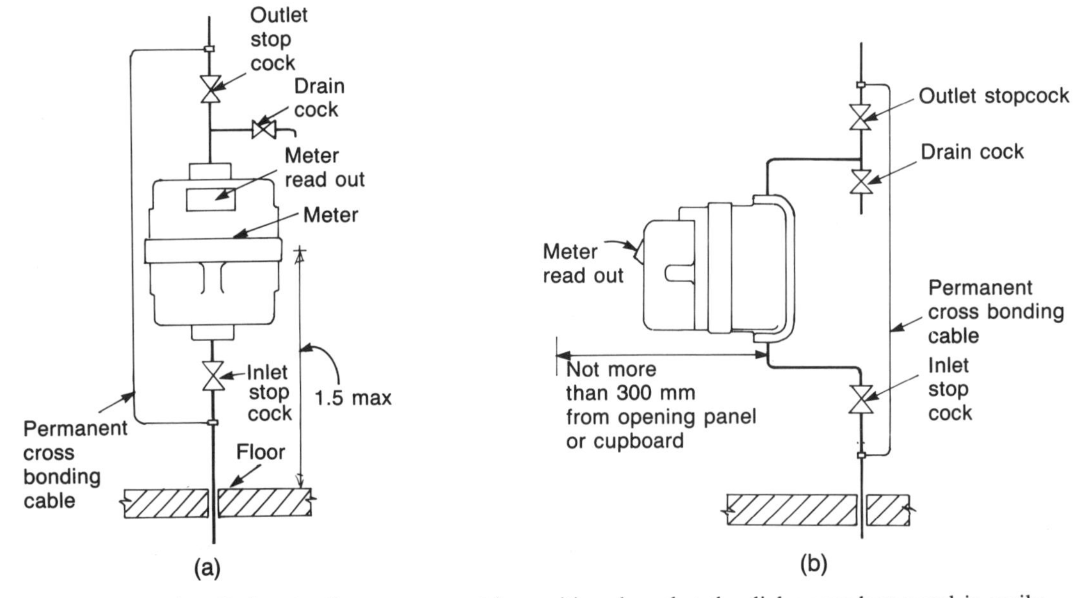

Fig. 5.30 Water meter: installation details: meters must be positioned so that the dial or readout panel is easily visible; all pipework must be adequately supported and stop cocks provided for servicing the meter; the meter must not be situated where it may be affected by frost, vibration or mechanical damage.

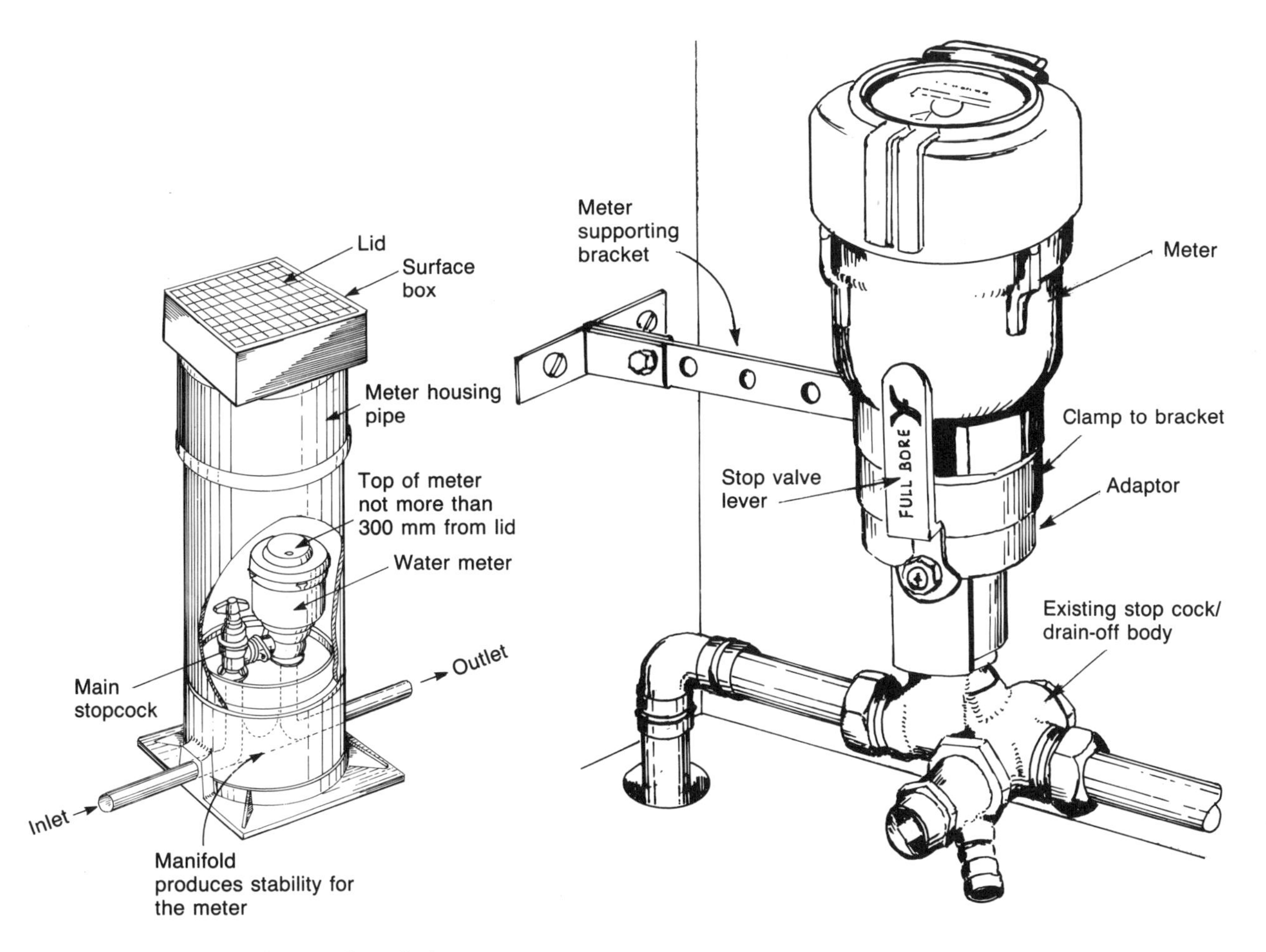

(c) Underground meter installation.

(d) 'Aquadapt' in line meter adaptor

This adaptor, marketed by IMI Yorkshire Fittings Ltd can be fitted to an existing stop cock body after removing the headwork. It can be used in both the horizontal and vertical positions. It is essential that the meter is adequately supported. The adjustable bracket shown is obtainable from the manufacturer. (Courtesy of IMI Yorkshire Fittings Limited.)

Fig. 5.30 *Continued*

as expanded polystyrene, glass fibre or vermiculite to a thickness of 75 mm. If pipes are run under suspended ground floors, these should be at oversite level and well insulated to avoid cold draughts from air bricks.

Generally speaking, pipe runs in a building should be readily accessible. If pipes are run in intermediate suspended floors, it may be necessary to notch the joists, but this should be avoided where possible as it weakens the joists. When notching must be resorted to, the joists should be cut just large enough for the pipe to move without chafing the side of the notch (see Fig. 5.31(a)). Tight notching is one of the most common causes of noise, especially with hot water systems where there is more movement of the pipes.

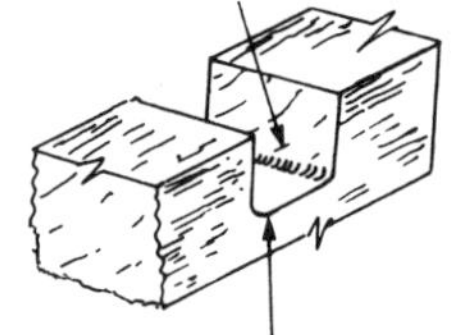

(a) Notch in joist

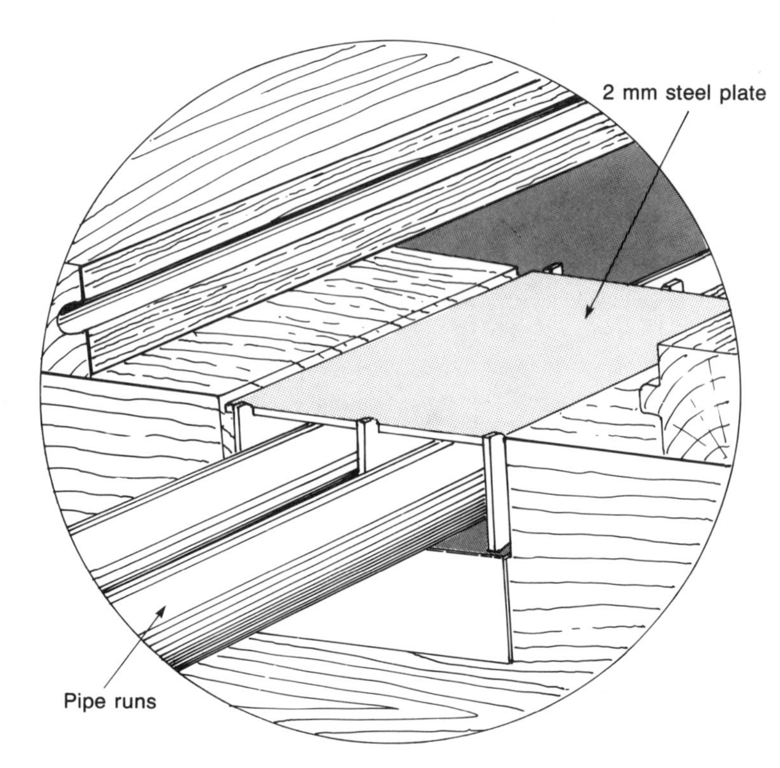

(b) Pipe guard

The use of these components in notches cut in joists allow for expansion of the pipes and prevent accidental damage from floor fixings.

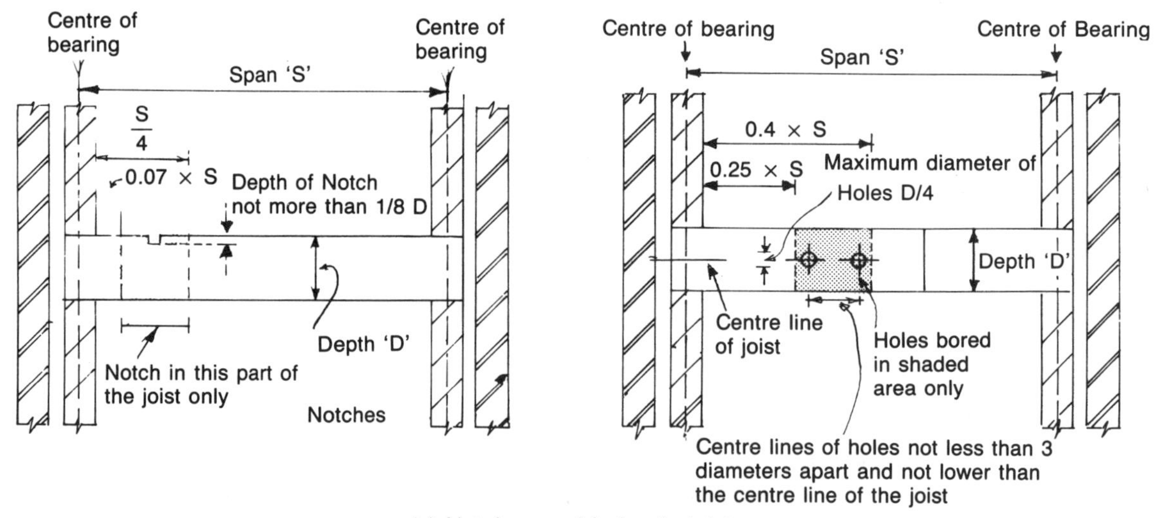

(c) Notches and holes in joists

Examples in a floor joist 200 mm in depth having a span of 4 m. The zone in which notches are permissible would start 0.280 m from the inside face of the wall, with a maximum distance away from the wall of 1.000 m. The depth of the notch in this case must not exceed 25 mm. If holes are bored the zone would start 1.000 m from the inside face with a maximum distance away from the wall 1.600 m. The diameter of the holes in this case must not exceed 50 mm.

Fig. 5.31 Pipe runs under suspended floors

Figure 5.31(b) illustrates a component which solves many of the problems of pipework installed under suspended floors. Its use ensures that the pipes do not bind on the woodwork, and the steel plate prevents penetration of the pipe by nails or screws when the floor is laid. The position of the notches is most important as to notch a joist in its centre would create a seriously weakened point. Figure 5.31(c) shows the limitation of notches and holes in joists to comply with the Building Regulations 1985. If these limitations are adhered to, no serious weakening of the joist will occur, especially if the corners are radiused.

Floor boards over pipe runs should be screwed down to permit removal for maintenance purposes. When flooring is removed to facilitate new pipe runs, the nails should be punched down and in the case of tongued and grooved boards, the tongue should be cut through with a pad or circular saw to avoid damage to the edges of the board when they are lifted. If possible, board cuts should be over joists but if this is not possible, the edge of the joist may be extended by using a stout cleat nailed to it (see Fig. 5.32(a))

In existing premises where chipboard flooring has been used, it is much more difficult to remove than wooden floor boards in order to gain access to the space below. Because of this problem it may be easier to run the pipework in suitable ducts above the floor level. It this is impracticable a section of chipboard may be removed and replaced as shown in Fig. 5.32(b). A circular saw with the blade set at the thickness of the flooring is the best way of making the saw cuts – but beware of nails!

Always remember that replaced flooring must be made good in a safe and professional manner.

It is not advisable to run pipes in solid floors where this can be avoided unless they are fitted in properly constructed ducts with removable covers. They are expensive to install and are seldom found in domestic properties. If it is absolutely necessary to run pipes under such floors, a proper channel

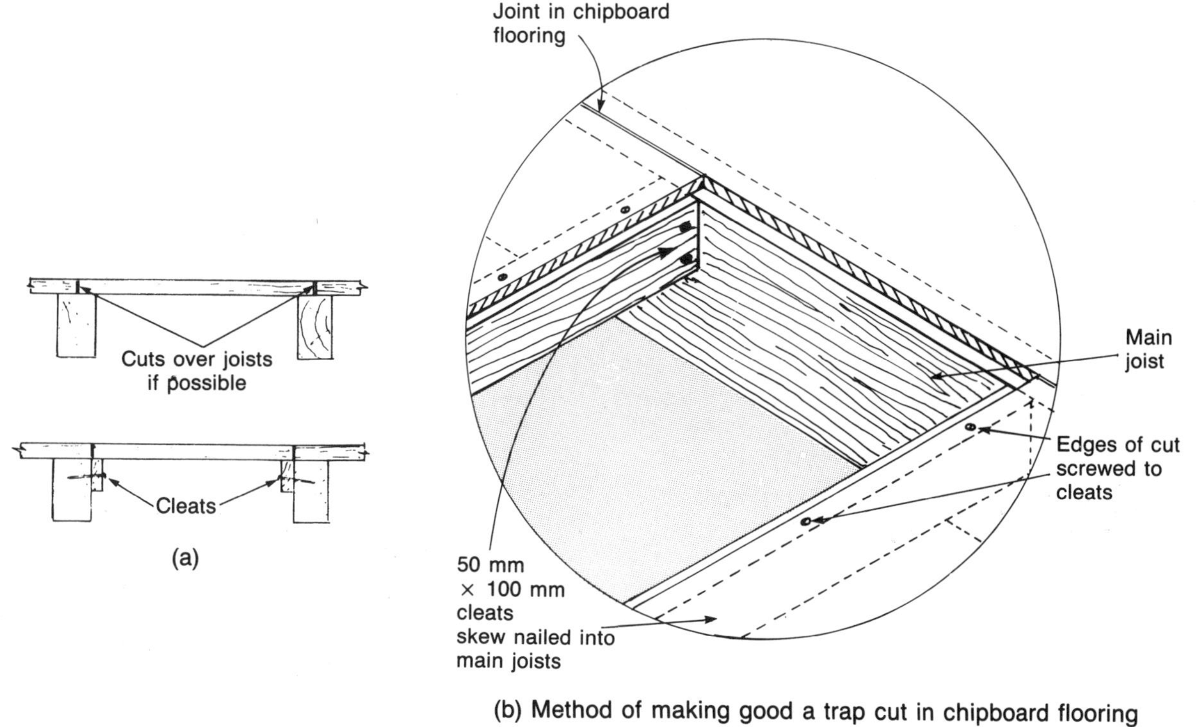

(b) Method of making good a trap cut in chipboard flooring

Note: As the joints in this type of flooring are normally tongued and grooved the panel may be damaged when it is removed and will require replacement.

Fig. 5.32 Replacing wooden flooring

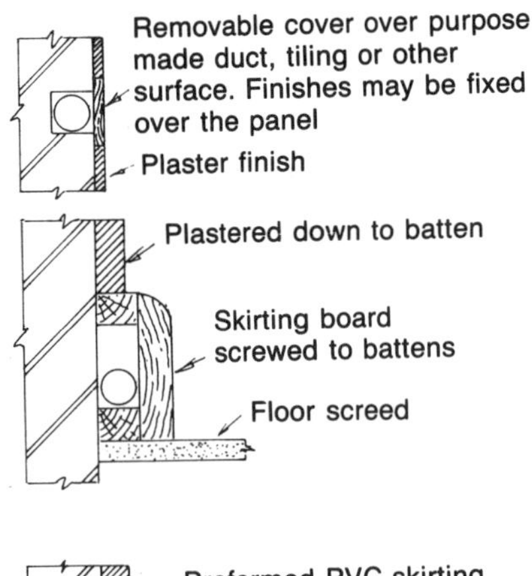

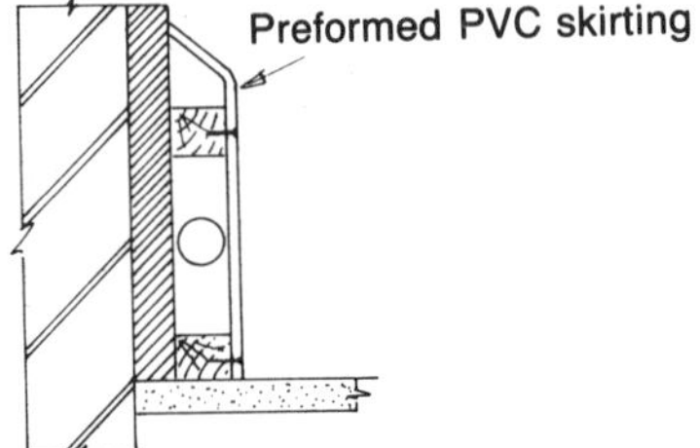

(a) Accessibility of pipe concealed in or adjacent solid walls

Several systems of purpose made PVC skirting are available for both pipe and cable concealment. Most ranges include internal and external corners and special fittings providing for pipe exits through the skirting to permit for example connections to radiators

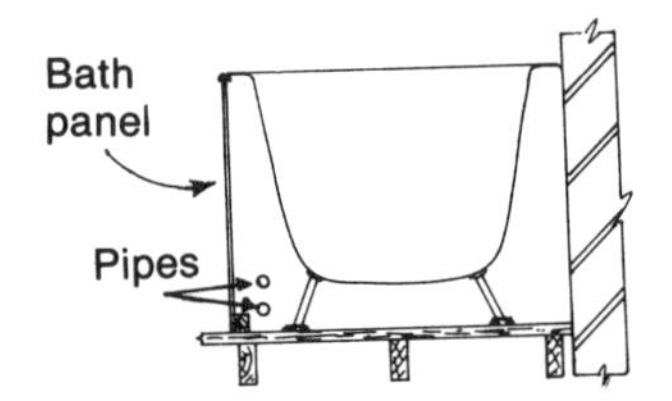

(b) Accessibility of service pipes under baths

Pipes to be fixed on the panel side of the bath and adequately supported

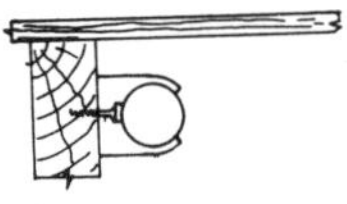

(c) Pipes under intermediate suspended floors

Pipes to be suitably supported with provision for access at intervals of not more than 2 m and at every joint. Such pipes are not normally exposed to temperatures of less than 0 °C and in such cases insulation may only be required if they are part of a hot water installation.

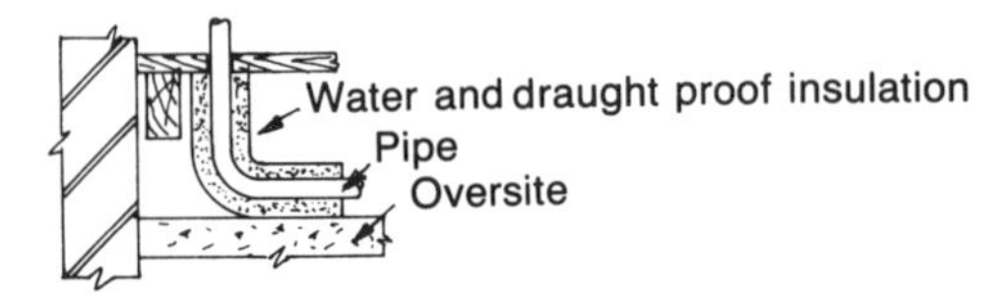

(d) Pipes under suspended ground floors

Pipes should be well insulated and preferably be laid on the oversite to avoid any cold draughts. Risers must be fitted as far as possible from any air bricks. The recommendations for accessibility in intermediate floors also apply here

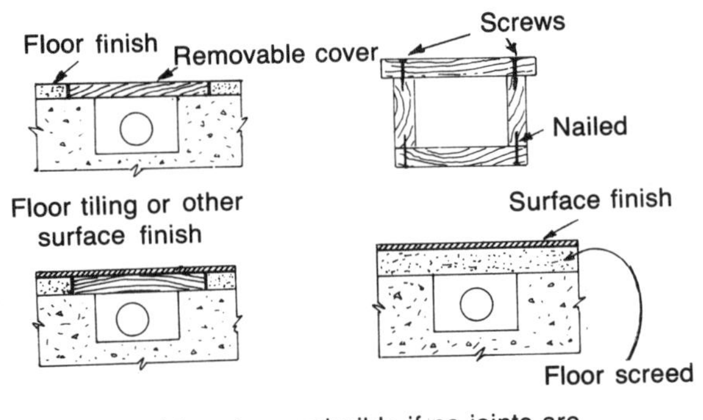

This method is only permissible if no joints are enclosed and the pipe can be withdrawn for inspection

(e) Accessibility of pipes in or under solid floors

Suggested construction of wooden box let into oversite duct. Timber to be treated against decay and coated externally with bituminous paint. Purpose made galvanised steel ducts are also manufactured for installation in floors. All hot water or heating pipework in solid floors must be suitably insulated to prevent heat loss.

Fig. 5.33 Approved methods of running pipework in the building fabric. It should be noted that it is not recommended to run pipes in external walls.

should be provided by the builder. A little planning is necessary here and the plumber should notify the builder where the channel is to be provided by means of a dimensioned sketch. Before it is laid, the pipe should be well insulated to allow it freedom of movement and to enable it to resist any corrosive effects of the floor screed or finish. Figure 5.33 illustrates some of the approved methods of running pipework in the building fabric.

Protection from frost damage

CP 6700, 1987, deals with precautions against frost damage. The water bylaws also lay specific emphasis on this subject as burst pipes can result in large quantities of water being wasted, quite apart from the damage caused to a householder's property. The best protection against frost is to keep the premises warm, but this may prove difficult when dealing with cisterns in a roof space, supplies fitted outside the building such as standpipes, or cattle feeding troughs. No amount of insulation will prevent freezing, it simply delays it, the thicker and more effective the insulation, the longer the delay. The following deals generally with frost precautions. Reference should also be made to the section on insulation in Chapter 6.

Storage cisterns in roof spaces should be situated, if possible, over or near the airing cupboard. If this is impossible, siting near a chimney is the next best choice. The underside of the cistern should not be insulated; heat from the rooms below must be allowed to keep it warm. Both cisterns and pipes should be situated away from draughts and well insulated. Pipe runs in the roof should be as short as possible and in no circumstances fitted near the eaves. Never run a pipe immediately under a roof as shown in Fig. 5.34 as it will be very exposed to frost damage and difficult to maintain.

All water pipes, both hot and cold, should be fitted in such a way that the whole system can be drained in the event of the building being unoccupied during periods of frost. In those buildings fitted with fully automatic heating, it is suggested that the thermostat is turned down and the time switch overridden so that the heating system is operational during the owner's absence.

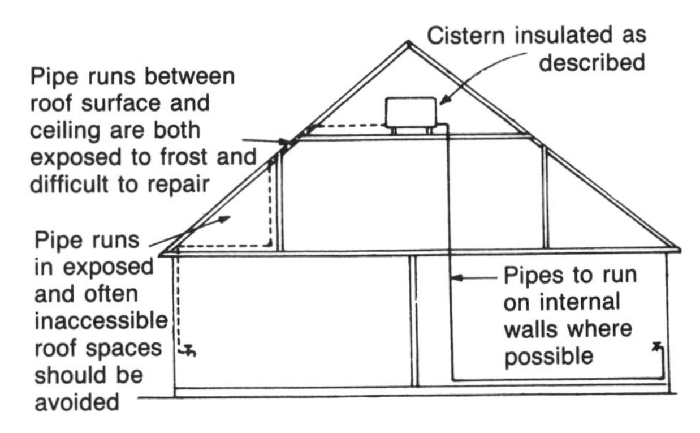

Fig. 5.34 Avoiding frost damage to pipework

This ensures that the premises are kept warm and there is no danger of the heating pipes becoming frozen, as it is a fact that in many cases it is almost impossible to fully drain these systems.

Outside taps should be fitted in such a way that they are controlled by a stop cock inside the premises. The pipe should be insulated where it passes through the outside wall as shown in Fig. 5.35(a). Standpipes away from the building should be fitted in an insulated box with a stop cock and draw-off fitted at low level. Figure 5.35(b) illustrates the treatment for outside draw-offs. Much the same applies to external WC apartments which are exposed to frost damage. Do not forget to instruct the householder or client about the precautions to take during frost. Figure 5.35(c) shows a hose union bib tap having integral check valves which comply with the water bylaws.

Automatic control devices

In the case of large or industrial buildings where the water supply must be kept working during severe weather, several alternatives to insulation are available for protection against frost damage. In cistern or tank rooms of buildings which are centrally heated, a radiator fitted with a thermostatic valve may be used. The sensory element on the valve is set at the frost position and any drop in temperature below this will cause the valve to open and keep the temperature of the room just above freezing point. An alternative is to insulate the cistern with materials containing an electrical heating element, rather similar in fact to an electric blanket. Pipework can be traced with an electrical heating element in tape form (see Fig. 5.36) which is bound to the pipe with adhesive tape, after which the whole is insulated with a suitable material. To avoid the possibility of anyone unwittingly cutting the heating element, warning notices should be attached to the insulation stating that the pipework is electrically traced. Control of the heating element can be manual – i.e. switched on during low temperatures – or automatic by means of a frost thermostat. These operate in the same way as a room thermostat for central heating, but they work over a lower temperature range, usually −10 °C to 20 °C. The position of the frost thermostat

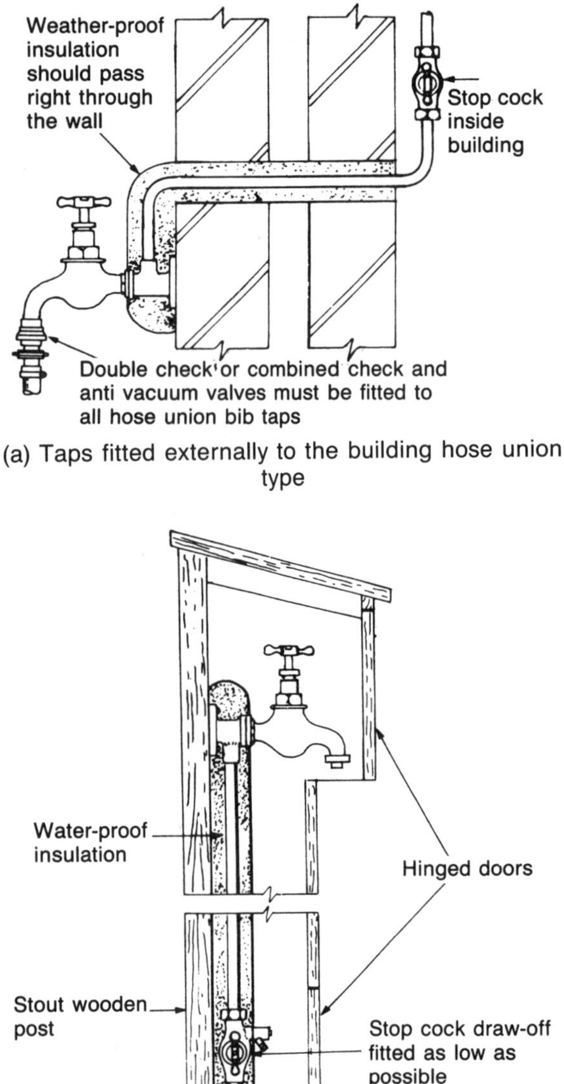

(a) Taps fitted externally to the building hose union type

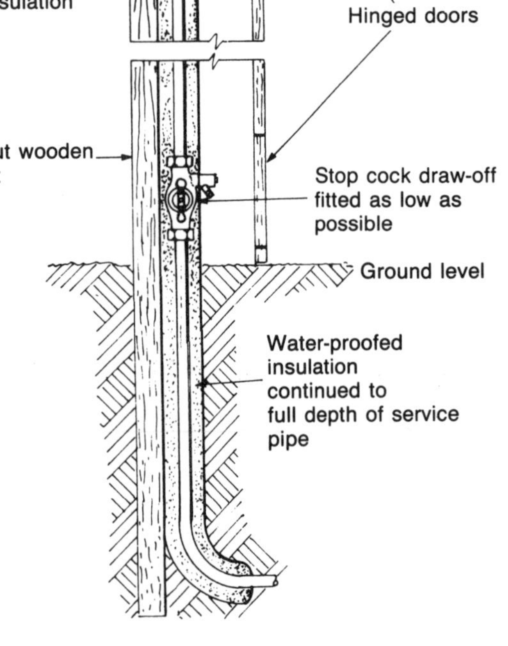

(b) Standpipes

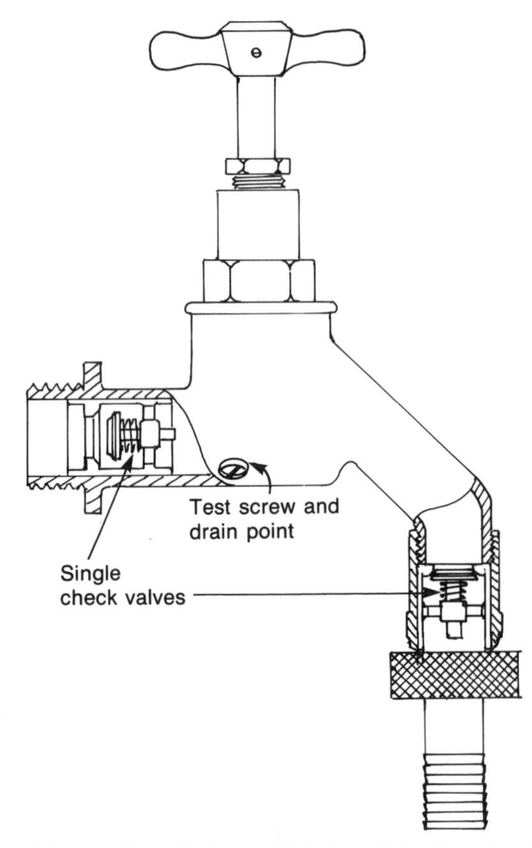

(c) Hose union bib tap with integral double check valves

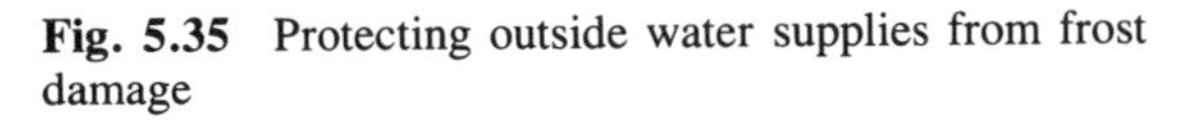

Fig. 5.35 Protecting outside water supplies from frost damage

should be carefully chosen to ensure that it operates correctly. If, for example, the thermostat is situated on a wall facing east, the heat from the early morning sun may well cause the heating element to be de-energised when freezing conditions still prevail in other parts of the building. A suitable thermostat setting is 2–3 °C.

The water regulations permit the installation of electric immersion heaters in cold water storage cisterns to prevent the contents freezing. This form of protection is well worth consideration in situations such as where bulk water storage is installed in unheated areas. Control of these heaters may be effected in the same way as an ordinary heater but having a special thermostat which works, like a frost thermostat, over a lower

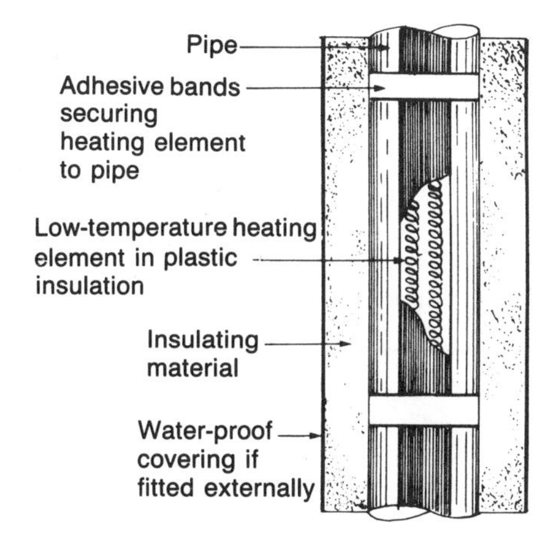

Fig. 5.36 Use of low-temperature heating element for tracing exposed pipework

temperature range. A suitable thermostat setting would be 2 °C. This ensures the water temperature would never drop below this point, above 2 °C the heater would be inoperative.

In the case of a feed cistern which supplies water to a hot water scheme only, the temperature setting may be up to, but not in excess of, 39 °C.

Finally, do not forget that any insulation fitted outside a building *must* be waterproof. If it is allowed to become wet the result will be worse than having no insulation at all.

Pipe supports

Selecting the correct type of fixing methods to be used in plumbing installations is a very important part of a plumber's work. Lack of adequate support to pipework and fittings can result in their damage and possibly subsequent failure under service conditions, or to the development of faults such as air locks. In the case of large installations, the type of fixings are specified by the designer, but on smaller jobs the plumber often has to make his own decisions. The factors to be considered include the type of construction and surfaces into which fixings must be made, the mass of the component to be fixed and whether, in the case of a pipe run, it is fitted horizontally at high level, low level (when it may be prone to damage) or in a vertical position. Pipework fitted in the latter position generally requires fewer fixings. The pipework material will also have some influence on the number of fixings required, i.e. polythene pipes require more support than copper or steel pipes. To avoid possible corrosion problems, the support or clip should, if possible, be made of the same or similar material as that of the pipe.

Pipe clips and supports vary considerably both in cost and effectiveness, and a variety are illustrated in Fig. 5.37. Clips or other supports in schools, hospitals and factories should be strong, robust, and to achieve maximum support, should be built into the fabric of the building. The holderbat bracket shown in Fig. 5.37(a) is typical of such a bracket for building into a wall. These brackets are made of cast brass or malleable iron and are suitable for copper and steel pipes. Holes may be made in a wall to accept these brackets by using a large tungsten tipped drill or by using a chisel of the type shown in Chapter 2, Fig. 2.35(c). In the case of brick walls, if the 'perpends' or upright joints can be located, a fixing is often more easily achieved as these joints are not always filled so solidly with mortar. The techniques for 'making' the brackets into the wall are the same as those described in Chapter 7 for fixing cantilever brackets.

Saddle clips (Fig. 5.37(b)) are made for pipes of steel, copper and semi-rigid plastic such as PVC. The main objection to these clips is that they clip a pipe back tightly to the wall making painting difficult and allowing an accumulation of dust and dirt to build up between the pipe and the wall surface. For these reasons it is considered good practice when fixing rigid pipes such as steel, copper and semi-rigid plastic pipes such as PVC, to use 'stand off' or spacing clips, of which many patterns are available. One of the most common of these types of clip is shown in Fig. 5.37(c). They are made of polypropylene or PVC and have the advantage of being secured by just one screw which saves fixing time. These clips are suitable for copper, stainless steel and PVC pipes which 'snap' into the clip after it has been fixed.

A similar clip for securing double pipe runs is shown in Fig. 5.37(d). The double pipe clip is made of PVC and is secured by one screw only instead of the two or three normally required. They also ensure both pipes are parallel throughout

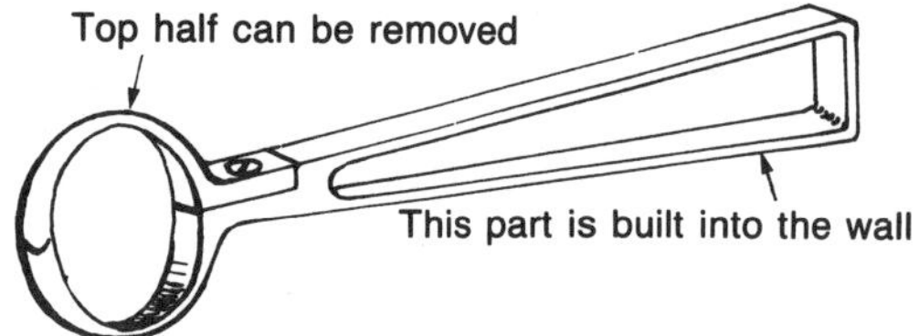

(a) Holderbat built-in bracket

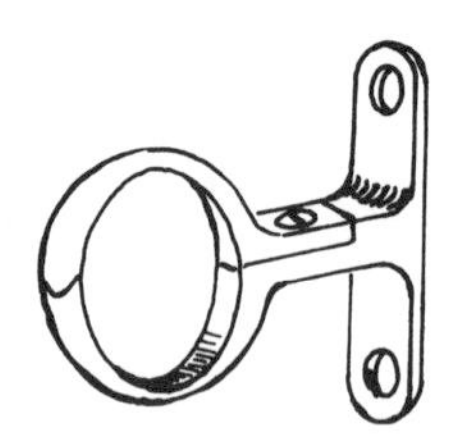

(f) School board bracket

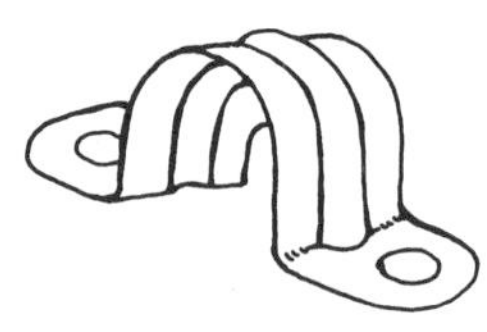

(b) Saddle clip

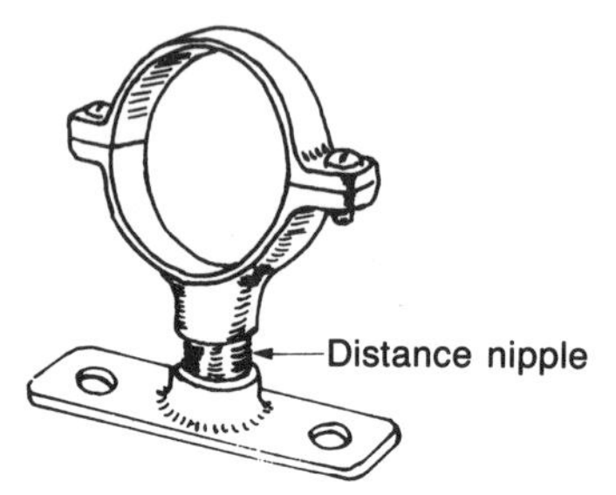

(g) Pipe or Munson ring fixing

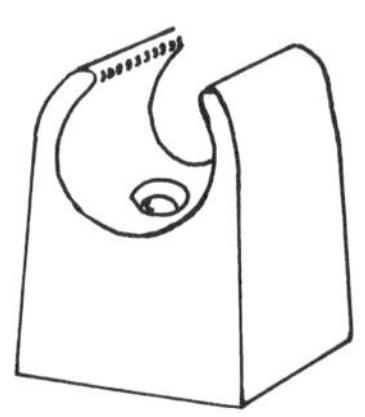

(c) Thermoplastic 'snap on' clip for copper or plastic tubes

Pipe stands off wall with this type of clip

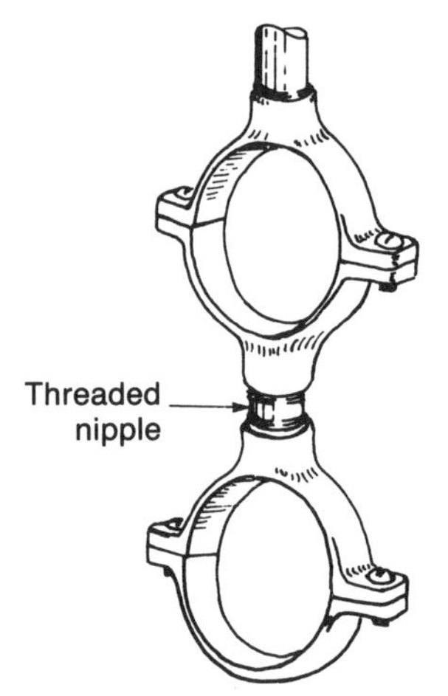

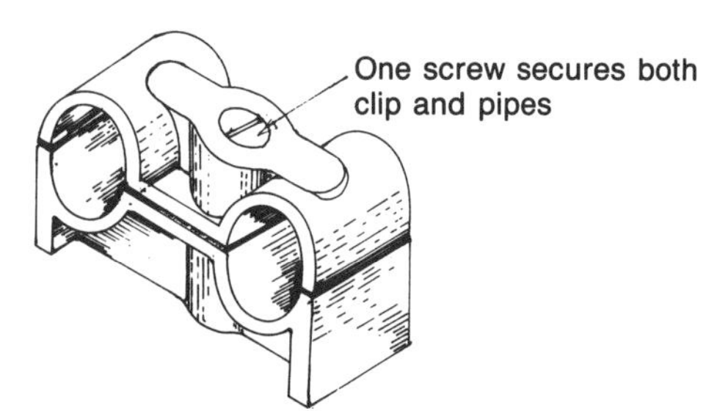

(d) Double pipe clip for copper or PVC pipes

15 and 22 mm OD only

(h) Series of pipe rings supported by a hanger

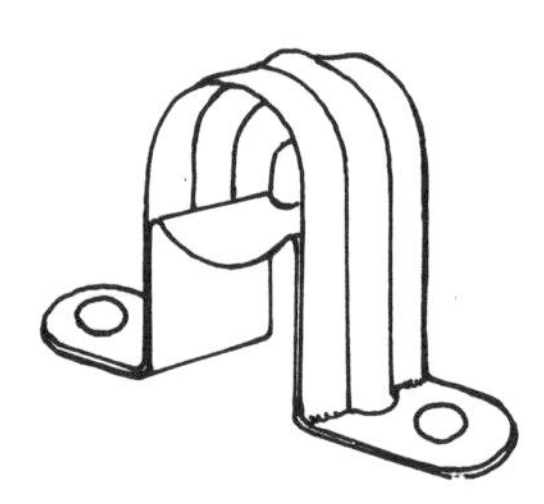

(e) Two-piece spacing clip

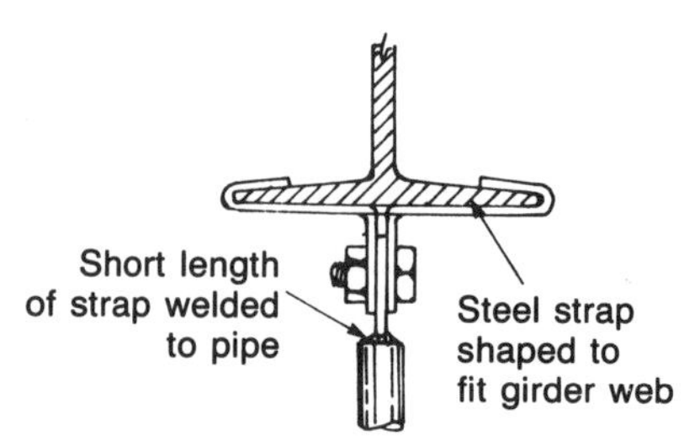

(i) Fabricated girder clamp for pipe rings

Fig. 5.37 Pipe clips and supports

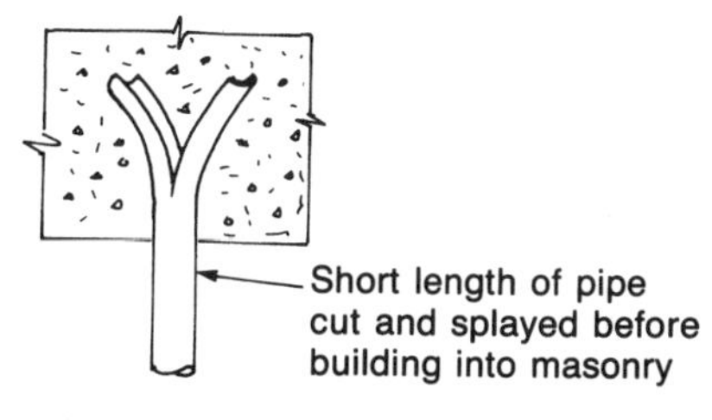

(j) Masonry fixing for pipe rings

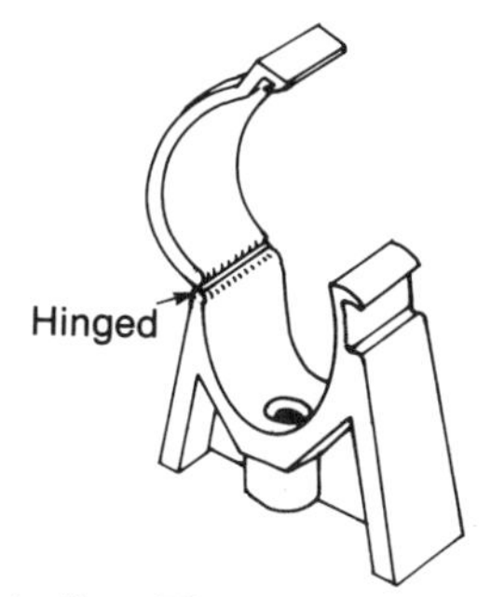

(k) Hinged clips. The type shown is suitable for polybutylene and cross-linked polythene. More robust clips of this type are made for copper tubes

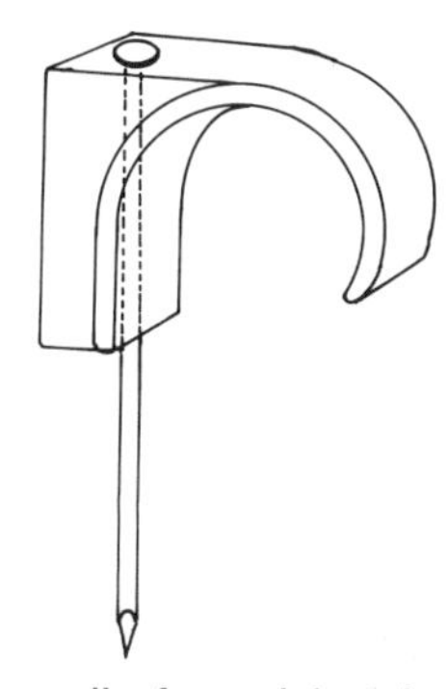

(l) Shows a nail on clip for polybutylene or cross-linked polythene pipes

Fig. 5.37 *Continued*

their length. Extra care is necessary when fitting the pipework as like all pipe fixings made of plastic, this material has insufficient strength to 'pull' a badly fitted pipe into the correct position. A good rule to follow is to carefully set out and fix the clips prior to fitting the pipework.

The clip shown in Fig. 5.37(e) is made of pressed copper and prior to the introduction of the plastic type was the most popular for copper tube fixing in domestic work. None of the 'stand off' clips shown in Fig. 5.37(c), (d) and (e) are very strong and their use is not recommended in public buildings, especially for skirting level work. A stronger, more rigid type of screw-on bracket is shown in Fig. 5.37(f). With the exception of the base plate, they are similar to built-in holderbat brackets, although the name does vary according to the locality.

Another type of bracket, referred to as 'pipe' or 'Munson' rings, is very adaptable and this bracket can also be used as a floor mounted support as shown in Fig. 5.37(g), or as a double-ended type as shown in Fig. 5.37(h). The use of a nipple permits a variation of the centre-line distance between the pipe and the floor, or pipe to pipe and gives a considerable degree of flexibility. In the case of Fig. 5.37(h) a series of pipes can be secured from one fixing such as the girder clamp shown in Fig. 5.37(i). Another alternative for fixing directly into masonry is shown in Fig. 5.37(j).

Pipe or Munson rings are tapped with a $\frac{1}{4}''$ or $\frac{3}{8}''$ BSP thread, the latter size being used for pipes of over 25 mm nominal diameter. As with school board and built in brackets, those used with steel pipes are made of malleable iron while those used for copper or stainless steel are made of cast or pressed brass. Figure 5.37(k) and (l) show two types of clip for polybutylene pipes. Type (k) is similar to a hinged top clip, sometimes used for copper tubes and is mainly used for surface fixing. Type (l) is a simple nail on clip similar to those used for electric cables and is mainly used for concealed work.

Good general recommendations for the distance between pipe supports are given in CP 6700, from which Tables 5.3 to 5.5 have been extracted. It will be seen from the tables that fewer supports are needed on vertical pipes. No recommendations are given for extra support for pipes run at low level so the plumber uses his own judgement here. Copper pipes especially need to be well fixed in this position as they are more easily damaged than pipes made of low carbon steel. Much the same applies to plastic pipe services and some thought should be given to the possibility of fitting such pipes in a position where possible damage to them

Table 5.3 Spacing for copper tube supports.

Diameter of pipe (mm)	*Interval for vertical runs (m)*	*Interval for horizontal runs (m)*
15	1.800	1.200
22	2.400	1.800
28	2.400	1.800
35	3.000	2.400
42	3.000	2.400
54	3.000	2.700
67*	3.600	3.060
76	3.600	3.000
108	3.600	3.000

*Note that 67 mm is a non-standard size

Table 5.4 Spacing for low carbon steel pipe supports.

Nom. diameter of pipe (mm)	*Interval for vertical runs (m)*	*Interval for horizontal runs (m)*
15	1.800	1.800
20	3.000	2.400
25	3.000	2.400
32	3.000	3.000
40	3.600	3.600
50	3.600	3.600

Table 5.5 Spacing for plastic pipework supports.

Polybutylene and cross-linked polythene complying to BS 5955: Part 8.		
Nom. diameter of pipe (mm)	*Interval for vertical runs (m)*	*Interval for horizontal runs (m)*
10	0.3	0.5
15	0.3	0.5
22	0.5	0.8
28	0.8	1.0
Low-density polyethylene complying with BS 1972 (colour black for above ground use)		
Nom. diameter of pipe (inches)	*Interval for vertical runs (m)*	*Interval for horizontal runs (m)*
$\frac{1}{2}$	0.800	0.400
$\frac{3}{4}$	0.800	0.400
1	0.800	0.400
$1\frac{1}{4}$	0.900	0.450
$1\frac{1}{2}$	0.900	0.450
2	1.100	0.550

is minimised. Always remember one clip too many is better than one too few.

Manufacturers of plastic pipes for discharge pipework always include directions for fixings in their freely available technical literature. The tables of fixing distances shown here relate to water supply pipework only.

Fixing of pipe supports

Whatever type of anchor, clip or pipe support used, its effectiveness will depend entirely upon how well it is fixed to the wall, ceiling or floor of the building. Fixings have to be made to a variety of structural materials in modern buildings and there are many differing types of equipment available which enable this to be achieved effectively.

Wood is the easiest material to which a fixing can be made, wood screws usually being used for this purpose. Generally they should be of sufficient length and stout enough for the purpose for which they are used. They can be made of brass, low carbon steel or aluminium alloy, the latter often being used by the plumber for the fixing of aluminium sheet fixings and cleats. Screws are made with countersunk, round or mushroom heads (see Fig. 5.38).

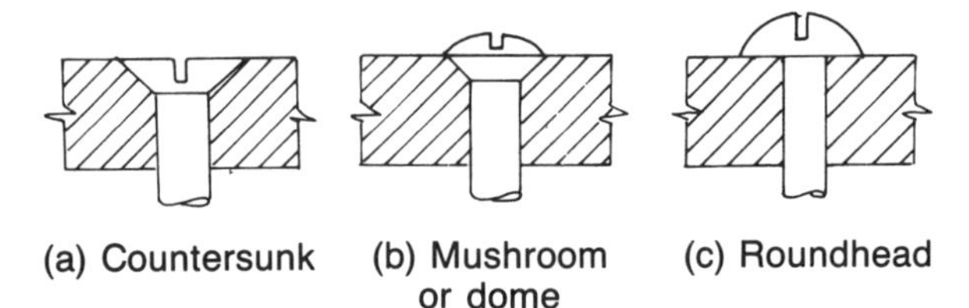

Fig. 5.38 Types of screw heads

The mushroom-headed screw is also countersunk and is used for fixing equipment such as towel rails and chromium-plated bathroom ware. Countersunk screws are used where the screw head must be flush with the surface of the work. Typical examples of this type of screw are found in the fixing of floor boards and wooden ducts over pipe runs, and also for fixing down wash basin pedestals and WC pans. Round and dome head screws are used for securing materials which are not thick enough to be countersunk or those in which no countersunk recess has been provided.

Brass and alloy screws have less tensile strength than those of low carbon steel and if the hole into which they are screwed is in hardwood or other dense materials they may shear off when screwed home. To prevent this a hole should be drilled in the material equivalent to the root core or minor diameter of the screw (see Fig. 5.39). A little oil or grease on the screw thread will assist the

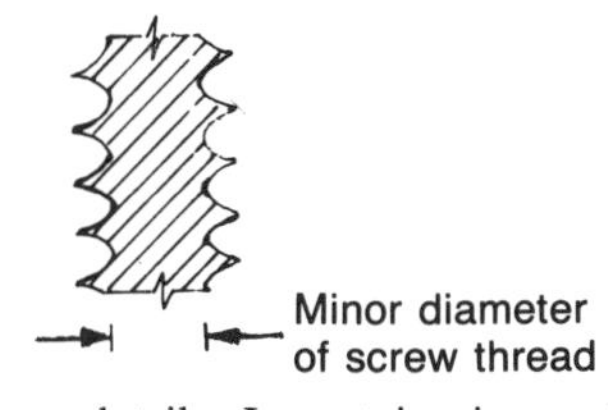

Fig. 5.39 Wood screw details. In certain circumstances it is necessary to drill a hole into the material before inserting a wood screw.

screwing operations into wood. The use of a lubricant also has the advantage of protecting the screw from corrosion, and enables easy withdrawal if and when required.

The original method of screwing into masonry was to use wooden plugs or wedges. However, as wood shrinks over a period of time due to loss of moisture, fixings of this type often become loose, and fibre or plastic plugs are used as they are much more reliable and save time. Fibre plugs are made in a variety of sizes to suit the diameter of the screw. If the top of the plug is flush with, or only just below, the wall surface, then the shank of the screw will expand the plug causing it to crack the surface of the work, therefore it is very important, especially when fixings are made to ceramic tile surfaces, to sink the plug into the hole clear of the wall surface. Careless fixing can result in cracking a tile which is time consuming to replace. The correct position of the plug in the hole is shown in Fig. 5.40(a).

When fixing to materials such as hollow blocks or steelwork, the spring toggle shown in Fig. 5.40(b) is typical of the many types of fixing available which enable the fixing load to be spread over a large area.

For lighter fixings the rubber nut is a cheap convenient alternative. Figure 5.40(c)(i) shows the fixing in position before the screw is tightened,

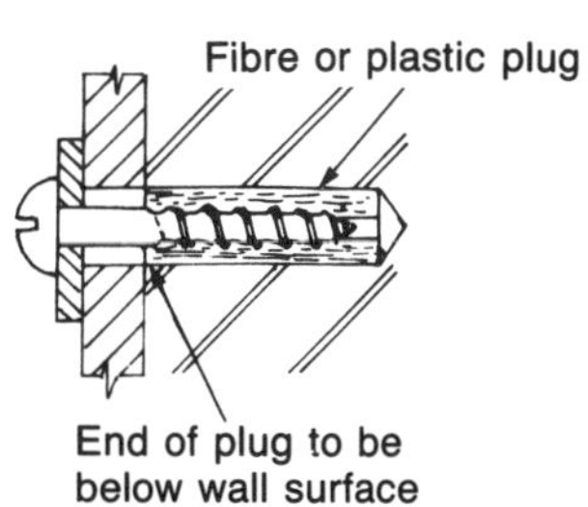

(a) In masonry using fibre or plastic plugs

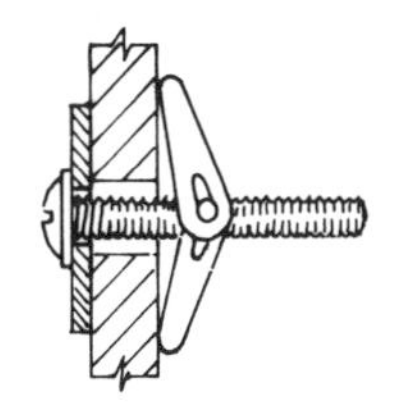

(b) Spring toggle fixing for steelwork or hollow blocks

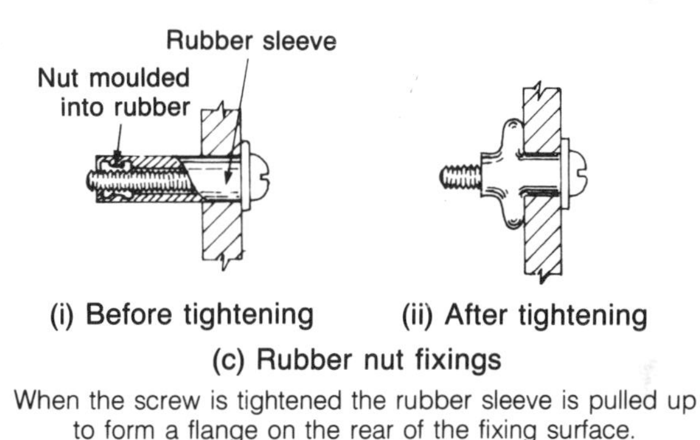

(c) Rubber nut fixings

When the screw is tightened the rubber sleeve is pulled up to form a flange on the rear of the fixing surface.

Fig. 5.40 Screw fixings

while Fig. 5.40(c)(ii) shows how the rubber is compressed to form a washer at the rear of the surface when the screw is tightened.

It should be recognised that there are limitations on the weight that can be supported by, for instance, plasterboard, which in itself is not a very strong material. In buildings under construction where partition walls are built of hollow blocks or plasterboard on studs, careful planning at an early stage can ensure good fixings. In the case of hollow work, a breeze or slag based block may be inserted where fixings are likely to be required for, say, a radiator, at a later stage of the job. In the case of studwork partitions ‘noggins’ should be fixed across the studs and nailed in appropriate

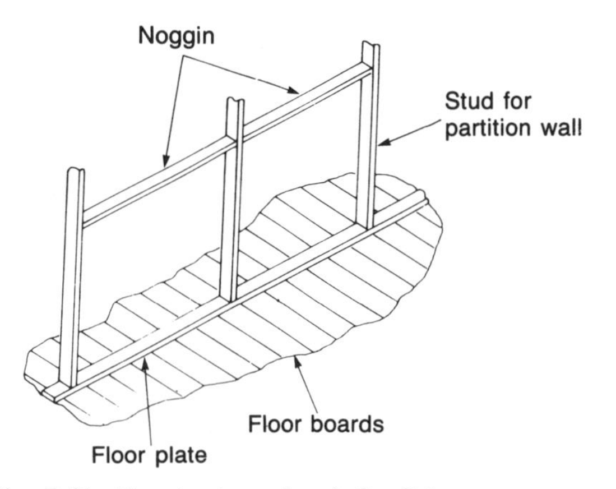

Fig. 5.41 Noggins in studwork for fixings

positions as shown in Fig. 5.41 before the plasterboard is fixed.

Where it is necessary to provide fixing for a series of pipes from the floor to the ceiling, i.e. in an airing cupboard, it is good practice to provide a board on which to fix the pipes. Its use will save time and reduce the possibility of damage to the finished plaster. The board can be easily fixed to lightweight blockwork using cut nails, but if fixings are to be made to brickwork it is recommended that the wall is plugged and screwed. Figure 5.42 shows details of this type of fixing.

There are two main methods of making heavy fixings in dense building materials such as concrete. The simple rag bolt shown in Fig. 5.43(a) can be positioned and built into the structure as it is being constructed, if sufficient information is available at an early stage in order to locate the rag bolt correctly. Alternatively, they may be built in at a later stage by first drilling a hole in the wall and then cementing or caulking in position using lead wool. The only problem with this type of fixing is that a period of time must be allowed for the cement mortar securing the bolt to dry out or ‘cure’.

The other method of fixing which is used in a similar way to the rag bolt is where the fitting and fixing can be made in one operation, the only prerequisites being the correct location of the bolt and an accurate hole in the structure affording a push fit for the bolt. One of the many variations of this type of fixing is shown in Fig. 5.43(b)). After

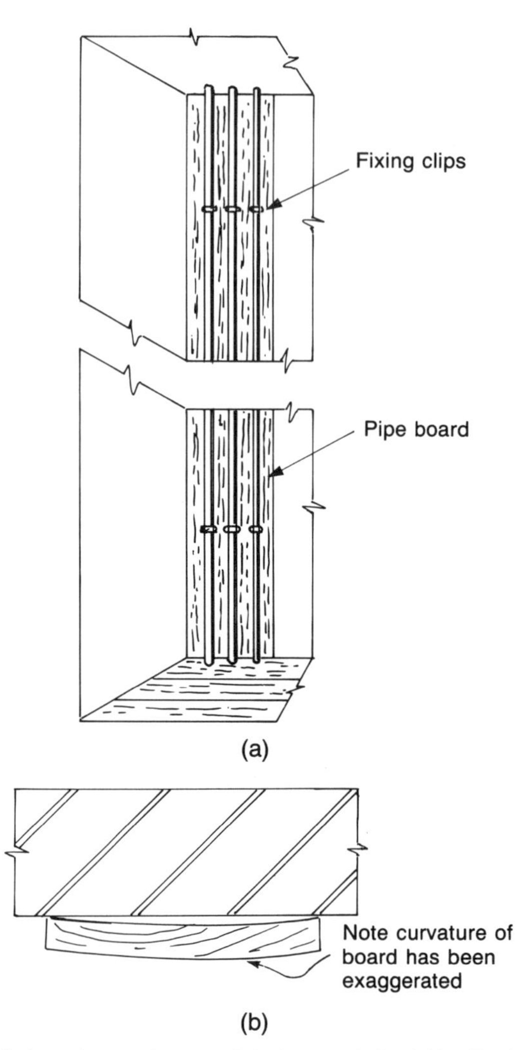

Wide boards may have a slight bow and should be fixed as shown. One or two fixings in the centre of the board will pull it back to the wall

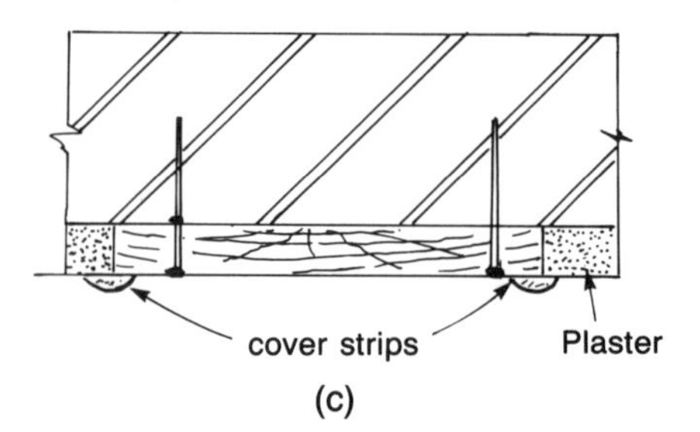

(c)

Section of board showing fixings. Board should be nominally 25 mm thick. Cover strips may be used to seal the shrinkage gaps between the board and plaster where the board is exposed.

Fig. 5.42 Pipe fixing boards

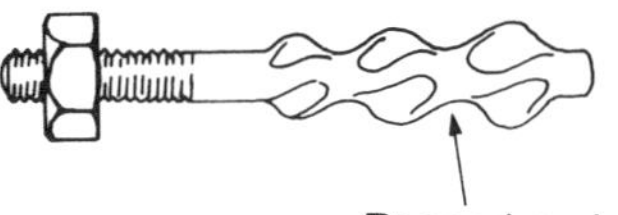

(a) Rag bolt

The ragged end of the bolt is built into concrete, etc.

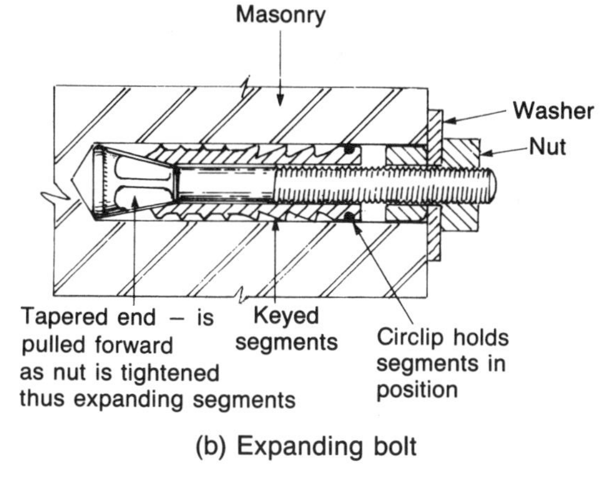

(b) Expanding bolt

Fig. 5.43

the bolt has been pushed into the prepared hole the nut is tightened and the tapered end of the bolt drawn outward, opening up the four keyed segments against the sides of the hole. Not only must the hole be accurate in size, but the structure itself should be fairly dense to use these bolts to their full effect. In old buildings constructed with soft bricks laid in mortar joints, a 'built in' rag bolt is often a more satisfactory fixing.

In most cases the fixings made by a plumber are either screwed or bolted but occasions do arise when light fixings can be nailed directly into the structure. Nails made of high tensile steel are available, being strong enough to be driven directly into masonry without bending. By using short positive hammer blows, they can be driven to a depth of approximately 20–25 mm into masonry, depending upon its density. These nails may be brittle and could shatter if struck with a glancing blow, therefore care should be taken to ensure that no other person is working in the vicinity where they are used and the operator should always wear suitable goggles.

One of the most versatile and labour saving devices for fixing is the cartridge tool. These

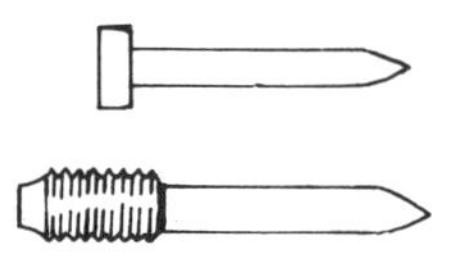

Fig. 5.44 Typical pins fired by explosive charge into masonry, concrete and steel

actually fire a pin or bolt into the structure with a force capable of penetrating low carbon steel, and this enables them to be used for fixing into dense brickwork or concrete with the minimum of effort. Two types of pin are used, those having a head similar to that of an ordinary nail and those with a threaded end to which fixings can be bolted, both types are illustrated in Fig. 5.44. Needless to say the tool that fires these pins is virtually a firearm and as such it is very important it is used only by a qualified person. Most companies manufacturing this type of equipment also offer short courses in their safe use and the potential user would be well advised to take advantage of such a course.

Cold water storage

In domestic premises where all the cold water supply, except that of drinking water is supplied by a cistern, the minimum storage capacity should be 114 litres. If the same cistern is also used as a feed cistern for hot water supply, the minimum capacity should be 227 litres. The materials from which these cisterns are constructed are described in Chapter 3.

Special care, however, is needed when polythene or polypropylene cisterns are installed. As they are made of a non-rigid material, they must be fitted on a flat area which fully supports the base.

Some consideration must also be given to the siting of cisterns in the roof space where it is not possible to position them over a load-bearing wall. If the mass of water in the cistern is supported over too few joists or trusses, or the bearers are insufficiently strong, the building fabric will be damaged. Cistern platforms should be constructed of a material that will not distort due to dampness which may cause the water connections to the system to leak. Plywood or a suitable alternative not less than 25 mm thick is suitable for this purpose. Chipboard is unsuitable due to the fact

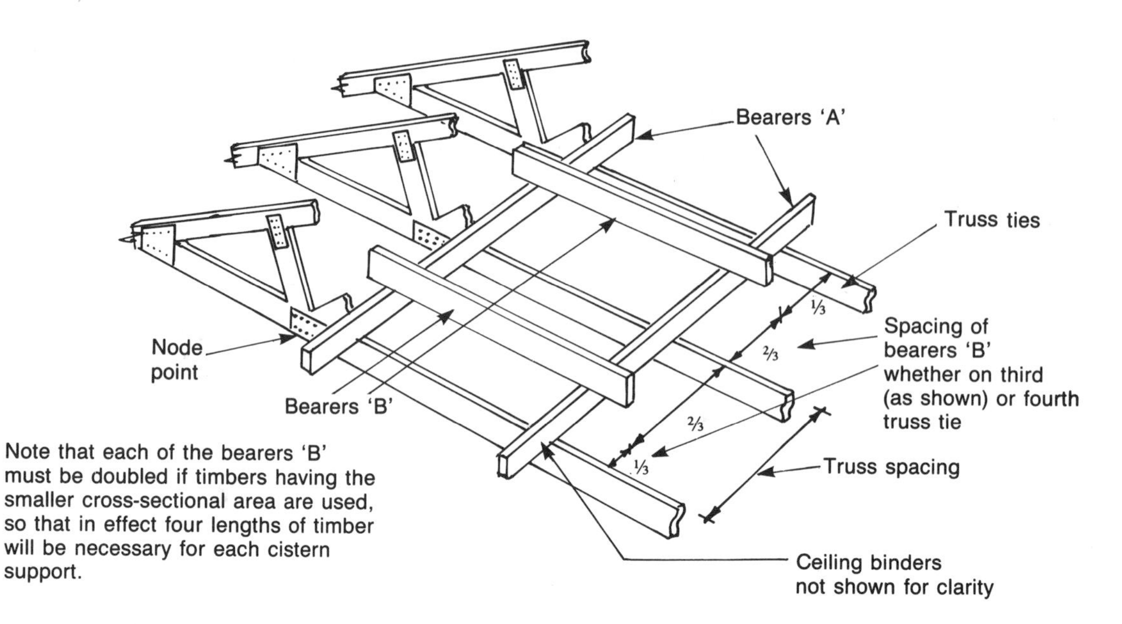

Size of each bearer 'B' relating to truss span and cistern capacity

Truss span (m)	*300 litres actual capacity*	*230 litres actual capacity*
6.5	Two* 38 × 100 mm or One 50 × 75 mm	One 50 × 100 mm
9.0	Two 38 × 100 mm or One 50 × 150 mm	Two 38 × 100 mm or One 50 × 125 mm
12.0	Two 38 × 150 mm	Two 38 × 125 mm or One 50 × 150 mm

*This indicates that two lengths of timber must be laid for *each* bearer

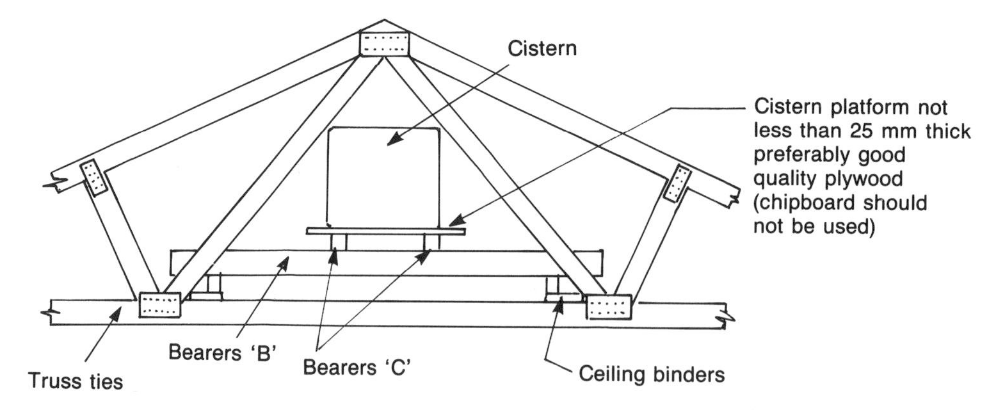

Fig. 5.45 Recommendations of Building Research Establishment for supporting storage cisterns on roof trusses. Cisterns of 300 litres actual capacity must be carried on not less than four trusses. Those of 230 litres may be supported on three as shown. Bearers 'A' must be fixed as close as possible to the 'node' points. Bearers 'A' and 'C' should be constructed of timber not less than 50 × 75 mm. The trussed rafter designer should be informed of the cistern capacity and its location in the roof space. (*Source*: BRE Defect Action Sheet DAS 44, 1984)

that it will soften and distort in damp conditions. Figure 5.45 illustrates the main points relating to cistern support on roof timbers or trusses.

On all but the smallest of these cisterns a metal or plastic reinforcing plate is supplied with the cistern to reinforce the area through which the float-operated valve passes giving the side of the cistern more rigidity against the upthrust of the float. Holes in these cisterns should be made with a proper cutter – those shown in Chapter 2, Fig. 2.2 are satisfactory for this purpose and ensure a hole of the correct size to accept a standard BSP thread. The practice of heating a piece of copper tube to melt out the hole is most unsatisfactory as it can give rise to subsequent cracking at its edges due to possible overheating. This has been known to cause complete failure of the cistern after a very short period of time. Care must also be taken to use a suitable packing material on any connection made in the cistern. Traditional materials such as paint, putty or paste jointing must not be used as the oil they contain has the effect of softening most thermoplastics and early failure of the material may occur with disastrous results. It has been suggested that polythene washers should be used but unless they are of a very good fit, a slight dampness or drip may be discernible around the joint when it is tested. Satisfactory joints can usually be obtained by the use of PTFE tape in conjunction with polythene washers or rubber seals which can be cut from suitable sheet.

Where holes are cut in galvanised cisterns, it is important to remove all the swarf as failure to do this can often result in serious corrosion of the cistern due to electrolytic action being set up. All cisterns should be tested before any holes are cut in them as manufacturers will not normally entertain claims made for defects after a cistern has been fitted.

Glass fibre reinforced plastic (GRP) is a thermosetting material and has been found satisfactory for the manufacture of cisterns. It has a greater degree of rigidity than those made of thermoplastics and is suitable for the construction of large cisterns that are built up by using bolted sections. The remarks relating to the support and fitting of thermoplastic cisterns apply equally to those constructed of GRP.

Figure 5.46 illustrates the requirement of the Water Bylaws 1986 regarding storage of water and it should be noted they are much more stringent than hitherto.

It is not uncommon in old installations to find the water polluted with dust from the atmosphere, flies and other small insects and even dead birds and vermin. In many cases in the past, bylaws regarding the use of a cistern cover have been ignored.

Suitable covers must now be used and in many cases they are bolted or screwed down on a rubber seal. Atmospheric pressure in the cistern is maintained by a vent in the cover which like the overflow pipe, must be screened. These fine mesh screens prevent access of insects to the stored water. They should be made of stainless steel or other equally suitable material. Building owners should be made aware that these screens should be cleaned regularly, as, if they become blocked, especially the overflow, serious flooding could take place. Figure 5.47 illustrates a cistern and its associated components which is available as a complete unit complying with the bylaws. It should be noted that any replacement of existing cisterns must comply with the new bylaws.

Access to float-operated valves The requirements of BS 6700 relating to working space above cisterns are shown in Fig. 5.48. Sufficient room must be allowed to enable maintenance of the valve, inspection, and means of cleaning to be carried out. The measurements shown are the minimum and should be increased where possible.

The term 'nominal capacity' of a cistern relates to the quantity of water it contains when it is full to the brim. When the float-operated valve and overflow are fitted, the amount of water is reduced to 'actual capacity'. To ensure the maximum actual capacity, the valve and overflow must be carefully positioned both in relation to each other and in relation to the top of the cistern (see Fig. 5.46). This ensures that the valve cannot be submerged if the cistern is overflowing, which would be a health hazard due to the possibility of back-siphonage.

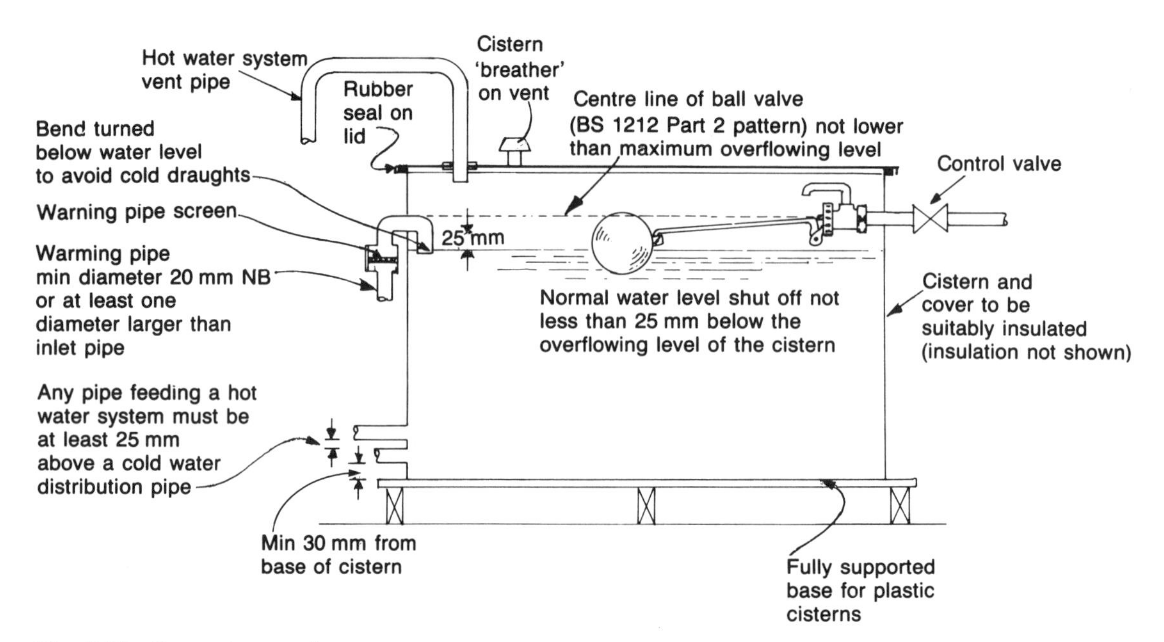

Fig. 5.46 Connections to cisterns

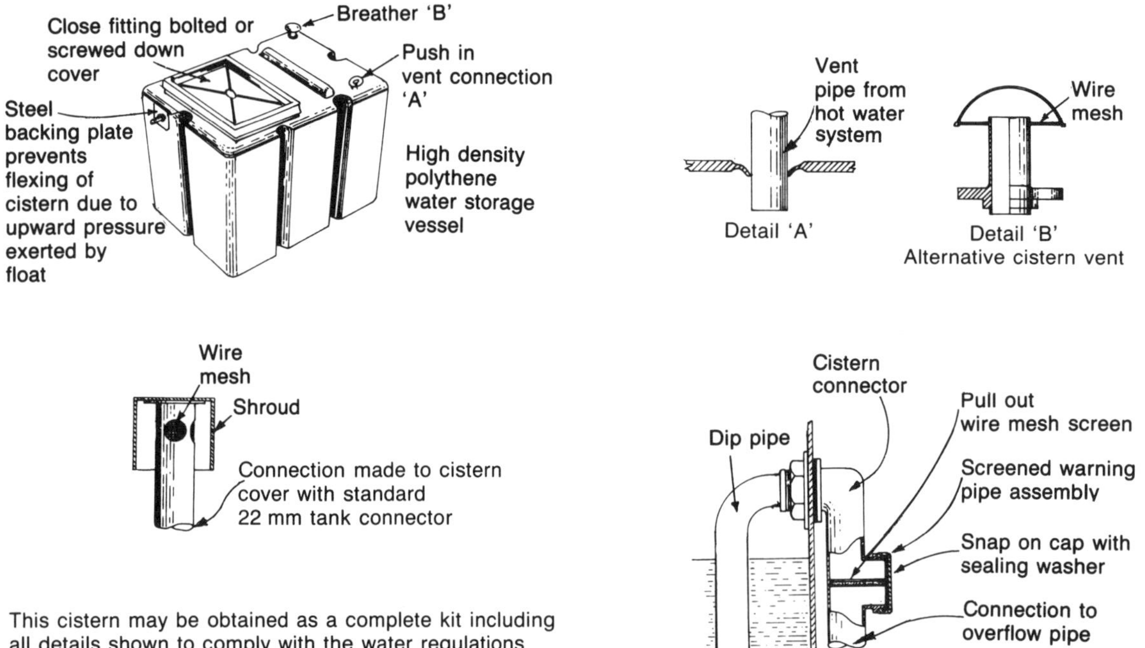

This cistern may be obtained as a complete kit including all details shown to comply with the water regulations requirements relating to screening with wire mesh any openings such as overflows through which vermin (insects etc) might obtain access to the stored water

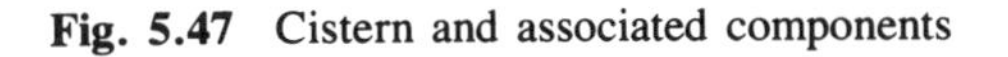
Fig. 5.47 Cistern and associated components

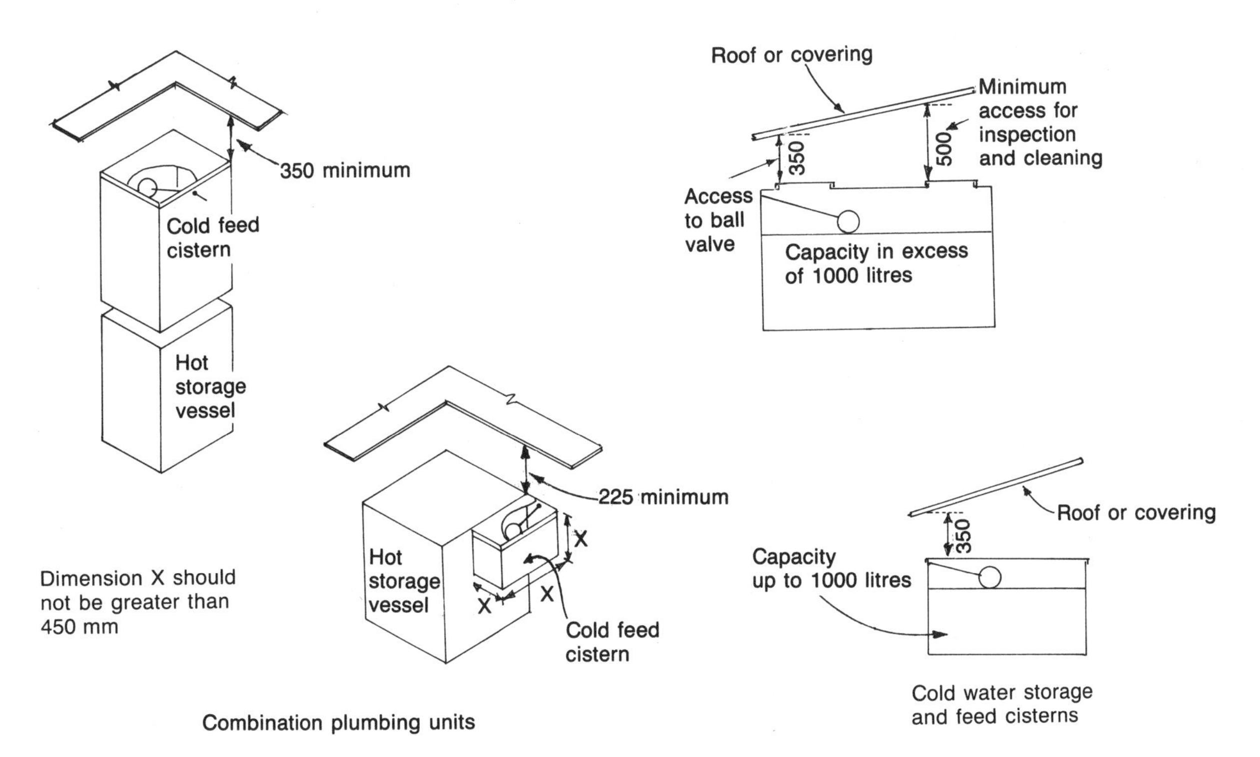

Fig. 5.48 Recommendations for providing access to ball valves in cisterns and combination units

Further reading

Much useful information can be obtained from the following sources, which are available from BRE Bookshop, Building Research Establishment, Garston, Watford, Herts, WD2 7JR.

BRS Digest 83: Plumbing with stainless steel.

Defect action sheets

DAS 43: Cistern Supports.

DAS 109: Hot and cold water systems, protection from frost.

DAS 120: Solid floors, water and heating pipes in screeds (design).

DAS 121: Solid floors, water and heating pipes in screeds (site).

Codes of Practice and British Standards

BS 6700: Specification for design, installation, testing and maintenance of services supplying water for domestic use within buildings and their curtilages.

BS 1010: Specifications for draw-off taps and stop valves for water services (screw-down pattern).

BS 1212:Parts 1, 2, 3: Float-operated valves.

Water storage vessels

Polytank Ltd, Naze Lane East, Freckleton, Preston, PR4 1UN.

Tanks and Drums Ltd, Bowling Iron Works, Bradford, BD4 8SX.

Water services

Edward Barber & Company Ltd, Paxton Road, London N17 OB3.

F.W. Talbot & Company Ltd, Winnall Valley Road, Winchester, Hants, SO23 8LL.

Wavin Plastics Ltd, Parsonage Way, Chippenham, Wiltshire, SN15 5PN.

Water meters

Kent Meter Ltd, Pondwick Road, Luton, Bedfordshire, LU1 3LJ.

Self-testing questions

1. State the requirements for the installation of a water meter in a cupboard under a sink unit.
2. Describe the effects of water containing dissolved carbon dioxide on chalk and limestone.
3. What will be the effect on a volume of water if its temperature is raised or lowered above or below 4 °C?
4. State the causes of burst pipes in frosty weather.
5. (a) List two types of hardness in water.
   (b) Describe the problems which can arise from using hard water in a hot water system.
6. Describe the dominant features of all screw-down taps.
7. State the minimum and maximum depths of underground water services.
8. Explain why it is advisable to fit lock-shield taps in public buildings.
9. State the purpose of 'breaking' a tap prior to fitting into a sanitary appliance.
10. Explain the action of an equilibrium float operated valve and state the circumstances in which you would recommend its use.
11. List three advantages of diaphragm type ball valves.
12. Explain the reason for a gooseneck connection on a water service pipe to the main ferrule.
13. Describe the type of fixing to be used for a steel pipe run at low level in a horizontal position in a hospital or a school.
14. State the minimum working space between the top of a cistern (not exceeding 1,000 litres) and the ceiling or underside of a roof.

6 The hot water supply

After reading this chapter the reader should be able to:

1. Identify the effects of heat and understand how heat quantity is expressed.
2. Demonstrate a knowledge of the forms of heat transmission and how they apply to hot water systems.
3. Calculate the boiler power necessary to heat a quantity of water through a given temperature rise.
4. Recognise possible causes of defects and faults in hot water systems and explain how they should be avoided or remedied.
5. Select materials suitable for boilers, hot water storage vessels and cisterns.
6. Recognise possible dangers in connection with hot water supplies.
7. Understand the working principles of hot water supplies.
8. Identify by name the components and pipes in hot water systems and describe their function.
9. State the purpose of secondary circulations, recognise where they are necessary, and describe methods of control and limiting heat losses.
10. Understand the need for effective insulation of hot water pipes and components.
11. Select and apply various types of insulation for a given purpose.

The introduction of hot water supply in domestic premises is a comparatively recent innovation. Prior to the period of the First World War 1914–1918, only the very wealthy were able to afford piped hot water in their homes, while one of the most attractive selling features of houses built for speculation during the period 1920–1930 was the introduction of proper bathrooms and hot water supply. One of the problems encountered with houses built during this period was the poor design of the schemes that were installed, using long exposed pipe runs often in uninsulated roof spaces resulting in frozen feed pipes and cisterns. In houses built since the 1945 period the subject of domestic hot water has been given greater thought, with the result that in many cases bathrooms are generally built adjacent to or over the kitchen, enabling sanitary fittings to be grouped together with shorter pipe runs. It has been realised too that if the hot storage vessel is situated in a position close to the boiler, the circulating pipes will be shorter with correspondingly less heat loss.

The design of hot water systems has also improved due to a better understanding of the basic physical principles involved and it is essential for the plumber to have a thorough grasp of these principles if he is to be able to install hot water systems efficiently.

Heat

It is difficult to describe heat. Like electricity it cannot be seen, although its effects can be appreciated. It can also be felt, as anyone will know who has accidentally touched, for example, a newly soldered joint. The simplest definition of heat is that it is a form of energy, usually created by chemical change or friction.

Temperature

The *intensity* of heat is termed temperature, and

can be measured with a variety of instruments. As the temperature of domestic hot water and heating systems seldom exceeds 85 °C, an ordinary thermometer is satisfactory for most purposes in connection with hot water supply.

One of the effects of heat is that as the temperature is increased almost all forms of matter, i.e. solids, liquids and gases, expand. This is a very important principle because it has many differing applications to the work and equipment used in the plumbing industry.

A simple example which proves the fact that liquids expand when heated is shown in Fig. 6.1. It also demonstrates the basic working principles of mercurial and alcohol thermometers. A flask is filled with a coloured, heat-sensitive liquid, usually alcohol, which expands rapidly on a rise in temperature. The flask is closed with a rubber bung through which is passed a capillary glass tube. When the hands are placed on the flask, heat from the body raises the temperature of the fluid, causing it to expand up the tube. So that the effects can be clearly seen, a piece of white card is placed behind the tube.

A thermometer works in exactly the same way, the liquid used being mercury or alcohol, both these materials being very sensitive to temperature change. A fixed scale is used in a thermometer, the boiling and freezing points of water being indicated as fixed points with suitable degree graduations between the two.

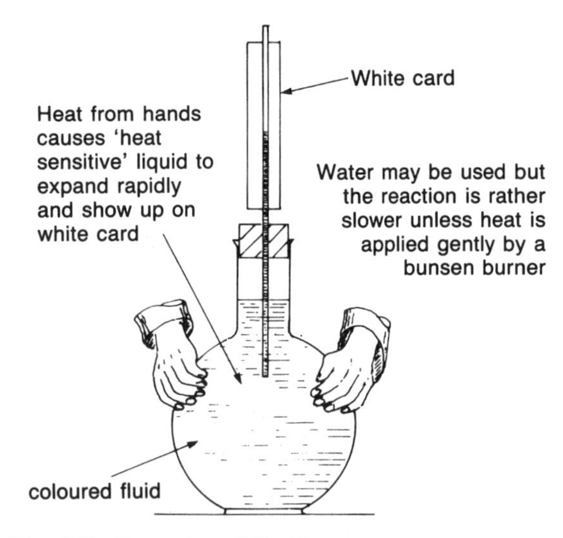

Fig. 6.1 Expansion of liquids

Long lengths of hot water pipe are affected to a considerable degree by increased temperature and the accompanying expansion so precautions must be taken to prevent the effects of the expansion causing a fracture or break-down in the joints and fittings. Additionally, one of the most common sources of noise in pipework systems is due to the expansion and contraction of hot water pipes laid in wooden joists or passing through floors without sufficient space for thermal movement.

Coefficient of linear expansion

If a piece of material 1 m long expands by 1 mm when its temperature is raised through 1 °C, it can be said to have expanded by one thousandth of its length. This number written as a decimal would read 0.001 and relates to the expansion of that material per degree Centigrade temperature rise. This fraction is referred to as the *coefficient of linear expansion*.

The foregoing has only been used as an example but reference to the materials listed in Table 6.1 will show the coefficients of linear expansion of many of the materials with which the plumber is commonly in contact.

To calculate the increase in length of a given piece of material due to thermal expansion, three facts must be known.

(a) The original or normal length of the material.
(b) The increase in temperature through which it is raised.
(c) The coefficient of expansion for that material.

Table 6.1 Coefficient of linear expansion.

Material	*Coefficient* $°C^{-1}$
Brass 66% copper; 34% zinc	0.0000189
Cast iron	0.0000102
Copper	0.0000167
Invar steel	0.0000009
Lead	0.0000291
Low carbon steel	0.000011
Polythene	0.00018
PVC	0.00005–0.00006
Stainless steel	0.00001
Zinc	0.0000258

For instance, if the temperature of a 6 m length of copper pipe used on a central heating system is raised through 70 °C, it will expand by approximately 7 mm. The calculation is made using the following simple formula.

$$\begin{aligned}&\text{Coefficient} \times \text{Original length of material}\\&\qquad \times \text{Temperature rise}\\&= \text{Increase in length}\\&= 0.0000167 \times 6\,\text{m} \times 70°\\&= 0.007014\,\text{m, or approximately 7 mm}\\&\quad \text{increase in length}\end{aligned}$$

It will be seen that the materials listed have widely differing rates of expansion. It is worth noting that invar steel has a very low increase in length when it is raised in temperature and for this reason it has wide applications in the field of thermostatic devices.

Expansion joints

Provision is made for the effects of expansion in pipework by using expansion joints or bends. Predetermined points on the pipe run are 'anchored' or firmly fixed so that any increase in length must be taken up by the expansion joints. A typical example of this is shown in Fig. 6.2 which illustrates a run of pipe between two rooms with an expansion loop formed in the dividing wall.

A typical method of anchoring pipes in ducts is shown in Fig. 6.3. It is important to realise that any pipe clip or support between the anchors must permit some degree of movement to the pipe. Two types of purpose made expansion joints are shown in Fig. 6.4.

To allow for continuity of movement, pipes should not be 'built in' to walls or solid floors, but sleeved as shown in Fig. 6.5. A short length of pipe of larger diameter is used as a sleeve, any gaps between it and the pipe passing through being sealed with heat-resistant rope or similar fireproof

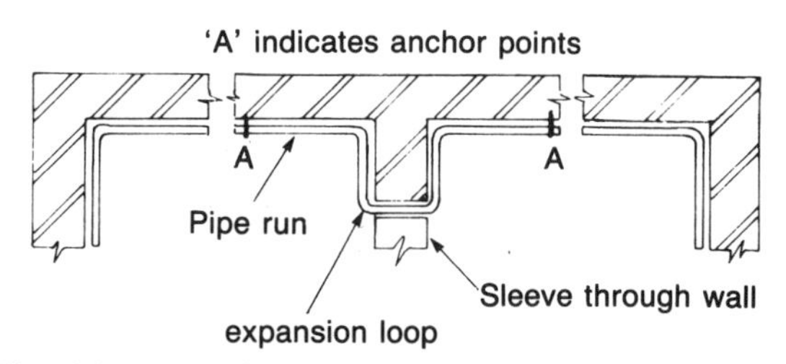

Fig. 6.2 Expansion loop

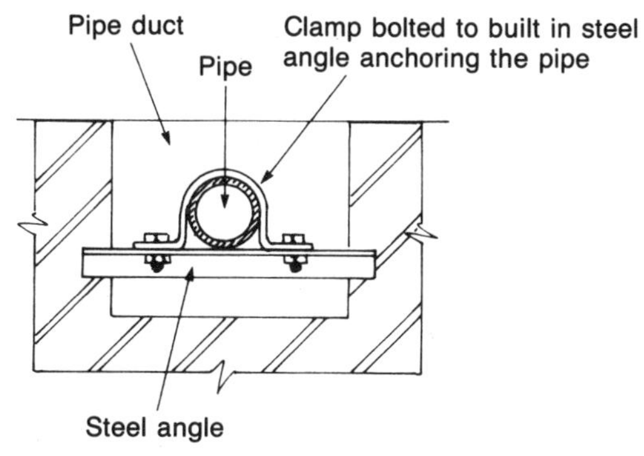

Fig. 6.3 Anchors

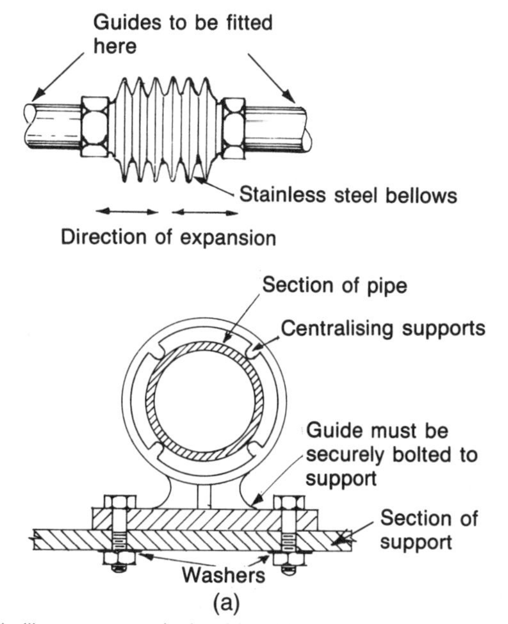

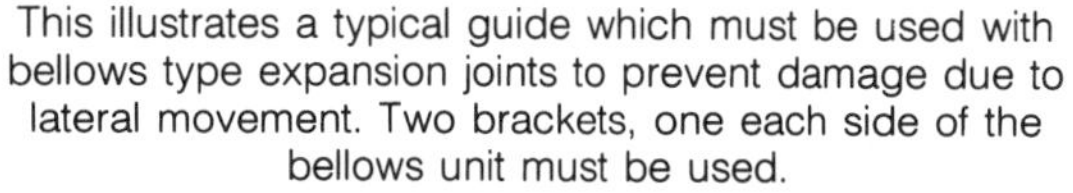
This illustrates a typical guide which must be used with bellows type expansion joints to prevent damage due to lateral movement. Two brackets, one each side of the bellows unit must be used.

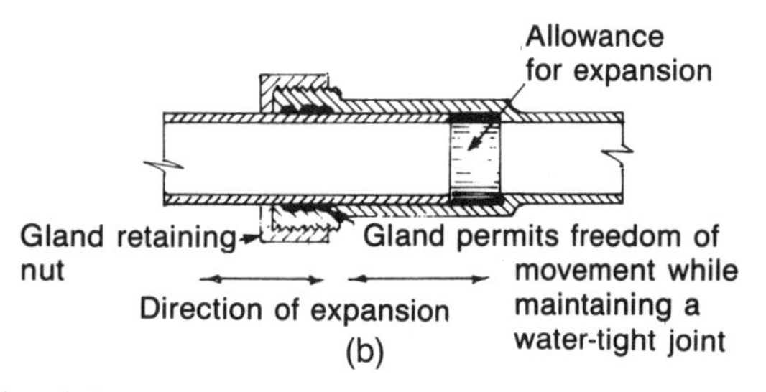

Fig. 6.4 Expansion joints

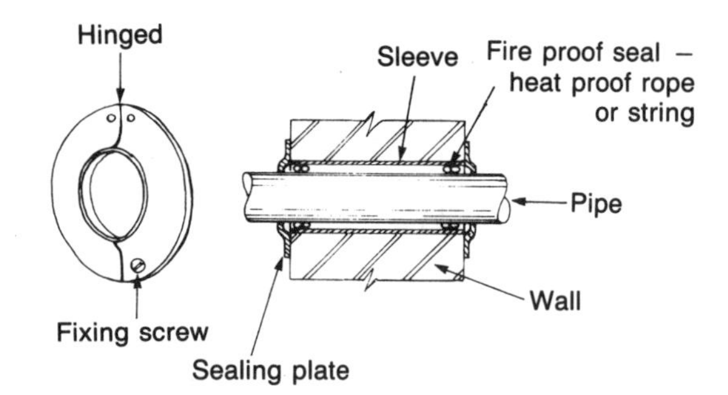

Fig. 6.5 Sleeving through walls to allow for expansion. Detail of sealing plate which provides a neat finish to sleeves in walls and floors

material. To provide a neat finish on the face of the wall or floor, a sealing plate is used, made of metal or plastic, usually in two halves to enable them to be fitted after the pipe is installed.

Quantity of heat

The difference between the intensity of heat (temperature) and its quantity is not always understood and it is often wrongly assumed that the higher the temperature of a substance the greater the quantity of heat it contains. This is not necessarily so as the volume of the heated substance must also be considered.

The basic metric unit of heat quantity is the *joule* (J). It is a very small unit, only 4.186 J (usually approximated to 4.2) being required to raise the temperature of one gram of water through 1 °C. A gram of water is also a very small quantity, some idea of its size will be apparent if a cube, each side measuring 10 mm, can be imagined – something a little smaller than a lump of sugar! To raise the temperature of 1,000 grams (one kilogram) of water through 50 °C, i.e. 42 × 1,000 × 50 would require 209,300 J. This is a very large number for a fairly small amount of heat, so to avoid ridiculously large numbers in calculations the usual practice is to work with *kilo*-joules and *kilo*grams:

1,000 grams = 1 kg
1,000 joules = 1 kJ

(Note: 1 litre of water is equivalent to 1 kg of water.)

The following simple calculation will show the difference between intensity of heat (temperature) and quantity. Assume two vessels are taken, one containing 1 litre of water which is heated from 10 °C to 60 °C (a 50 °C increase in temperature), the other containing 10 litres which is heated from 10 °C to 20 °C (only a 10 °C increase). The following calculations show that the former, although hotter at 60 °C, contains less heat than the 10 litres which is at 20 °C.

1 litre × 50 × 4.2 = 210 kJ heat content
10 litres × 10 × 4.2 = 420 kJ heat content

From this it will be seen that temperature alone does not indicate the quantity of heat in a substance, but its volume must also be taken into consideration.

Transfer of heat

Some knowledge of the ways in which heat is transferred is necessary to understand fully the working principles of heating and hot water systems. There are three methods of heat transfer: conduction, convection and radiation. Each of these and its applications will be discussed separately.

Conduction

If a soldered capillary fitting is made on a piece of copper tube, it will be noticed that although only the area adjacent to the joint is heated, the tube will be hot some distance on either side of the joint, the heat having travelled along the pipe. This heat transference is known as conduction and it occurs in all solids, liquids and gases, the two latter generally being poor conductors of heat. Conduction takes place due to the increased vibration of the molecules of the material when it is heated. The vibrations are passed on to the adjacent molecules, those which are actually subjected to the source of heat being the most active, and gradually diminish as they move further along the material. This rate or speed of conductivity varies with different materials. Most metals are good conductors of heat while plastic, ceramics and timber have a poor conductivity rate, these latter often being the basic materials from which thermal insulation is manufactured. Where it is necessary to reduce heat losses from pipes and vessels made of metal, they should be insulated with a poor conductor such as fibre glass, synthetic foam rubber or a PVC sleeve.

Metals themselves have differing rates of thermal conductivity, and a simple experiment as shown in Fig. 6.6 gives a good guide to the conductivity rates of some of the more common plumbing materials. It consists of an open topped container into which are fixed a series of rods of differing materials. The upper end of these rods are inside the vessel while the major length of the rods are outside. Loose fitting clips are placed over the rods all at the same level, and to prevent these from falling by their own weight, each rod is coated with wax – candle wax is suitable for this purpose. The container is then filled with hot water which heats the upper end of the rods. As heat is conducted through the rods, so the wax is melted, causing the clips to slide slowly downwards. Assuming one of the rods is made of copper, it would be seen that it has a high conductivity rate, as the clip secured to it would fall over a longer distance in comparison to the clips on the rods made of iron, lead or plastic. The distance through which the clips fall gives a general idea of the thermal conductivity of a particular material. There are, of course, precise figures for the thermal conductivity values of different materials.

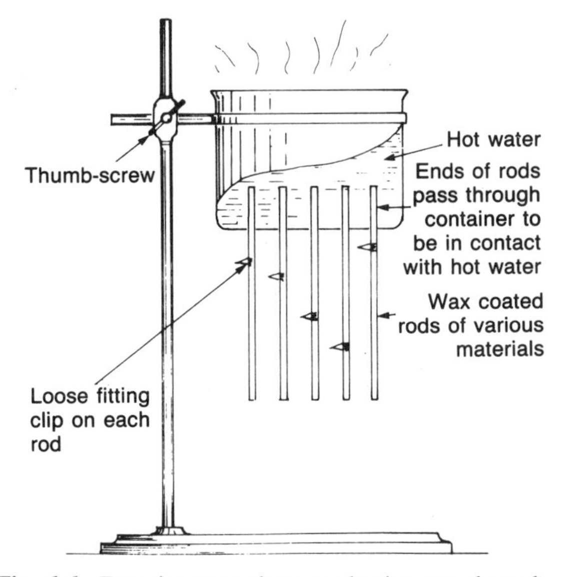

Fig. 6.6 Experiment to show conduction rate through various solids

A practical application of conductivity is the use of the copper coil or annulus in an indirect cylinder, as copper has a higher thermal conductivity rate than, for example, steel.

In years gone by boilers were often made of copper, not only because it has a high conductivity rate, but due also to its resistance to corrosion, especially in soft water areas. However, due to the high cost of copper and the fact that corrosion problems can be overcome by using indirect systems of hot water supply, iron or steel is now used almost exclusively as a boiler material. A hot water storage vessel made of copper needs to be well insulated otherwise the high conduction rate of the copper would result in considerable heat loss.

Convection

This form of heat transference occurs due to the physical movement of the molecules within a material. As fluids and gases are the only substances which are able to move freely at ordinary temperatures, this form of heat transference takes place only in these types of materials.

Liquids and gases expand when heated, and the molecules nearest the source of heat expand more quickly than those further away. On expansion they become lighter or less dense and are consequently pushed upward by the colder and heavier molecules surrounding them, these sinking to lower levels thereby creating circulation of the liquid. A simple illustration of what happens when a pot or kettle containing water is heated is shown in Fig. 6.7. The arrows indicate the upward flow of molecules as the water is heated. This movement is maintained throughout the period of heating, the water becoming progressively hotter.

The effect of heat on gases is precisely the same, illustrated in Fig. 6.8. This shows the effect of heating air in a sealed glass fronted box. The heat emitted by the electric lamp heats the air which surrounds it. The heated air rises and passes out of the box through chimney 'B', causing

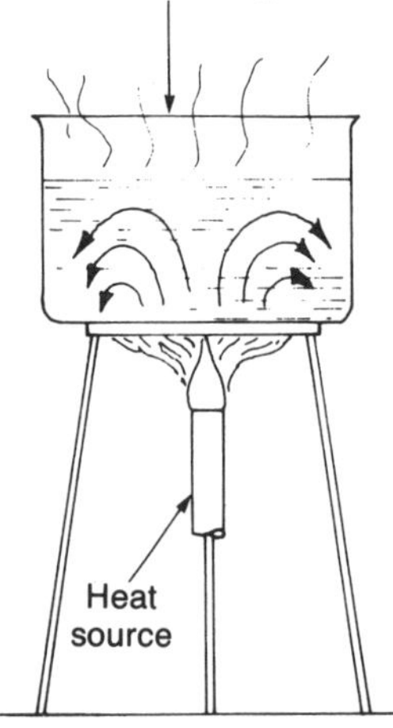

Fig. 6.7 Convection currents in fluids. Heated molecules rise and cooler molecules sink to lower level in vessel

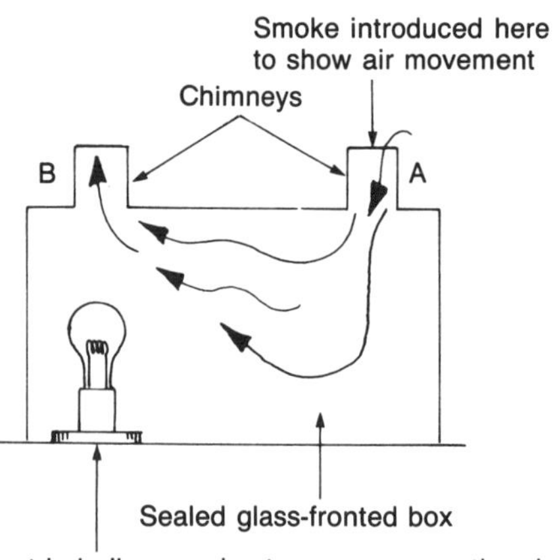

Fig. 6.8 Convection of gases

further air to be drawn into the box via chimney 'A'. The air movement will be seen very clearly if a little smoke is introduced to chimney 'A'. This experiment illustrates how the air in a room with an open fireplace will be heated by the fire and pass up the chimney, and how heat from a radiator causes the circulation of warm air in a room. Figure 6.9 shows how the heated air from the radiator may well be drawn up the chimney and so wasted. It may also be noticed that the wall above a radiator is darker than the surrounding areas.

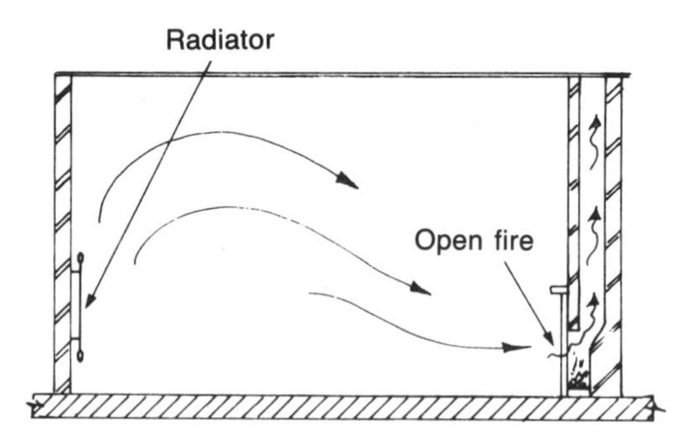

Fig. 6.9 Loss of warm air in room via open fireplace. The air warmed by the radiator may be drawn up the flue by the updraught of the open fire. This could result in a serious loss of heat

This is caused by dust particles adhering to the wall as the air, due to convention, moves upward. This disfiguration of the decorations is called pattern staining, and when a radiator is fitted on a wall as illustrated, it should have a shelf fixed above it to deflect the warm air into the room.

Convection is the principle which causes water, when heated in a boiler, to flow upward into the hot storage vessels. The same movement of water occurs as was shown in Fig. 6.7, but in this case the top of the vessel is closed and the hot water rises upward through the flow pipe being replaced by colder and heavier water moving downward in the return (see Fig. 6.10). These two pipes are called the primary circulators.

Water has a unique property in that when it falls below or is raised above the temperature of 4 °C, it expands. It therefore occupies the least space (i.e. is at maximum density) at 4 °C. A rapid and considerable expansion takes place when the temperature is lowered, but expansion does not take place so quickly or by so much when the temperature is raised. In fact, water only expands by approximately 1/24th of its volume through a range of 4 °C to 100 °C, (shown pictorially in Fig. 6.11). (The fraction is often changed to 1/25 for easy calculation purposes.) This very small amount of expansion means that the hotter water has less mass (i.e. is lighter) than the cooler water which creates the pressure head and causes circulation of the water by gravity. This is called the 'circulating pressure'. To express this in another way, the

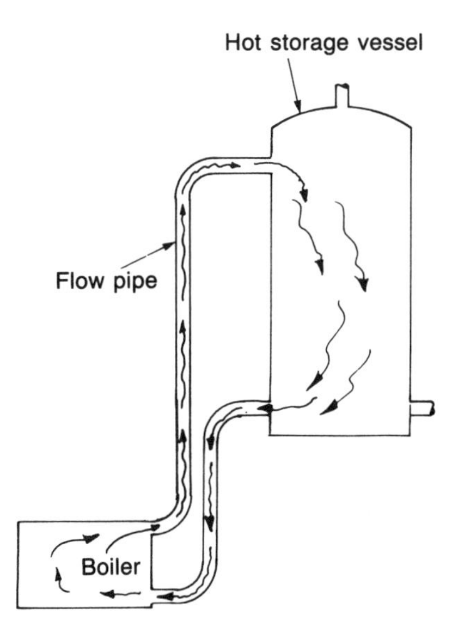

Fig. 6.10 Primary circulation in domestic hot water systems. The circulation of water between the boiler and hot storage vessel is due to convection currents

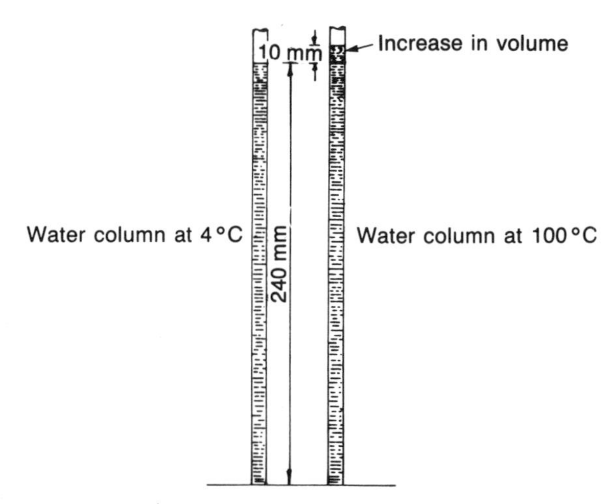

Fig. 6.11 Difference in height of water columns at 4 °C and 100 °C

mass of 1 m^3 of water at 4 °C is 1,000 kg while at a temperature of 100 °C the same quantity has a mass of only 958.5 kg.

In the average hot water scheme the difference in densities is much less. Normally water enters a system at approximately 10 °C having a density of 999.7 kg and assuming its temperature is raised to 70 °C, its density is then reduced to 977.7 kg. The difference in mass is only 22 kg at these widely varying temperatures, and as the temperature difference becomes progressively less, so the available circulating pressure head that causes circulation by gravity is reduced even more. For this reason the primary circulation pipes must assist the circulation as much as possible by being of adequate size, as short as possible, and any changes of direction made with easy bends.

It is often thought that the pressure exerted by the head of water in the feed cistern causes the circulation of water in the primary flow and return. This is quite wrong as the pressure head exerted from this source is the same on both pipes.

Radiation

Heat transfer by both conduction and convection requires a medium through which to travel; radiant heat requires none. It may be defined as a form of energy which travels in straight lines through both air and space. The heat from the sun is a good example of radiant heat as it travels through millions of miles of space to reach the earth. It will pass through the air without appreciably warming it, but any solid object obstructing the rays will become warmed by them. Dark matt surfaces tend to absorb more radiant heat than those of a light colour, which reflect the heat. For this reason people living in hot climates often wear white clothing and live in buildings which have a light-coloured exterior.

One of the most promising new developments in the field of energy conservation is the possibility of using heat from the sun's rays for water heating. Collecting plates consisting of a series of pipes fixed upon a dark heat absorbent surface are situated in a position, usually a roof, where they can receive the maximum exposure to the sun's rays. The dark surface absorbs the heat from the sun's rays and transmits it by conduction through the pipes to the water they contain, the heated water then being pumped to a suitable heat exchanger. Heat transmitted in this way is used to augment that provided by traditional heating appliances. Although solar heating is at present hardly a commercially viable proposition, the continuing shortage of fossil fuels and their alarming increase in cost in recent years has

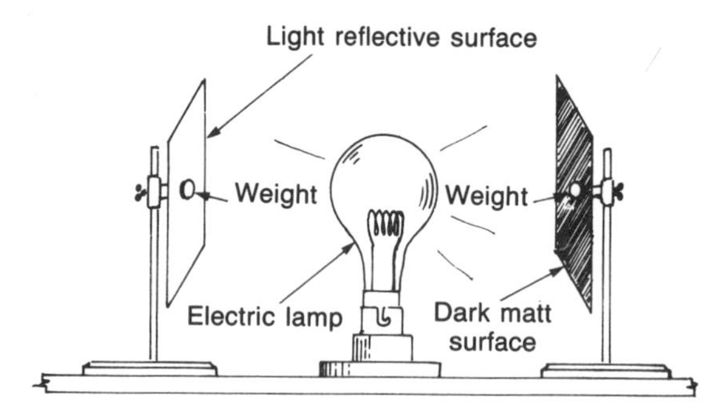

Fig. 6.12 Effect of radiant heat on light reflective and dark matt surfaces

rapidly narrowed the gap between what is, and what is not, an economic proposition.

A simple experiment to illustrate the effect of radiant heat on differing surfaces in shown in Fig. 6.12. Two pieces of sheet metal of the same size are taken, one being highly polished, the other having a dull matt, preferably black, surface. Into the centre of each plate is secured, by wax, two small pieces of lead, both of the same weight. An electric lamp is used as a heat source and is placed exactly between the two metal plates. The wax securing the weight to the matt surface will quickly melt, releasing its lead weight, clearly indicating that this surface has absorbed more heat. Bright surfaces reflect the rays of heat and for this reason do not absorb it so quickly, a well-known principle used in many forms of insulation.

To reduce heat losses in buildings by radiation plasterboard is obtainable with one surface covered with aluminium foil and, when used with stud work partitioning, radiant heat loss through this material is very low. A futher lining of mineral based felt backing to insulate against conducted heat will result in even further savings. The wall surface behind radiators is sometimes covered with aluminium foil. While this may be effective in preventing radiant heat loss, its use in preventing conduction is limited.

Some heating appliances, especially gas and electric fires, convert the heat generated by a flame or hot element to radiant heat. In the case of electric fires, the heat generated by the glowing element radiates on to a bright shining surface and is reflected out into the room. In the case of a gas

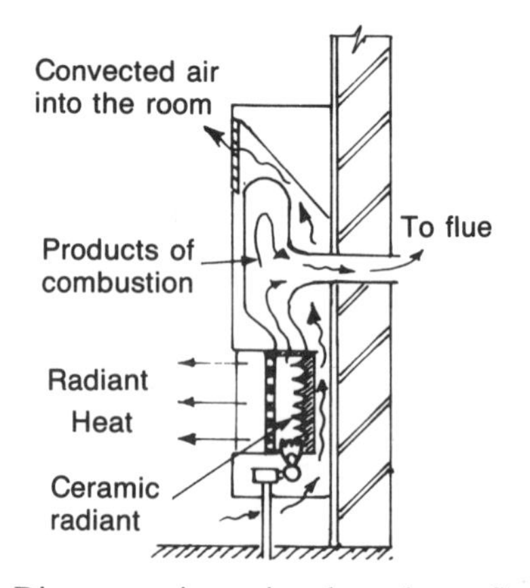

Fig. 6.13 Diagrammatic section through gas fire convector. Modern gas fires are very efficient, providing heat by both convection and radiation.

fire the ceramic radiants are made having many 'nodules' which increase the surface area when they are heated by the flame and become almost white hot. Heat from these radiants is directed into the room by radiation in the same way as that of an electric fire (see Fig. 6.13). All modern gas fires are designed in such a way that they produce convected heat as well. However, it is an interesting fact that the heating equipment we call radiators in fact impart about 90 per cent of their heat into a room by convection and only 10 per cent by radiation (see Fig. 6.14).

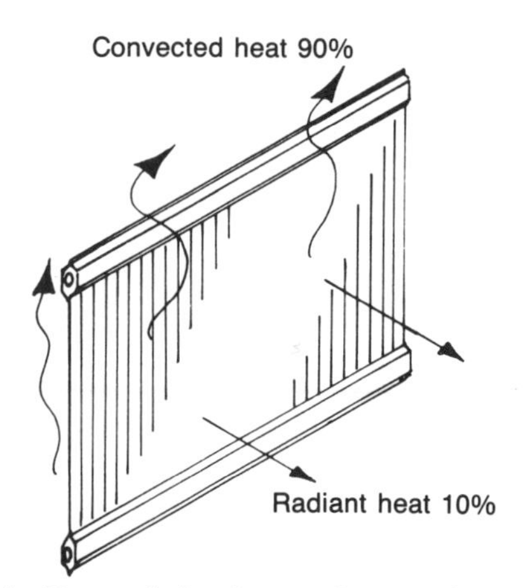

Fig. 6.14 Heat emission from radiators. The name radiator is misleading as heat is provided mainly by convection creating circulating currents of warmed air.

Simple domestic hot water systems

Having obtained some idea of the basic principle of hot water supply, some consideration can now be given to the component parts of the schemes used.

Hot water systems as we know them are the result of many years of careful thought and in some cases, trial and error. Until the last century the only form of hot water supply in many homes was a pot filled with water suspended over an open fire. When cast iron cooking ranges became popular, some enterprising manufacturer fitted a water container with a loose fitting cover on one side of the fire. The cover was removed to top up the container after the heated water had been drawn off by the tap in the base of the container. It was found that the production of hot water sometimes exceeded the requirements of the moment but often fell woefully short so the possibilities of hot water storage were explored, coupled with the need for a supply of hot water directly over the sink. A scheme known as the 'tank system' was eventually evolved, similar in some ways to modern systems, the main difference being that all the hot draw-offs were situated on the flow pipe. This stemmed from the thought that any draw-off should be taken from the hottest part of the system, i.e. near the boiler. Unfortunately what in fact happened was that when the supply of water in the boiler was exhausted, cold water in the return pipe became mixed with that drawn off from the flow. The result was that after a small amount of very hot water was drawn off, the remainder was only lukewarm.

Surprisingly this inefficient system with all its defects survived for many years, only gradually being replaced by that used today known as the cylinder system to identify it from the tank system. The cylinder system owes its name to the cylindrical hot water storage vessel generally used, its main feature being that hot water is drawn from the top of the storage vessel instead of the primary circulation, thus ensuring that water at the maximum temperature is drawn off.

Boilers

The heart of any hot water system is the boiler, the design of which depends largely on the type of fuel used. Until the period 1950–60, hot water supply in most small dwellings was derived from a small back boiler or an independent boiler, suitable for domestic hot water supply only, and heated by solid fuel.

In the early 1950s cheap oil became a popular fuel, and shortly afterwards cheap gas, both of which brought about a substantial change in home heating. As a result of this and the development of small bore heating, many homes are now fully or partially centrally heated, and the small boiler capable of heating hot water for domestic use only is becoming very rare.

For the purpose of this book only the domestic types of solid fuel boilers will be discussed. Unlike gas or oil appliances, they are not on/off boilers and, as such, a gravity circulation of water between the boiler and hot store vessel is always recommended to allow for the dissipation of hot water should it become overheated for some reason. This is sometimes called a 'heat leak'.

Back boilers The term back boiler is given to those boilers situated behind a fireplace and built into a chimney opening. The original back boilers were made to supply domestic hot water only, a typical example being shown in Fig. 6.15(a). A similar type of boiler is used in a variety of cooking ranges (Fig. 6.15(b)). Due to the demand for domestic central heating the type of solid fuel boiler used for this purpose has changed considerably. The independent (free-standing) boilers which were once very popular, have been superseded by back boilers having very high outputs, this being achieved by increasing their heating surface.

For example, the area in a fireplace once occupied by fire bricks is now part of the boiler waterway. Two types of high-output boilers are illustrated in Figs 6.15(c) and (d), and it will be seen that in comparison to the small boilers previously discussed, the area of boiler exposed to the heat source is considerably increased. The result has been that these boilers are capable of producing sufficient hot water for both domestic and space heating in the average domestic dwelling. It should be noted that both of these boilers operate with a traditional open fire.

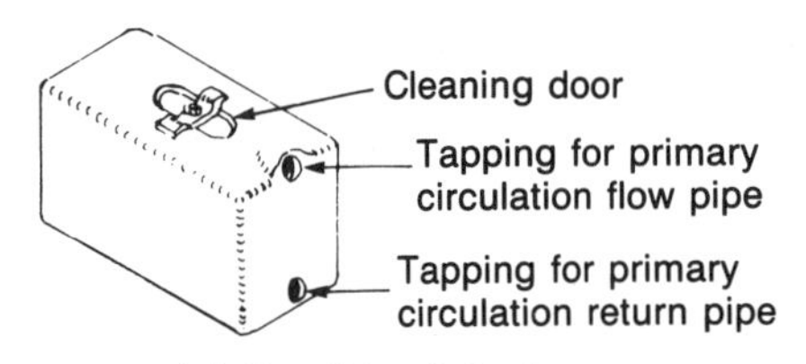

(a) Small back boiler
These boilers are sited in a cast iron frame which forms the boiler flue. They are suitable for domestic hot water supply only.

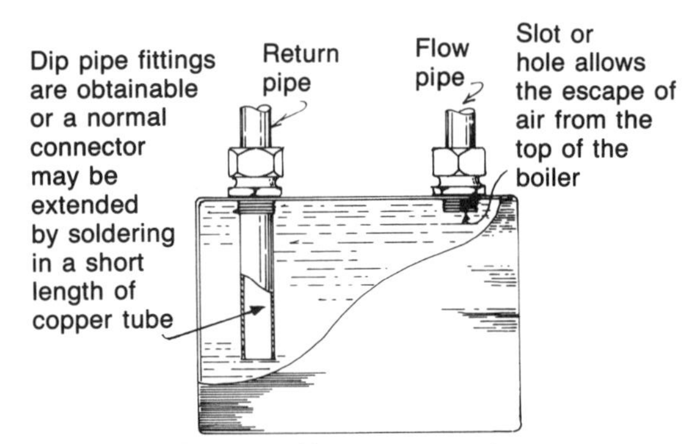

(b) Boilers with top connections
Some boilers fitted in cooking stoves have top connections. The return pipe must be extended into the boiler to ensure positive circulation

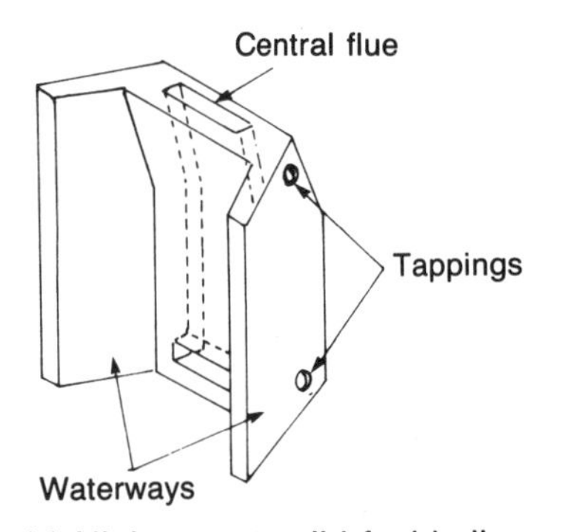

(c) High output solid fuel boiler
The waterways in the cheeks of the boiler and those surrounding the flue increase the area in contact with the fire.

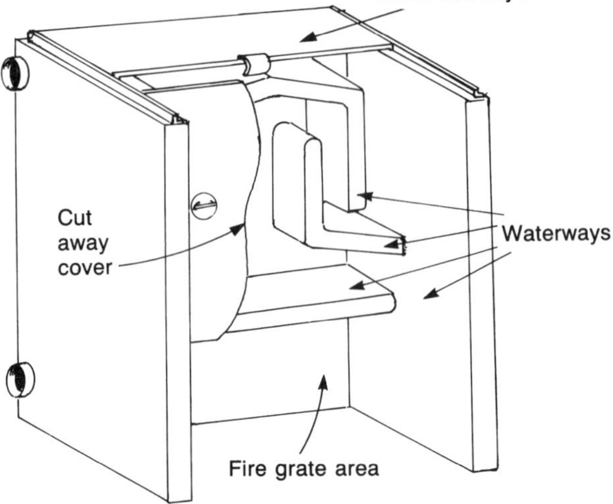

(d)(i) Cut-away view
One of the two plates which cover the flueways is shown cut-away. The periodic removal of these plates enables the flueways to be cleaned

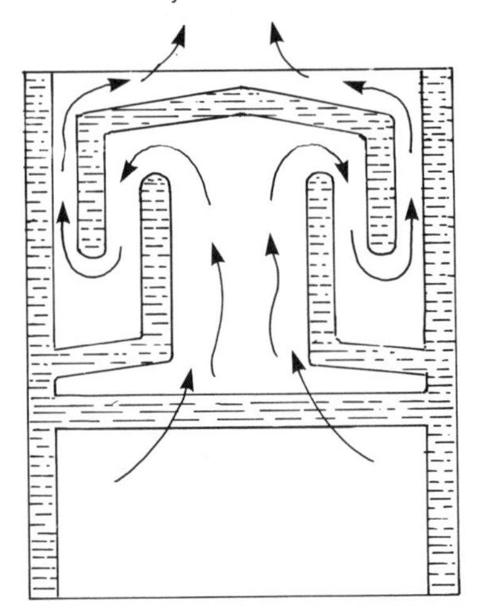

(d)(ii) Sectional view
From front of boiler – with the flueway covers removed the path of the hot flue gases is shown

Fig. 6.15 Back boilers

So that the range can be built into a limited sized opening and to provide access for making the flow and return connections, such boilers often have top tappings. Figure 6.15(b) illustrates such a boiler showing the necessity of fitting a dip pipe in the return to ensure a good circulation and a slot or hole in the flow connection so that any air that may collect in the top of the boiler can escape through the flow pipe.

Convector fires are another alternative using solid fuel and are very efficient, as being totally enclosed, the burning rate of the fuel can be controlled effectively. It should be noted one of the problems with open fires is that, due to the convective effect of the flue, warm air from the room is often drawn up the chimney with consequent loss of efficiency. These sealed convector heaters largely overcome this problem. Figure 6.16(a) and (b) illustrates a typical example of an installation of this type of heater. It will be seen that the boiler and fire unit, made of iron castings, is situated in a sealed concrete container. Air is drawn from the room under the front casing where it picks up heat from the exterior of the boiler and fire unit before being discharged through the louvres at the top of the casing. This convected heat is normally sufficient to heat the room in which it is situated, indeed, if the room is very small the doors may have to be left open to dissipate the warm air. The boilers used with this type of appliance have varying outputs, the larger types, like those used with open fires, being capable of whole house heating. The burning rate of the fuel is controlled thermostatically and is shown in Fig. 6.17.

As with all modern heating equipment it is essential they are installed to the manufacturer's specification. If these instructions are not complied with the result will almost certainly be an inefficient installation and a dissatisfied customer. One of the most common complaints with concrete units is that, due to the lack of care during installation, the air seals are not properly made and warm air escapes up the chimney resulting in a cold room. It is also important that the correct type of fuel is used, as recommended by the manufacturer.

Controls for solid fuel boilers The two main types of control used by most small back boilers are:

(a) a sliding damper
(b) thermostatic control

Some boilers are provided with both. The sliding damper is made in such a way that it causes the heated flue gases to either pass around the boiler or directly into the flue. A simple non-electric thermostatic device, which controls the supply of air for combustion, is shown in Fig. 6.17 and is suitable for all small solid fuel appliances except for open fires. It operates by reducing the air (i.e. the oxygen) supply needed to burn the fuel. The thermostat consists of a loosely fitted metal disc on a flexible steel arm, pivoted at its upper end allowing it to swing inward so that the disc covers the air inlet. A phial containing a heat sensitive gas is situated in a pocket in the waterway of the boiler and is connected to the bellows unit by a small-bore copper tube. As the water temperature increases, the gas in the phial expands and opens the bellows which in turn exerts pressure on the flexible arm closing the air inlet. As the air inlet closes so the fuel has a reduced air supply, less heat is generated and consequently less hot water is produced. Variations of temperature are achieved by the control which lengthens or shortens the bellows by means of a screw thread.

Independent boilers An independent boiler is one which is entirely free-standing, requiring only a flue. This term applies to boilers using all types of fuel. The smaller solid fuel independent boilers, once very common, have been superseded by the larger back boilers previously described. There are, however, installations which require greater heat output and the most effective method using solid fuel in such cases is the use of hopper-fed independent boilers.

Hopper-fed boilers The main feature of these boilers is they have an integral fuel store which avoids the necessity of stoking the boiler every few hours. They are designed to burn small, graded fuel known as anthracite peas. Fuel of this size will flow from the hopper by gravity on to the firebed. An electric blower, which is controlled by a thermostat, provides air for combustion. A typical boiler of this type is illustrated in Fig. 6.18. They are very efficient due to their design, the quality of the fuel they burn, and the forced draught which provides sufficient oxygen for complete combustion of the fuel.

Boiler sizes are specified by their rated output in watts per hour, this being based on the quantity of heat that will be passed through the boiler plates to the water. Solid fuel boilers can be made to work harder and produce more hot water than their rated output, but this leads to waste of fuel and damage

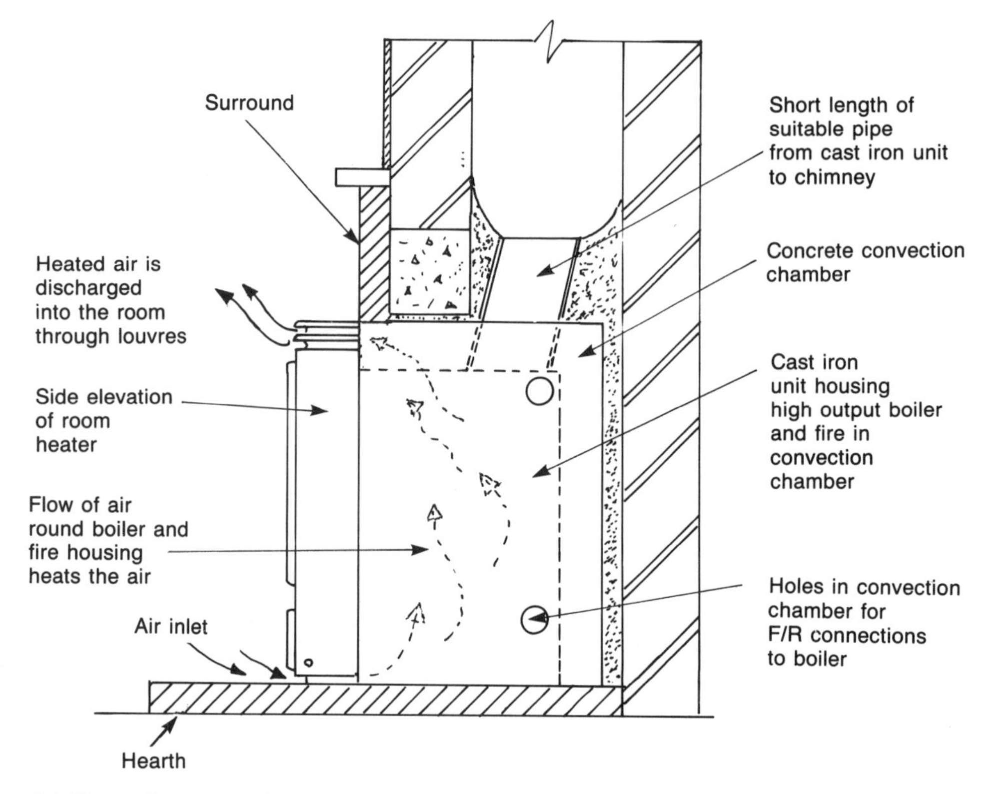

(a) Illustrating general principles of solid fuel convector heaters with high output boilers

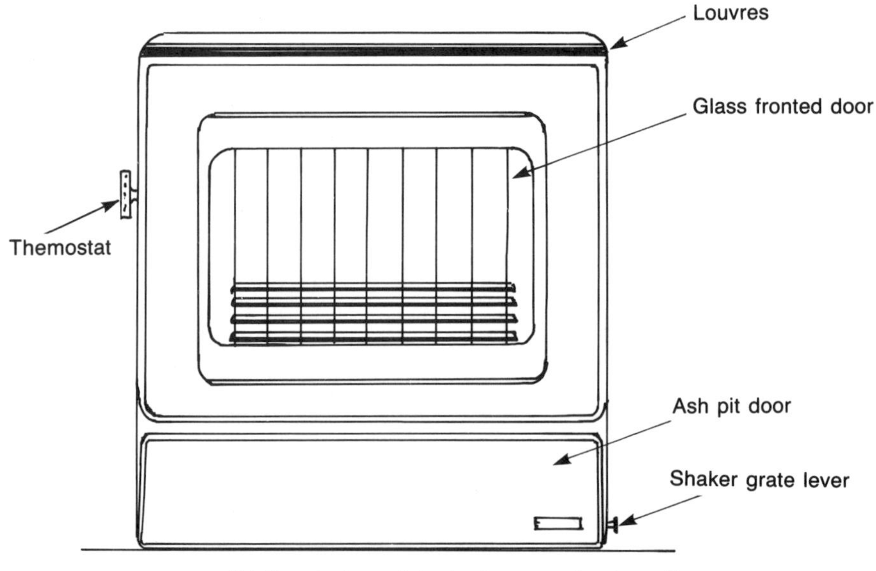

(b) Front view of typical convector type fire

Fig. 6.16 Convector heater

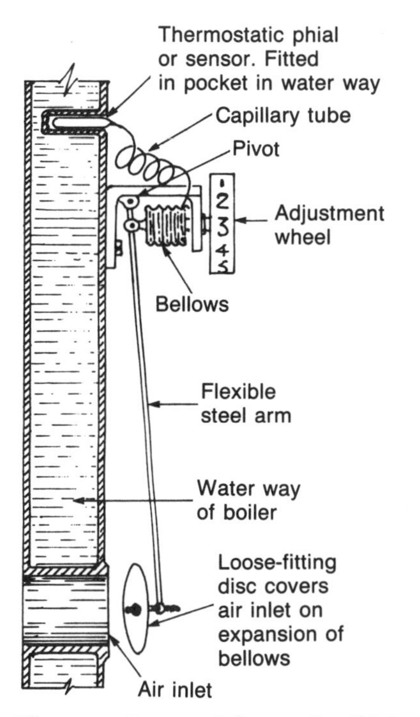

Fig. 6.17 Thermostatic control for small solid fuel boilers and room heaters

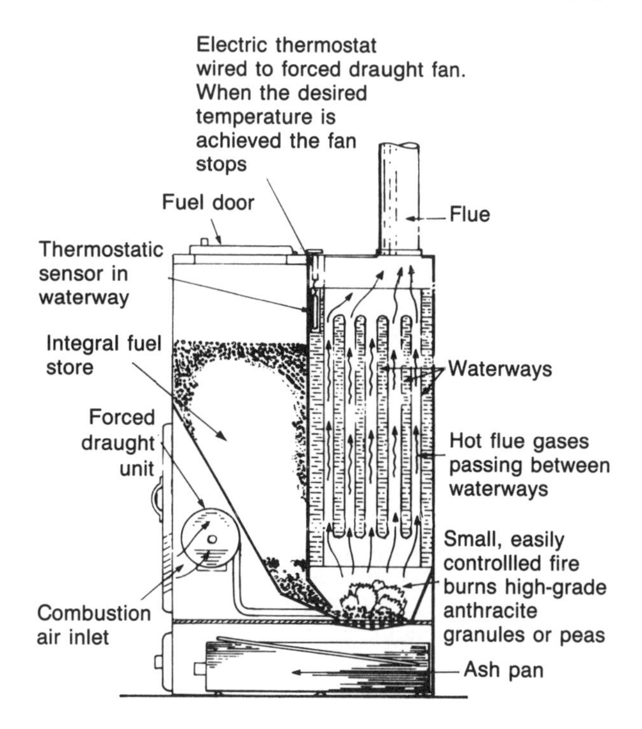

Fig. 6.18 Automatic hopper-fed solid fuel boiler

to firebricks and enamelled surfaces. Most boiler manufacturers are aware of this fact and base the output on standard laboratory tests.

For hot water supply in domestic premises having a storage capacity of approximately 120 litres, it is usual to allow 3 kW boiler power (i.e. 3,000 watts per hour). If the scheme is combined with space heating a suitable allowance must be added.

It is common practice to oversize solid fuel boilers by about 20 per cent to avoid the necessity of attending to the fire every two or three hours, as having a larger fuel capacity, the air inlet can be adjusted so the fire will keep in for a much longer period. This also makes provision for very cold periods when the air temperature may fall below 0 °C for short periods of time. For the same reason it is advisable to increase the power of both gas and oil boilers by approximately 10 per cent.

Oil- and gas-fired boilers are generally more efficient than those of solid fuel due to the relative purity of the fuel they burn. A certain amount of deposit from the products of combustion is left on the boiler flue ways but provided it is removed when the boiler is serviced, little loss of efficiency will result. These two fuels also have the advantage in that they can be controlled electrically and can therefore be operated automatically to a greater extent than solid fuels. They do, however, require much more specialist attention and unlike solid fuel boilers involve the householder in an annual maintenance charge.

Oil- and gas-fired boilers are both dealt with more fully in Volume 2 of this series.

Oil-fired boilers The actual boiler construction used with atomising burners is very similar to that used for gas boilers. Figure 6.19 illustrates the flueways of a typical cast iron boiler, which, like many gas boilers, are constructed of a series of sections screwed together with right and left hand nipple or tie bolts. Welded LCS plates are also used for the construction of oil-fired boilers. Modern boilers using oil as a fuel employ atomising burners for firing and, like gas, can be adapted for balanced flue operation.

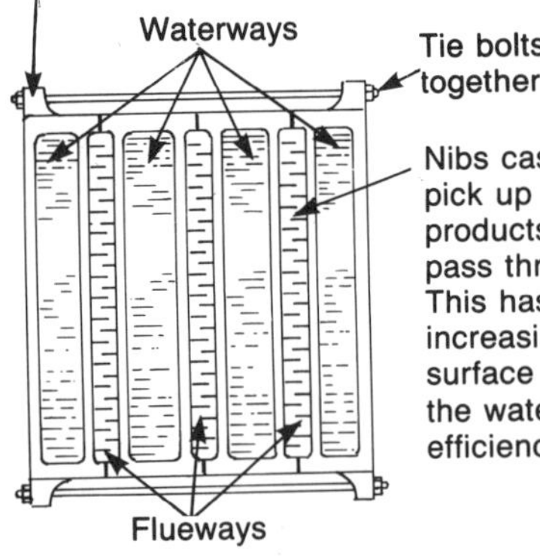

Fig. 6.19 Plan view of section of gas- or oil-fired boilers showing the flueways. Because these fuels are cleaner the flueways are much smaller than those of solid fuel appliances. It illustrates the need to ensure that fuel/air ratios are carefully adjusted as failure to do so will cause the rapid formation of carbon deposits blocking the flueways.

Gas-fired boilers Gas is a very popular and versatile fuel and can be burned in a variety of boilers all of which embody the same basic principles. The original gas boilers were of the free-standing independent type and very early models required only a supply of gas, all the controls being operated by the gas supply itself, with the exception of the clockwork timer switch which was wound manually. With the advent of electrical control this type of timing device was quickly replaced by those operated electrically, and electrical thermostats and solenoid valves have replaced the older gas thermostats and relay valves to control the burning operation. Gas boilers have also been developed to be fitted behind a fireplace on the front of which is mounted a gas fire that heats the room in which it is situated, normally making the installation of other space-heating appliances unnecessary. Wall hung boilers are now very popular due to their advantage in saving floor space and being suitable for fitting almost anywhere – in a roof space if necessary – providing access for servicing is available.

Condensing gas boilers These boilers have been developed to increase the efficiency of gas water-heating appliances. They are likely to become increasingly used in the near future especially for the replacement of older boilers of the open type.

Many modern gas appliances do not now require a chimney, a distinct advantage as building costs can be reduced. These appliances are fitted with a balanced flue and are known as 'room sealed'. They do not take air for combustion from the room in which they are situated, both the air inlet and outlet for the combustion products being situated in a special grille which passes through the wall to the outside. The fact that air for combustion is not taken from the room obviates the cold draughts at low level that are often noticeable with conventional appliances of all types.

Competency in gas work It should be noted that only plumbers and fitters who are qualified in the installation and servicing of gas appliances are allowed to carry out gas fitting, as poor quality workmanship and lack of knowledge can result in dangerous situations, the only exception to this being where gas work is carried out on a DIY basis. However, all work whether carried out by certified gas fitters or DIY enthusiasts must comply with the Gas Regulations which are enforceable by law. Most plumbers and fitters who undertake gas work are members of a body known as the Confederation of Registered Gas Installers (CORGI), and to ensure that standards are maintained CORGI representatives make regular inspections of the registered installer's work.

Flues

The effectiveness of the boiler flue has a considerable bearing on the efficiency of any boiler whatever type of fuel is used. Flues should be lined to prevent condensation forming, a very important consideration when oil or gas appliances are used, due to the low temperatures of the flue gases.

The formation of the throat of a flue in the case of a solid fuel back boiler must be carried out observing the manufacturer's recommendations in order to obtain the maximum efficiency from the appliance without waste of fuel. The throat is the portion of the flue which is narrowed or reduced

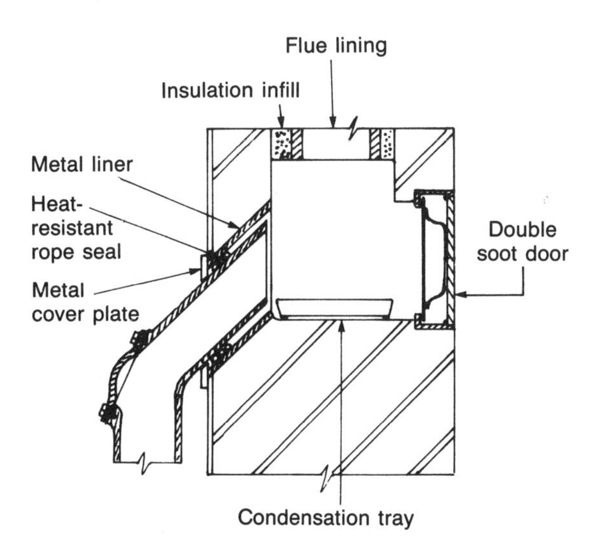

Fig. 6.20 Connection of flue pipe from independent boiler to chimney

in size immediately above the appliance, creating a positive updraught.

The connection of the flue pipe of an independent boiler to the stack should be made as shown in Fig. 6.20. This method allows the flue pipe to expand and avoids cracked plaster where it enters the chimney. A soot door should also be provided externally to enable the chimney to be swept.

Hearths

Before fitting a boiler consideration should be given to its position in relation to the suitability of the surrounding building fabric as a boiler support and its fire resistance. Reference must be made to the Building Regulations 1985 for details of the requirements for hearths and flues.

Hot storage vessels

Storage vessels should be large enough to contain sufficient hot water at suitable temperatures to supply the maximum anticipated demand over a specified period of time, and be strong enough to withstand the pressure of the water they contain as well as that exerted by the head of water in the feed cistern. The Water Bylaws 1986 permit the use of unvented hot water systems. Due to the higher pressures involved stronger storage vessels having more sophisticated controls are necessary. Details of these systems are dealt with in more detail in Volume 2 of this series. Storage vessels should be sited as near as possible to the boiler to reduce heat losses from the circulating pipes and 'dead legs' should be as short as possible. A dead leg is a run of pipe from the hot storage vessel to a drain-off point and most draw-off pipes in small domestic properties are in fact dead legs. Long dead legs will result in waste of water and heat due to the amount of water having to be drawn off before hot water reaches the tap and the subsequent heat losses from the hot water remaining in the pipe gradually cooling when the tap is closed.

Hot water storage vessels may be cylindrical or rectangular in shape (see Figs 6.21(a) and (b)), although the use of the latter is confined mainly to replacement of existing vessels. Both types of storage vessel are made of galvanised steel, although the smaller sizes of cylinders for domestic use are usually made of copper, due to its corrosion-resistant properties and the fact that most domestic supplies are also run in copper tubes. It is not good practice to mix galvanised components and copper in the same system if it can be avoided due to the possibility of electrolytic corrosion.

It is important to make provision for air movement under the storage vessel by standing it on battens. Failure to do this may result in the formation of condensation, causing damage to wooden or chipboard flooring.

Some manufacturers produce units which combine the cold water storage with that of the hot (see Fig. 6.21(c)). These are suitable for small domestic properties such as flats and old age pensioners' dwellings where the demand for hot water is not so high. Their main advantage is the limited space they occupy, and the fact that being a unit the vent and feed pipes are built in which reduces the installation costs. They are also very useful in schemes involving the upgrading of older properties where space is limited.

Indirect cylinders So far all the storage vessels discussed have been of the 'direct' type. This is to say that water heated in the boiler is circulated in

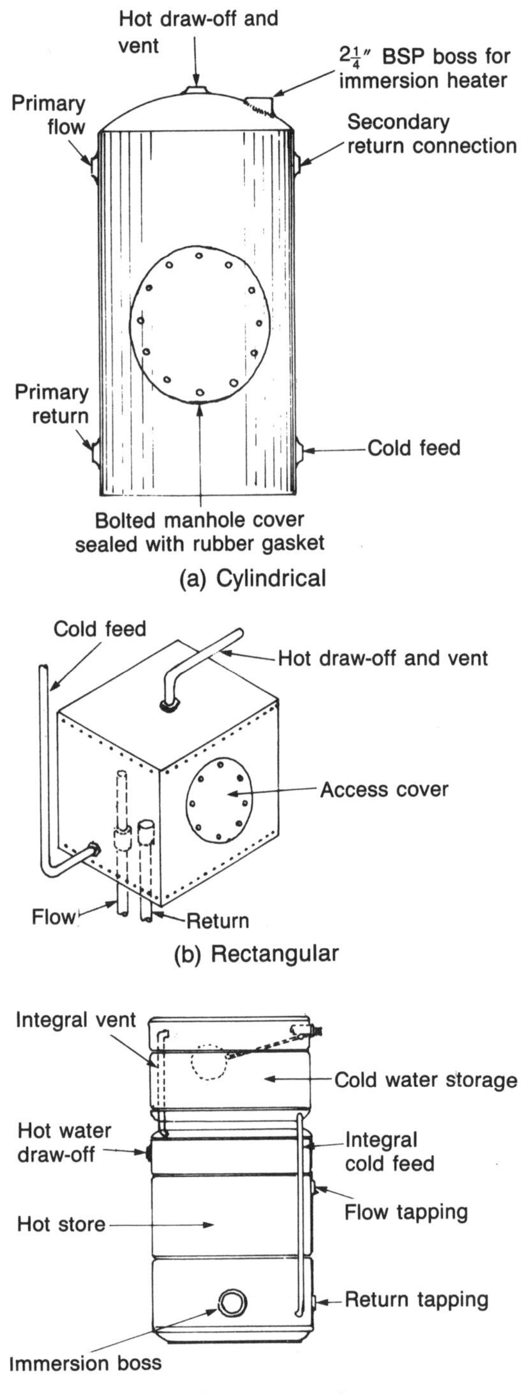

Reference should be made to Chapter 5 regarding accessibility of the float valve in this unit.

Fig. 6.21 Hot water storage vessels

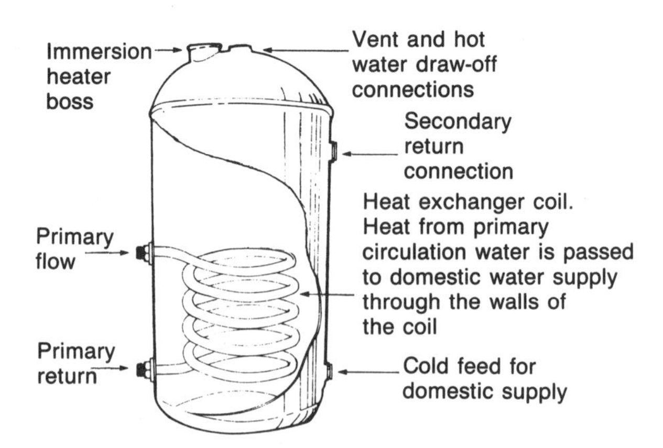

Fig. 6.22 Indirect hot water cylinder. This type of hot storage vessel is used on indirect systems of hot water supply to reduce corrosion and scale formation in hard-water areas. They must always be used in conjunction with combined hot water and heating systems.

the storage vessel and will be eventually drawn off through the taps, resulting in a continuous change of water. While this is not serious in most cases of domestic supply, if the scheme is combined with central heating, an indirect cylinder must be used (see Fig. 6.22).

The object of using an indirect cylinder is to separate the water in the boiler, radiators and associated pipework, the primary circulation, from that actually drawn off through the taps. The only reason water has to be supplied to the primary circulation after the initial filling of the system is to replace losses due to evaporation. The use of an indirect cylinder avoids the build up of fur in the boiler and primary flow and return pipes caused by the precipitation of salts from temporary hard water. It also avoids any corrosion problems that could occur with continually changing water coming into contact with radiators made of ferrous metals. Heat is imparted to the water in the storage vessel by a coil or annulus (see Fig. 6.22), the coil being preferred due to the higher pressures developed in fully pumped and pressurised systems of combined hot water and heating supplies. A coiled tube is deemed more efficient than an annulus and so less heating surface is required for the same performance.

Feed cistern

This is the term given to the cistern which feeds

the domestic hot water supply system to distinguish it from the feed and expansion cisterns used in indirect hot water supplies and from cold water storage cisterns. In most domestic properties, however, where the cistern feeds the cold water services it also serves as a feed cistern for the hot supply. Its minimum nominal capacity, for hot water supply only, should not be less than 114 litres plus the amount required for cold water storage if it is used as a dual purpose storage vessel.

The materials from which cisterns are constructed will be found described in Chapter 5 on cold water supply.

Hot storage vessels are graded in relation to the thickness of the material from which they are made and the internal pressure which they can withstand, as can be seen from Table 6.2.

Table 6.2 Grading of hot storage vessels.

Vessel	*Grade*	*Test pressure (kN/m²)*	*Max. working head (m)**
Copper cylinder to BS 699 (1972)	1	365	25
	2	220	15
	3	145	10
Galvanised steel cylinders to BS 417	A	483	30
	B	276	18
	C	138	9
Galvanised steel tanks to BS 417	A	69	4.5
	B	52	3.0

*The maximum working head is vertical distance between the base of the storage vessel and the water line of the feed cistern.

Hot water storage capacities Table 6.3 shows suggested hot water store capacities for domestic properties based on the number of bedrooms. It is assumed here that 3 kW is available for water heating and the temperature of the stored water is 60 °C. The recommendations do not take into account the fact that larger properties may have two bathrooms, in which case extra storage may be necessary, or increased storage temperatures and higher heat input. In special cases where a more accurate assessment of hot water storage requirement is necessary, reference should be made to CP 6700.

Table 6.3 Hot water storage capacities for small dwellings.

No. of bedrooms	*Storage in litres*
2	120
3	144
4	166

Boiler power Having decided on the capacity and type of storage vessel, the next step is to ascertain the boiler power required. It used to be thought the boiler should be capable of heating the contents of the storage vessel in one hour. This is unnecessary of course as the only time during the day when all its contents are required will be the 'peak' period, this being the time during which the heaviest demands will be made upon the stored water. Two to three hours are allowed for the boiler to heat the stored water, this being termed the firing period. Using this method of boiler sizing, there is still sufficient water at the right temperature to meet the incidental demands of the kitchen sink and wash basin at off peak times.

To give the reader some idea of calculating the boiler power required to heat a domestic storage vessel containing 120 litres of water, the following example is shown. It is assumed the temperature of cold water entering the system is 10 °C and it is to be raised to 65 °C. Water at 65 °C is hot enough for culinary purposes and bathing, and by keeping the maximum temperature at 65 °C, the loss of heat to the surrounding air is minimised and excessive scale in hard water areas is reduced. In this example the water has to be increased in temperature by 55 °C and assuming 4.2 kJ is needed to raise the temperature of 1 kg (litre) of water through 1 °C, the following formula is used:

$$\text{Quantity of water} \times \text{Temperature rise} \times \text{Specific heat of water}$$

where the specific heat of water (the heat required to raise 1 litre of water through 1 °C) is 4.2 kJ. Therefore:

$$120 \times 55 \times 4.2 = 27{,}720\,\text{kJ}$$

It should not be forgotten that the time scale on which this calculation has been made is only one second, but boilers are rated in kilowatts (kW) per hour so the kilojoules figure has to be divided by 3,600, the number of seconds in one hour, to convert it to kilowatts per hour, i.e.

$$27{,}720 \div 3{,}600 = 7.7\,\text{kW}.$$

The boiler power required is therefore 7.7 kW per hour, but assuming a firing period of three hours:

$$7.7 \div 3 = 2.56\,\text{kW}.$$

Just over $2\frac{1}{2}$ kW will be required.

It is seldom necessary to go through these calculations for small domestic systems as all but the smallest boiler will produce sufficient power for this purpose. It does, however, serve to illustrate the method used to calculate boiler power in comparatively simple terms.

Domestic hot water pipework systems

A simple 'direct' hot water system of this type is illustrated in Fig. 6.23. The primary flow and return pipes should not be less than 25 mm in diameter, except in areas where the water is known to have a high temporary hardness (unless an indirect system is to be used) when one size larger is often used so the build-up of fur and scale does not have immediate effects on the flow of water. The minimum recommended diameter of the cold feed pipe on a system having only three or four draw-off points is also 25 mm. While it cannot be ignored that in many cases a feed of 19 mm diameter appears to function quite well, it should be borne in mind that hot water can only be drawn from the storage vessel as quickly as it is replaced. Thus if two draw-offs are opened simultaneously, it is likely that the supplies to both will be diminished. This can result in air being drawn in through the vent and produce a mixture of air and water at the draw-off, as shown at Fig. 6.24. If a hot water draw-off is fixed at a level too near the feed cistern water level a similar situation occurs, and in such cases the only remedies are either to increase the head pressure on the system by raising the cistern or to increase the diameter of the feed pipe.

As shown in Fig. 6.23, a valve is fitted on the cold feed to isolate the hot water system for repair work. While in the past stop cocks have often been used for this purpose, a gate valve is now used as they are classified as full way and offer little or no frictional resistance to the passage of water, an important point to remember when dealing with the low pressures involved. No other supplies or draw-offs should be connected to the cold feed as this could lead to starvation of the hot water supply by reducing the flow of water into the hot water storage vessel. In a small system supplying only a bath, basin and sink, a draw-off pipe of 19 mm diameter is normally sufficient. This is also the minimum size of the vent, as smaller pipes would be more prone to obstruction. The vent is a very important part of the system, its

Fig. 6.23 Simple domestic hot water supply system

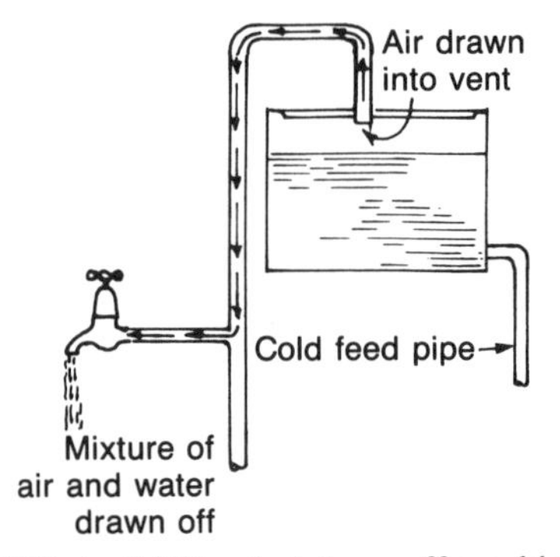

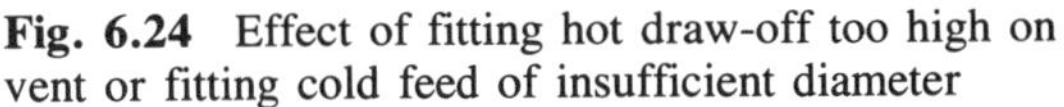

Fig. 6.24 Effect of fitting hot draw-off too high on vent or fitting cold feed of insufficient diameter

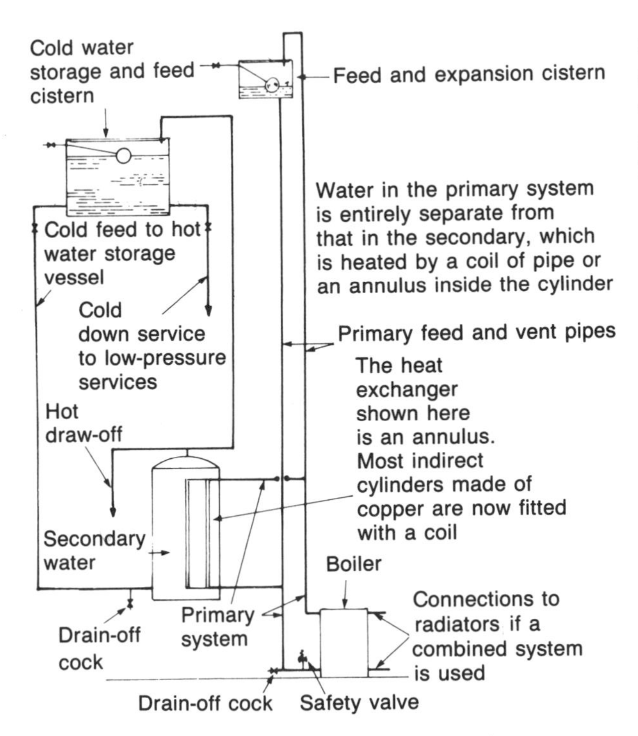

Fig. 6.25 Indirect system of hot water supply

main functions being to maintain atmospheric conditions in the pipework. It permits the escape of air which has entered the system and more rarely in the event of the water becoming overheated, allows it to discharge over the feed cistern.

Indirect systems of hot water supply

An indirect system is designed in such a way that the water in the boiler and primary pipework is never changed, the only loss of water being in the supply cistern through evaporation. When the water is initially heated, any gases it contains are given off, including the oxygen and carbon dioxide, which, as stated earlier, are largely responsible for corrosion and scaling respectively. This is not to say that an indirect system is entirely immune from corrosion, but its incidence is considerably reduced. A study of Fig. 6.25 shows that the primary water is quite separate from the secondary water, and may be defined as that contained in the boiler, primary flow and return pipes, the annulus or coil in the cylinder, and any radiators that are fitted. The secondary water is that drawn from hot storage vessels to supply the hot taps. It is heated by conduction, being in contact with the annulus or coil.

It is important that indirect cylinders conform to BS 1566 which specifies the minimum heating surface area of the annulus or coil. Hot water vessels which do not comply with this standard often have an insufficient primary heating surface which results in slow heat up of the water, wastage of fuel and higher running costs. The cylinders, too, are often made of thinner copper sheet than that specified, resulting in a shorter working life. The moral of this is to ensure that all components carry the British Standard kite mark.

The primary part of the system is fed by a feed and expansion cistern which must be big enough to accommodate the expansion of the water in the system when it is heated. For small indirect systems a cistern of 50–70 litres actual capacity is normally of sufficient size. An accurate method of sizing may be necessary if a *combined system* is fitted. This term applies to systems where one boiler serves both domestic hot water and space-heating appliances such as radiators.

Although hot water expands by approximately only $\frac{1}{24}$ in volume when raised from 4 °C to 100 °C, it is usual to make provision for an expansion of $\frac{1}{20}$ when sizing feed and expansion cisterns; this is to say that a system filled with water containing 100 litres when cold should have provision in the expansion cistern for a further 5 litres. This is a generous allowance as the average temperature of the cold water supply is 10 °C and it is seldom necessary and, indeed, is wasteful to heat domestic hot water to more than 70 °C. Temperatures higher than this give rise to the formation of fur in the secondary part of the system and excessive heat loss from the stored water. Indirect systems of this type are sometimes called ‘closed circuits’ which can be confusing as, unlike fully pressurised systems which have no outlet to the atmosphere, the system under discussion has a vent and feed pipe open to the atmosphere.

Failure of the boiler to heat the domestic hot water is usually caused by evaporation of the water in the expansion cistern, while the ball valve, because it is seldom required to open, becomes corroded and remains in the closed position. When the water level falls below the flow connection on the cylinder, no circulation can take place in the annulus, with the result that the secondary water remains cold.

It is rare but not unknown for a leak in the annulus or coil to allow the mixing of the primary and secondary water. This may remain undetected for a long time if the surface of the water in the feed and expansion cistern is level with that of the feed cistern. The first indication that something is wrong is the discharge of discoloured water from the hot taps, or worse, corrosion in a radiator causing it to leak. For this reason it is recommended that the cisterns are fitted at different levels so that persistent overflowing of the lower cistern will indicate the defect.

Single feed indirect systems

This type of system employs the use of a hot storage vessel known as a self-priming cylinder. The difference between this and a normal indirect cylinder is that both the primary and secondary sections of the system are fed with water from a common cold feed, a special heater admitting water to the boiler and any radiators that may be fitted. These cylinders are slightly more expensive than the normal indirect types, the main object in their use being to reduce the amount of pipework involved in the installation and to render a separate feed, vent and expansion cistern unnecesary thus reducing fitting costs. They are, in fact, fitted in the same way as a direct cylinder, the primary and secondary water being kept separate by what is best described as an air lock which is contained in the single feed heater situated inside the cylinder. This heater contains two hemispheres, under one of which air is entrapped. The movement of air between these two hemispheres allows for the expansion of the primary water when it is heated. A study of Fig. 6.26 shows how both the primary and secondary components are filled with water and charts the movement of the air when the primary water expands. It will be seen that the space containing the air in the hemisphere is limited, and if the volumetric expansion of the water exceeds that of the space available, some air will be forced out and the space it occupied replaced by water. If this continues, a slow but sure change of water in the primary part of the system will take place with all the undesirable characteristics that the use of an indirect system is supposed to prevent. This state of affairs is usually brought about by an excessive quantity of water in the primary circulation which, when it expands, the hemisphere is unable to accommodate. For this reason the heater units are made in more than one size, the volume of the hemispheres being larger for cylinders fitted to schemes having a larger quantity of primary water. In all cases when fitting single feed cylinders the manufacturers' recommendations regarding the primary capacity of the system must be observed to avoid costly mistakes.

Cylinders of this type must be fitted with a drain-off cock on the cold feed like that of ordinary indirect cylinders, as they can only be partly drained by the boiler drain-off cock. Due to their design, they must always be installed in the upright position, otherwise no air will be entrapped in the hemispheres.

When self-priming cylinders are drained down and refilled, the heater unit sometimes becomes air locked resulting in the secondary water not heating. This is usually due to a blockage in the inverted U vent fitted to the heater unit. Corrosion products sometimes obstruct the hole where the vent is connected to the heater unit. If this happens the problem can sometimes be solved by draining the cylinder completely and allowing the feed cistern to fill before fully opening the gate valve on the cold feed. The resulting sudden rush of water into the cylinder often removes the corrosion products allowing the cylinder to function normally. Single feed cylinders must not be fitted with valves controlling the temperature of the hot water as if the boiler overheats the air seal may be displaced. A hot water system using a single feed cylinder is shown in Fig. 6.27 and it will be seen that the connections are the same as those for a direct installation.

Hot water systems with a secondary circulation

As has previously been described, when a hot tap

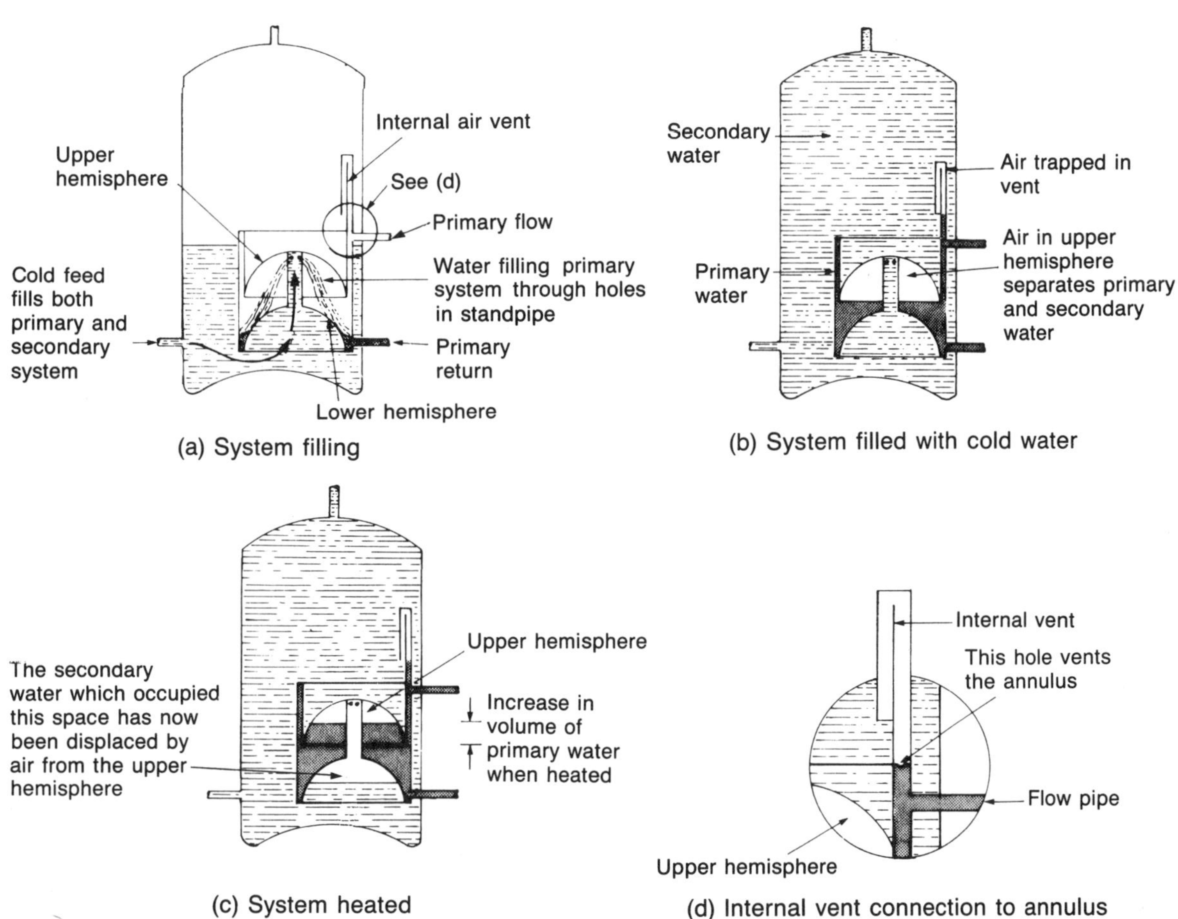

Fig. 6.26 Single feed self-priming cylinders

is closed after use, the water that remains in the pipe will cool and when the tap is next opened, this cooled water must be first drawn off before a further supply of hot water reaches the tap. Draw-off pipes of excessive length are therefore a cause of waste of both heat and water. Water authorities are aware of this fact and the old model water bylaws specify that the maximum length of a dead leg does not exceed that shown in Table 6.4. It is unlikely that any new legislation will have any significant effect on the lengths indicated.

In cases when a longer or larger draw-off than the maximum permissible is required, a secondary circulation will become necessary. Such a circulation is only advisable in very large houses and public buildings where they are absolutely essential. The diagram at Fig. 6.28 shows a direct hot water system incorporating a secondary circulation. It will be seen that its main function of providing an immediate supply of hot water to the draw-offs is accomplished, but the following two disadvantages should be noted. The long lengths of

Table 6.4 Permissible length of dead legs BS 324.

Nominal diameter	*Length of dead leg*
Pipes of up to 19 mm	12.000 m
Pipes of 20–25 mm	7.500 m
Pipes of over 25 mm	3.000 m

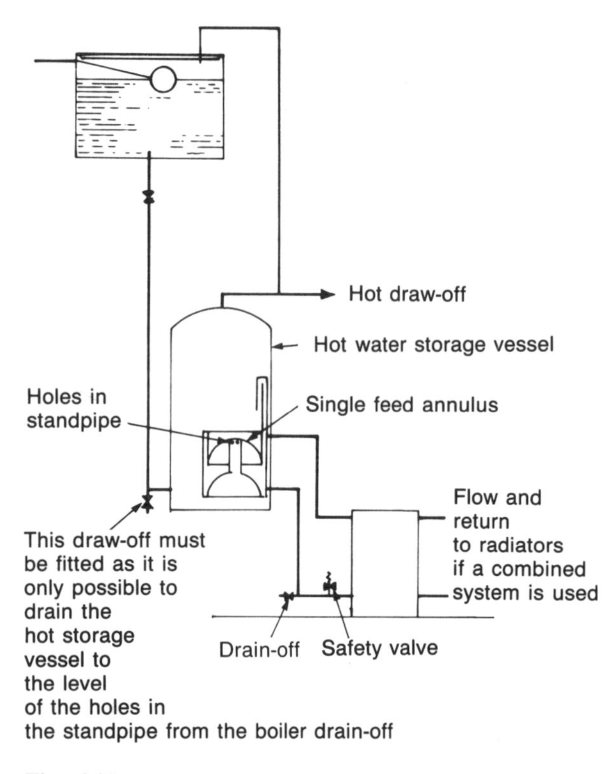

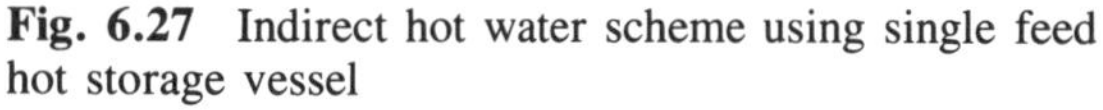

Fig. 6.27 Indirect hot water scheme using single feed hot storage vessel

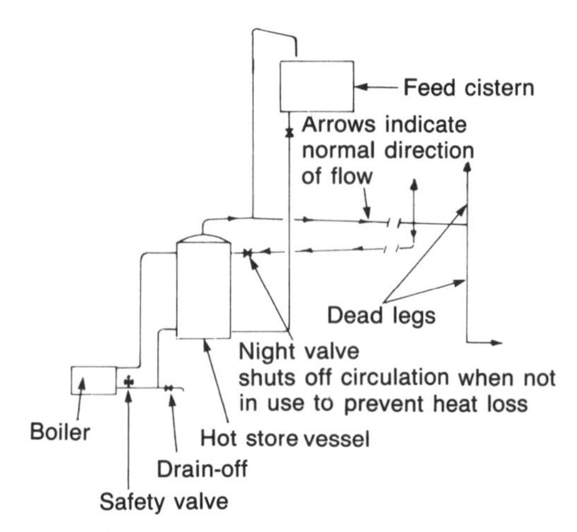

Fig. 6.28 Domestic hot water system with secondary circulation. This type of system is not normally required in small houses as the draw-offs should be kept as short as possible by careful planning. Note that if a secondary circulation is fitted the hot storage vessel must always be lower than the pipe runs if it is to operate by natural or gravity circulation.

circulating pipe must be well insulated to prevent excessive heat losses and a night valve must be provided to prevent overnight circulation and subsequent cooling of the stored hot water if the boiler is shut down. These used to be manually operated gate valves, but with the availability of motorised valves and time switches, automatic shutdown at predetermined times is now possible.

So far it has been assumed that the secondary circulation operates on a gravity basis. Where such an arrangement is not possible due to the position of the storage vessel, or head losses are excessive due to very long pipe runs, then a pump must be provided. Due to the long lengths of pipe usually associated with secondary circulations, heat loss can make such systems very expensive to run and effective insulation of the circulating pipe is essential.

Secondary circulations in public and commercial buildings

While secondary circulations were originally designed to function by gravity the design and size of modern buildings often make a pumped circulation necessary. If, for instance, the hot storage vessel is at a high level in the building as shown in Fig. 6.29 a gravity circulation would not operate due to the lack of circulating pressure. In large buildings the length of the secondary circulation would make it extremely sluggish and, in view of the fact that the recommended temperature drop between the flow and return should not exceed 22 °C, a pump would definitely be necessary. Those used for this purpose must be constructed of non-ferrous metals or the working parts will quickly seize up. If the system is arranged to circulate only when the pump is functioning, a time switch can be used to determine when the circulation is operative thus avoiding the necessity of fitting a night valve.

When a draw-off on a secondary circulation is opened, water is drawn from both the flow and return (see Fig. 6.30). If the return connection to the hot storage vessel is too low, this will result in cooler water from the base of the storage vessel being drawn off producing a lukewarm supply, especially when large quantities of water are required such as when hot water for a bath is

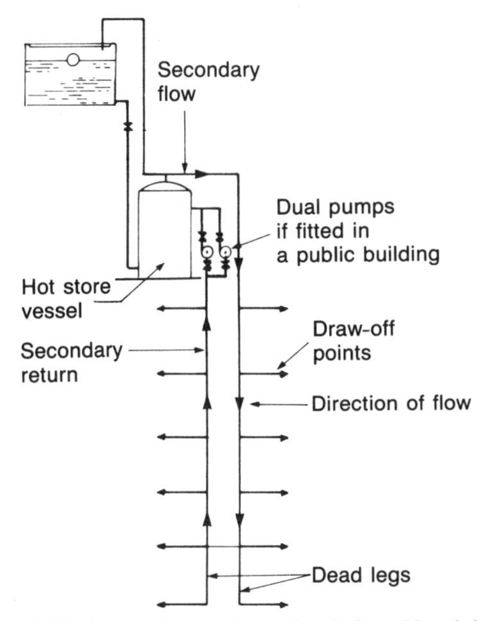

Fig. 6.29 Pumped secondary circulation. No night valve will be necessary on this system as it will only circulate when the pumps are operating. Pumps can be fitted on the secondary flow if more convenient.

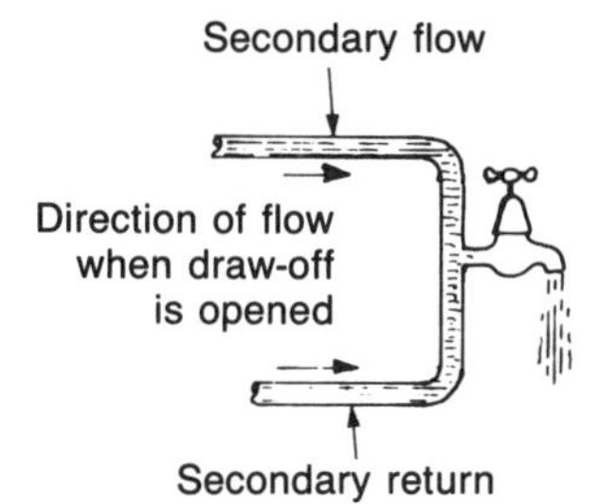

Fig. 6.30 Effect of opening draw-off on secondary circulation

drawn. For this reason the position of the secondary return tapping in the storage vessel must be as high as possible, and certainly not further down the cylinder than one-quarter of its total height.

Defects in hot water systems

Broadly speaking, there are two main causes of faulty hot water supply apart from the effects of frost. There are those that happen naturally and are usually the result either of corrosion caused by the use of dissimilar metals and unsuitable fittings, or scale formation in areas of temporary hard water. There are also the mechanical faults that are often built into a system resulting in air locks and noise. In most cases these defects are the result of lack of knowledge and appreciation of basic principles.

Air locks

Air locks are a very common cause of trouble with both hot water and heating systems. Most air locks are caused by unventilated arches formed in badly fitted pipework. An air lock is a small quantity of air trapped in a pipe which due to the very low circulating pressure available prevents water passing through the pipe. Even if an air lock does not completely stop the flow of water it can reduce the flow considerably. Air locks are very often the result of pipes sagging, or not being laid truly horizontally or to appropriate falls. Figure 6.31(a) shows how an air lock occurs in a pipe run, although they are not always as obvious as this.

Another common cause of air locks is shown in Fig. 6.31(b) where the cold feed, instead of falling away from the cistern, rises causing an unventilated arch in the bend.

Air may also be locked in a boiler due to defective tappings as shown in Fig. 6.31(d)(i) but the boiler manufacturers are aware of this and usually the flow connection is in the position shown in Fig. 6.31(d)(ii). Probably the worst effect of defective tappings would be a rather noisy boiler, but if this were further aggravated by an air lock in the flow pipe, reversed circulation could take place causing the hot water to be forced up the return pipe instead of the flow. Reversed circulation sometimes corrects itself as the water temperature increases, but the noise caused by this problem is alarming and should be investigated.

Care must be exercised when installing boilers and primary circulating pipes to avoid collections or pockets of air. Yet another common cause of air locking is shown in Fig. 6.31(c) where the hot draw-off has been badly fitted causing it to sag and create an air trap at the cylinder connection.

One of the most common methods of removing air from a hot water system is to fit a hose from a hot draw-off tap to a high-pressure cold supply tap and then turn both taps on, where the high

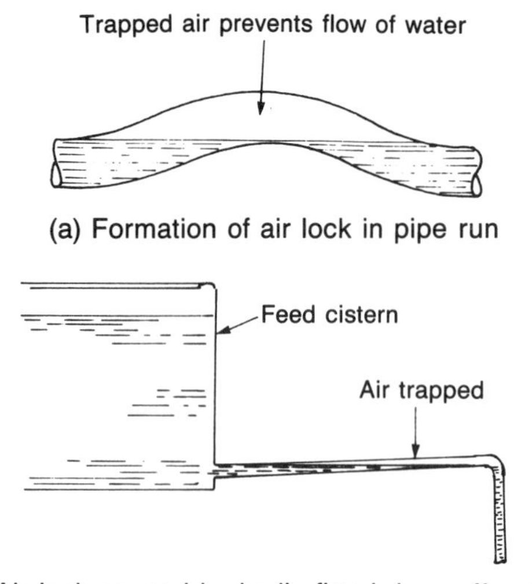

(a) Formation of air lock in pipe run

(b) Air lock caused by badly fitted draw-off or feed pipes. This is possibly the most common cause of air locks in small domestic hot water systems

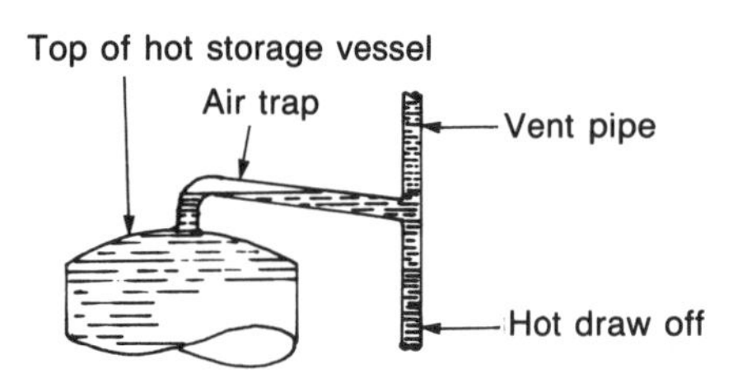

(c) Badly fitted hot draw-off pipe causing air lock

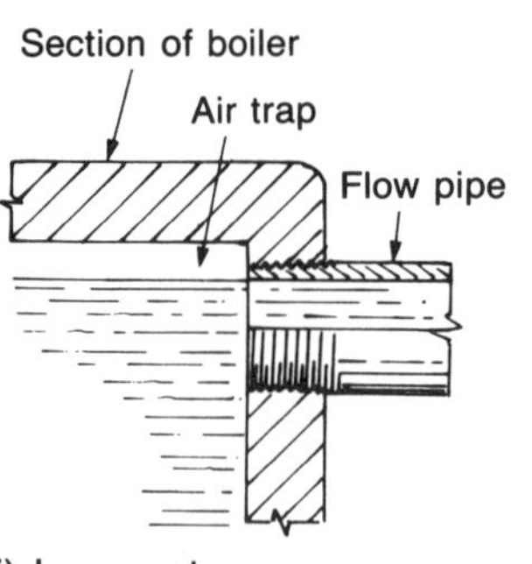

(i) Incorrect

A boiler with the flow tapping incorrectly fitted is likely to be the cause of noise.

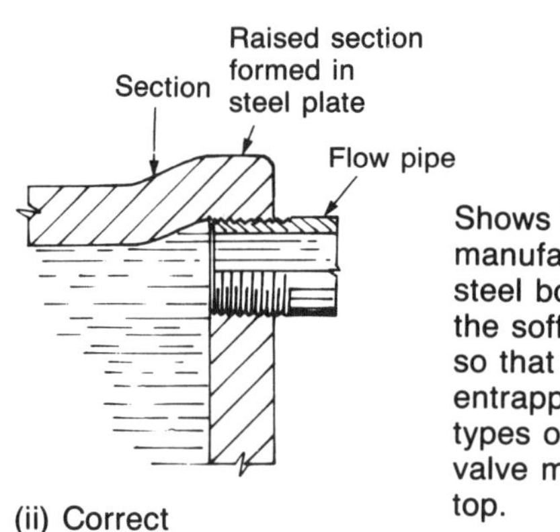

(ii) Correct

Shows how some manufacturers form the steel boiler plates to raise the soffit of the flow pipe so that no air can be entrapped. With certain types of boilers an air valve may be fitted in the top.

(d) Flow tappings in boilers

Fig. 6.31 Common causes of air locks

pressure will force out the air. While this method is usually effective, it is not to be recommended as the whole of the hot water system is subjected to water at high pressure and should the vent and feed pipes be obstructed (possibly in freezing weather by ice) it may result in bursting the storage vessel.

Experience has shown that air locks usually occur in the cold feed pipe. The safest method of removing the air is to connect a force pump to the cold feed connection and force water through the air lock. If the air lock is in the hot draw-off, the force pump can be connected by a hose to a hot tap, pumping out the air as described. It must be stressed, however, that before taking any measures likely to exert pressure on the system, ensure that both the vent and feed pipes are not obstructed in any way.

Boiler explosions

Boiler explosions are fortunately rare, but the result of such a happening can be disastrous, especially when a cast iron boiler is involved. An explosion will only occur if both of the primary circulating pipes are obstructed. This can come about for two reasons, the first being most common in domestic buildings. If a building is left without heat in freezing conditions, the whole of the system including the flow and return pipes may become frozen. If the boiler is fired under these conditions, the heating of the water will cause it to expand and burst. If it is suspected that the system is frozen, no attempt should be made to fire the boiler until it is certain any ice has thawed.

The other cause of boiler explosion is where both primary circulating pipes have been 'valved' for some reason. If both valves are closed the same situation will exist as previously described, with the same result. It should be quite unnecessary for valves to be fitted on domestic installations, although it has been known. On large installations having a series of boilers, it is of course usual to valve both primaries on each boiler to enable servicing or repairs to be made without shutting down the whole system. On installations of this size, however, the valves will be operated by a qualified person who would be aware of the dangers. Even in these circumstances, when a job is completed on such a system, the valves must be locked in the 'open' position.

Scale deposits

The formation of scale deposits in temporary hard water areas can obstruct the primary circulating pipes, but this is unlikely to happen to both pipes simultaneously as the flow pipe carrying the hotter water tends to scale up first and the resultant noises in the boiler would prompt an early investigation.

Safety valves

It has always been considered good practice to fit a safety or pressure relief valve on a hot water system, but in order to cut down on cost they are often omitted. While no legislation exists to enforce their provision, it is good practice to fit them. It is interesting to note that regional gas boards will not undertake to maintain a gas-fired boiler unless it is equipped with a suitable pressure relief valve.

A typical spring-loaded safety valve suitable for domestic use is shown in Fig. 6.32. Any excess pressure in the system lifts the valve against the spring, the excess escaping through the holes in the valve casing. Safety valves should be fitted directly on the boiler if possible, and some manufacturers provide extra tappings for this purpose. In the case of back boilers this arrangement may not be possible and the usual practice is to fit the valve on the return pipe as close as possible to the boiler. There is less likelihood of scale formation in the return which could cause the valve to adhere to its seating.

Safety valves should be removed periodically and checked to ensure they are in working order.

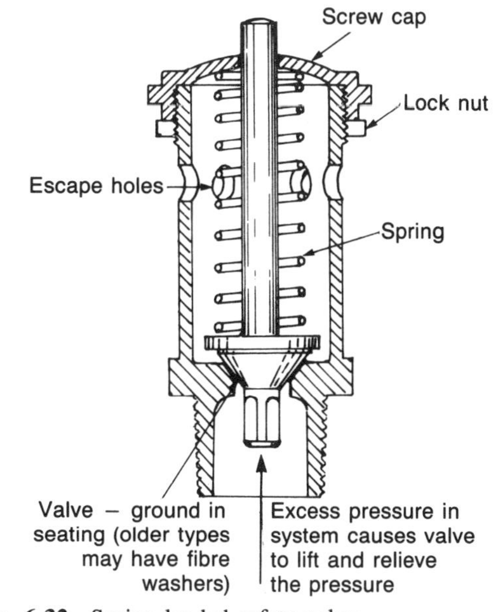

Fig. 6.32 Spring-loaded safety valve

The best practical way of adjusting a safety valve to function satisfactorily is as follows. The cap should be unscrewed, relieving tension on the spring until a drip of water is seen to be escaping from the holes in the casing. It should then be screwed down half a turn and locked in this position with the lock nut.

Cylinder collapse

Cylinder collapse is due to the pressure on the inside of the cylinder becoming less than that of the atmosphere. A simple experiment is shown in Fig. 6.33 which illustrates quite clearly the effect of atmospheric pressure. A small amount of water is put into a can which has a close-fitting airtight lid. This lid is removed and the water is heated until it boils and gives off steam, when the source of heat is removed and the lid firmly replaced on the can allowing the contents to cool. When the steam condenses and the air in the can cools, it occupies less space, consequently lowering the pressure inside the can. The atmosphere exerts a pressure of approximately 100 kN on an area of one square metre, and since this pressure will be far greater than that inside the can, the walls of

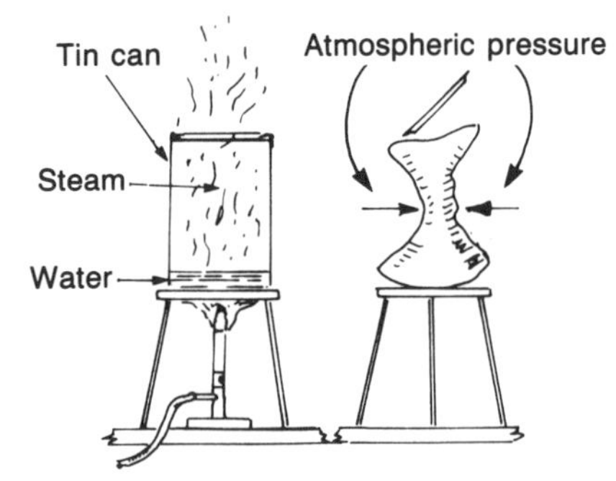

Fig. 6.33 Experiment illustrating cylinder collapse

the can will be forced inward until the can has been completely deformed.

Various situations can occur whereby similar conditions can arise in hot water storage vessels causing their collapse. It almost invariably happens in freezing weather when the vent pipe, often in an exposed position, becomes frozen. There will be no indication that all is not well, as any expansion of the water will take place via the cold feed pipe to the storage vessel. As the water temperature increases, water vapour may form, pushing more water back up the cold feed. In freezing weather the air temperature drops overnight causing the cold feed to freeze as the fire dies down, the system then becomes closed to the atmosphere, and a situation now exists in the cylinder similar to that of the tin filled with steam, and it will have the same result when the boiler is shut down and the water in the storage vessel cools. The cooling water will contract in volume and reduce the pressure inside the cylinder which will slowly deform and partially collapse.

Another cause of cylinder collapse is when a hot water draw-off at a lower level is opened with both the vent and cold feed frozen. Water will flow from the tap but as none can enter the cylinder to replace it, a siphon is started. As the water is withdrawn the pressure inside the cylinder becomes less than that of the atmosphere so the cylinder sides are crushed inwards. In this case the cylinder may collapse fully and become completely flattened.

If only a slight collapse has occurred, the cylinder can often be salvaged by filling it with water and pressurising it with a force pump. Providing the damage is not too severe, the cylinder can be made serviceable again after thorough testing for leaks.

The incidence of cylinder collapse is much less in modern homes equipped with central heating systems and where the roofs are made draught-proof by sarking felt under the tiles. It should not be forgotten, however, that many older properties often have very exposed roof spaces where in severe weather it is not unknown for snow to be blown in. Where cisterns and pipework are exposed to such conditions, it is essential they are well insulated and draught-proofed.

One-pipe circulation

The positions of the connections on a storage vessel are most important. The hot water draw-off is taken from the top of the vessel as the temperature of the water will be at its maximum at this point. It is for this reason that when a secondary return connection is made it should be at a high level in the storage vessel.

To avoid 'one-pipe' circulation and subsequent waste of heat, the vent and draw-off should be fitted as shown in Fig. 6.34(a). Convection currents will occur in the vent pipe (see Fig. 6.34(b)) if it rises directly from the storage vessel.

Vent pipe termination in cisterns

The height of the vent above the water level in the feed cistern depends on the distance between the surface of the water level in the cistern and the base of the hot store vessel and may be calculated as follows. Working on the assumption that water will expand by 1/25 fom 4 °C to 100 °C, a column of water 1 m high, if raised through this temperature, would expand by 40 mm. On this basis 40 mm is allowed for every one metre in height of the system. As it is not desirable for water to be continually discharging from the vent, a distance of 150 mm is added. To give an example, supposing the water level in the cistern is 2.500 m above the base of the storage vessel, the vent should extend above it by 100 mm + 150 mm = 250 mm (see Fig. 6.35). The allowances that have been made are quite adequate, but failure to provide a vent of sufficient height will result in water discharging into the cistern as the water

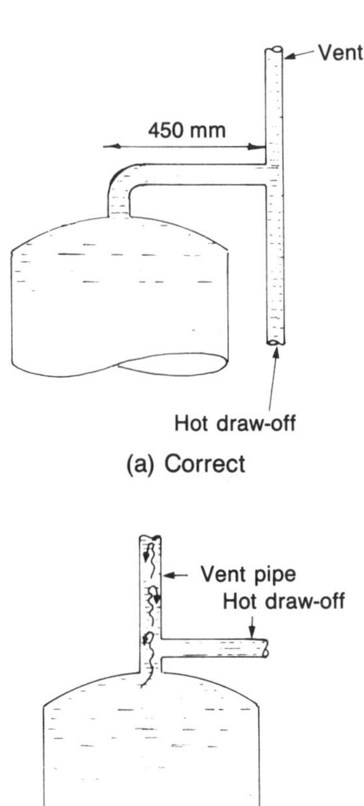

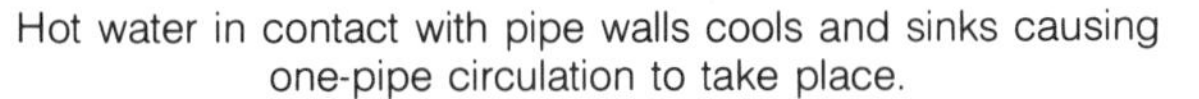
Hot water in contact with pipe walls cools and sinks causing one-pipe circulation to take place.

Fig. 6.34 Connecting vent and hot draw-off to avoid one-pipe circulation

temperature increases. This not only causes a waste of fuel but may also result in luke-warm water at the cold water draw-off points if the cistern is used for hot water feed and cold water storage.

The other important factor to observe in relation to vent pipes is to make sure they are trimmed off well above the water level. Should the vent become submerged, cold water could be siphoned out of the cistern when a hot draw-off is opened.

Corrosion

The subject of intermetallic or electrolytic corrosion is dealt with later in this book in relation to weathering and roofing. Surprisingly, many plumbers who are well aware of the disastrous effects of mixing dissimilar metals in a roofwork

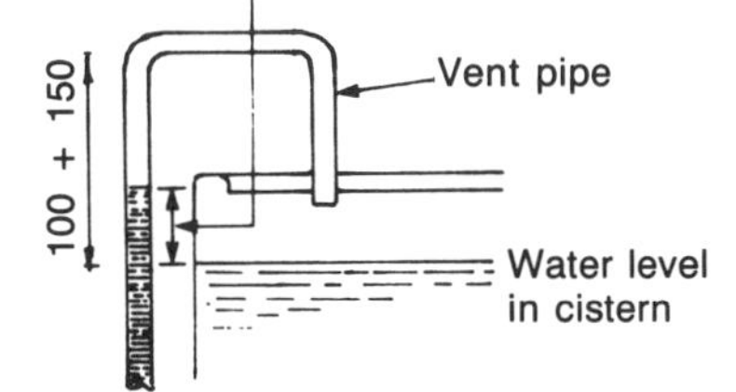

Fig. 6.35 Allowance for expansion of water in vent pipe

situation think nothing of using combinations of copper pipes and iron bushes! In hard water areas a film of scale often protects the iron from any serious corrosion, but this is not the case with soft water when the result is often the rapid deterioration of small ferrous metal fittings such as bushes. To reduce this danger, such fittings should be made of bronze or brass with a high copper content, especially when fitted to direct systems of supply.

Another form of corrosion which arises in both hot and cold water systems is known as 'dezincification'. For many years the manufacturers of copper fittings have used brass with a high zinc content, partly to economise on copper, but mainly to produce an alloy which can easily be pressed or stamped. (These terms relate to a method of manufacture where a component is pressed out into a mould instead of being cast, the latter method being more expensive.) Unfortunately, brasses having a high zinc content are very unstable alloys in the presence of acidic waters and tend to break down into their constituent parts. The zinc content of the alloy is reduced to a basic carbonate and loses all its metallic properties, causing it to increase in volume and obstruct the waterway of the fitting. The copper content of the alloy, if examined under a microscope, looks like a sponge and does in fact become porous. This form of corrosion is usually identified by a gradual lessening of the supply of water delivered through the draw-offs, and a white growth on the exterior surface of the fitting.

The use of brass as a material for copper pipe fittings has diminished, having been replaced by fittings made of copper or bronze. Quite recently fitting manufacturers have claimed to have overcome the problem of dezincification of brass by subjecting it to a form of heat treatment. Fittings of this type are, however, only available in the compression range, as the effect of soldering capillary joints has damaging effects on these specially heat treated alloys.

Copper cylinders which were often supplied with brass bosses and brazed with an alloy containing zinc are now specified as being zinc free. Galvanised storage cisterns should on no account be used with waters likely to cause dezincification unless they are suitably protected by painting with a non-coal tar paint approved by the local water authority.

Prevention of heat loss

The effective insulation of both hot and cold water service pipes is important, both to protect from frost and, in the case of hot water, to avoid heat losses. The theory of insulation is very simple and consists of covering the pipework and components of plumbing systems with a material which has a low heat conductivity rate, and which at the same time entraps bubbles or small pockets of air, air being a poor conductor of heat. The reflective properties of a bright surface are also sometimes employed to resist heat loss, a typical example being foil-backed plasterboard which if used for ceilings offers considerable resistance to heat flow into the space above.

The conductivity of materials for insulation vary considerably depending on the type used. The flow of heat through an insulating material is indicated as the heat conducted per hour in watts, per square metre, per degree Celsius temperature difference. This is usually abbreviated to W/m °C and it is important to understand this so that a comparison can be made between the various types of insulating materials. Another fact that should also be understood is the thickness of the insulation often differs in manufacturers' lists as to whether protection against frost damage or the prevention of heat loss (in the case of hot water pipes) is required. The reason for this is that the insulation thickness suitable for preventing heat loss on hot water services may not be sufficient to delay the effects of frost. The materials used for insulation vary considerably, some being most suitable for a specific purpose than others, but a good insulating material should be of adequate thickness and have the following characteristics:

(a) it should not be flammable
(b) it should be vermin-proof
(c) it should be draught-proof and impervious to moisture if it is fixed externally or likely to be subject to damp conditions, i.e. under suspended ground floors
(d) it should be sufficiently robust for its purpose

Pipe insulation

Hair felt was one of the most commonly used insulating materials, mainly because of its low cost in comparison to alternatives. The original material only met one of the foregoing requirements in that it was a poor conductor of heat. Better qualities of this type of insulation are now manufactured which comply with the rot and fireproof requirements, but because of its nature cannot be said to be waterproof and should only be used in dry, moisture-proof situations. It can be obtained in both sleeve or strip form the latter being rolled around the pipe as shown in Fig. 6.36(a). Its performance is considerably improved by overwrapping with a polythene sheet which renders it draught-proof and improves its damp-proofing qualities.

A far superior form of insulation is expanded synthetic rubber sleeving (see Fig. 6.36(b)) which, although expensive, is a very good material. A high-quality insulation of this type should be fire resistant, but it is advisable to check that this is so as some types of formed plastic and rubber are highly flammable and give off poisonous fumes when burning.

This type of insulation is supplied with a slit throughout its length so that it can be opened up and fitted to existing pipework. If there is no slit, it must be fitted as the pipework is installed. Any cuts may be made using a very sharp knife, the joints being sealed with an adhesive painted on to the surfaces to be joined, or a self-adhesive tape, both of which are available from the manufacturer.

Copper pipes are available with a preformed PVC sleeve which fits tightly around the pipe as shown in Fig. 6.36(c). Like most plastics, PVC is a poor conductor of heat, and the extrusion is made to incorporate the insulating characteristics of still air. Copper tubes fitted with this type of sleeve can be bent with a spring or in a machine with a special former and back guide. It should be noted that sleeved tube of this type is ideal for fitting under solid floors or in aggressive soils, as it also resists attack by acids and alkalis.

Some preformed insulation for pipes is moulded in two halves. Various materials are used, calcium silicate, expanded polystyrene and glass fibre being typical examples. The latter is commonly used for hot water work, having a linen wrapper which acts as a hinge allowing it to be fitted over the pipe as in Fig. 6.36(d). The linen overlap may be sealed with an adhesive or stapled, and to clamp it securely on the pipe, black Japanned steel bands are fitted at 450 mm intervals.

One problem that will be encountered with all types of performed insulation is the difficulty of obtaining a neat finish at changes of direction and at tee joints, but most types can be cut and fitted around these obstacles using a fine-toothed hacksaw, a sharp knife and a little care, making sure the joints are well fitting and jointed where applicable with an appropriate adhesive. Figure 6.37 illustrates a typical example of mitred preformed sectional insulation around a tee joint.

Protection of insulation

In situations where external pipework is insulated the insulation must be waterproof. The traditional method of achieving this is to wrap it in bituminous felt, sealing the joints with an adhesive. To protect it from mechanical damage the felt is overwrapped with galvanised wire mesh (see Fig. 6.38(a)). A modern alternative to this method is the application of polyisobutylene sheet cut into strips, allowing for overlaps of 40–50 mm, which is sealed using white spirit as an adhesive (see Fig. 6.38(b)). Where insulation is

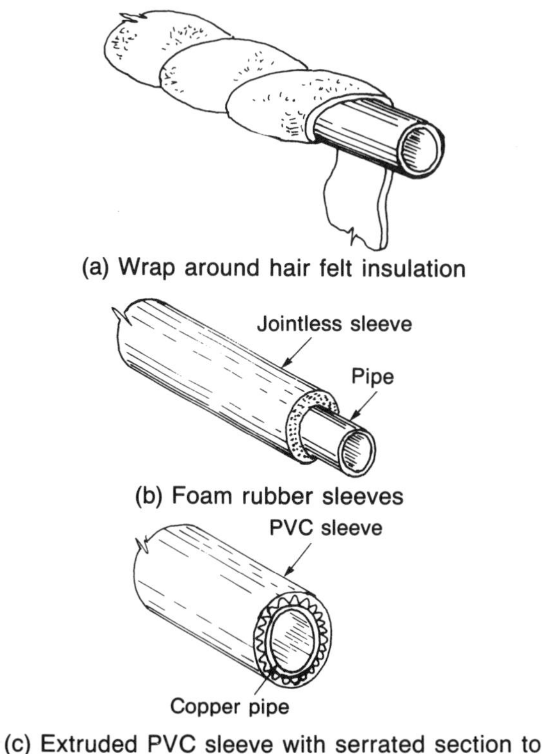

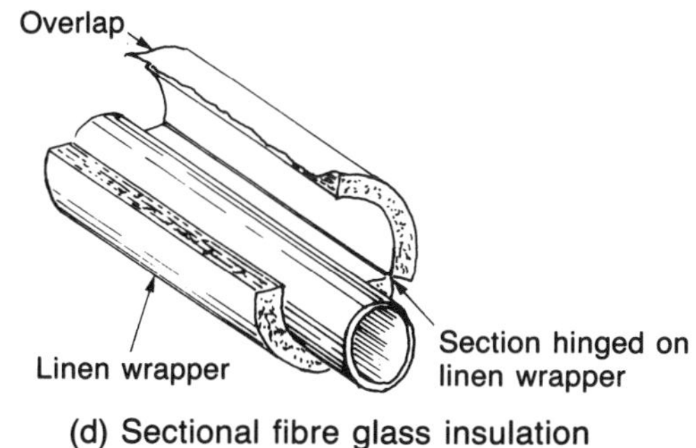

Fig. 6.36 Pipe insulation

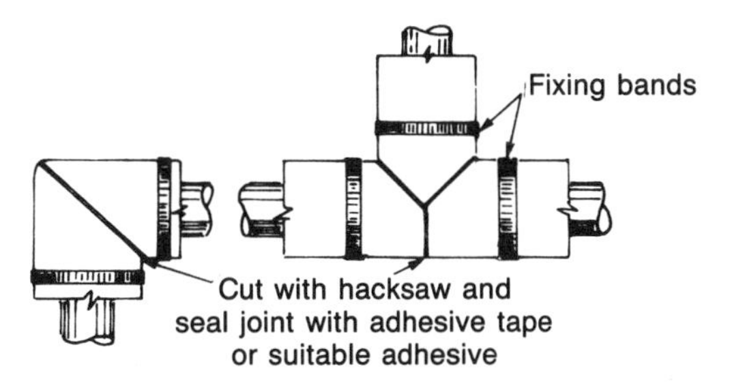

Fig. 6.37 Fitting pipe insulation. Cutting sectional insulation round tees and elbows

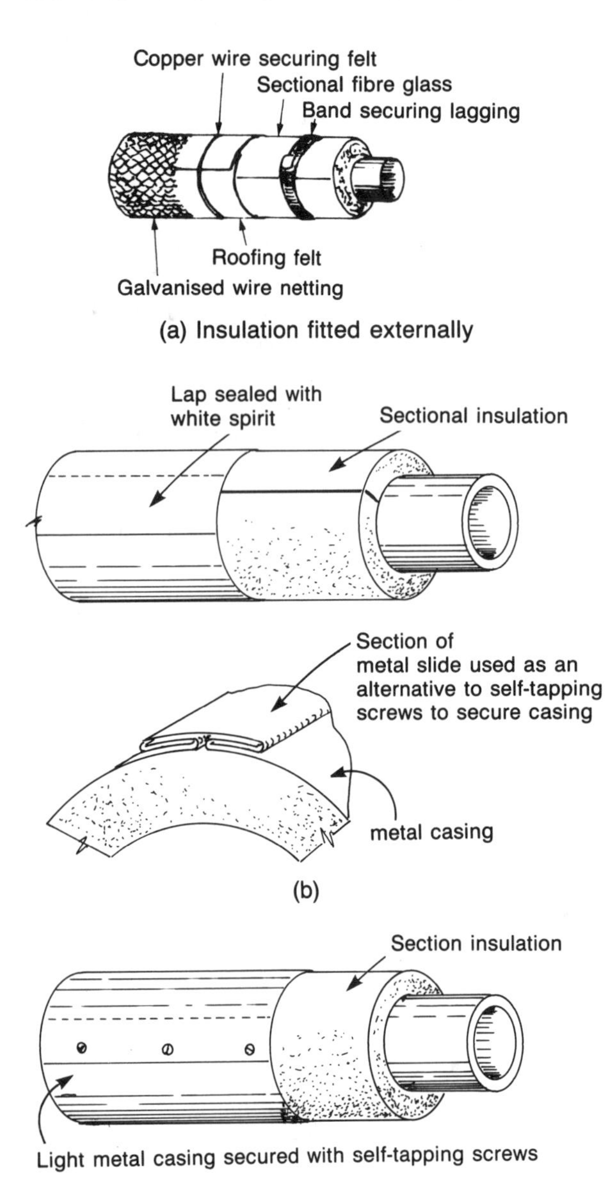

Fig. 6.38 Protection of insulation

Table 6.5 Maximum recommended lengths of uninsulated circulating pipes.

Outside diameter of pipe (mm)	*Maximum length (m)*
12	20
Over 12 and including 22	12
Over 22 and including 28	8
Over 28	3

used in situations where it is likely to suffer mechanical damage, for example at low level in boiler rooms, it can be covered with a light gauge galvanised sheet which is fixed as shown in Fig. 6.38(c).

The maximum recommended length of uninsulated circulating pipes, except those which contribute to the space heating of a building, is listed in Table 6.5 and applies to both primary and secondary circulations.

Insulation of hot storage vessels and cisterns

Insulation of domestic hot water cylinders is essential to keep fuel costs to a minimum. One of the most effective methods is the use of foamed polyurethane which is sprayed on by the manufacturer. An alternative for those vessels which are not treated in this way is the use of sectional insulation which is secured by tapes or metal bands, as illustrated in Fig. 6.39.

The sections are made of quilted glass wool or fibreglass with a canvas or PVC cover. When

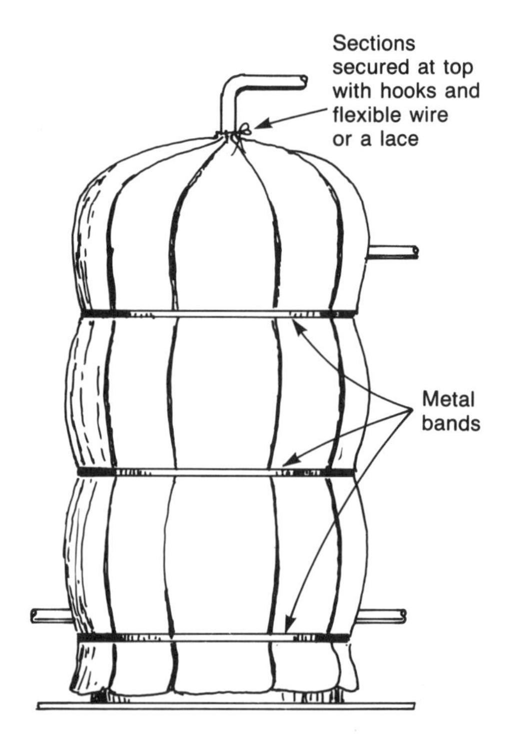

Fig. 6.39 Sectional insulation jacket for hot water storage vessel

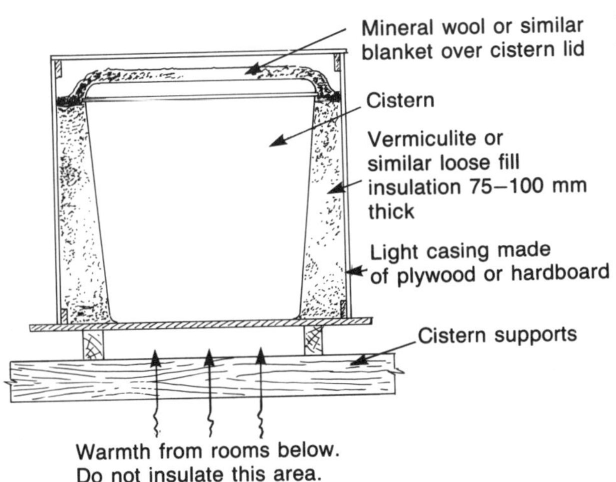

(a) Insulation of cisterns in a roof space

ordering a jacket of this type, the measurements of the storage vessel must be specified.

A similar type of jacket can be obtained for tanks and cisterns. Unfortunately it is not always possible to make a snug fit due to the pipework connections, which may result in undesirable gaps between the sections being unavoidable. An alternative is to use one of the insulating materials marketed in sheet form, such as rigid sections of mineral wool or foamed rubber. Yet another alternative is to construct a casing out of light timber or hardboard, allowing about 100 mm all around which is filled with a loose fill material such as vermiculite or expanded plastic granules. When insulating cisterns in roof spaces, if the cistern rests on the ceiling joists, it may be advisable to leave the underside free so that warm air from beneath the ceiling provides a little warmth in very cold weather. Figure 6.40(a) and (b) illustrates the foregoing methods.

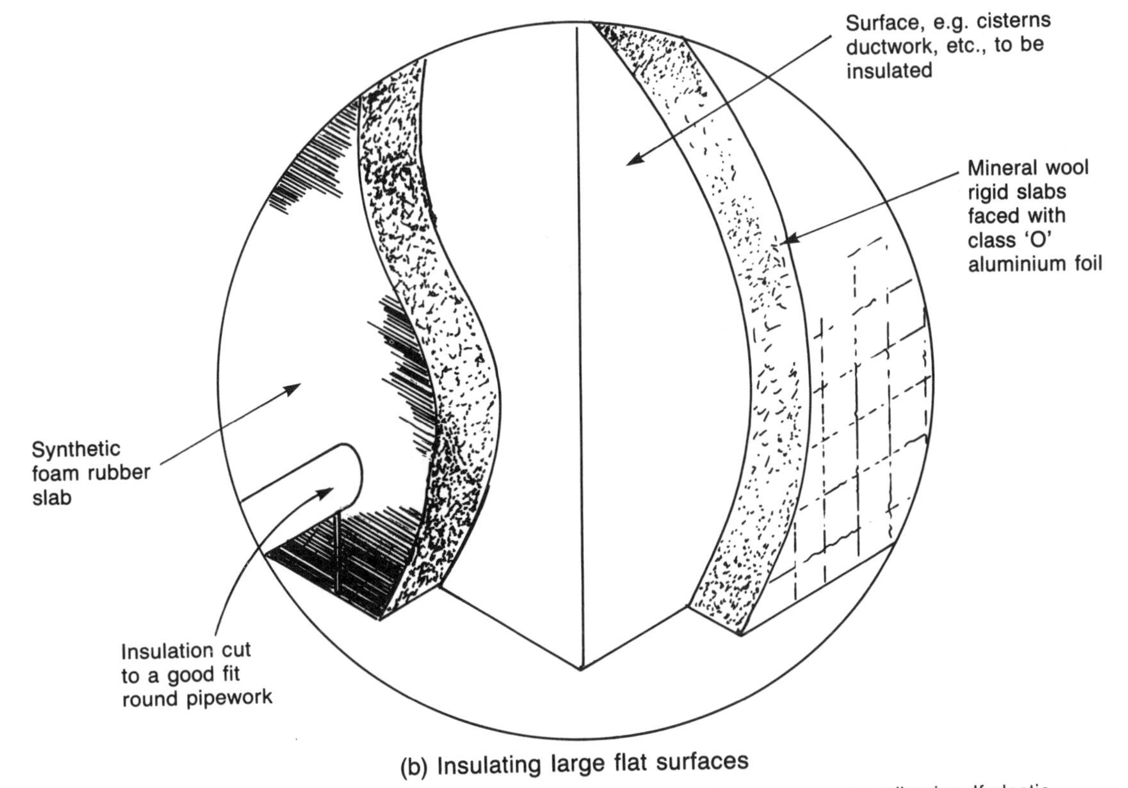

(b) Insulating large flat surfaces

Insulation of the types shown is usually fixed with purpose-made bands or straps or an adhesive. If plastic cisterns are to be insulated check that the adhesive is suitable and will not soften the plastic.

Fig. 6.40 Insulating cisterns

The insulation of large installations is usually the province of specialist firms who use modern techniques to speed up the work of lagging large pipes and surfaces. Asbestos-free mineral fibres mixed with a suitable binder is the usual material used for this type of work and it is often sprayed on. The fuel savings effected with this type of lagging are quoted as almost 50 per cent by one manufacturer and while it may not be possible to show savings of such magnitude on smaller systems, it is a fact that in most cases the cost of insulation pays for itself in a year or two, and is an investment well worth making.

Further reading

Much useful information can be obtained from the following sources:

The Building Regulations
BS 6700 (1987): Design, installation, testing and maintenance of services supplying water for domestic use within buildings and their curtilages.
BS 1566: Specification for double feed indirect cylinders. Part 1: 1984.
BS 699: Specification for copper direct cylinders for domestic purposes, 1984.
BS 5422: Use of thermal insulating materials.

All the foregoing are obtainable from HMSO.

Defect Advisory Sheets
Obtainable from BRE Bookshop, Building Research Establishment, Garston, Watford, Herts, WD2 7JR.
No. 126: Domestic chimneys – Solid fuel flue design.
No. 127: Domestic chimneys – Solid fuel flue installation.

Solid fuel heating appliances
Parkray Ltd, Park Foundry, Belper, Derby, DE5 1WE.
Grahamston Iron Foundry, PO Box 5, Gowan Avenue, Falkirk, FK2 7HH.

Insulation
Pilkington Insulation Ltd, PO Box 10, St Helens, Merseyside, WA10 3NS.
Armstrong World Industries Ltd, Armstrong House, 38 Market Square, Uxbridge, Middlesex, UB8 1NG.

Copper hot water vessels
Range Boilers Ltd, PO Box, Stalybridge, Cheshire.

Self-testing questions

1. State the three forms of heat transfer and how they apply to hot water systems.
2. Describe the methods used to accommodate expansion in long runs of hot water pipes.
3. Describe the function of the primary circulation pipes in a hot water system.
4. Assume that a hot water system is to be installed using copper pipework. State the materials you would recommend for the boiler, hot storage vessel and feed cistern. Give reasons for your choice.
5. Over a period of 2 hours 240 litres of water are to be raised by 50 °C. State the boiler power required in kW.
6. State the fundamental requirements of a well-designed domestic hot water scheme.
7. Explain the reasons for fitting a secondary circulation of hot water supply.
8. Explain the causes of air locks and their effect on hot water systems.
9. Describe two sets of conditions under which a boiler explosion could occur. Name the components that should be used on all heating schemes to prevent a build-up of pressure in the system.
10. State the requirements of good insulating materials.

7 Sanitary appliances

After reading this chapter the reader should be able to:

1. List the materials from which sanitary fittings are made.
2. State the fixing heights of baths, wash basins and sinks.
3. Select appropriate methods of connecting sanitary fitments to various pipework materials.
4. Describe the differences between bidets having an over-rim supply and those fitted with a douche attachment.
5. State the main principle of siphonage and understand its application in connection with flushing cisterns.
6. Describe the methods of support and fixing for domestic sanitary fittings.

Materials for sanitary appliances

The materials from which appliances are manufactured are dependent upon the type of fitment and the use of the building into which it is to be installed. In general terms the materials must be non-corroding, non-absorbent and easily cleaned. In factories and schools materials must also be capable of withstanding rough treatment, but for domestic use this requirement can be discounted.

Metal

Baths, shower trays and flushing cisterns were commonly made out of *cast iron*, although this is being superseded to a large extent by the use of plastics. The interior surfaces of cast iron baths and shower trays are *vitreous-enamelled* to provide a smooth, hard wearing surface that is corrosion resistant. Articles manufactured from cast iron are very heavy and brittle and great care must be exercised when handling them. Due to the high cost of iron sanitaryware, its use is limited to public buildings and commercial use, where its robust characteristics ensures a long working life.

Vitreous-enamelled pressed steel is also used for the manufacture of baths and sink units. It is a cheaper material but does not have the long lasting qualities of cast iron. Manufacturers of *porcelain-enamelled ware* always attach a label to the appliance giving instruction on its care and methods of cleaning. Always make sure that this is intact when the job is complete.

Stainless steel is also extensively used for a wide range of sanitary appliances including WCs, sink units, urinal stalls and wash-hand basins. Its popularity derives from the properties it possesses:

(a) clean, pleasant appearance
(b) non-corrosive
(c) hard wearing
(d) has no vitreous-enamelled surface to chip
(e) easily cleaned

Ceramic

Ceramicware or *potteryware* has always been a popular material for such appliances as WCs, wash-hand basins, urinals, sinks and shower trays. The term ceramic means a substance made by firing clay and includes various forms of pottery such as fireclay, stoneware and vitreous china. Fitments manufactured from fireclay and stoneware would be porous, i.e. would absorb moisture, unless coated with vitreous enamel.

Appliances made from *fireclay* and *stoneware* are strong and heavy and this makes them particularly useful in situations where hard-wearing qualities are essential, i.e. factories and sanitary annexes. Butler's sinks, urinal slabs and stalls, shower trays and WCs are all made from these two materials.

Vitreous china is a special type of earthenware which, as result of high firing temperatures, is made impervious, i.e. will not absorb water. Its hard surface coating serves to aid cleaning and to improve its appearance. The strength of vitreous china allows fitments to be manufactured with a very thin section reducing their weight. Wash-hand basins and WCs are the most common appliances made of vitreous china.

Plastic

Comparatively recently, plastic materials have become increasingly used for sanitary fittings, especially baths and flushing cisterns.

Two types of plastics are used in the manufacture of sanitary appliances: *thermosetting* and *thermoplastic* materials.

Thermosetting plastics are generally harder and have a greater degree of rigidity than thermoplastics. This fact makes them suitable for the manufacture of such components as WC seats and flushing cistern shells. Conversely a thermoplastic material called methacrylate is one of the most common materials used for the manufacture of baths and shower trays for domestic use. It should also be noted that most of the discharge pipework used for sanitation is made of thermoplastics. Although these materials are cheap and light to handle they have the distinct disadvantage of being softer than metals and thus easily damaged. Baths also suffer from movement when in use because thermoplastics are flexible. This makes it essential to provide a wooden cradle to give the necessary stability.

The specific heat, or heat capacity, of plastics is lower than that of metals, so less heat is absorbed from the hot water by the appliance, which is an advantage especially in relation to baths.

Sanitary appliances

Butler's sinks

These are usually specified as 'London' or 'Belfast' sinks. London sinks were originally less deep than Belfast types, but the only difference nowadays is that the Belfast pattern has an integral overflow while the London has not. Both types are made of glazed fireclay and are capable of very hard wear. They are made in a comprehensive range of sizes, but those for domestic use usually conform to the measurements shown in Fig. 7.1.

Due to the thickness and the rounded top edge it is difficult to make a watertight joint between such a sink and the wall. The best way of overcoming this problem is to cut away the plaster behind the sink so that it abuts the brickwork. The joint can then be made good with cement mortar and tiled across its top edge to make it watertight (see Fig. 7.2).

The recommended height from the top edge of a butler's sink to the floor is 865 mm.

Cleaners sinks

These are seldom necessary in small domestic households, but do differ from ordinary sinks. The main difference is that cleaners sinks are fitted with either a galvanised or brass-hinged grating on which buckets may be rested while being filled. They are usually fitted at a low level to avoid unnecessary lifting of the bucket and a hardwood face is fitted on the front edge to reduce damage

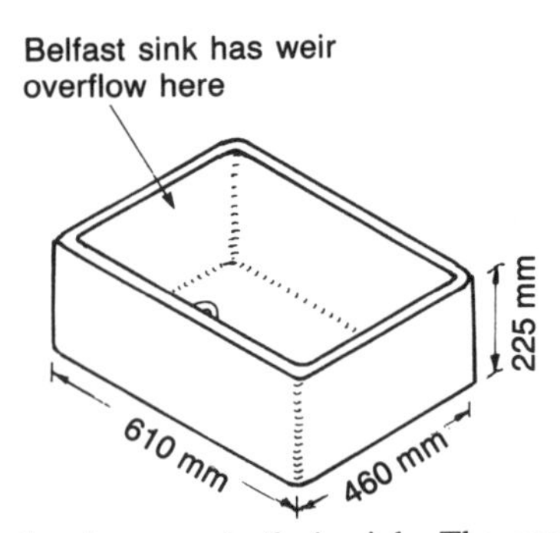

Fig. 7.1 London type butler's sink. The general appearance of both London and Belfast sinks is the same excepting that the Belfast pattern is fitted with a weir overflow. Both types are constructed of glazed fireclay and are made in various sizes. The dimensions shown are representative of the most common size.

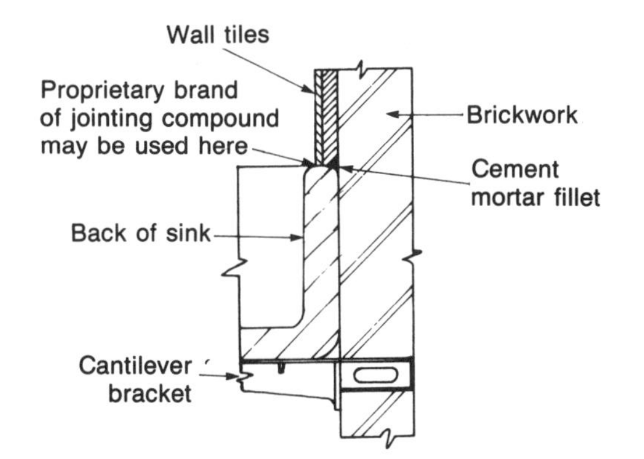

Fig. 7.2 Making a watertight joint between a butler's sink and the wall surface. Plasterwork must be cut away behind a butler's sink so that the wall tiles finish over the rounded top of the sink. Failure to do this will lead to damp and insanitary conditions behind the sink.

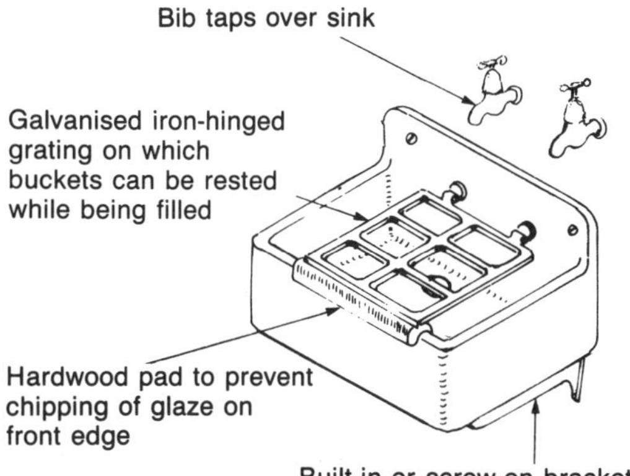

Fig. 7.3 Cleaner's sink. These sinks are usually fitted between 300 and 450 mm from the floor level to facilitate easy removal of the bucket from the grating.

to the glaze. A typical example of a cleaner's sink is illustrated in Fig. 7.3. They are normally supported on cantilever brackets with the upstand at the back screwed to the wall with large brass screws to prevent it sliding forward. Hot and cold taps are fitted at a suitable height over the sink.

Sink units

This term applies to units where the sink and draining boards are pressed or moulded as a complete component, rendering unnecessary the often insanitary wooden draining boards sometimes associated with butler's sinks.

The most popular materials for construction of sink units are stainless steel and vitreous enamelled steel. Both these materials are easily pressed into shape by modern manufacturing processes. They are almost invariably fitted to a cabinet which determines the height of the front edge, usually 900–915 mm from the floor. The sink should be a good fit on the cabinet and secured to it firmly by the clips provided. The cabinet should also be screwed to both the wall and floor.

There are many differing combinations of sink units to suit a variety of purposes, some of which are illustrated in Fig. 7.4.

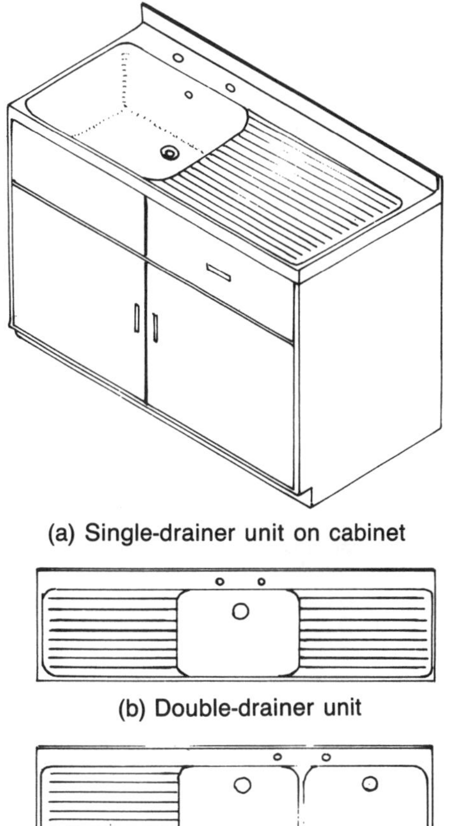

(a) Single-drainer unit on cabinet

(b) Double-drainer unit

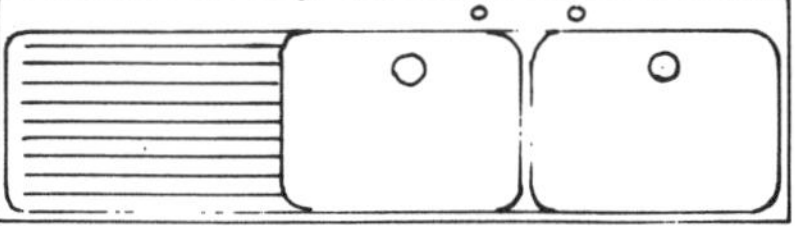

(c) Double sink single-drainer unit

Note position of tap holes for mixer taps.

Fig. 7.4 Sink units. Standard widths of sinks is 500 or 600 mm, depending on the type of cabinet used.

Baths

Baths vary little in their basic shape whether they are made of cast iron, steel or acrylic, but refinements in design are available beyond the standard rectangular type to include such features as soap sinkings, hand grips or dropped front edges (see Fig. 7.5).

Baths made for corner fittings are also becoming increasingly popular. The measurements shown in Fig. 7.6 apply to standard baths, but longer and wider baths are available if required, the full range of measurements being quoted in manufacturers' handbooks.

Baths of cast iron have adjustable feet so that the bath height can be varied, but it is generally recommended that they are fixed as low as possible so they can be used more easily by the elderly and young children. Most modern baths are designed to be panelled in, some being supplied complete with a panel which determines the fixing height. Panels should be constructed and fitted in such a way that they are easily removable for maintenance purposes. Similarly, pipes should not be run in inaccessible positions behind baths.

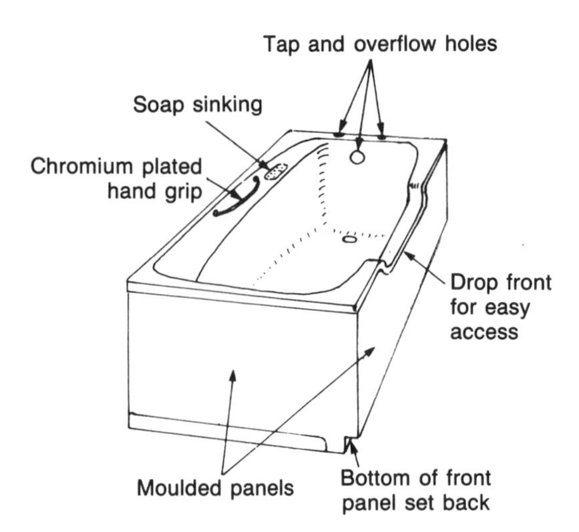

Fig. 7.5 Typical moulded acrylic bath

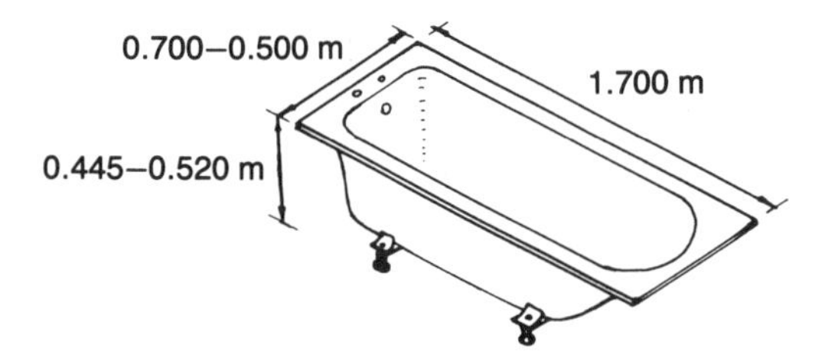

Fig. 7.6 Measurements of a standard size cast iron bath. The height of the bath is variable depending on the adjustable feet. Generally they should be as low as conveniently possible, a hole being cut in the floor to accommodate a deep seal trap.

Wash basins

These are bowl-shaped fittings used for ablutionary purposes. They are made of glazed earthenware, vitrified china or when they are required to be particularly strong, durable, and vandal-proof, stainless steel or cast iron. Sizes of basins vary considerably, the standard size conforming to BS 1188 (see Fig. 7.7(a)). Smaller basins for hand washing only (Fig. 7.7(b)) are often fitted in WC apartments where space is limited. Angle basins (see Fig. 7.7(c)) can be fitted in similar circumstances to save space.

Tap holes in basins vary considerably depending on the choice of taps required. Basins having one centre hole only, or three holes, are available to accommodate the wide variety of mixer taps that are currently produced. To avoid special orders many basins are made now with provision for three 'knock out' holes. Having selected the holes necessary to accommodate the appropriate taps, the glaze covering the hole *must* be tapped out with a ball pein hammer from the 'face' side of the appliance, not the underside. Failure to do this may result in chipping the surface glaze in such a way that it is not covered by the flange on the tap.

Wash basins of all types are supported by a variety of methods which are dealt with later in this chapter. The pedestal shown in Fig. 7.7(a) is most popular in domestic dwellings for the larger type of basins. The basin should be well secured to the pedestal to prevent any movement, and for the same reason, the pedestal must be firmly screwed to the floor via the holes in its base. Two alternative methods of securing the basin to the pedestal are illustrated in Fig. 7.8(a) and (b). The former shows a bracket which fits round the exposed waste fitting below the basin which is adjustable to meet any variation in the centre line of the waste to the wall.

Another type of fitting is illustrated in Fig.

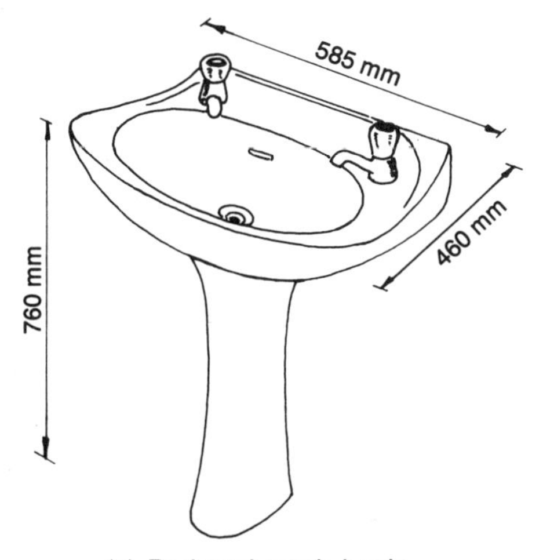

(a) Pedestal wash basin

The height of the basin should not vary substantially from that shown whether brackets or a pedestal is used. The measurements shown are generally representative of this type of basin

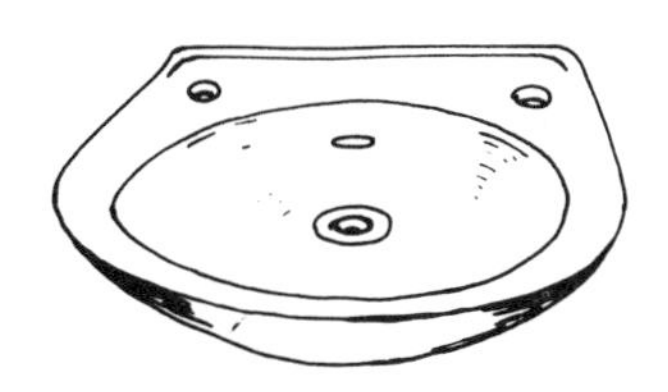

(b) Small hand washing basin suitable for WC apartments

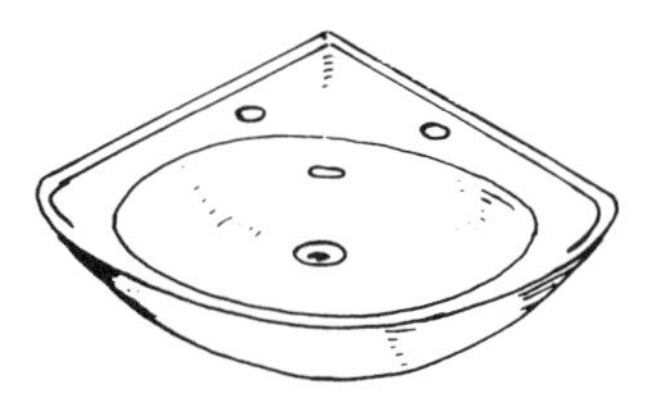

(c) Angle or corner basin

These basins are useful in bathrooms or toilets where space is limited.

Fig. 7.7 Wash basins

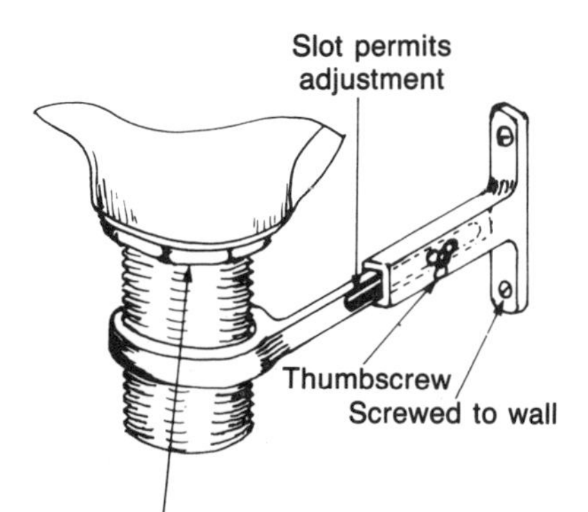

(a) Adjustable basin waste bracket for pedestal basins

This type of bracket is used to prevent movement of basins fitted on pedestals

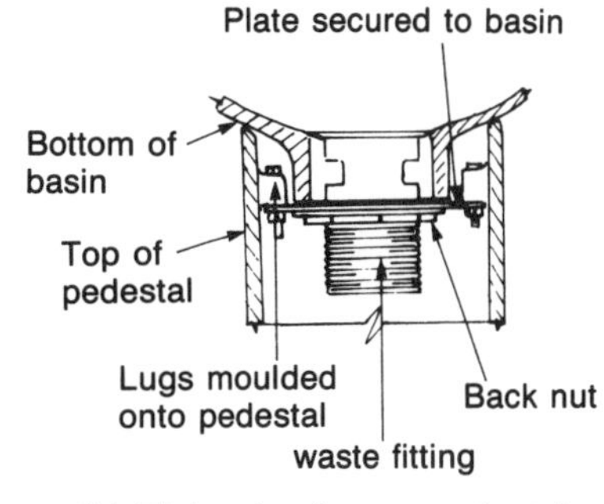

(b) Fixing basins to pedestals

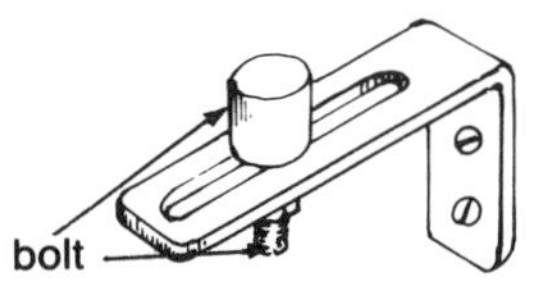

(c) Wall fixing for pedestal basins

This raised portion of the bolt is adjustable and is made to engage into two holes under the side edges of the basin. It effectively ties the basin back to the wall

Fig. 7.8 Fixing pedestal basins

7.8(b). It consists of a flat plate which is fitted into position when the waste is fitted into the basin and is attached by means of metal screws and wing nuts to the pedestal. This firmly fixes the basin to the pedestal, but the basin must also be firmly screwed back to the wall through the holes on the underside.

Always remember a little grease applied to mechanical fixings of this type enables them to be fitted much more easily and, if necessary, removed at a later date. All excess jointing paste must also be removed from threads to facilitate removal after fixing. Angle brackets (Fig. 7.8(c)) are often supplied with pedestal basins to enable the basin itself to be fixed back firmly to the wall. Basins are normally fitted on the face of tiled walls and should be made watertight by a suitable mastic joint between basin and wall.

Vanitory units

Although used for the same purpose, a vanitory unit is not a wash basin in the strict sense of the word. It is simply a bowl, made of vitreous china, stainless steel or vitreous enamelled steel, and fitted into a prepared cabinet with a wooden top, this being cut away to admit the bowl. The main consideration when dealing with these units is to ensure an absolutely watertight joint between the woodwork and the bowl to prevent seepage of water into the cabinet. Manufacturers supply the bowl with the necessary jointing compound and sufficient clamps to fix the bowl securely into the woodwork.

Water closets

Modern water closets may be of the wash-down or siphonic type. The former relies on the momentum of the flush of water from the cistern to remove the contents of the trap while in the latter case the contents are removed by siphonic action.

The design of the wash-down closet has not varied a great deal since it was first produced. The integral trap has a 50 mm depth of seal and an outgo diameter of approximately 90 mm. Figure 7.9(a) illustrates a typical WC which can be converted to an 'S' trap or side outlet using a bent WC connector made of plastic material, but two-piece WCs are available which permit the trap to be fitted to any angle. These are very useful for connecting to the non-standard outlets which are often found in older properties, both P and S outlets being available. The one disadvantage with this type of WC is the difficulty of making the swivel joint. Great care must be exercised to ensure a smooth internal bore throughout the trap (see Fig. 7.9(b)).

Another type of WC, very popular in industrial and public buildings, is the corbel type which has the advantage of leaving the floor area clear for cleaning. Shown in Fig. 7.9(c) this type is designed for use with a plumbing duct. The water and waste services are all installed behind the panel forming the duct, which is an advantage in appearance and in situations such as hospitals,

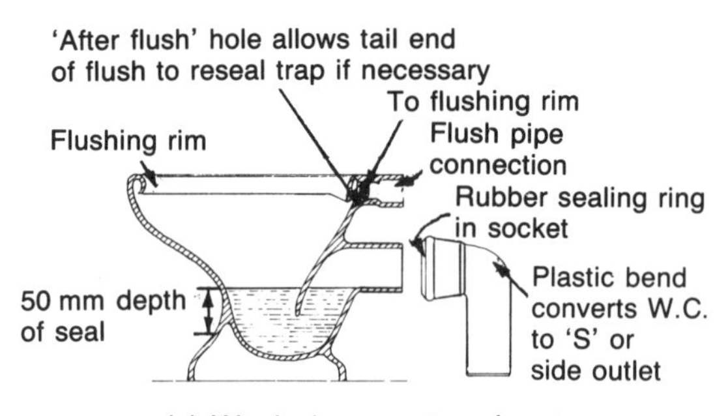

(a) Wash-down water closet

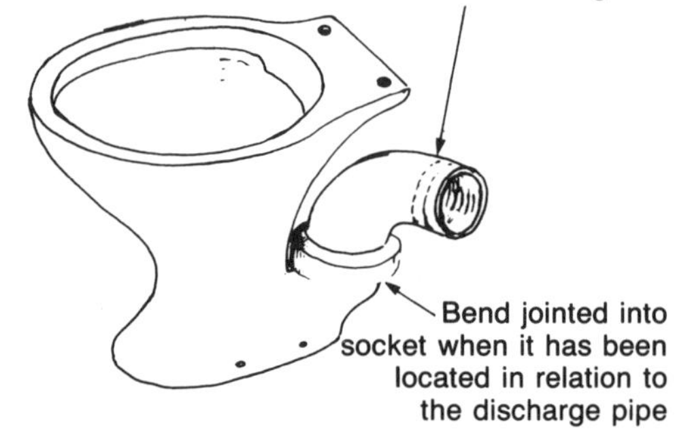

(b) Two-piece water closet

The use of these WCs had two main disadvantages: (1) the difficulty of removing any surplus jointing material from inside the WC and (2) the joint is permanently below the water seal.

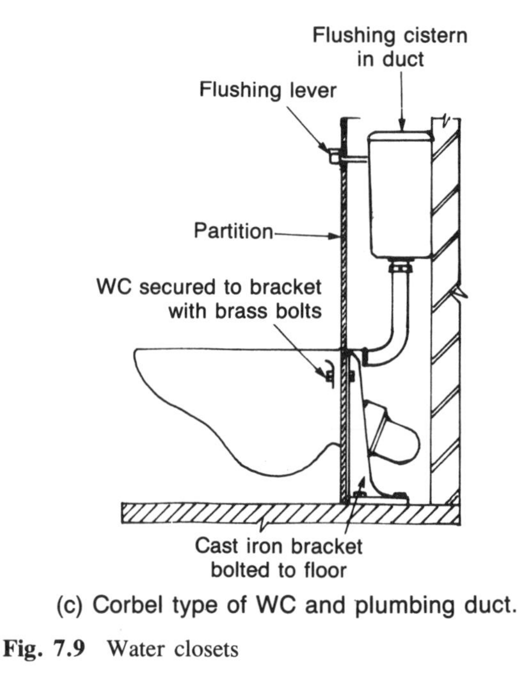

(c) Corbel type of WC and plumbing duct.

Fig. 7.9 Water closets

schools and offices since the wall and floor area around the WC can be easily cleaned. The ducting also reduces the possibility of vandalism in public conveniences.

The imposed strains that occur when the WC is in use are taken by a cast iron frame situated in the duct to which the WC is secured. Although the plumber seldom constructs the ducts and partitioning, he should ensure that adequate access is available for future servicing of the installation.

Two good tests may be applied to WCs to ensure that the whole of the bowl surface is reached and scoured by the flushing water. First, a suitable water-bound paint (thinned out emulsion may be used for this purpose) is applied to the inside of the bowl. When the cistern is flushed all the paint should be washed off or at least wetted. The second test relates to the ability of the flush to clear the trap. Six to eight sheets of toilet paper are placed in the bowl and the cistern flushed. If the paper is not removed with one flush, then check the flush pipe joint for obstruction.

Some plumbers use putty or mastic, in most cases unnecessarily, to seal the plastic flush pipe connection to the WC. If too much of such jointing material is used it may obstruct the holes leading to the flushing rim and reduce the flushing action. If a check on this is negative an investigation should be made on the joint between the WC and the drain which also could be partly blocked due to unsatisfactory jointing.

The minimum diameter of flush pipe for high-level cisterns is 32 mm nominal diameter.

Low-level WC suites The practice of fitting cisterns at a lower level originated with the need for a better appearance, the loss of height being compensated for by a larger flush pipe, usually 40 mm nominal diameter. A further development of this idea is the close coupled suite where the cistern is fitted directly to the WC without a flush pipe (see Fig. 7.10).

A larger capacity flushing cistern would possibly solve the problem, but this is no longer permissible. Although close couple units with wash down closets are made, it will generally be found that a siphonic WC will be more effective in clearing the contents of the trap.

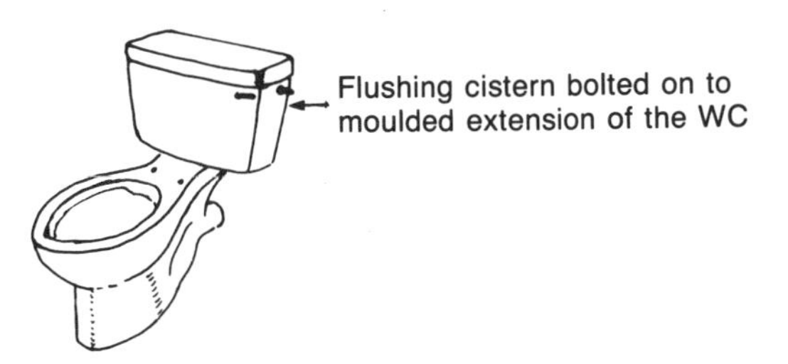

Fig. 7.10 Close-coupled WC

Siphonic water closets

The siphonic closet has many advantages over the wash-down closet. When the cistern is flushed, a more positive action is exerted on the contents of the WC bowl, due to the siphonic action set up in the closet itself. In the case of wash-down closets, the contents of the bowl are emptied solely by the momentum of the flushing water. This is effective, although rather noisy, where high-level flushing cisterns are used. Low-level and close-coupled suites are increasingly used these days and are subject to the problems previously described, this being due to the loss in head or pressure of the flushing water.

Siphonic closets have many other advantages from both health and sanitary points of view. They are quiet in action, and as the surface of the water seal covers a larger area than that of wash-down closets, there is less possibility of fouling the bowl. In performance tests with wash-down and siphonic closets, it was found that wash-down closets, when flushed, cause small quantities of water in the form of a fine spray to be thrown over the edge of the bowl. This is due to the more violent action of the flush and could result in what is in effect an unseen bacteria-laden aerosol spray being deposited on the floor adjacent to the WC.

There are two main types of siphonic closet, the single-trap type and the double-trap type. The single-trap type is no longer manufactured, as a reduction in the bore of the outgo was necessary to start siphonic action. Due to misuse this was the cause of many blockages with this type of WC. The double-trap type, however, does not have this disadvantage, having full bore throughout.

Double-trap siphonic closet The siphonic action in

this type of closet is started by a reduction of the air pressure between the two traps. The way in which this is achieved can best be understood by looking at the principle of a filter pump as shown in Fig. 7.11. A filter pump is a simple device in which a small jet of water is discharged from the inlet nozzle (A) to the outlet (B), causing a lowering of the air pressure in area C. Atmospheric pressure on the surface of water in the reservoir (D) expels the air from the pipe (E) as it forces water upwards into the outlet (B).

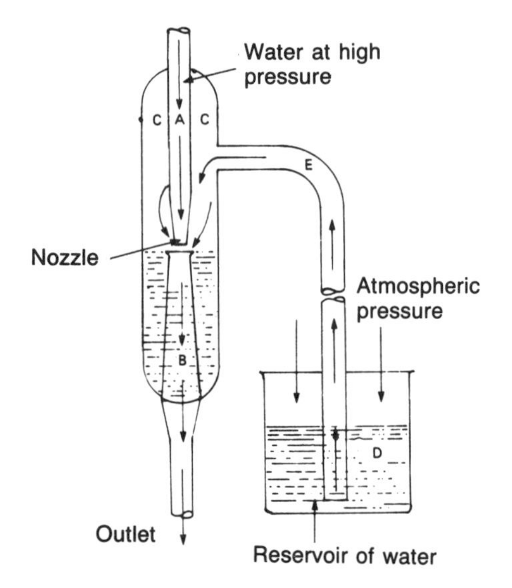

Fig. 7.11 Principle of a filter pump. Water at high pressure discharged from the nozzle 'A' spreads out in the shaped tube 'B', abstracts air from the pump body 'C' causing a reduction of pressure. Atmospheric pressure on the surface of the water in the reservoir 'D' forces it upward to discharge through the outlet tube 'B'.

Figure 7.12 shows a section through a typical double-trap siphonic closet in which, for it to act effectively, a pressure less than atmospheric pressure must be set up in the space between the two traps. Trap B has a full 50 mm seal depth; trap C is shallow, its purpose being simply to retard the discharge of water when the cistern is flushed to allow sufficient time for air to be withdrawn from the space between the two traps shown as A. The principle used to achieve this is almost the same as that of the filter pump: water flowing through the pressure-reducing fitting in the flush pipe abstracts air from area A thus lowering the pressure between the two traps. As in all cases of siphonage, atmospheric pressure acts on the free surface of water (in this case on the water in the WC bowl) forcing it into the outlet and forming a solid plug which not only removes the contents of the bowl, but that of trap B as well. This is replenished by the after flush water retained in the reservoir (D).

Figure 7.13 shows details of the components used to lower the pressure between the two traps.

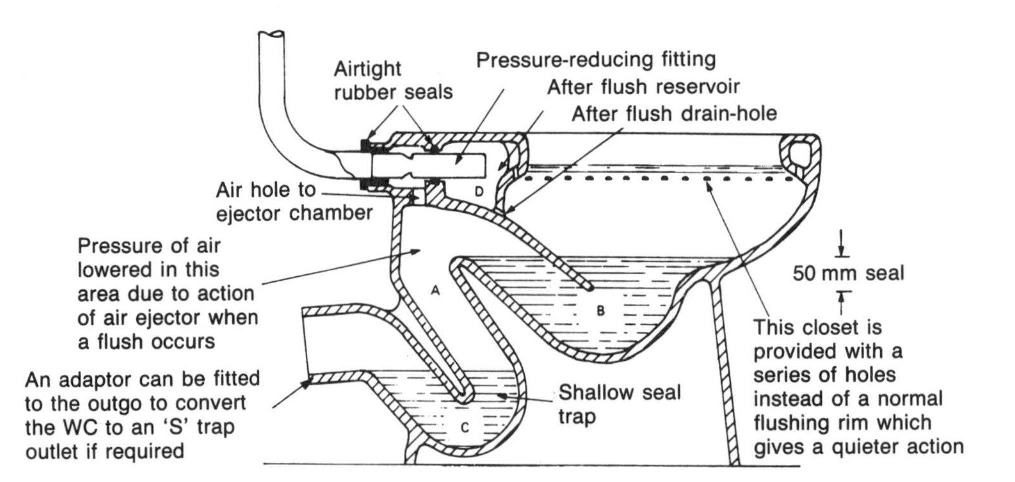

Fig. 7.12 Double-trap siphonic closet

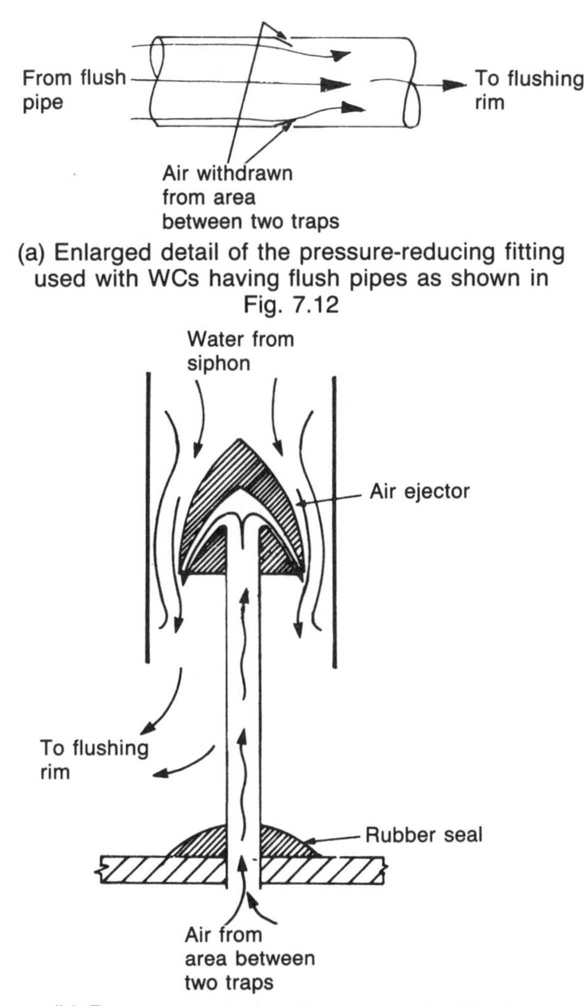

Fig. 7.13 Pressure-reducing fitments used with double trap siphonic closets

Figure 7.13(a) illustrates the special adaptor for an appliance fitted with a flush pipe, and Fig. 7.13(b) shows a special pressure-reducing fitting used with close-coupled appliances which have no flush pipe. When fitting these components, care must be taken to ensure that all the rubber seals are correctly positioned and air-tight. Failure to observe this and any special fitting instructions supplied by the manufacturer will result in inefficiency or failure in the performance of the appliance.

Flushing cisterns

The type exclusively used employs the plunger type of siphon and is suitable for both high- and low-level arrangements. The older Burlington or well-type cast iron cisterns are now obsolescent, and when it becomes necessary, they are replaced with a modern type of cistern.

Working principles of flushing cisterns

The function of these appliances is to flush away the contents of soil fittings to the drain. Their basic action involves the discharge of a body of water, by siphonic action into the WC. Before discussing the operation of the cistern, it is important to understand the principle of the siphon and the part played by atmospheric pressure in its action.

Siphons Air, like everything else on the earth, is pulled towards the earth's centre by the force of gravity. The pull of gravity on the atmosphere, which we usually refer to as atmospheric pressure, is 101.3 kN/m^2 at ground level. Since commercial pressures are normally indicated by the 'bar', it will be seen that the pressure of the atmosphere is approximately 1 bar. To give a practical illustration of atmospheric pressure so that some idea of the force it exerts can be imagined, the pressure created by a column of water approximately 10 m high will be nearly the same as that exerted by the atmosphere. There is a slight variation of pressure due to the contours of the earth. For example, the atmospheric pressure will be less great on a mountain than at sea level but this can be ignored as far as siphonage in plumbing is concerned.

The simple experiment shown in Fig. 7.14 shows how a siphon works. In cistern A the pressure of the atmosphere is the same inside the bent tube as that on the surface of the water before siphonic action commences. If, however, air is withdrawn from the tube, the atmospheric pressure exerted on the water in A will be greater than that in the tube and will force the water up and over the bend to discharge into the lower cistern B.

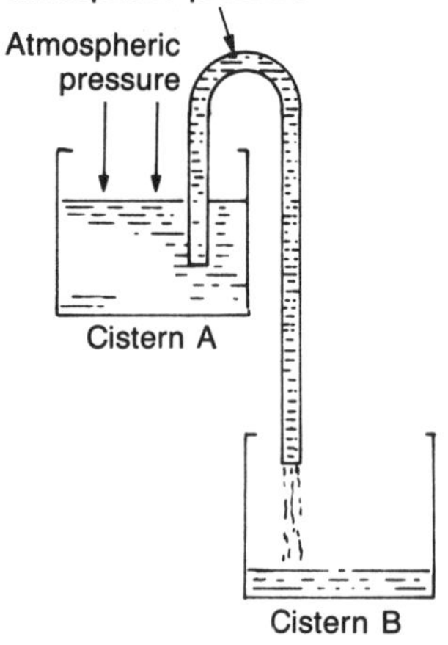

Fig. 7.14 Experiment showing how a siphon works. The basic action of the siphon has many applications in plumbing where, for instance, it may be necessary to empty a hot storage vessel due to a defective or non-existent drain-off cock.

This principle has many applications in plumbing, one of the most common being the flushing cistern.

A similar arrangement to that shown is often employed to empty hot storage vessels in hot water systems fitted with defective drain-off cocks. The action of siphonage also affects the working principles of discharge pipe systems as will be seen in Chapter 8.

Flushing cistern syphons The 'plunger' or 'piston' type shown in Fig. 7.15 is now used for all manually operated flushing cisterns. Its working principles are not new and from the early period of modern sanitation were in use with better-class flushing cisterns. This type of siphon was often fitted into a lead or copper-lined wooden box before cast iron cistern shells became popular. It also fits very conveniently into modern types of ceramic or plastic flushing cisterns as it does not require a well, thus giving the cistern much cleaner lines.

The siphon apparatus for these cisterns was originally made of brass, copper or a lead alloy. They are now made principally of polythene which is both cheap and corrosion resistant.

When the handle is depressed it lifts the plunger or piston carrying the water in the cylinder over the siphon bend. As with the well type of cistern, when the water falls down the flush pipe it carries away the air in the siphon bend, lowering the air pressure in the outlet leg of the siphon. The atmospheric pressure on the surface of the water in the cistern is then greater than that in the siphon causing the water to flow through the piston, which is not solid but spoked like a wheel. The piston simply supports a rubber or plastic washer which gives the effect of a solid piston only on the upstroke. When the piston has reached its maximum length of travel in an upward direction, water continues to pass through it, pushing past the piston and the flexible washer until the cistern is emptied and air is admitted to the underside of the piston, halting the siphonic action. A sketch showing some of the details of this action is shown in Fig. 7.16(a) and (b).

Defects with this type of cistern are uncommon unless the cistern is misused. The washer supported by the piston may distort or perish after a long period of time but it is easily replaced when the siphon is removed. The hole through which the piston rod passes will wear in time, especially with those siphons made of brass. When this occurs, the cheapest form of repair is to renew the complete unit. Make sure the correct type is replaced, carefully measuring the height and the diameter of the piston. The simplest solution is to take the defective siphon to a builder's merchant and match it exactly.

Most flushing cisterns are reversible, i.e. the ball valve can be fitted into either the left-hand or right-hand side of the cistern which means both the inlet and the overflow are at the same level. To ensure the inlet discharges into the cistern at a level above that of the overflow, diaphragm ball valves with top outlets which meet the requirements of the Water Bylaws are supplied with new cisterns.

Water conservation

All flushing cisterns, except those that operate automatically, must be designed in such a way that in the event of an overflow the appliance will create such a nuisance that the defect will be

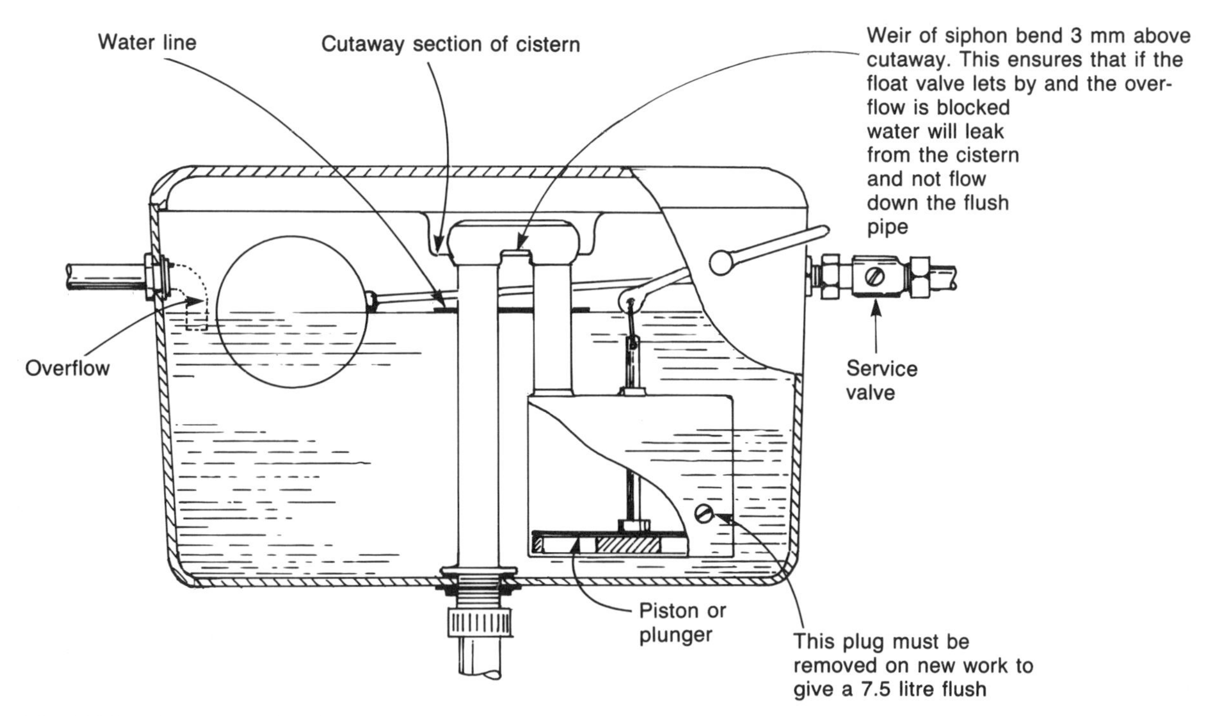

Fig. 7.15 Plunger-type flushing cistern

repaired quickly, thereby limiting the waste of water.

Water authorities insist on the cistern overflow pipe discharging in an obvious location to prompt speedy repair. The old original 'valve type' cisterns which are no longer in use were often not fitted with an overflow pipe and allowed overflow water to discharge down the flush pipe into the WC. As this did not create a nuisance many people did not bother to effect the necessary repairs which resulted in considerable wastage of water. Water authorities will now only permit the waste water preventing type of flushing cistern to be used. A close inspection of Fig. 7.15 shows a cut-away area falling below the invert of the siphon bend. This arrangement prevents the entry of water to the flush pipe by any means other than manual operation of the siphon.

Dual flushing cisterns were also introduced in an effort to conserve water, but due to the fact that most people did not understand how to use them properly, more water was wasted than was saved. Because of this the 1987 Water Bylaws specified that after 1 January 1993, in all new installations, flushing cisterns will be single flush only, but made to deliver 7.5 litres instead of the 9 litres hitherto permissible. This bylaw does not apply to the replacement of those cisterns fitted before 1993. Rather than reduce the actual capacity of flushing cistern shells, which would involve manufacturers in considerable expense, this problem has been overcome in a very simple way. All siphons now produced are fitted as shown in Fig. 7.15, which must be removed on all new installations. This will permit only 7.5 litres of water to discharge when the cistern is operated. When an existing cistern is replaced the plug is left in position thus allowing for a full 9 litre flush.

Flush pipe joints Flush pipes are joined to the WC by a flexible rubber or plastic connector, the most common type being shown in Fig. 7.17(a). To be effective they must be tight fitting and as such are sometimes difficult to pull over the WC nozzle. This operation will be simplified if the surfaces are wetted with water or a little washing-up fluid.

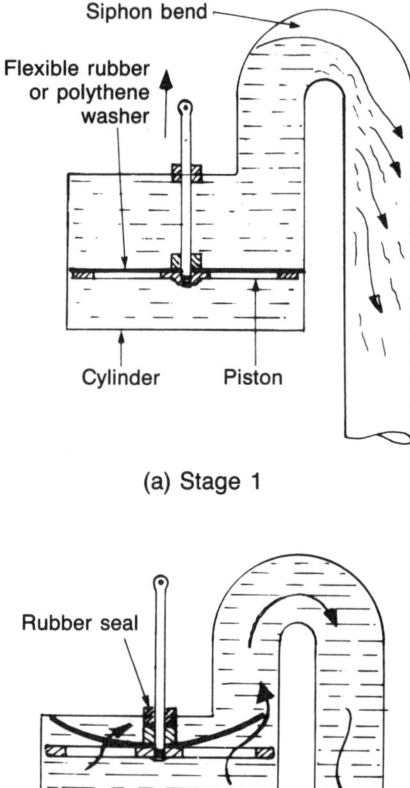

(a) Stage 1

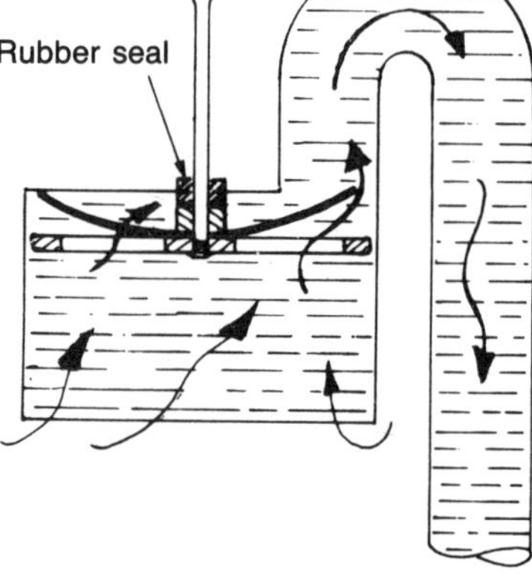

(b) Stage 2

When the piston reaches the end of its travel, water continues to flow through the piston and the flexible washer. As it does so siphonage will continue until the cistern empties and air is admitted under the edge of the cylinder thus breaking the siphon action

Fig. 7.16 Flushing action details

Another type of connection between a WC and the flush pipe is shown in Fig. 7.17(b), this type being usually provided with the WC and flush pipe as a set. They are not suitable for all WCs, being either too tight or too loose. When fitting flush pipes, make sure they do not enter the nozzle of the WC too far, as this very often results in a defective flush.

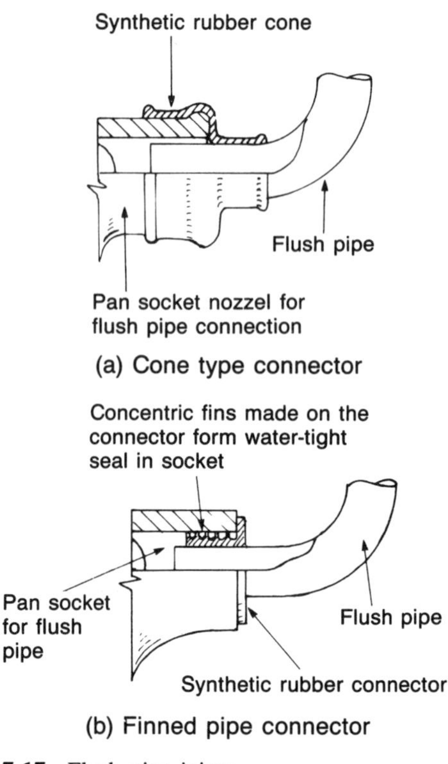

Fig. 7.17 Flush pipe joints

WC joints to the drain The joint between the WC and the drain will depend upon the material of which the drain is constructed. A very versatile joint marketed under the trade name of 'Multiquick' can be used with nearly all materials, including plain ended pipes, and is illustrated in Fig. 7.18.

They are especially useful in cases where a WC connected to a clayware drain is to be replaced. It is often found the clayware socket is cracked, or if this is not so it is almost impossible to cut away the existing cement joint without cracking it. The easiest way to overcome this problem is to cut the socket away completely until it is flush with the floor. The 'Multiquick' can now be simply inserted into the drain. Care must be taken to seal the hole with paper or rags so that chippings are not allowed to enter the drain where they may contribute to a blockage.

Connections of WCs to clayware or cast iron socketed pipe are illustrated in Fig. 7.19. If the WC is fitted on a solid floor where there is no

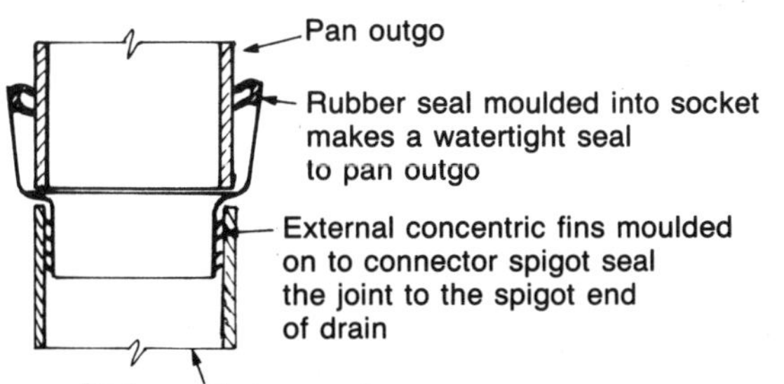

(a) Section through 'Multiquick' pan connector

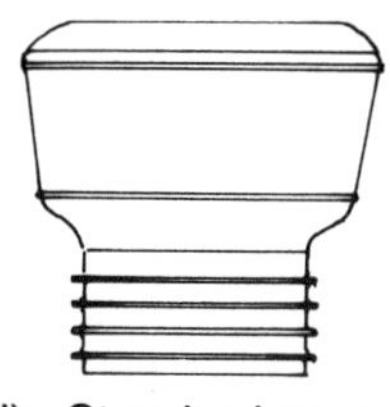

(i) Standard type

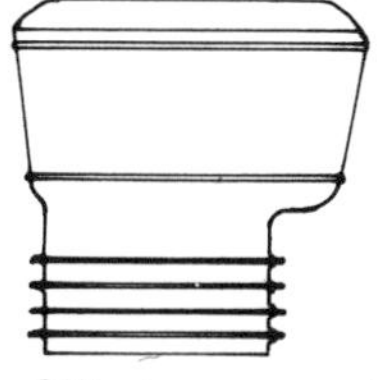

(ii) Offset or eccentric

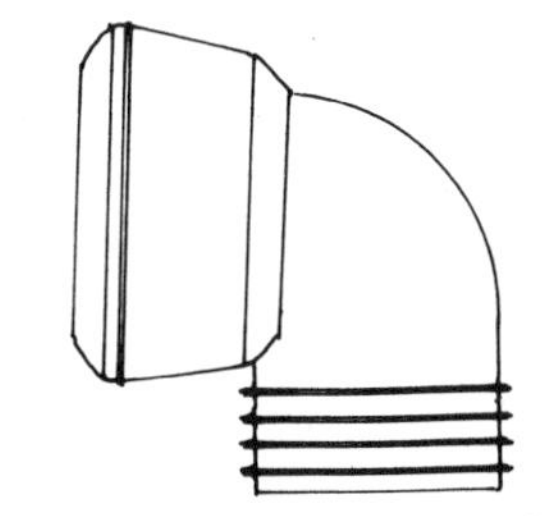

(iii) Bent connector can be used in some circumstances to convert a 'P' trap WC to side or 'S' outgo

(b) Typical 'Multiquick' connectors

Fig. 7.18 'Multiquick' joints

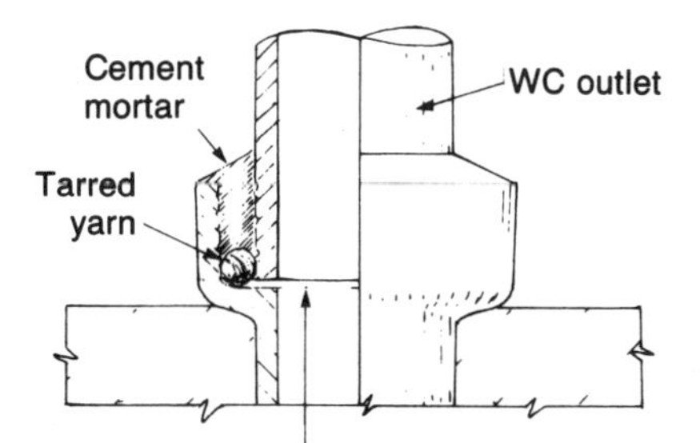

(a) Connection to earthenware socketed pipe

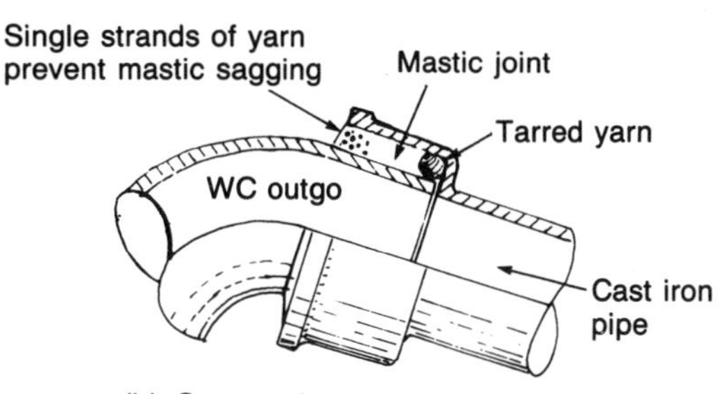

(b) Connection to a cast iron pipe

This type of joint must be used where the WC is sited on a suspended floor. The joint, being soft, will give slightly to accommodate any shrinkage of the timber.

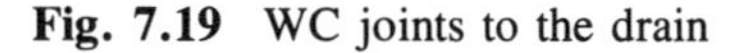

Fig. 7.19 WC joints to the drain

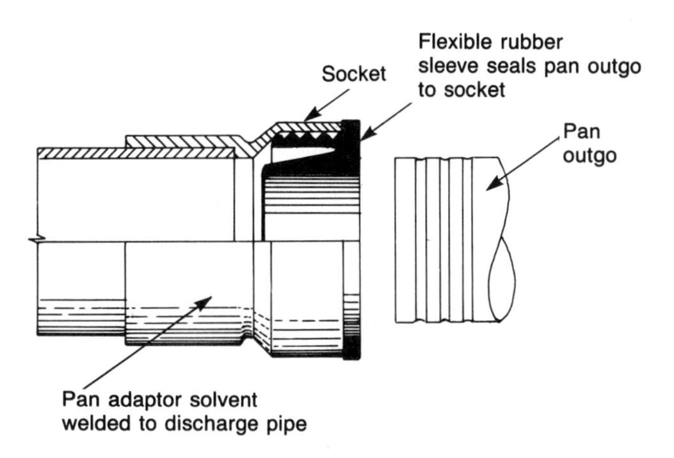

Fig. 7.20 WC connector for PVC discharge pipe

likelihood of movement or shrinkage, and assuming the drain pipe to be either cast iron or earthenware, the outlet joint can be made with yarn and cement mortar as shown in Fig. 7.19(a), or by a soft joint using yarn and a proprietary mastic. The advantage of using the latter is that the WC can, if necessary, be removed without breakage (see Fig. 7.19(b)).

There is a wide range of connections available for connecting WCs to plastic discharge pipes. Figure 7.20 illustrates a typical example having a synthetic rubber seal, common to all such connections, which requires no other materials to make a sound joint. When making rubber-sealed joints, using a lubricant will make the task much easier. As these joints are inherently flexible, any shrinkage of a timber floor is unlikely to result in a broken WC.

Fixing WCs to floors When WCs are fixed to wooden floors, especially new ones, a considerable amount of shrinkage can be expected and the type of joint illustrated in Fig. 7.19(b) should always be used to avoid damage to the WC. A WC must be rigidly fixed on all types of floors with suitable brass screws to facilitate their easy removal. Do not overtighten the screws and make sure they enter into the floor at the same angle as the holes in the WC. Failure to observe these two points often results in a cracked pan.

Bidets Bidets are an ablutionary fitting designed for cleansing the excretory organs, or for use as a foot bath. Figure 7.21 shows a bidet without a spray nozzle normally associated with these fittings. It is equipped with pillar taps and a normal waste fitting, the same in fact as those used for wash basins.

The bidet shown in Fig. 7.22 is fitted with a mixing tap set, pop up waste and spray jet. The mixing valve is constructed so that a mixed supply of hot and cold water can be delivered through the rim so that it is warmed prior to use. The same water fills the bowl for washing purposes. Alternatively, the mixed supply can be diverted to the spray jet for the purpose of douching. Because the spray jet is situated below the flood level of the fitting there is a high risk of water pollution which must be taken into account when the hot and cold supplies are fitted, and water authorities have stringent regulations concerning these connections.

Due to its bowl shape and its use, the waste from a bidet should be treated in the same way as that of a wash basin with the same restrictions on its length, diameter and fall. Discharge pipes to bidets fitted at the ground-floor level, if not connected to the main discharge stack, should discharge directly to the drain or be fitted to a back inlet gulley in case of possible misuse.

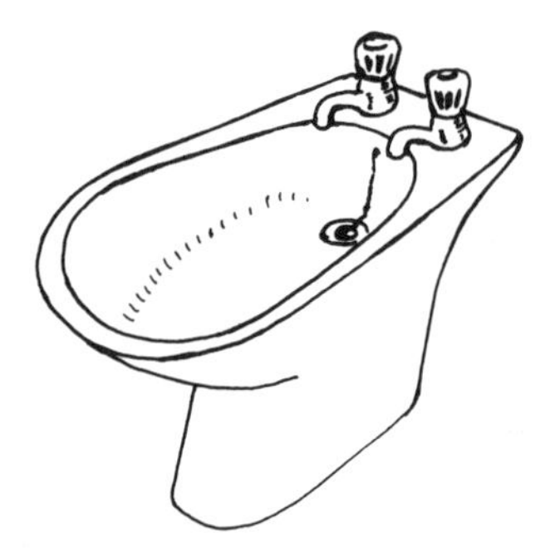

Fig. 7.21 Bidet without spray jet below flood level of the fitting. There is no risk of back siphonage with this type of bidet

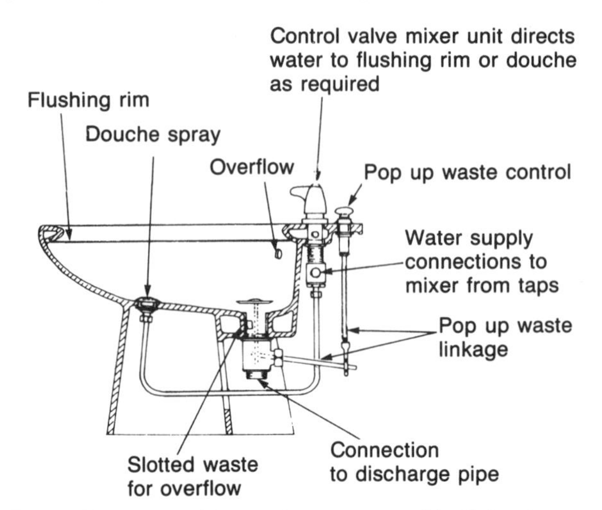

Fig. 7.22 Bidet with douche spray. As this fitting has a water inlet below its flood level, the cold supply must never be connected directly to a mains water supply, i.e. it must be fed from a cistern

Diagrammatic details of the integral pipework of a bidet are illustrated in Fig. 7.23 and show how a mixed supply of water is distributed to the rim and douche jet. Two main points should be borne in mind when fitting bidets with a douche attachment. They must be supplied with cold water from a cistern, not from the mains, and both hot and cold supplies must be of equal pressure and quite separate with no branches to other fittings. Further details of the Water Regulations relating to bidets are dealt with in Book 2 of this series.

Overflows to sanitary fittings

Overflows are provided with most sanitary fittings with the exception of some types of sinks and shower trays. They may be fitted separately as in the case of flushing cisterns and baths, or they

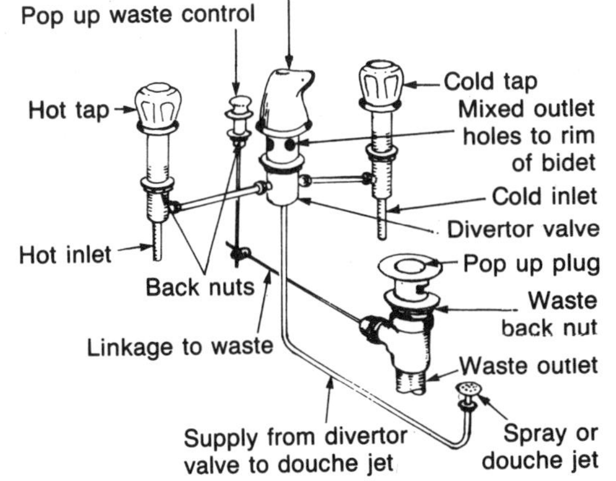

Fig. 7.23 Integral pipework and waste linkage for bidets

may be an integral part of the appliance as in the case of basins and sinks. Those described are illustrated in Fig. 7.24.

Weir overflows

This type of overflow illustrated in Fig. 7.24(a) is capable of being cleaned to some extent, and as such is often accepted by sanitary authorities. They are used with Belfast sinks and some types of wash basin. Hospital sinks do not have overflows, thus avoiding inaccessible areas that are difficult to keep clean.

Slotted overflows

A type of integral overflow, often referred to as the 'slotted' or 'secret' overflow, is incorporated in the design of most modern wash basins and bidets (see Fig. 7.24(b)). The main objection to its use is the difficulty of effective cleaning.

Separate overflows

Fittings such as baths and flushing cisterns have separate overflows which discharge externally from the building, except in high rise dwellings where a common overflow is often employed. One of the problems associated with bath overflows which discharge externally is that they admit cold draughts into the building and a method of overcoming this is shown in Fig. 7.24(c). In the case of flushing or water storage cisterns the

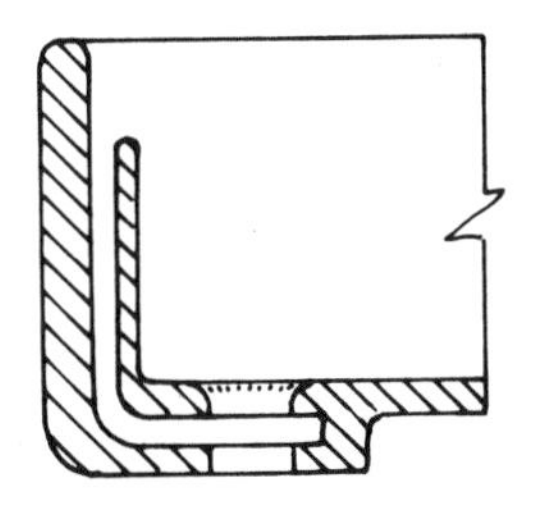

(a) Weir overflow

This type of overflow is usually fitted to butler's sinks and has the advantage of being easily cleaned

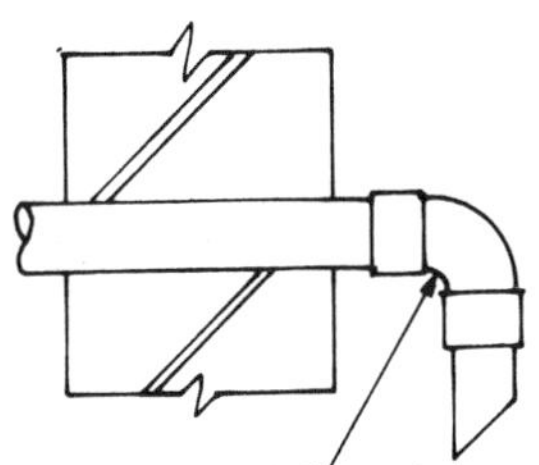

(c) Draught deflection bend for external overflows

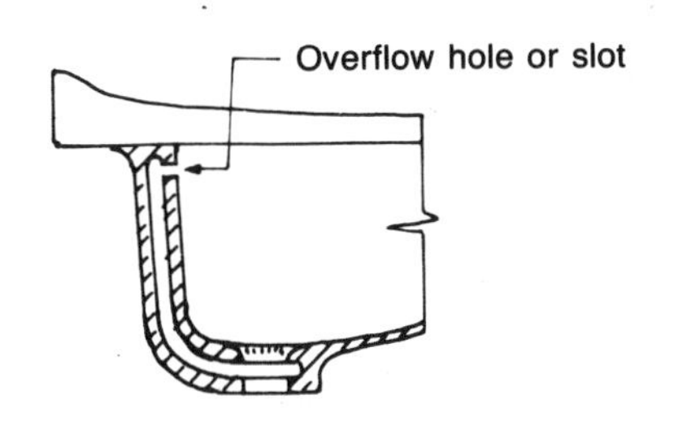

(b) Basin fitted with secret or slotted overflow

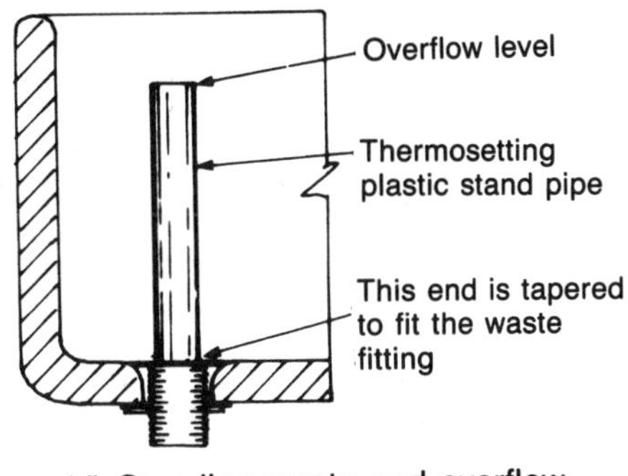

(d) Standing waste and overflow

Fig. 7.24 Types of overflow

overflow should be turned down beneath the water level as shown in Fig. 7.15.

Standing waste and overflow
The type of overflow shown in Fig. 7.24(d) is most commonly used in conjunction with hospitals and laboratory sinks, but these can be used with specially designed basins having a space at the back to accommodate the extension of the waste. This prevents the standpipe being disturbed when the basin is in use. The standpipe is usually made of 'vulcanite', a type of thermosetting plastic. It has a machine taper on one end to enable it to seal the waste effectively in the same manner as a plug. The advantage with this type of overflow is the ease of cleaning and lack of maintenance it requires.

Waste fittings

These are all basically of the same type with the exception of their length and diameter which vary depending upon the type of fitting with which they are used. A bath, for example, having no integral overflow and constructed of thinner materials than basins or sinks, requires a waste with only a short length of thread. The other difference is that some wastes are slotted or have a cut away portion to coincide with entry of integral overflows (see Fig. 7.25(a)). When a waste of this type is fitted it is important to check that the slot in the waste is in line with the entry of the overflow. It is also necessary to ensure that the slots are not obstructed by excess jointing material (see Fig. 7.25(b)).

Wastes are fitted into sanitary fittings using putty and paint, or mastic, as a jointing medium. In the case of the former, the paint provides a key for the putty, which on its own is unsatisfactory. A word of warning here: some wastes are made of plastic materials and on no account should paint or putty be used as it can damage thermoplastics. A washer, either purpose-made polythene or one made of lead, should be used between the fitting and the back nut to provide a cushion into which the nut can be tightened without damage to the fitting. Some manufacturers provide rubber washers for waste fittings but these are sometimes

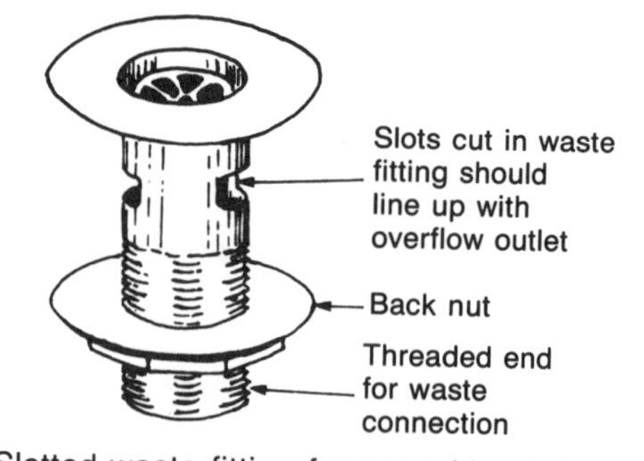

(a) Slotted waste fitting for use with appliances having integral overflows

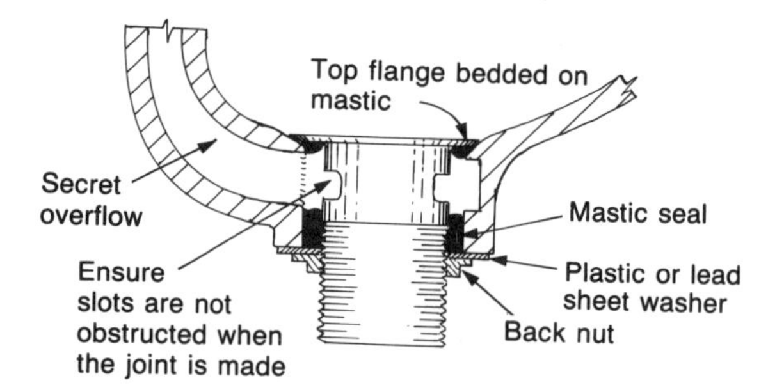

(b) Sealing slotted waste fittings into sanitary appliances

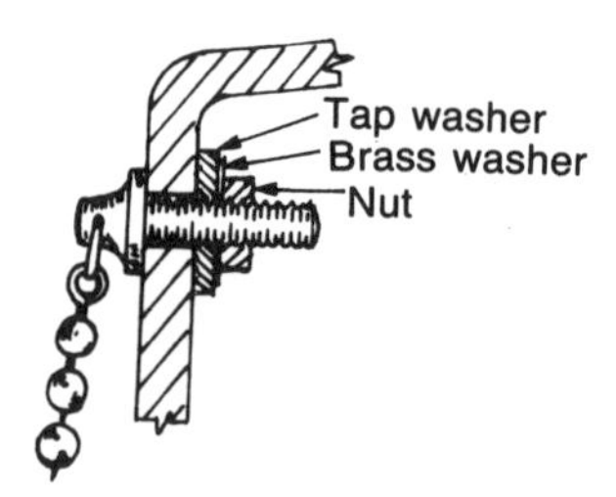

(c) Method of securing chain and stay to a basin

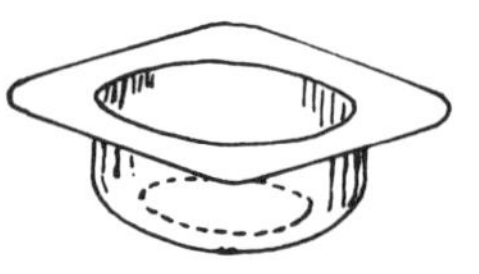
(d) Special plastic washer used with back nuts to secure pillar taps to sanitary fittings

Fig. 7.25 Waste fittings

quite useless as when they are tightened, the rubber distorts and squeezes out of shape causing leakage. This is because the jointing surfaces of ceramic and GRP sanitaryware are seldom suitable for this type of washer, due to the methods of manufacture. Older plumbers often refer to this distortion as 'spewing', a very descriptive term for what actually happens. Yet another word of warning; many modern baths and shower trays are made of a type of plastic called acrilic resin and only silicone-based mastics are suitable for this material if the rubber washers are unsuitable. When the waste has been satisfactorily tightened, make sure all the surplus jointing material has been removed and the fitting is wiped clean.

Basin plugs are normally secured by a stay (or stud) and chain, the stay passing through a hole in the basin. A method of securing the stay and ensuring a watertight joint is shown in Fig. 7.25(c).

Prior to fixing taps into sanitary fittings they should be 'broken' as described in Chapter 5 (see p. 101). The joint between the appliance and the tap must then be made watertight, especially in the case of basins and sink units, and to ensure this the taps are bedded down on suitable mastic with the back nuts tightened against a specially flanged washer made of polypropylene (see Fig. 7.25(d)). These washers are usually supplied with the taps as a complete set. The back nut should be tightened sufficiently to stop the tap turning in the hole.

Brackets and supports for sanitary fittings

The type of support required varies with the type of appliance. Fireclay sinks, for instance, having considerable weight, require stronger fixings than those for wash basins. It is, therefore, important to select the appropriate type of support.

Brackets

Brackets, a common method of supporting appliances, generally fall into two main groups: they are either 'built-in' or screwed to the wall, the 'built-in' type being chosen for the heavier loads.

The most common support for butler sinks are cantilever brackets, which are designed to be built into a wall with strong cement mortar. These must be carefully levelled at the correct height, as once the mortar has set their position cannot be altered. One important factor must always be remembered when any built-in fixing is to be made: the hole into which the bracket fits must always be wetted. Failure to do this will cause the moisture in the mortar to be absorbed too quickly by the wall causing it to shrink, with the result that the brackets move. The dotted line shown on the bracket illustrated in Fig. 7.26(a) indicates the cut-away portion that is required to accommodate a fitting having an integral overflow, such as a Belfast sink. Ensure the cut-away bracket is built in at the end at which the waste is to be fitted.

A similar type of 'built-in' bracket is available for supporting single wash basins in schools or factories where screwed-on fittings may be pulled off the wall (see Fig. 7.26(b)).

Pedestals

The usual method used to support wash basins in modern domestic premises is by a pedestal, as illustrated in Fig. 7.7(a). Where it is desirable to keep the floor area clear under a basin, towel-rail brackets which are screwed to the wall surface may be used (see Fig. 7.26(c)). The screws must be of sufficient length to enable secure fixing to be made and purpose-made fibrous or plastic wall plugs should always be used. The practice of using wooden plugs leaves much to be desired as it is invariably damp on a building site and plugs made of damp wood will eventually shrink causing the brackets to loosen.

All the brackets that have so far been mentioned are made of cast iron and protected from corrosion by painting or porcelain enamel. Brackets made of steel are usually galvanised.

Stands

When fixings are to be made to soft brickwork walls with mortar joints, a reliable fixing for screws cannot always be obtained. In such cases, either built-in brackets must be considered or, in the case of wash basins, a chromium-plated tubular stand can be used of the type shown in Fig. 7.27. When these stands are used, the floor takes most of the weight, the fixings to the wall only being

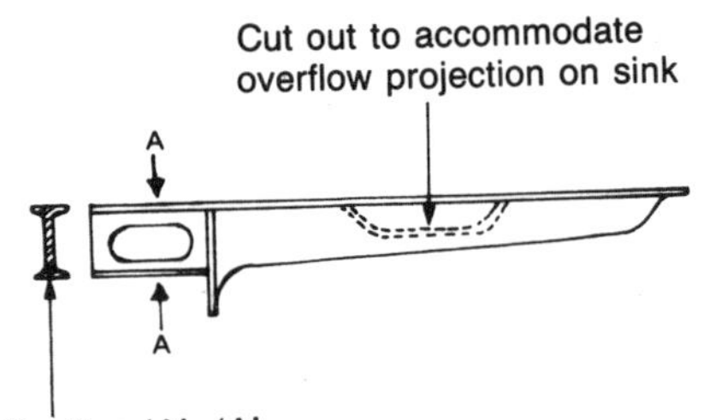

Section 'A'–'A'

(a) Cantilever bracket

These brackets are built into a wall and provide a strong fixing for heavy earthenware sanitary fittings.

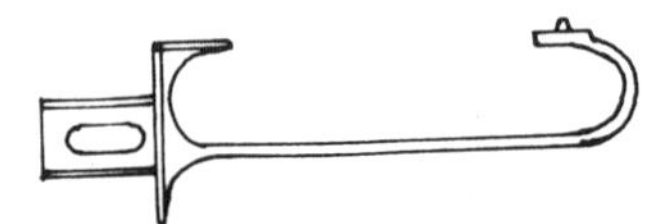

(b) Built-in cantilever towel-rail bracket for wash basins

These brackets are used in schools, hospitals, factories where strong fixings are essential.

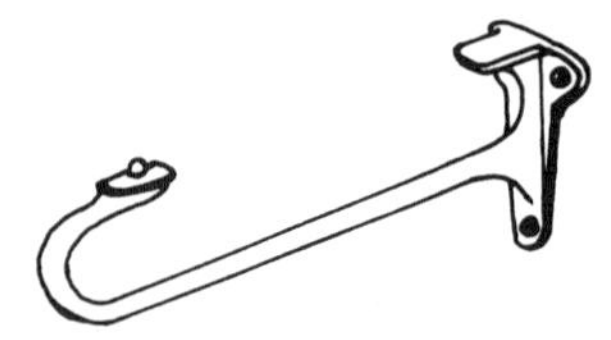

(c) Screw-on towel-rail bracket for wash basins

Used for normal domestic dwellings.

Fig. 7.26 Brackets for supporting sanitary fittings

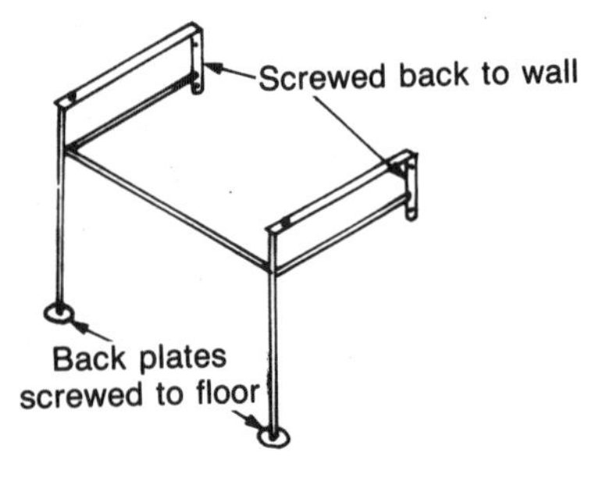

Fig. 7.27 Chromium-plated tubular basin stand. These stands form a strong attractive support for basins and can be used instead of pedestals or brackets.

necessary to prevent the stand from falling forward. A similar type of frame made of cast iron is available for the support of many types of combination sinks and drainers when other types of support are not suitable or desirable.

Legs

Many sanitary fittings are supplied with adjustable legs or feet so they can be levelled on uneven floors. Figure 7.28 shows such a bracket used to level in and adjust the height of cast iron baths.

Cradles

Both acrylic and pressed steel baths, being less rigid than those of cast iron, require the use of a special cradle for their support. Figure 7.29 shows a typical cradle with adjustable feet for acrylic

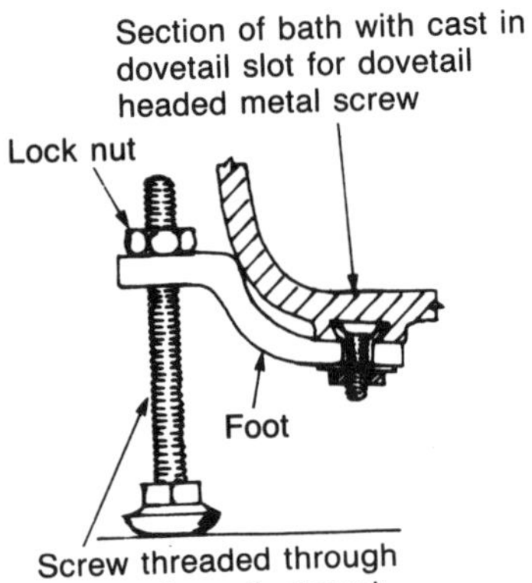

Fig. 7.28 Adjustable bath foot for cast iron baths

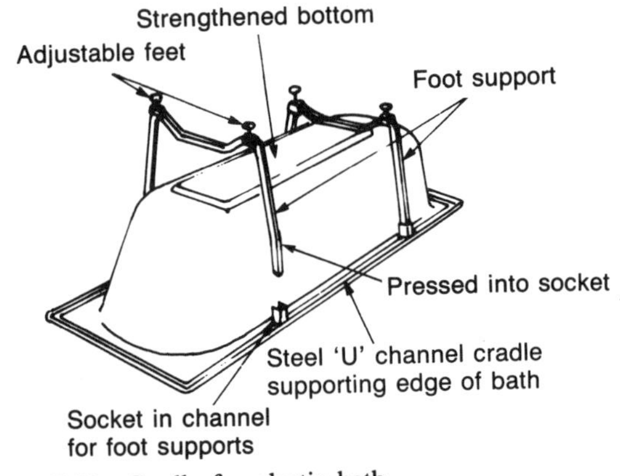

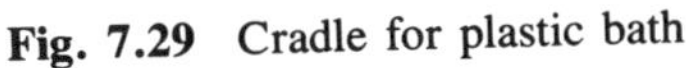

Fig. 7.29 Cradle for plastic bath

baths. Care must be taken to ensure that these cradles are correctly fitted to avoid subsequent damage to the fittings.

Fixings for flushing cisterns
The fixings for flushing cisterns vary depending on their type. Cisterns made of synthetic plastics vary considerably. Those of good quality are usually provided with two back fixing plates wich are screwed to the wall (see Fig. 7.30(a)). The threaded screws passing through the plate are made to fit through the holes in the back of the cistern case, which is then secured to the plates by brass nuts and suitable washers. The very light flushing cisterns are normally screwed through the back of the cistern directly on to the wall, and brass round-head wood screws of sufficient length are required. Cisterns made of ceramics, being heavier, require more positive support. They are usually supplied with two galvanised steel angle brackets which support the cistern, the top being held back to the wall by two brass screws fixed through the holes provided (see Fig. 7.30(b)).

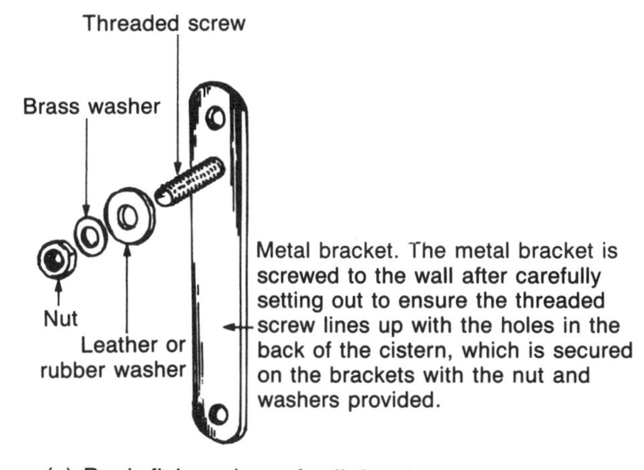

(a) Back fixing plates for light plastic flushing cisterns

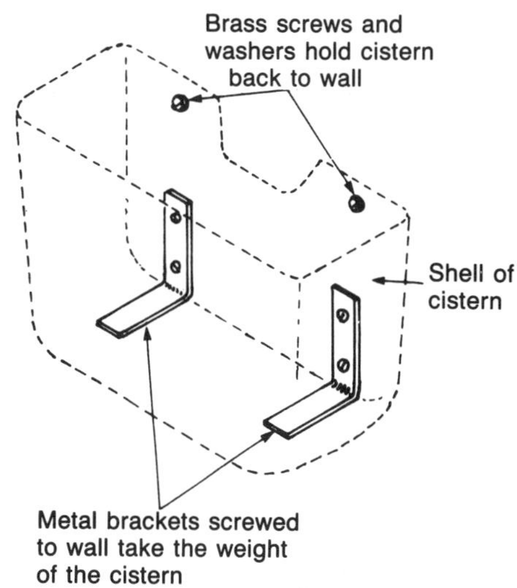

(b) Angle bracket and screw support for cisterns constructed from ceramic materials

Fig. 7.30 Fixings for flushing cisterns

Further reading

Relevant British Standards for sanitary fittings are as follows.

BS 1125: Specification for WC flushing cisterns.
BS 1189: Specification for cast iron baths for domestic purposes.
BS 1206: Specification for fireclay sinks: dimensions and workmanship.
BS 1213: Ceramic wash down WC pans: dimensions and workmanship.
BS 1244: Specification for metal sinks. Part 2: Metric units.
BS 1390: Sheet steel baths for domestic purposes.
BS 3380: Specification for wastes for sanitary appliances and bath overflows.
BS 3943: Specification for plastic waste traps.
BS 4305: Specification for baths for domestic purposes made from cast acrylic sheet.
BS 5572: Sanitary pipework.

Information on sanitary appliances is obtainable from the following companies:

Ideal Standard Sanitary Ware Ltd, PO Box 60, National Avenue, Hull, HU5 4JE.
Armitage Shanks Ltd, Rugeley, Staffs, WS15 4BT.

Self-testing questions

1. State the essential differences between glazed fireclay ware and vitreous chinaware.
2. Identify the essential difference in design between a Belfast sink and a sink of the London pattern.

3. State the recommended fixing height for a wash basin fixed on towel-rail brackets.
4. State two tests that can be applied for checking the efficient working of a WC.
5. List the advantages of using soft joints for joining a WC to either discharge pipes or to drains.
6. State the two basic factors which are necessary for siphonic action to take place.
7. Explain the action of siphonic closets.
8. Explain how water is prevented from entering the flush pipe in the event of failure of the float-operated valve and a blocked overflow.
9. State the precautions to be taken when fitting a slotted waste to a wash basin.
10. Explain why special cradles are necessary to support plastic and sheet steel baths.
11. State why purpose-made plugs, instead of those made of wood, are recommended for most types of fixings.

8 Sanitary Pipework

After completing this chapter the reader should be able to:

1. Understand the need for the highest standards of quality in workmanship and materials for sanitary pipework.
2. Define the causes of seal loss in traps.
3. Describe the evolution of modern above-ground soil- and waste-discharge systems.
4. Identify the main design factors relating to waste-disposal systems.
5. Define the need for means of access in discharge pipe systems.
6. State the special points to be considered when installing washing machine and dishwashing equipment to discharge systems.

Introduction

The Great Plague of London during the seventeenth century was the direct result of poor sanitation and overcrowded living conditions. It swept large areas of the country and caused many deaths, but it should not be assumed that this was the only outbreak of its type. Such epidemics have occurred throughout history and are still rife in underdeveloped countries, though fortunately nowadays they are fairly localised.

It was not until the early Victorian era that forward-thinking people associated the terrible toll of human life with the lack of effective sanitation. In 1875 the London County Council first laid down a system of bylaws or rules to which the installation of sanitary systems had to conform. Other boroughs and local authorities quickly followed, each producing its own local bylaws. While in most cases these bylaws were very similar, the differences which did occur were due largely to their interpretation by individual local authorities. So it will be seen that the sanitary pipework systems as we know them today are the product of necessity and legislation by successive governments and local authorities in comparatively recent years.

Building Regulations

The Building Regulations of 1965 brought a degree of uniformity to the local bylaws and these now apply to all forms of construction throughout the country. They have been modified and amended since they were first introduced to permit the use of new materials and techniques. The latest version of these regulations became operative in 1991. The other important document that applies especially to the plumbing industry in relation to sanitary pipework is Code of Practice 5572. Codes of Practice are not legal documents but give a good general guide to the minimum standards of workmanship and materials to be used.

The general arrangement of a well-designed discharge-pipe system should embody the following features. It should permit the speedy removal of waste to the underground drainage system and must prevent the ingress of foul air into the building by efficient trapping and maintenance of the trap seals. Adequate provisions should be made for access for the removal of blockages in the system and materials should be chosen to ensure a trouble-free and corrosion-resistant life.

Terminology

To avoid confusion, some knowledge of the terms

relating to sanitary pipework is necessary.

The terms 'soil pipe' and 'waste pipe' are both derived from the original two-pipe system of sanitation where a soil pipe was connected to a WC and a waste to an ablutionary fitting. These terms are not generally used now, both being designated *discharge pipes*.

A *stack* relates to a main vertical pipe whether a discharge 'stack' or a ventilating 'stack'. In certain cases a discharge stack can be used to convey rainwater where a combined system of underground drainage is permitted by the local authority.

A *ventilating* pipe may be a branch or main ventilating pipe, its main function being to maintain atmospheric pressure inside the whole of the discharge and drainage system.

A pipe carrying waste water from a fitting or group of fittings to the main discharge pipe is called a branch discharge pipe.

Traps

Traps are an integral part of a modern sanitary system, being designed to retain a small quantity of the waste water from the discharge of the fitting to which they are attached as a barrier to prevent foul air entering the building. Traps should be self-cleaning, that is to say, they should be designed so that their walls are scoured by the discharging water. Due to the necessary bends in their construction, traps are prone to blockage and adequate access for their cleansing must be provided. One of the advantages of modern traps constructed of plastic materials is the ease with which they may be dismantled for cleaning.

P and S traps

Traps are designated either P or S, the difference being the position of their outlet. The P trap outlets are in an almost horizontal plane, while the outlet leg of an S trap is in a vertical position. A slight angle, usually 2½°, is in fact provided to the outlet of P traps to ensure a slight fall.

Figure 8.1(a) illustrates a tubular trap. These are made of pipe and to reduce the space they occupy, the bends are made to a very small radius. The parts of the trap are indicated and also the depth of seal, which is the measured length from the soffit of the return bend and the weir or invert of the outlet.

A trap of a different design called a bottle trap is shown in Fig. 8.1(b). These are not considered self-cleansing by many but due to their neat appearance they are commonly used. They are always made with a P outlet, but purpose-made bends are available to convert them to S outlets if required. The effective depth of seal is measured from the base of the dip pipe to the weir.

Trap seals

The depth of the trap seal varies with the size and usage of the pipework to which it is connected. If the diameter of the waste is 50 mm or less and the fitting is connected directly to a main discharge pipe or drain, the seal must be 75 mm,* yet outlets of larger diameter traps have seals of only 50 mm. The reason for this apparent anomaly is that large-diameter traps are unlikely to receive a full bore discharge which is often the cause of seal loss. Typical examples of these larger traps may be seen in WCs and gulleys.

Traps with a seal depth of 38 mm are permitted only if the appliance discharges over a trapped gulley, because the gulley provides an air break between the drain and the appliance should the trap seal be destroyed.

Loss of trap seal

Loss of trap seals will result in objectionable odours entering the building which, if persistent, may cause poor health. For these reasons it will be seen that the water seal of a trap must be retained in all circumstances. The three main reasons for seal loss due to bad design of the discharge pipe system are dealt with fully later in this chapter.

Trap seals, however, are also lost by natural causes, most of which are rare, but of which the plumber should be aware.

Evaporation This is the most common natural cause of seal loss. The warmer and drier the air

* CP 5572: 1994 suggests that bath traps may have a seal of only 50 mm owing to the difficulty of fitting one having a seal of 75 mm under a bath.

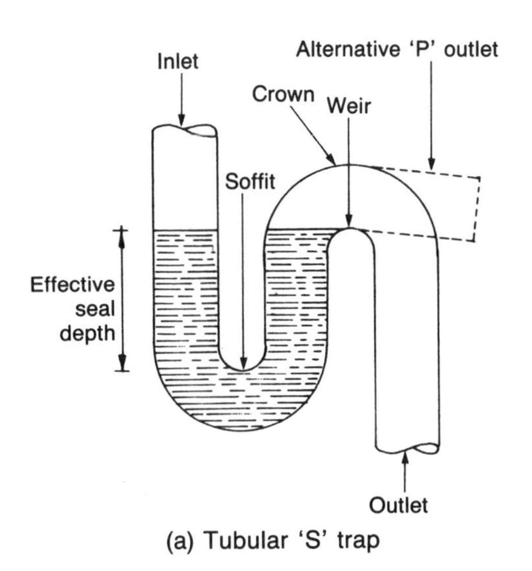

(a) Tubular 'S' trap

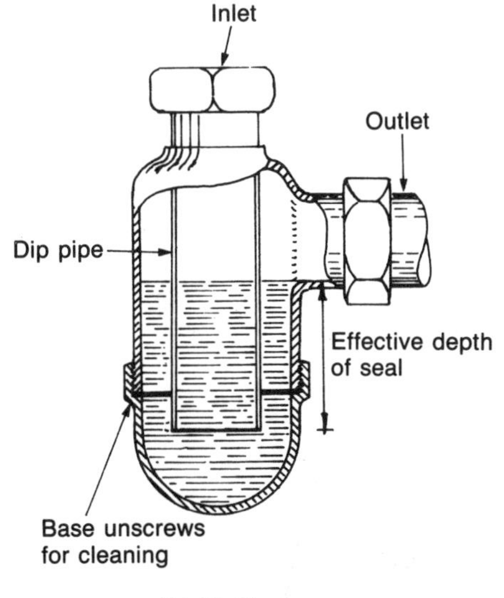

(b) Bottle trap

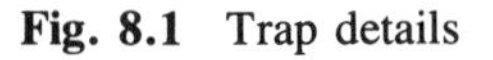
Fig. 8.1 Trap details

becomes, the greater will be the rate of evaporation. It is unlikely, however, that traps having a seal depth of 75 mm would evaporate completely, even in very warm weather, providing the appliance was used occasionally. It has been calculated that even during a very hot summer it would take a period of about ten to twelve weeks to evaporate a trap seal.

Capillary action Capillarity can result in the rapid destruction of a seal due to particles of dish-cloths being deposited over the weir of the trap (see Fig. 8.2(a)). This is, however, a rare occurrence which only happens in S traps. A similar effect will be seen if a cloth is left hanging over the edge of a bucket of water.

Wind The effects of wind currents and pressures at the vent terminal can cause 'waving' out of traps especially those of larger diameter. Unstable atmospheric conditions in the vent pipe due to this cause sometimes produce a wave movement of the water in the trap allowing it to gradually wash over the weir. This is not a common occurrence and cannot easily be foreseen. It can sometimes be remedied by extending the vent upward or terminating it with a bend as shown in Fig. 8.2(b).

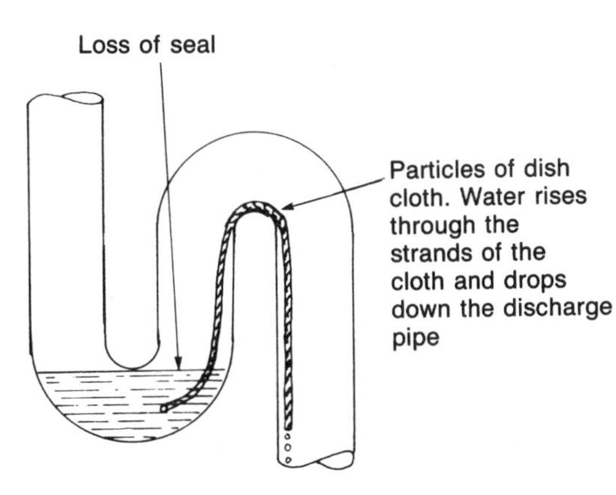

(a) Effect of capillary attraction on tube trap

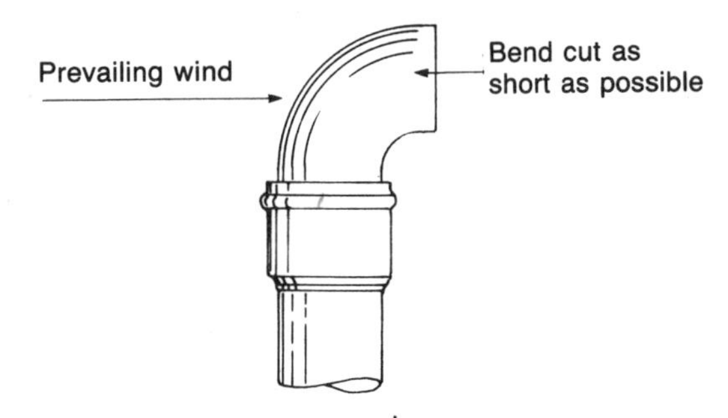

(b) Prevention of seal loss by 'waving out'

Fig. 8.2 Loss of trap seal

Momentum Momentum is the term used to describe the effect of a sudden discharge of water from a bucket into a WC or gully trap. Due to the velocity of the discharge, it sometimes also carries away the water which should form the trap seal. When this occurs it can easily be seen and the householder can flush the cistern or empty more water into the trap to reseal it.

Access

Due to the low velocity of water flow at which most modern discharge-pipe systems operate, access is necessary to remove any silt which may build up and cause a blockage. Access is sometimes provided by an easily dismantled trap, a typical example being shown in Fig. 8.3. The swivel joint also permits some degree of adjustment to be made to accommodate the direction of the discharge pipe connection.

Special types of trap

Variable outlet traps This type of trap is illustrated in Fig. 8.4. Their unusual design permits them to be connected to a wide variety of plumbing layouts and they are therefore useful in situations where, for instance, a sink unit is to be replaced and the waste outlet on the new unit differs from that of the old. Due to the limit of flexibility with an ordinary two-piece trap, it might be necessary to reposition the waste with all the associated problems of cutting away the fabric of the building and making good. As the illustration shows a variable outlet trap has a telescopic inlet allowing adjustment for height. The swivel joints enable it to be used as a 'P' or 'S' trap with the outlet at varying angles, while the centre section is reversible giving a choice of seal depth between 40 mm (shallow seal) or 75 mm (deep seal).

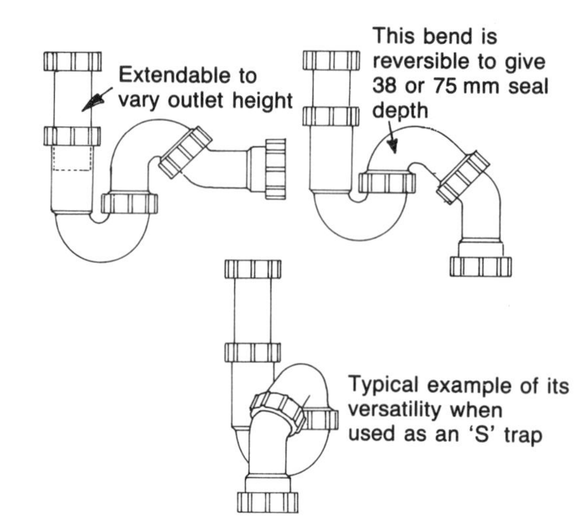

Fig. 8.4 Variable outlet trap

Straight through trap Figure 8.5 illustrates a straight through trap which can be used in some

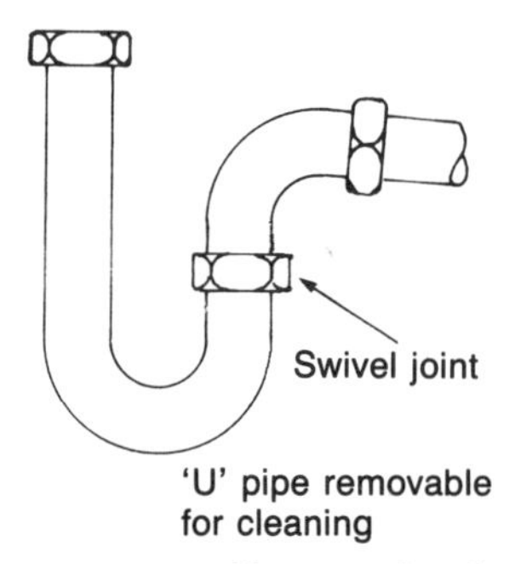

Fig. 8.3 Access to traps. Traps made of copper and plastics are made so that they can easily be dismantled for cleaning.

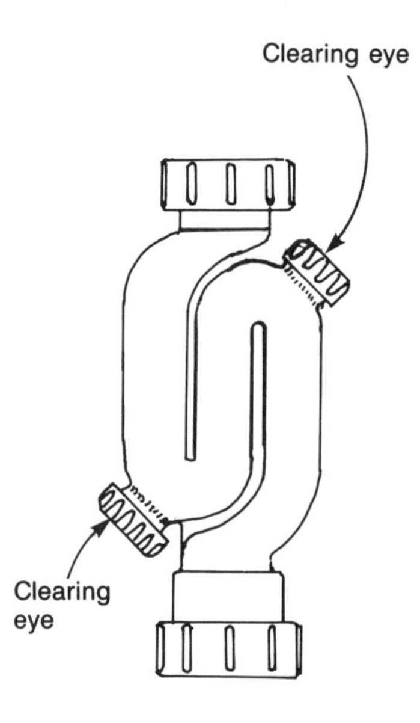

Fig. 8.5 Straight through trap (originally called bag trap). These can usefully be employed instead of an 'S' trap behind a basin pedestal.

instances instead of an 'S' trap. Its main disadvantage is the two relatively tight bends which slow up the discharge of water. These traps are especially useful for pedestal basins, as they can be concealed more easily than an 'S' trap behind the pedestal. As they are moulded in one piece and cannot be dismantled easily, adequate provision for clearing is provided.

Running trap Figure 8.6 shows a running trap which is nowadays only used in situations where it is not possible to use a 'P' or 'S' trap for some reason. Typical cases would be trapping a sink macerator or possibly a washing machine.

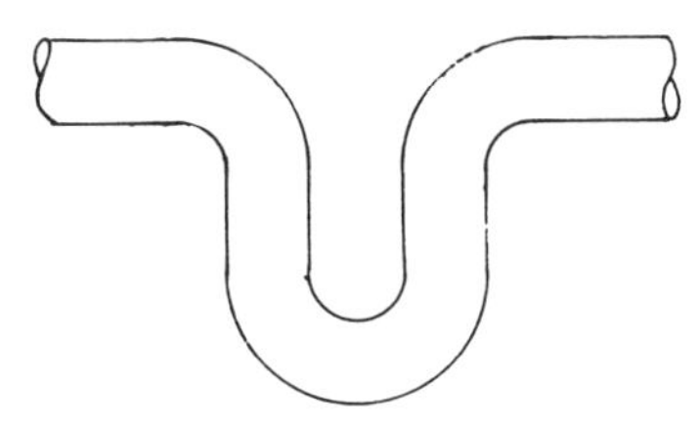

Fig. 8.6 Running trap

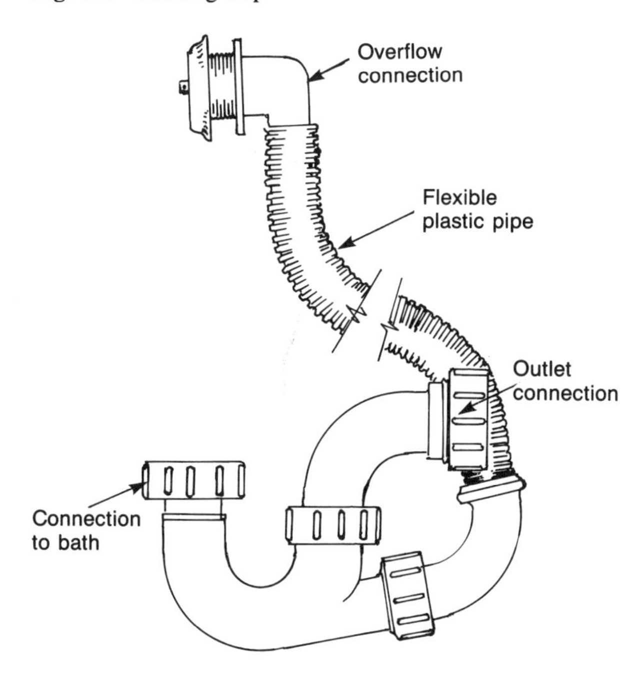

Fig. 8.7 Combination bath trap and overflow. Bath traps previously required a 75 mm seal, but the revised Code of Practice now allows a 50 mm seal.

Combination traps Another special trap is shown in Fig. 8.7, this being used on a bath waste avoiding the use of a separate overflow. There is a tendency for the waste water to wash back up the overflow pipe, having the effect of depositing soapy waste in its bore which is a disadvantage with this type of fitting. Most traps of this type, however, are made so that they can easily be dismantled for cleaning.

Resealing traps Later in this chapter the design features which are embodied in a good discharge system to prevent seal losses by siphonage will be discussed. Where these conditions cannot be met, however, a *resealing* trap may be used with advantage. These traps are designed so that a water seal remains as a barrier to foul odours even after siphonage has occurred, this being achieved by the reservoir of water which reseals the trap after conditions causing siphonage have ceased. Figure 8.8 shows a typical example of this type of trap.

Another type of resealing trap is shown in Fig. 8.9. This trap incorporates a bypass arrangement which admits a flow of air into the discharge pipe to maintain equilibrium of pressure inside the trap with that of the atmosphere. Figure 8.9(a) shows the trap with the water at rest; Figure 8.9(b) shows the effect on the water seal when siphonage is taking place (note the bypass action permitting air to enter the discharge pipe); Figure 8.9(c)

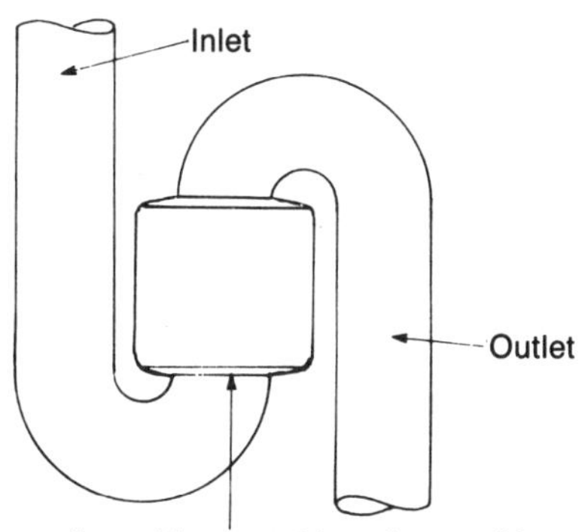

Fig. 8.8 Reservoir type resealing trap

shows the effect on the water seal when siphonage is taking place (note the bypass action permitting air to enter the discharge pipe); Figure 8.9(c) shows the effect on the water seal after siphonage has occurred. It will be seen that the seal has been maintained but is diminished in depth.

The two main objections to the use of resealing traps of all types is that they are noisy in operation due to the gurgling effect of the air–water mixture as it is drawn through the trap, and the fact that not all the traps are self-cleansing.

Another type of trap which limits the effect of siphonage incorporates a mechanical air valve in its crown. When pressures less than that of the atmosphere occur in the discharge pipe, the valve opens to admit air. These traps are quiet in operation, the only objection being that regular maintenance may be necessary to ensure the correct functioning of the valve.

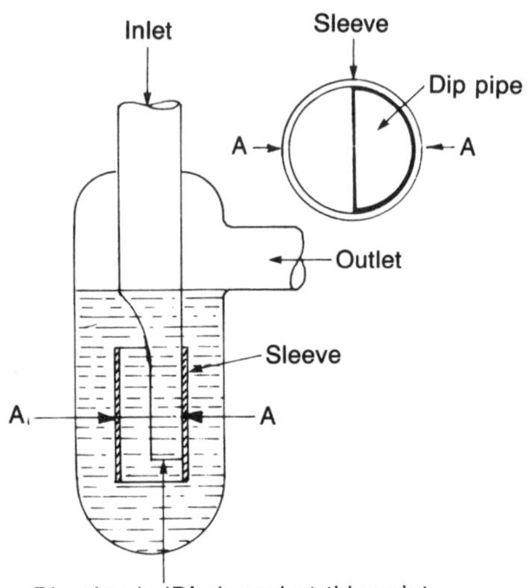

(a) Water at rest

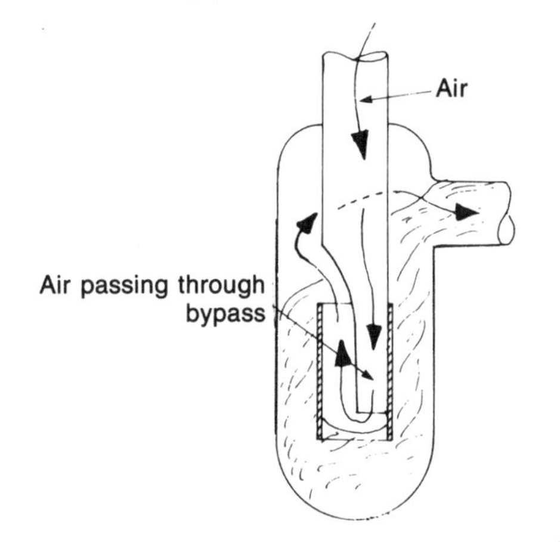

(b) Resealing trap during a discharge of the fitting

As the seal of the trap is lost during a discharge the end of the 'D' shaped outlet is exposed and air is allowed to enter via the bypass stabilising the air pressure in the discharge pipe

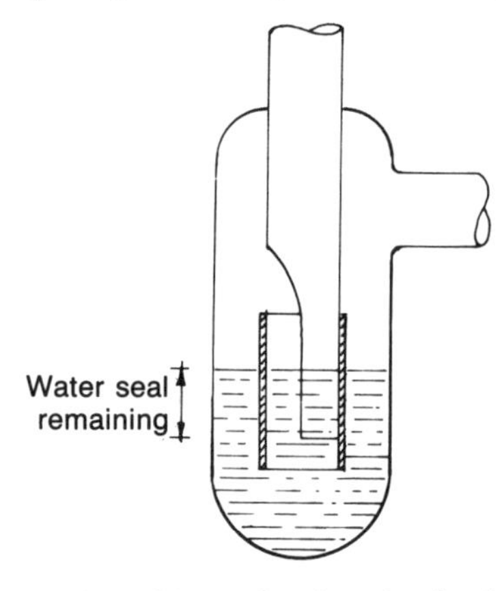

(c) Shows reduced but effective depth of seal after a discharge has taken place

Fig. 8.9 Resealing bottle trap

Discharge pipework systems

The two-pipe system

The original concept of pipework systems to comply with the London County Council requirements was the two-pipe system, the use of which is still permissible for modern buildings, although it has been largely superseded by more economic installations.

To meet the requirements of the two-pipe system, sanitary fittings are divided into two main groups. Those which are used to dispose excremental matter, such as WCs and urinals, are designated 'soil' fittings. Those used for ablutionary purposes such as baths and basins or those used for culinary purposes such as sinks are termed 'waste' fittings. The main feature of the two-pipe system is the completely separate disposal of the discharges from soil fittings and waste fittings requiring two independent above-ground pipework installations, both eventually discharging into the same underground drainage system. Only soil fittings are allowed to discharge directly into the drain; waste fittings must terminate over a gulley, preferably one of the back inlet type.

This type of system may seem peculiar by modern standards but it should be borne in mind that when it was first devised not all local authorities insisted on the trapping of waste fittings. The gully therefore often provided the only effective trap against drain air entering the building. When the two-pipe system is correctly installed it is an excellent method of soil and waste disposal; but unfortunately it is also an expensive one, due to the double set of pipes required.

What in fact happened during the 1920–40 period – the heyday of speculative building – was the installation of a hopper-head in small domestic buildings. This was used to collect the waste water from the first floor ablutionary (waste) fittings to avoid the cost of a properly constructed waste and vent pipe having proper branch connections for the branch wastes. While the use of hopper-heads for this purpose no longer conforms with the Building Regulations, many of these systems are still in use. For this reason an illustration showing a system of this type, with all its defects, has been included (see Fig. 8.10).

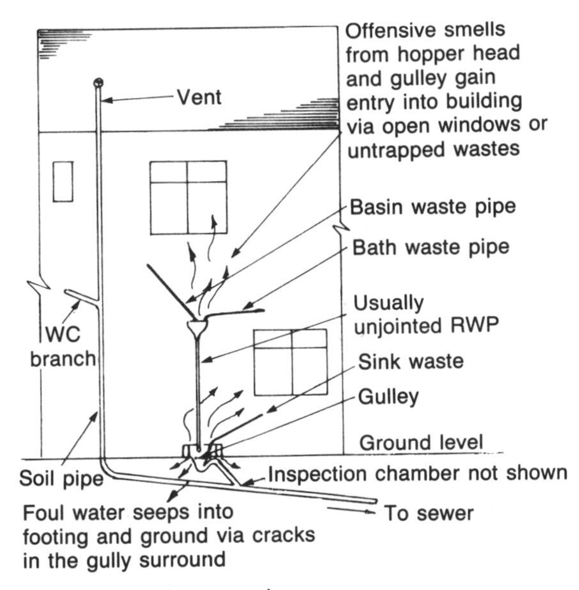

Fig. 8.10 Obsolete two-pipe system

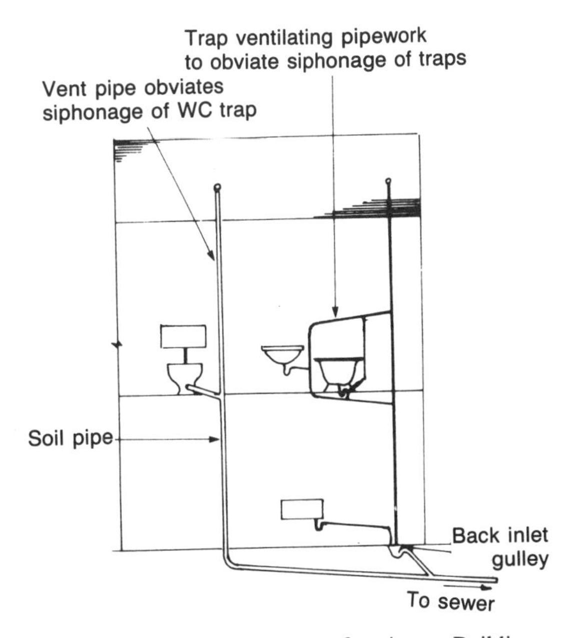

Fig. 8.11 Two-pipe system conforming to Building Regulations. Note that as with Fig. 8.10 soil and waste water are separated above ground level but discharge into same underground drainage system.

A two-pipe system that will meet the requirements of the Building Regulations is shown in Fig. 8.11. The only advantage of this type of installation is its suitability for buildings where, for some reason or other, it is not possible to group the sanitary fittings together around the main discharge stack. When it is compared with the single stack system it will be seen how uneconomic it is in terms of labour and materials, coupled with the difficulty of concealing the pipework inside the building. It is for these two reasons that it is seldom used, or indeed necessary, for most modern sanitary systems.

The one-pipe system

The one-pipe system, shown in Fig. 8.12, was an early attempt to economise in the labour and materials required for sanitary pipework systems yet maintain the high standards set by the two-pipe system. The main difference between the one- and two-pipe system is that with the former both soil and waste fittings are discharged into a common stack with the addition of some trap-ventilating

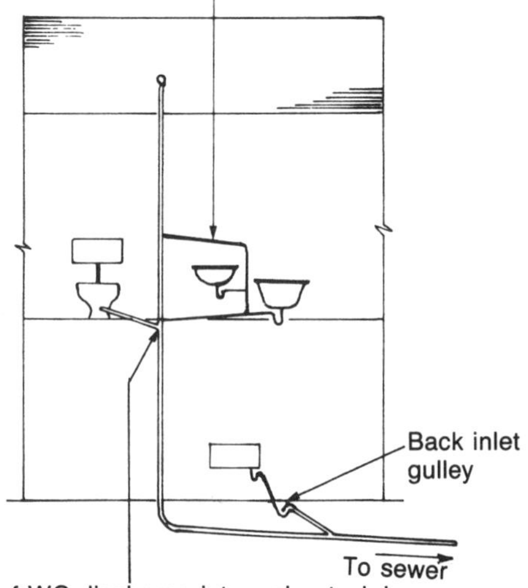

Fig. 8.12 One-pipe system (*note:* this is almost identical in design to the modern fully ventilated system). With this system both soil and waste discharges are connected to a common discharge pipe. All traps subject to risk of siphonage are fully ventilated.

pipes. Although this was a big advantage economically, it was not accepted by all local authorities for many years. The course of time has proved it to be an effective system and it was used successfully for multi-storey buildings such as blocks of flats in the immediate post-War period. it was seldom used for small buildings, the older two-pipe system incorporating a hopper-head being more popular at this time, not because it was a better system but because it was cheaper to install.

A distinct disadvantage with both the one- and two-pipe systems is the amount of pipework involved making such schemes difficult to conceal. Prior to the 1965 Building Regulations which require all discharge pipework to be fitted inside the building, the use of these systems often resulted in a maze of exposed pipework on the face of the building.

The single-stack system

With still greater economy in mind, the Building Research Station commenced a series of tests on fully ventilated sanitary pipework. It was found that many of the vent pipes which had previously been insisted upon were in fact unnecessary. The result of these experiments and tests revolutionised the thinking on waste-disposal systems and resulted in what is now called the single-stack system shown in Fig. 8.13(a) and (b).

The outstanding feature of this system is the complete absence of any trap-ventilating pipe, except in very special circumstances. This is accomplished by the main factors in the following list:

(a) limitations on the length of waste pipes
(b) limitation on their fall and in some cases an increase in their diameter
(c) the branch entry of a WC into the main stack and the positioning of waste pipes in relation to this

The initial concept of this system was to simplify sanitary pipework systems for both high- and low-rise housing. It has, however, been adapted to the needs of other types of buildings by introducing a limited degree of ventilation to permit its use with ranges of fittings.

Quite apart from economic reasons, this system met the 1965 Building Regulations which required

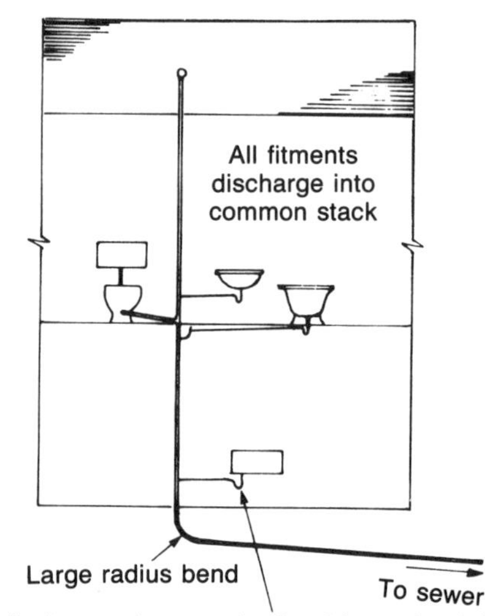

(a) Diagrammatic layout

Fig. 8.13 Single-stack system

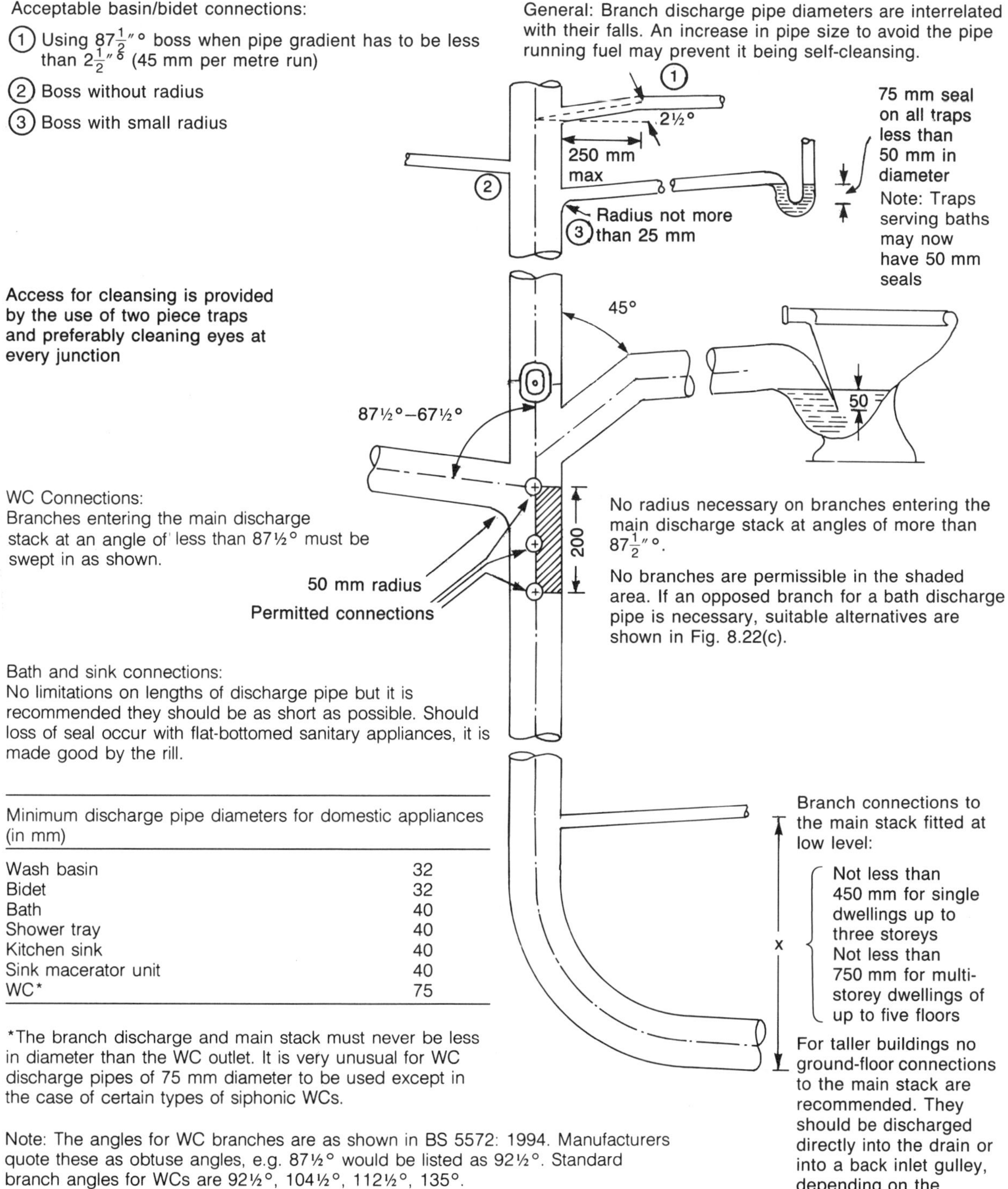

Minimum discharge pipe diameters for domestic appliances (in mm)	
Wash basin	32
Bidet	32
Bath	40
Shower tray	40
Kitchen sink	40
Sink macerator unit	40
WC*	75

*The branch discharge and main stack must never be less in diameter than the WC outlet. It is very unusual for WC discharge pipes of 75 mm diameter to be used except in the case of certain types of siphonic WCs.

Note: The angles for WC branches are as shown in BS 5572: 1994. Manufacturers quote these as obtuse angles, e.g. 87½° would be listed as 92½°. Standard branch angles for WCs are 92½°, 104½°, 112½°, 135°.

(b) Single stack system of overground sanitary pipework details (ref. BS 5572)

Fig. 8.13 *Continued*

all discharge pipework to be fitted inside the building; the absence of ventilating pipes rendering this system much more easily adaptable to this requirement. The single-stack system and its modified versions are the most likely types to be used nowadays on both new and conversion work. It is, therefore, necessary for the plumber to fully understand the working principles and limitations of the system, not forgetting that trap seals must always be retained.

A thorough understanding of how trap seals are lost due to bad design is a prerequisite to a full understanding of the single-stack system.

Seal loss due to bad design

Self-siphonage

Figure 8.14 shows what is probably the most common cause of seal loss. It is termed 'self-siphonage' and occurs when the discharge from a fitting unseals its own trap in the following manner. As water flows from the fitting down the discharge pipe, its progress is impeded by the resistance of the pipe walls until the water forms a solid plug, the technical term for this being the *hydraulic jump*. As the solid plug of water continues down the pipe, the air between it and the trap seal has to fill an ever increasing gap with subsequent loss of air pressure. The air in the inlet leg of the trap is at atmospheric pressure and therefore greater than that in the outlet side of the trap, and in an effort to equalise pressure in the pipe, pushes the water in the trap seal, out of the trap and down the discharge pipe. This is prevented in one- and two-pipe systems by fitting a vent pipe, sometimes called an anti-siphon pipe. Trap-ventilating pipes are only necessary on modern systems when for some reason the length of a discharge exceeds that recommended by CP 5572. In small buildings and dwelling houses having a well-designed drainage system this is normally unnecessary.

With the single-stack system, vents are used only in exceptional cases, the method of preserving trap seals being as follows. All discharge pipes must be of adequate size and in the case of certain fittings, i.e. wash basins, of limited length, all falls being kept within strictly confined limits. (A graph showing the relationship between slope and length of the pipe is given in Fig. 8.15.) The length of any discharge pipe should be as short as possible, and one of the most important design features of this system is the close grouping of the sanitary fittings.

The only fittings having a positive restriction on the length of discharge pipes are wash basins and bidets. Both these fittings are very prone to self-siphonage due to their bowl-shaped construction, which means there is no 'rill' or tail off in the final discharge of these appliances to reseal the trap if the seal is siphoned. If the trap seals of baths and sinks are siphoned they are usually resealed by the tail off of water due to their flat bottom. The graph in Fig. 8.15 shows the relationship of the length of a basin waste and its fall. To use the graph, the measured length of the waste from the weir of the trap to its junction with the main stack is related to its permissible slope.

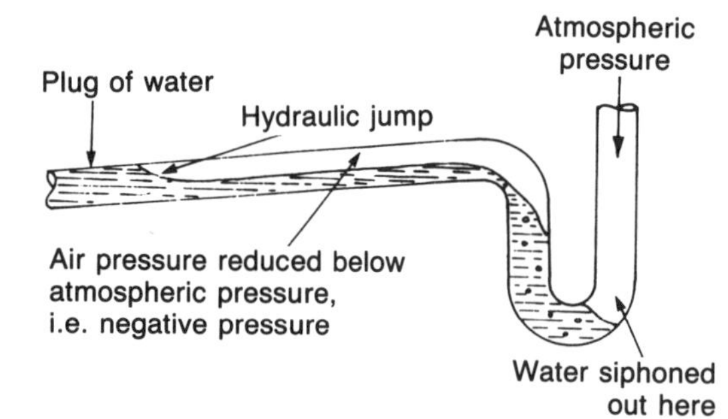

Fig. 8.14 Cause and effect of the hydraulic jump

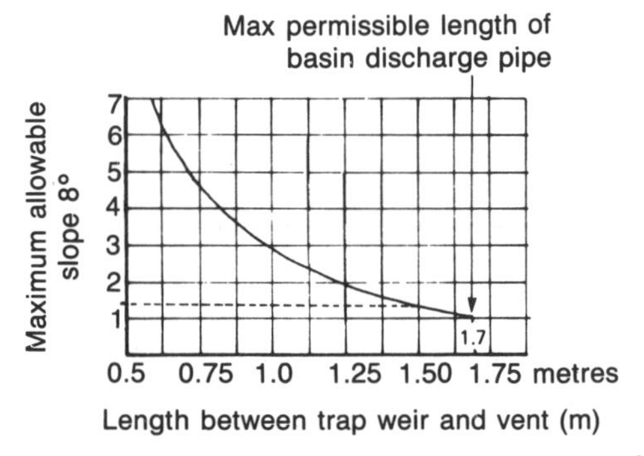

Fig. 8.15 Relationship between slope and length of discharge pipes

To give an example of its use, assume the measured length of a basin waste to be 1.500 m. Follow the vertical line upward from 1.50 until it intersects with the curved line. From this point, follow the dotted horizontal line across to the left and it will be seen to fall between 1° and 2°, in fact a little less than 1.5°. This is the maximum slope that can safely be permitted on a waste of this length.

The accurate measurement of these small angles is difficult under site conditions, but it can be achieved by the use of one of two purpose-made tools. The first is a simple incidence board which can be made by cutting a piece of timber to the required angle (see Fig.8.16(a)). A similar arrangement is often used for levelling a short length of underground drain to a predetermined fall. When the timber is placed on the pipe and levelled as shown, the pipe will be running at the correct angle. The angle cut on the timber is of course only suitable for one angle of fall, and to avoid cutting a new piece of timber every time a different angle is required, an adjustable incidence level can be made as shown in Fig. 8.16(b). A small bolt and thumbscrew passes through the middle of the two pieces of timber and when the correct angle of fall has been found a protractor is used to set the timbers. The thumbscrew is then tightened and with one timber at 180° (level) the other will indicate the correct slope of the waste.

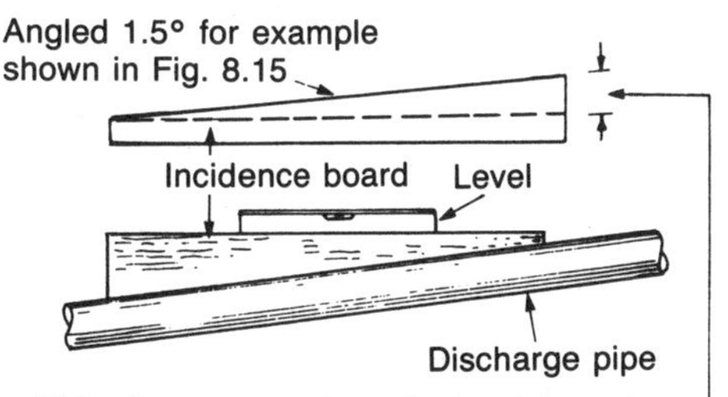

(a) Making and using a simple incidence board

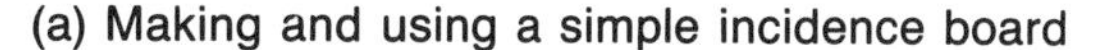

The board should be cut at the angle to which the waste is to fall and used as shown with a level.

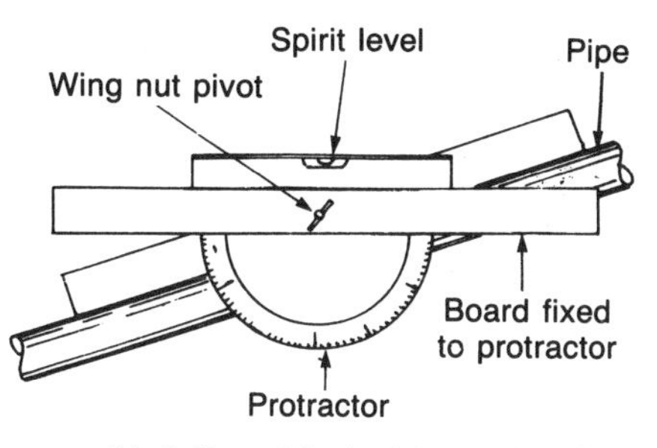

(b) Adjustable incidence level.

The advantage of using this tool is that it can be adjusted to any angle.

Fig. 8.16 Measuring the angle of fall in waste pipes

If it is impossible for a basin discharge pipe to conform to the maximum recommended length, three alternatives are possible. Resealing traps can be used, bearing in mind their disadvantages as described earlier, or a trap-ventilating pipe may be used. The third alternative is to increase the size of the discharge pipe, which has the effect of delaying the build up of the hydraulic jump. Unfortunately to overcome one problem, the measures taken can produce others. One of the biggest problems with a single-stack system is the fact that because the water moves through the pipe at a relatively low velocity, it is more likely to leave solid matter such as soap deposits on its invert. The larger the pipe diameter, the lower will be the velocity of the flow, usually resulting in larger deposits. Despite its disadvantages, this latter alternative is often the most viable proposition, but adequate provision must be made for rodding in case of blockages.

Induced siphonage

The second way in which trap seals may be lost due to poor design is known as *Induced Siphonage*, a typical example being shown in Fig. 8.17. This form of siphonage occurs due to the discharge of one fitting in a range destroying the seal of the trap of another fitting connected to a common discharge pipe. This form of siphonage rarely occurs in domestic properties as it is not normally necessary to connect two fittings to the same discharge pipe. In offices and factories, however, where several fittings are connected to a common waste some ventilation of the system may be required if induced siphonage is to be prevented.

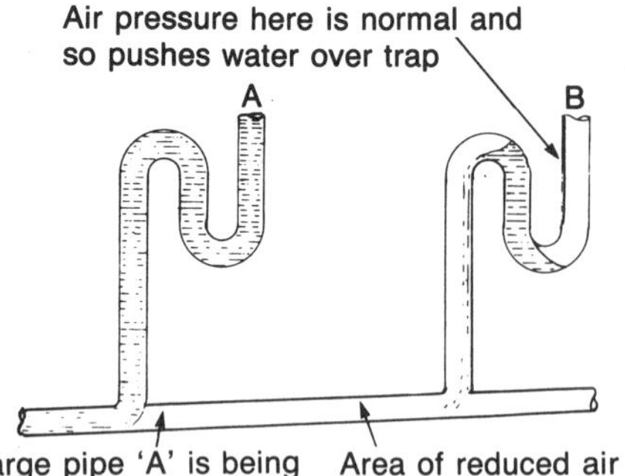

Fig. 8.17 Induced siphonage

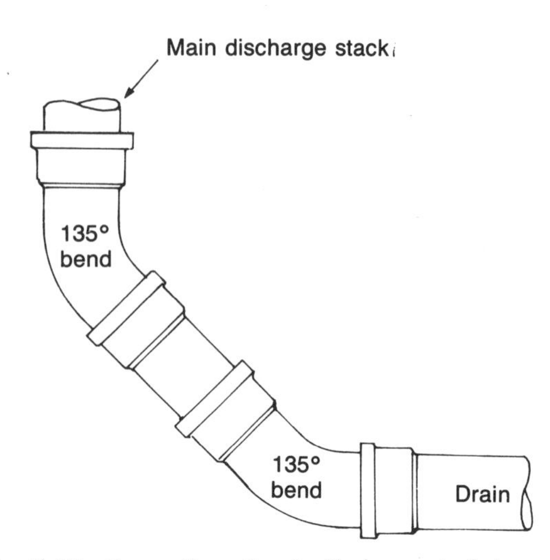

Fig. 8.19 Connection of main discharge stack to underground drain. The use of a large-radius bend or two 135° bends at the foot of a main discharge stack prevents 'compression' of air in the stack which would cause trap seals to be blown

Seal loss by compression

The remaining factor which must be guarded against is the possibility of the trap seal being blown back into the fitting by what is called *compression*, i.e. compression of the air in the discharge pipe. In many cases the water displaced will run back into the trap due to the shape of the appliance it serves, but foul air will have been released inside the building.

This problem only occurs in high-rise buildings where the discharge of sanitary fitments at high level compresses the air in the main stack as it falls (see Fig. 8.18). Sharp-radius bends at the foot of the stack are the most common cause of compression. For this reason it is usual to use a large-radius bend or two 135° bends as shown in Fig. 8.19. For the same reason, offsets or bends should not be used in the wet part of the main stack. If this is unavoidable the offset must be ventilated as shown in Fig. 8.20.

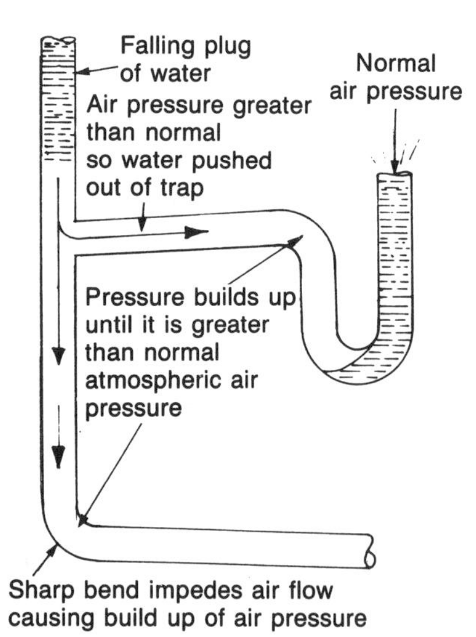

Fig. 8.18 Effects of compression

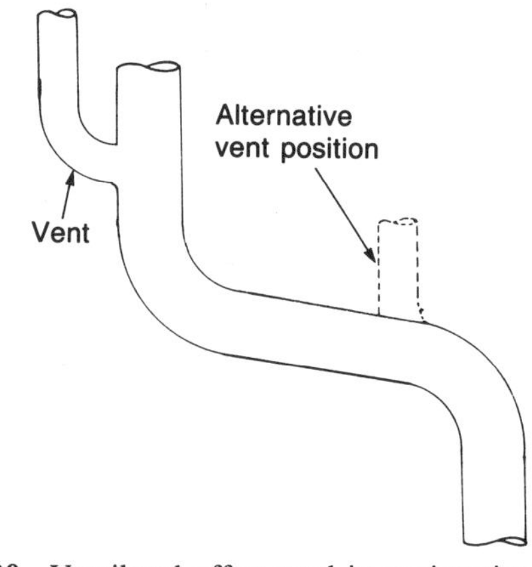

Fig. 8.20 Ventilated offset used in conjunction with the single-stack system. The use of offsets between discharging fittings should be avoided where possible when the single-stack system is fitted. If an offset must be used it should be vented as shown to avoid the build up of compression.

WC connections to discharge pipe systems

The usual outlet diameter for a wash-down closet is 100 mm, although 85 mm is quite adequate for low-rise buildings. Some types of siphonic closet require an outlet diameter of only 75 mm for their efficient functioning and reference should be made to the manufacturers' fitting instructions when these are installed. Generally speaking, WCs are less likely to be affected by loss of seal due to siphonage than other fitments as their outlet pipe to the main stack seldom runs full of water. The angle at which the WC branch joins the main stack is not critical, but a standard 104° branch is recommended. What is important, however, is the way in which the branch joins the main stack (see Fig. 8.21). The standard 50 mm radius on the invert of the junction leads the discharge from the WC into the main stack in such a way as to avoid the induced siphonage of other fitments sharing the same stack.

Cross flow

Another problem encountered in modern discharge-pipe systems is that of cross flow. When two opposing branches occur in a stack the discharge from one can shoot across the stack and into the other, causing fouled water to enter the other fittings (see Fig. 8.22(a)). Cross flow is more common with single-stack systems as the branch discharge pipes are not swept into the main stack

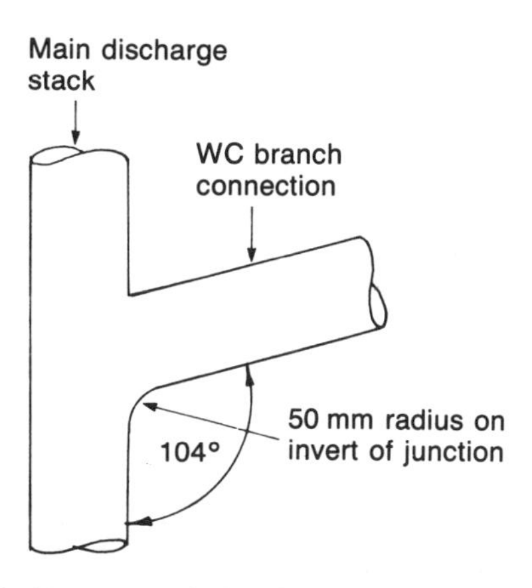

Fig. 8.21 Recommended WC branch connection to the main discharge stack

in the same way as those fitted to one- and two-pipe systems.

Swept branches, i.e. those with a large radius, on unventilated wastes can increase the possibility of siphonage. The maximum radius permitted at the junction of a branch discharge pipe and the main stack is 25 mm, as shown in Fig. 8.22(b). It will be found in practice that the WC branch would most conveniently enter the main stack almost directly opposite the bath discharge pipe connection, but as this would encourage cross flow, branches are not made in the main stack within the shaded area shown in Fig. 8.22(c). When necessary a short vertical branch is connected to the main stack below the restricted area in such a way that the maximum slope is not exceeded. The vertical branch is always one size larger than the discharge pipe from the bath, which prevents the formation of a solid plug of water in the vertical stub waste.

Ground-floor appliances

These may be discharged into the underground drain in several different ways. In low-rise dwellings the risk of compression is not great and it is, therefore, suitable to connect the pipe into the main discharge pipe, especially if it is within easy reach. If this is not possible it may discharge into a back inlet gulley.

A ground-floor fitting having a small diameter discharge pipe is sometimes connected directly to a drain using a special reducer (see Fig. 8.23) which is permitted by the Building Regulations, although the only reference made in CP 5572 refers to pipes of 100 mm in diameter which may be fitted vertically to a height of 1.5 m without venting. It is suggested that if, for example, a basin is discharged into a drain using the method illustrated, a check is made that the minimum permitted trap seal is retained after use as the vertical discharge may result in trap seal loss.

Termination of vent pipes

The termination of vent pipes of roof level is the same for all waste disposal systems. The most important point is to situate the terminal in such a

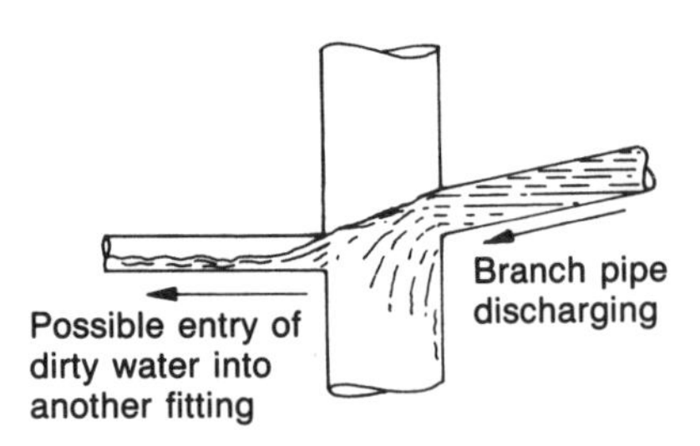

(a) Effect of cross flow

The discharge from opposing branch pipes can often cause cross flow, especially when branch pipes have only a slight fall

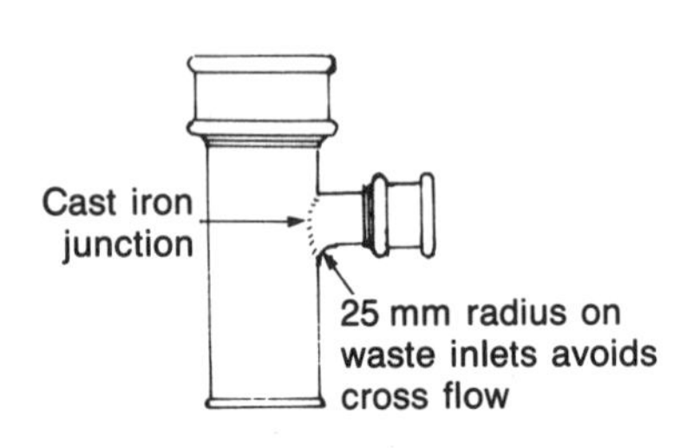

(b) Radiused inlet fitting to avoid cross flow

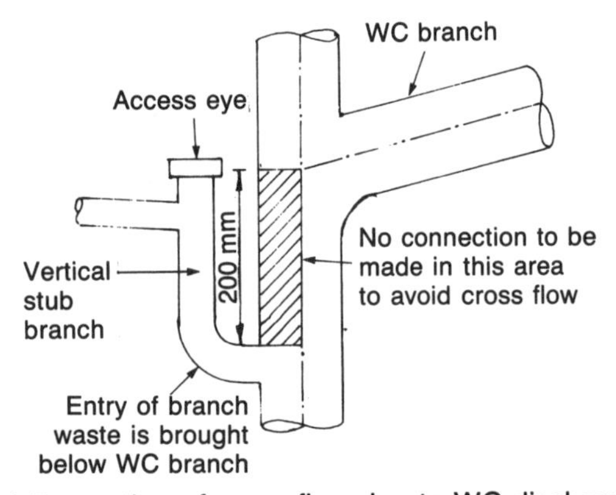

(c) Prevention of cross flow due to WC discharge

Bath and WC are usually at same level so their discharge pipes would most conveniently enter main stack opposite each other, but this would cause cross flow

Fig. 8.22 Cross flow

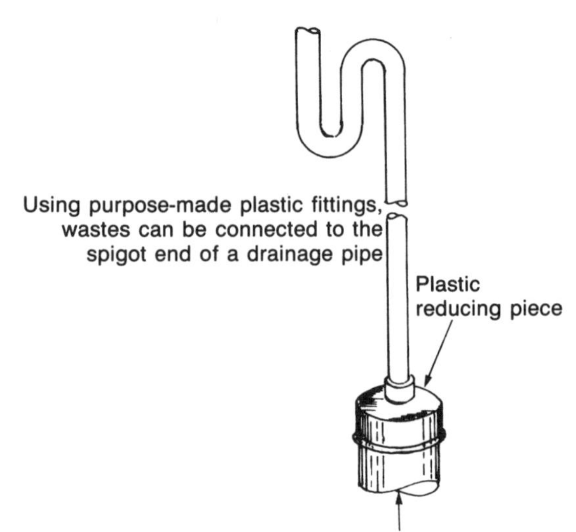

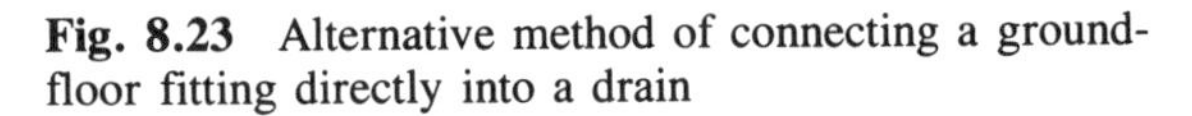

Fig. 8.23 Alternative method of connecting a ground-floor fitting directly into a drain

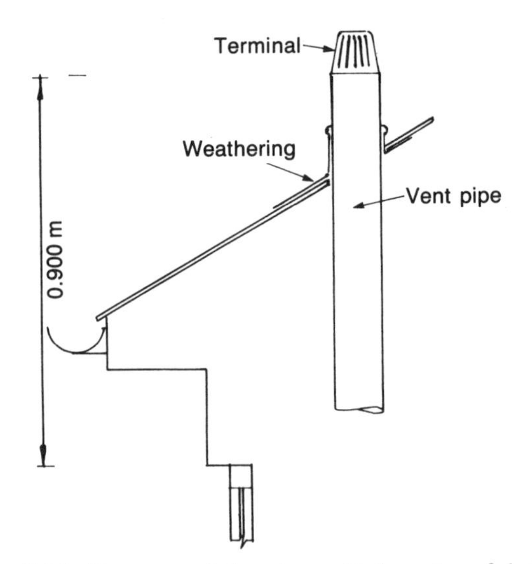

Fig. 8.24 Recommendations to avoid the entry of drain air to a building. If the vent pipe terminal is situated within 3 m of an opening window or ventilator it must be carried up to sufficient height to prevent drain air entering the building

way that foul air cannot enter the building via windows or ventilators. The recommendations given in CP 5572 are illustrated in Fig. 8.24.

The Building Regulations simply state that the vent should be carried upward to such a height as not to cause a health hazard or nuisance. If, as is usual, the main vent stack is fitted inside the building, the point where the stack passes through the roof should be weathered by a pipe flashing as detailed in Chapter 11. When PVC stacks are used, some systems include a special weathering component which avoids the necessity of fabricating metal weatherings. Some form of terminal should also be fitted at the top of the vent to prevent birds nesting in the stack and those constructed of high density polythene are adaptable for most piping materials.

Access points

The reason for providing means of access for clearing blockages has already been discussed. It is probably true to say that the traditional fully ventilated systems, especially those fitted to low-rise private buildings, seldom become obstructed, but modern waste-disposal systems are more prone to obstruction due to the low velocity of flow through the branch discharge pipes and the need for access is often very apparent. Generally speaking, access should be provided at the fitting and at the junction of the branch waste with the main discharge pipe. It was shown earlier in this chapter that easily dismantled traps can be used at the appliance, and clearing eyes at each junction or group of junctions in the pipework system (see Fig. 8.25(a)).

A greater number of access points are required in public and commercial buildings than private dwellings as blockages are more likely to occur due to the possibility of misuse. This is especially true of urinals, when due to the ammonia content of urine a hard surface coating of scale is built up on the inner walls of the pipes. The channel also becomes a depository for cigarette ends, matches and other small items of rubbish which, unless sufficient access is provided, can cause some unpleasant stoppages.

A screw cap type of access point is illustrated in Fig. 8.25(b), these being used at the end of a common waste serving several fittings. Figure 8.25(c) shows a cleaning eye bend suitable for branch discharge pipes while Fig. 8.25(d) illustrates a cleaning door in a length of pipe. These are usually used at ground level on a main discharge stack to permit access to the connection with the underground drain.

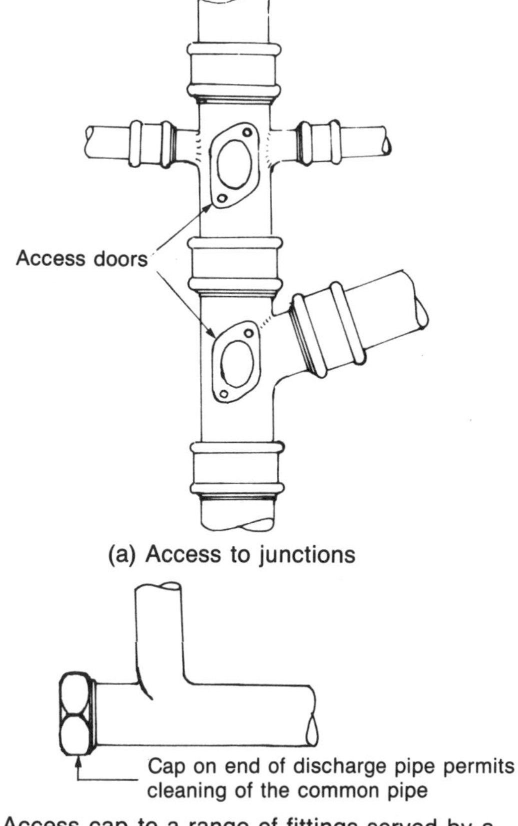

(a) Access to junctions

(b) Access cap to a range of fittings served by a common discharge pipe

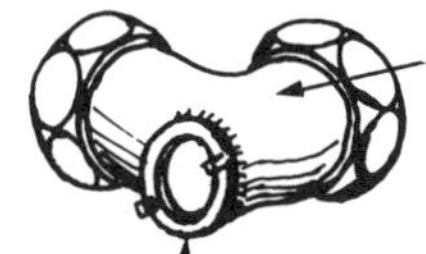

(c) Cleaning eye bend

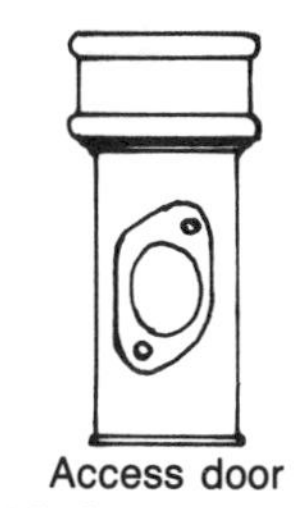

(d) Access pipe

This is usually fitted at a point where the main discharge stack connects to the underground drain. Some manufacturers of PVC pipe produce an access door which can be solvent welded into a pipe at any required point.

Fig. 8.25 Access points

Branch discharge pipe connections

The method of making additional connections to existing stacks depends upon the type of material being used. In the case of lead stacks, the traditional jointing methods of lead welding, solder wiping or, in the case of copper, bronze welding or brazing are used.

Where the stack is made of cast iron, one alternative is to cut out a section of the pipe and insert an appropriate junction. A form of slip collar will also be necessary if this method is adopted. Unless the additional connection is of a large diameter, e.g. a WC junction, the most convenient method is to use a 'Fixicon' type clamp as shown in Fig. 8.26(a). Adaptors are available which can accommodate a variety of pipework materials and sizes. The greatest difficulty with the use of this type of fitting is drilling a hole of sufficient size in the stack, especially in an awkward position. Great care must be exercised to ensure that the portion of pipe that is removed does not fall inside the pipe, as it may be very difficult to retrieve. Despite the difficulties, the use of these fittings can often save a lot of time which would otherwise be spent fitting a new junction into an existing stack, not an easy task even under good conditions.

The manufacturers of PVC systems employ several methods of making branch connections. Blanked off sockets may be moulded on to standard 100 mm junctions, the blank being cut out when the connection is required (see Fig. 8.26(b)). The other main alternative is the use of a fitting very similar to the 'Fixicon' bracket used with cast iron (see Fig. 8.26(c)). A hole is cut at the point where the branch is required using a suitable hole saw. The area round the hole and the face of the bracket are then treated with cleaning fluid prior to solvent welding the connection in position. The connection illustrated is provided with a moulded clamp which secures it in position until the solvent-welded joint has matured. Others are clamped in position by means of a special tool provided by the manufacturers.

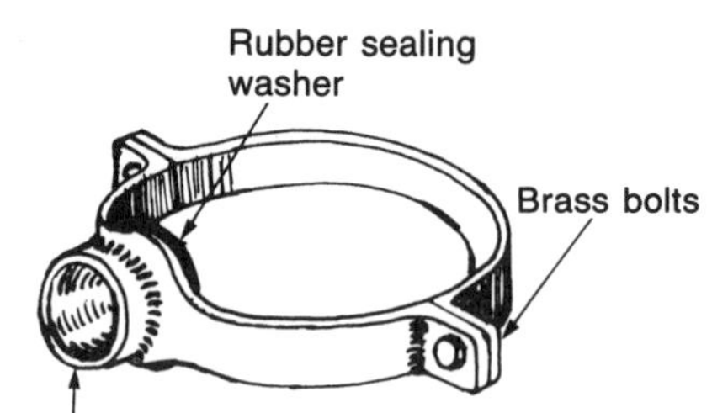

(a) Cast iron fixicon bracket

These fittings are clamped round the pipe over a precut hole.

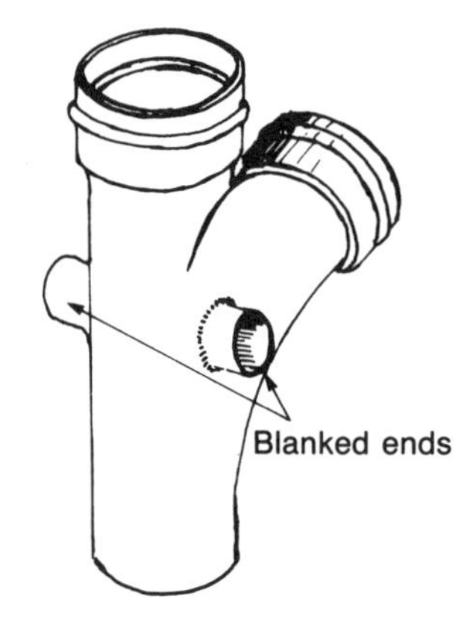

(b) PVC junction

This junction has moulded-on blank sockets for waste connections. Blanks are removed as and when needed.

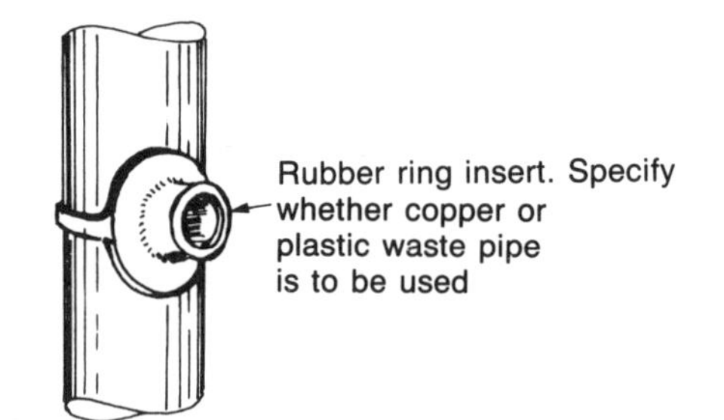

(c) Branch connection for PVC discharge stacks

Fig. 8.26 Connections to main stacks

Mechanical waste-disposal units

These fittings are becoming increasingly popular with householders and the plumber is required to install them in both new and existing properties. They are fitted beneath the sink and are designed to macerate waste materials such as vegetable peelings so they can be discharged into the drains without causing stoppages. They are electrically operated, and unless the plumber has competent electrical knowledge a properly qualified person should be employed to make the necessary electrical connections.

Before an installation is contemplated make sure an electrical supply is available! Such a warning may sound unnecessary, but experience proves otherwise. Another important point to remember is that the waste hole in many existing sinks is not large enough to accept such a unit, and in many cases a new sink will be required – a point he householder may not have considered. A hole of approximately 90 mm in diameter is required, which is far larger than that required for a normal BS 3380 waste fitting. Special cutters are available for enlarging the hole in certain types of existing stainless steel sinks and may be hired from the stockists of the equipment being fitted. Some of these stockists employ specialists who will fit the actual unit and make the necessary electrical connections, leaving only the waste connections to be fitted by the plumber.

It must be borne in mind that disposal units do not have an integral trap. A tubular trap must be fitted to the outlet and should be installed so that the invert of its outlet is lower than that of its inlet, which prevents water lying in the machine, Bottle traps must not be used as they tend to retain waste matter. The fall of the discharge pipe should be such that sufficient velocity of flow is maintained to prevent waste being deposited and causing a blockage. The absolute minimum fall is 1 in 10 or approximately 6° but some manufacturers recommend 15°. These relatively sharp falls preclude the direct connection of the equipment to a main discharge stack. They should always be connected to a back inlet gulley as illustrated in Fig. 8.27(a), as if they discharge over an ordinary gulley grating this will quickly become clogged by the waste particles. Suggested arrangements for fitting to a sink on an upper floor are shown in Fig. 8.27(b). The provision of a vent in the gulley is not normally necessary as it is very unlikely the drain will ever run full.

When this equipment is installed, the client should always be made to understand that a flow of water is required while in use. Failure to observe this will result in constant obstruction in the discharge pipe. Always make sure the client understands how to operate the unit and free it if it jams, a special tool being provided for this purpose. Leave the manufacturer's instructions and any guarantee forms with the client before leaving the job.

Washing machines and dish-washing equipment

The increasing use of domestic washing machines and dish washers has also provided the plumber with a new area of work. Before any machine of this type is fitted check that it is approved by the local water authority, as some types, especially older models, do not conform with the water regulations in respect of back siphonage.

Washing machine waste connections

The waste outlet on these units is a flexible rubber hose, approximately 20 mm nominal diameter. The waste water is discharged by a pump fitted to the

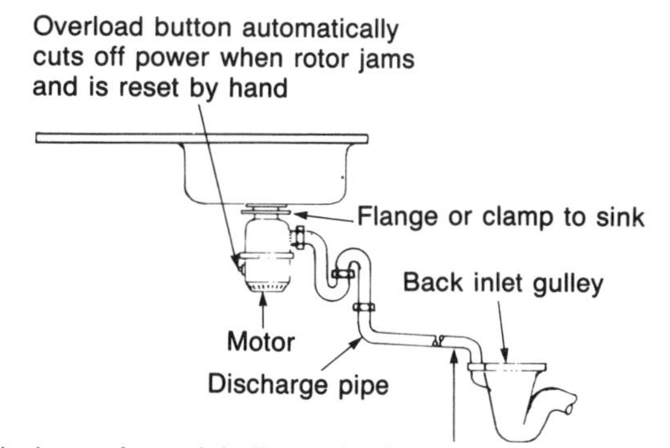

Discharge from sink disposal units must be connected under the gulley grating or via the back inlet to avoid fouling and blocking grating

(a) Fitted to a ground floor

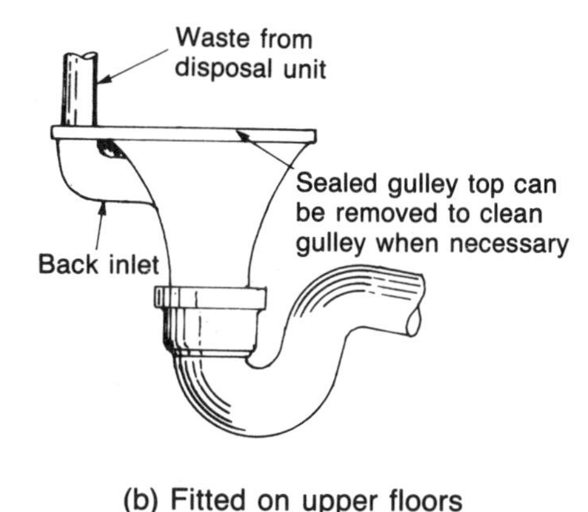

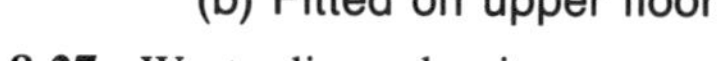
(b) Fitted on upper floors

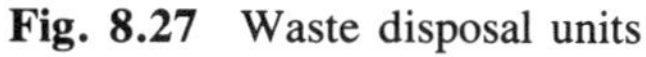
Fig. 8.27 Waste disposal units

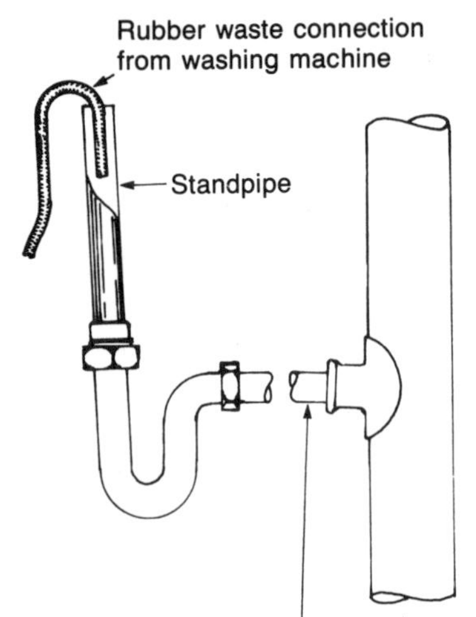

(a) Waste connection to discharge pipe

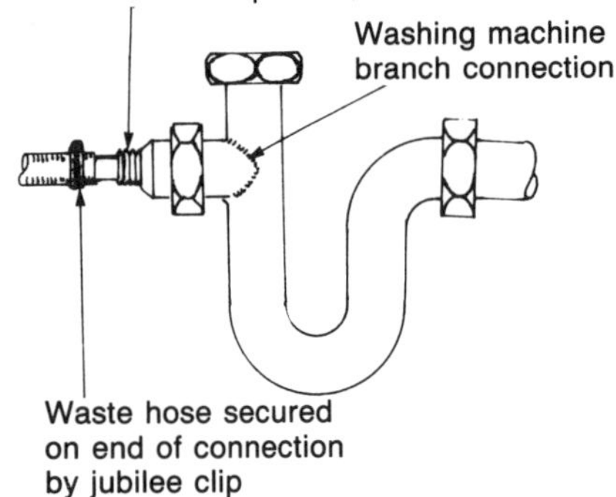

(b) Waste connection to sink trap

The branch connection enables the rubber waste hose to be discharged into the sink trap arrangement by providing a special trap as shown.

Fig. 8.28 Washing machine waste

machine. The original method of disposing of the waste was to connect the hose to a 22 mm pipe which terminated over a gulley, but this is now unacceptable for two reasons. It was found that this method often resulted in the pump becoming airlocked, necessitating a service call to the dealer. Secondly, due to the changeover to modern systems of sanitary pipework, there may not be a gulley available. Figure 8.28(a) shows a suitable arrangement for the connection of the waste from the machine to the main discharge stack, the rubber waste being hooked over the 40 mm riser. The air break so formed prevents siphonage of the pump and waste water back up into the machine.

If the machine is situated at a convenient point near a sink, it is often possible to replace the existing trap with one having a connection for the machine (see Fig. 8.28(b)). This connection is adaptable to a wide range of hose diameters; those not required can be cut off with a hacksaw and the hose secured with an adjustable clamp such as a 'Jubilee' clip.

Further reading

Much useful information can be obtained from the following sources:

BS 4118:1967: Glossary of sanitation terms.
BS 5572:1994: Sanitation pipework.
BS 1184: Copper and copper alloy traps.
BS 5943: Plastic waste traps.

Defect Action Sheets are obtainable from the Building Research Establishment, Garston, Watford, Herts WD2 7JR.

DAS 40: Plastic sanitary pipework – Storage and handling.
DAS 41: Plastic sanitary pipework – Jointing and support.
DAS 42: Plastic sanitary pipework – Site installation.

Plastic discharge pipe systems
See manufacturers lists, Chapter 3.

Self-testing questions

1. By means of a sketch show how the 'effective seal' of a trap is measured.
2. Explain why it is essential for traps of up to 50 mm diameter to have a 75 mm seal depth when in direct connection with a drain.
3. Name the three causes of seal loss in a trap due to the defective design of sanitary pipework. Briefly explain the reason for each cause occurring.

4. Describe the circumstances in which a resealing trap may be used.
5. Name the systems of above-ground discharge pipework.
6. State the minimum diameter of discharge pipes for baths, basins and sinks.
7. Describe how 'hydraulic jump' occurs in discharge pipes and how this affects the water seal in traps.
8. Explain why it is necessary to join a main discharge pipe to an underground drain with an easy or large-radius bend.
9. Define the term 'cross flow' in discharge pipes and give an example.
10. State why the discharge from sink waste disposal units should not terminate over a gulley grating. Sketch and describe the correct method.

9 Sheet-weathering materials

After completing this chapter the reader should be able to:

1. Name the materials used by the plumber for sheet roof coverings.
2. State the main physical and working properties of roof-covering materials.
3. Select materials with suitable properties for given jobs.
4. Describe the effects of various forms of corrosion on roof materials.
5. State the commercial sizes, types, thicknesses and qualities of sheet-roofing materials.
6. Define the terms 'work hardening' and 'annealing' and suggest suitable heat treatment.
7. Calculate the mass of roof coverings for specific roof areas.

Introduction

Before the introduction of piped water supplies the main activity of the plumber was roof weathering. Lead was the sheet material in general use and some idea of its applications can be gained by visiting historic buildings, many of which still retain their original roof coverings. During the last hundred years other materials have been increasingly used in this field of work and the plumber is required to have a thorough understanding of these materials and the working processes involved in their use. It should be noted that the plumbing NVQs cover only lead as a weathering material, but it is likely that this qualification will be extended to other roofing materials traditionally used by plumbers.

In recent years there has been an increase in the use of sheet metal weatherings, especially for large commercial, public and high-quality dwellings, and although the initial costs may be higher, the longer life of these materials over their alternatives has proved to be more economic in the course of time. Modern methods, coupled with the desire of many planning authorities to revert to traditional styles of building, have led to an increase in the use of sheet metal weathering, especially lead. While specialist roofing contractors are mainly involved in large-scale projects, the modern plumber should be competent to weather small canopies, dormers, gutters and chimney weatherings in sheet metal.

The sheet materials used by the plumber for weatherings are lead, copper, aluminium and zinc. Prior to discussing these materials it is necessary to consider certain terms and processes in order that the details given can be fully understood.

Physical properties

Colour and melting point are familiar terms to everyone but density and coefficient of linear expansion will require explanation.

Density

For commercial purposes the density of a substance relates to the mass (weight) of one cubic metre of the substance, for example the density of lead is 11,340 kilograms per cubic metre (11,340 kg/m^3).

Each material has a different density and this is often used as a means of its identification. Lead is an especially dense material, while aluminium has a low density of 2,720 kg/m^3, and both may be compared to water which has a density of 1,000 kg/m^3. (Note that when a comparison is made between the densities of various materials their volume, i.e. the space they occupy, must always be the same.)

When the density of a material is known then the mass of the sheet roof covering made from it may be found by calculation.

Example If lead has a density of 11,340 kg/m^3, calculate the mass of 1 square metre of BS code No. 6 lead sheet having a thickness of 2.5 mm.

First find the mass of one square metre of lead sheet which is 1 mm thick:

$$11{,}340 \div 1{,}000 = 11.34\,\text{kg}.$$

The mass of one square metre of sheet lead 2.5 mm thick will therefore be:

$$11.34 \times 2.5 = 28.35\,\text{kg}.$$

This simple formula can be used to find the mass (weight) of the total area of sheet lead, in square metres, necessary to weather a roof, information which would be needed by a building designer to make provision for sufficiently strong roof supports.

Coefficient of linear expansion

This subject has been dealt with in Chapter 6 (see pp. 132–3) where it was shown that an increase in temperature has the effect of increasing the length of pipe. The same principle also applies to metal sheet coverings for roofing and weathering which will be affected by variations in temperature ranging from snow in winter to heat from the sun in summer. Temperatures of up to 90 °C have been recorded on roofs in this country in the summer, so the matter of expansion (and subsequent contraction) of roofing materials demands careful consideration.

As it is so important to understand the need for restricting the areas and lengths of the materials used, a further example of this subject is included here.

Example One bay of an aluminium roof covering is 3 metres long and has been subjected to a temperature rise of 44 °C. Calculate the increase in the length of the aluminium. (Coefficient of linear expansion of aluminium is 0.000026.)

$$\begin{aligned}\text{Increase in length} &= \text{Length} \times \text{Temp. rise} \times \text{Coeff. of linear exp.}\\ &= 3\ \text{m} \times 44\ ^\circ\text{C} \times 0.000026\\ &= 0.0034\ \text{m}\\ \text{or} &= 3.4\ \text{mm}\end{aligned}$$

The bay will be 3.4 mm longer when its temperature has been raised by 44 °C. It must also be realised that this effect is reversed on cooling, i.e. when cooled through 44 °C to the original temperature, the aluminium will contract by 3.4 mm.

Expansion and contraction due to temperature changes must be allowed for when laying roofing materials or deformation and cracking will occur and the roof will leak. Methods of limiting damage from these effects will be considered in the next chapter.

Working properties

Consideration of mechanical and other properties will inform the plumber how a material will behave when being fixed and when in use.

Malleability

This term relates to the ability of a metal to be hammered or rolled into permanent shape without breaking. This property is required of metals that are to be formed by 'bossing'.

Ductility

The ability of a metal to withstand distortion without fracturing is indicated by the elongation that occurs while a sample is being tested to find its tensile strength. The term 'ductility' is used to denote this property. Neither malleability nor ductility can be measured accurately, the only test being to 'work' metals and then compare the results. Table 9.1 shows the order of malleability and ductility for cold-worked metals.

Tensile strength

The tensile strength of a material is its ability to resist being torn apart. The method of applying the test is to subject the sample to increasing loads

Table 9.1 Order of malleability and ductility.

Malleability	*Ductility*
Lead	Copper
Aluminium	Aluminium
Copper	Zinc
Zinc	Lead

until it breaks, the loading that causes breakage being the ultimate tensile stress.

Hardness

The hardness of a metal is its resistance to indentation, abrasion and deformation. Several tests can be used to determine hardness but they are all similar in principle and require that either a steel ball (Brinell) or a diamond point (Vickers and Rockwell) is pressed on to the metal surface and the indentation produced gives an indication of hardness.

A simple workshop test for hardness uses a steel ball bearing and a glass tube, and is illustrated in Fig. 9.1. When the ball is dropped into the tube it will hit the metal sample and then bounce upwards, the higher the bounce the harder the metal. By using a soft metal such as lead to provide a low mark and cast iron or steel to give a high mark the whole range of plumbing material can be tested to find how they relate to each other.

Temper

The temper of a metal is the amount or level of hardness and can vary between dead hard and dead soft. An example of the range of temper can be seen with copper where it would be of maximum hardness in a hammered or cold-worked condition, the dead soft state being when fully annealed.

The working properties of hardness, tensile strength and ductility are related so that if one property is changed, then changes will occur in the others. An example of this relationship can be seen when hardness is increased by work hardening: tensile strength also increases but ductility decreases. Similarly, as a metal is heated it becomes soft, its ductility increases but the tensile strength is lowered.

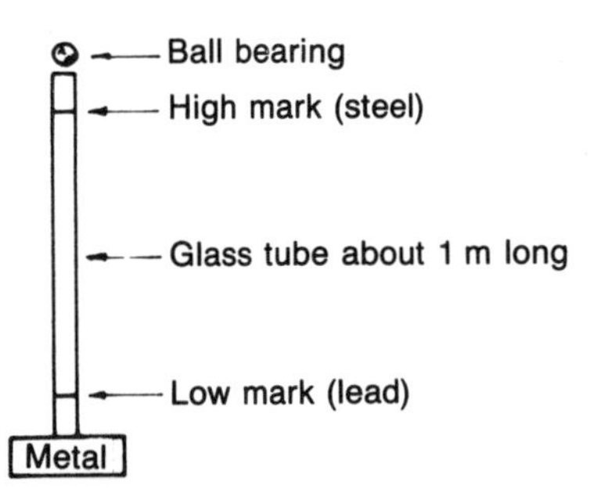

Fig. 9.1 Simple workshop test to compare the hardness of various metals. When the ball bearing is dropped in the glass tube it bounces on the metal below. The higher the bounce the harder is the metal.

Work hardening

Relates to an increase in hardness caused by cold working, e.g. bossing or hammering. All metals are formed of many grains or crystals which become deformed during 'working'; the effect is shown in Fig. 9.2. The deformation of the crystals reduces ductility which will prevent further cold working and could lead to fracture.

Annealing

This is the name of the heat treatment used to relieve the condition of work hardening. The metal is raised to a specified temperature; during the heating process recrystallisation takes place and the crystal grains return to normal. This treatment softens the metal, relieves internal stresses and allows further cold working to be carried out. Any attempt to anneal metal while it is being fixed to a wooden substructure will present a serious fire risk, so this practice must be avoided or undertaken with great caution.

Fatigue cracking

This is an example of work hardening that can be demonstrated by bending a strip of metal backwards and forwards in the same position. The strip will quickly work harden and if bending

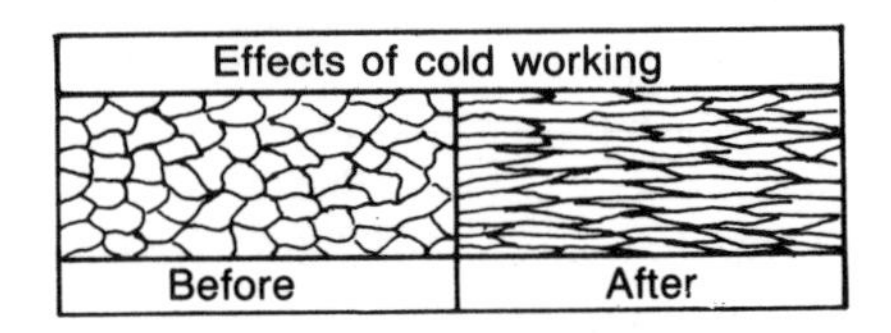

Fig. 9.2 The effects of cold-working metals

continues it will break, i.e. brittle fracture or fatigue cracking will occur.

Creep

The tendency of metals to stretch slowly over a period of time if under constant loading is termed 'creep' and it is an important consideration in sheet metal weathering. Creep is a form of plastic deformation and is illustrated in Fig. 9.3 where two different metal samples are seen to have lengthened due to the application of a load. When the loads are removed, metal 'A' has returned to its original length, i.e. the elongation to load was elastic deformation. Sample 'B', however, has suffered a permanent increase in length, i.e. plastic deformation, known as creep.

An experiment to compare the 'creeping' action of metals when heated is shown in Fig. 9.4. Samples of sheet metals 6 mm wide are cut and fixed at the top by a screw and at the bottom to a pointer. The initial position of the pointer is marked and steam is passed into the tube causing the strip to become hot and expand. The expansion causes the pointer to move down the scale and, upon cooling to its original temperature, to move up again. The difference between first and final positions of the pointer indicate amount of creep.

The effects of heat and mass will increase the tendency of sheet roofing metals to 'creep'. Hot metal has a lower tensile strength and the mass of metal will provide the load causing elongation, especially in the case of sheets which are fixed vertically or placed on steeply pitched roofs.

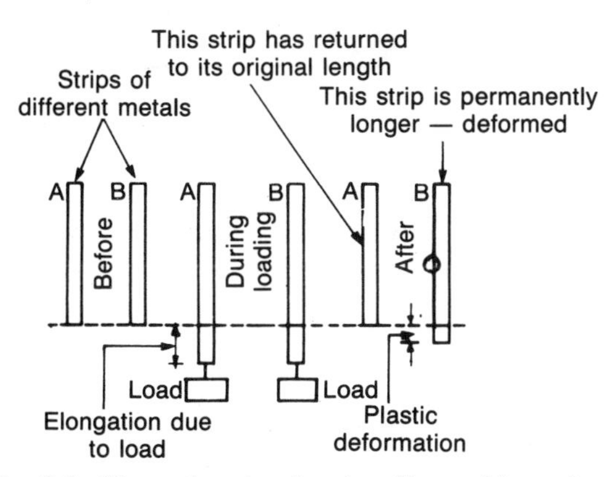

Fig. 9.3 Illustration showing the effects of 'creep' on materials

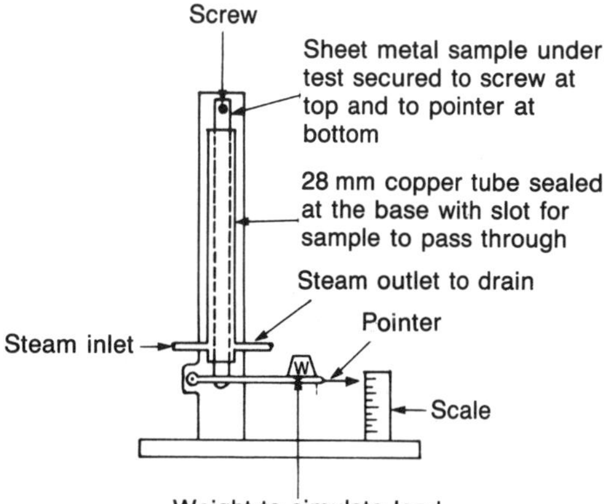

Fig. 9.4 Simple experiment showing how heating and loading a sample of material can increase the effect of creep

British Standards

The British Standards Institute was founded in 1901 to standardise such industrial activities as manufacture, design and installation practice. British Standards specifications refer to standards of manufacture, e.g. sheet aluminium is made to the specifications contained in BS 1470. British Standards Codes of Practice apply to methods of good practice in installation work, e.g. BSCP 143 deals with sheet roofing. By having standards which have been agreed by both manufacturers and industrial experts, the problem of getting the right materials and correct installation procedure is made easier. Designers need only to specify that material must conform to the relevant BS and that work is to be undertaken in accordance with the particular BSCP, to ensure materials and work are of a satisfactory standard.

Copper

Chemical symbol	: Cu
Colour	: Reddish-brown
Density	: 89,000 kg/m^3
Melting point	: 1083 °C
Coeff. of linear expansion	: 0.000016/°C
Tensile strength	: 210–390 MN/m^2

BS specification No. 2870
BSCP 143, Part 12

Copper is obtained from the ore copper pyrites, the USA being the biggest producers, but Zambia, Canada and Chile provide the majority of copper for Great Britain. In Roman times copper was in use as a roofing material, the dome of the Pantheon in Rome being roofed with bronze (an alloy of copper and tin) tiles under which was laid a sheathing of copper plates. An example of the early use of sheet copper for roofing in this country is the great dome of the British Museum which was covered in 1857.

Sheet copper is malleable and highly ductile so it can be easily worked, particularly when in a soft or fully annealed state. Work hardening, however, occurs with cold working so as little hammering as practicable should be done. Annealing must be undertaken as soon as the copper becomes hard and difficult to work, or brittle fracture will occur. Annealing is carried out by heating to a dull red and then either quenching (rapid cooling) in water or cooling naturally in the atmosphere, either of which will make the copper dead soft.

The physical properties of copper are such that it will not 'creep', and this means that it is suitable for fixing to steep slopes. It has a low mass and this factor is of great advantage when designing the roof structure that the sheet is to weather. If the dome of St Paul's Cathedral had been covered with sheet copper (as was proposed by Parliament), instead of cast lead, the mass of the roof covering would have been reduced by many metric tonnes. The reduction in load would have made maintenance of the roof support structure of Wren's masterpiece an easier task than it has been.

Copper is available in the form of rectangular sheet or in 'strip' form, i.e. long lengths, the latter usually being supplied in coils although it may be flat or folded. 'Strip' copper can be obtained in many different widths and lengths, the latter to a maximum of 90 metres. The standard strip widths are 114, 228, 380, 457, 533, 610 and 686 mm, while the length is determined by the thickness and mass of the coil which is usually limited to 25 and 50 kg in order to assist manual handling on site. 'Sheet' copper can be obtained in almost any size depending on the quantity required, but standard sizes of 1,830 × 910 mm and 1,830 × 610 mm are normally specified. Both strip and sheet is sold by mass.

The availability of a range of widths makes the use of strip sheet copper more popular for flashings, damp-proof courses and cavity wall flashings as there will be little waste. In deciding which sizes of strip or sheet are needed for a job it should be remembered that an extra charge is usually made for small quantities so it is important to limit the number of different sizes and thicknesses.

Two basic copper roofing techniques are in use, the 'traditional' and 'longstrip' methods. The latter is chiefly used on very large areas and is outside the normal range of plumbers' work. The former method, however, is suitable for weathering the small flats and canopies traditionally laid by plumbers. Laying techniques using this method are described in the next chapter. The temper of the copper sheet is dependent on the technique being used, but for the majority of 'traditional' type roofing and flashing work, the temper should be dead soft or fully annealed. Where a degree of rigidity is required such as for the weathering of window sills or along cornice or coping edges, a half-hard temper may be necessary. The recommended thicknesses of copper sheet for various situations is shown in Table 9.2.

Table 9.2 Thickness of sheet copper for various fixing positions.

Fixing position	*Thickness (mm)*
Cover flashings, gutters	0.45
Flat roofs, canopies, gutters	0.60 or 0.70

Lead

Chemical symbol	: Pb
Colour	: Blue-grey
Density	
Milled	: 11,340 kg/m^3
Cast	: 11,300 kg/m^3
Melting point	: 327 °C
Coeff. of linear expansion	: 0.0000293 °C
Tensile strength	: 18 MN/m^3

BS specification No. 1178

BSCP 143, Part II (Roofing)

Lead is obtained from the ore galena and is mined and smelted in many different parts of the world, the principal producing countries being Australia, Canada, the USA and the USSR. Lead is also mined in Britain but the amount obtained nowadays is very small. The plumber took his name from the Latin name for lead which is *plumbum*. In Roman times the use of lead in building was restricted mainly to sanitary and ablutionary installations, but during later years, particularly in the Norman period, the use of lead was extended to include roofing and weathering. There are many examples of the use of this material on English churches and cathedrals of all periods, notably Lincoln, Canterbury, York and St Paul's cathedrals. Sheet material used for these buildings was usually produced on site by pouring molten lead on to a flat bed of fine sand. These cast lead sheets were thicker than most of those used in modern practice and varied between 3.15 and 3.55 mm in thickness. It should be noted that one company producing lead sheet has developed a technique for continuous casting. It is obtainable in roll and strip form like milled sheet and has comparable thicknesses.

As can be seen from the stated densities, cast lead is less dense than the milled sheet in current use, the latter being compressed during manufacture by being passed between steel rollers to obtain the required thickness. Traditionally cast lead is little used nowadays except for ornamental details such as rainwater pipes and hopper heads. If cast lead weatherings are renewed using traditional methods the work is usually executed by specialist firms with the necessary equipment and craftsmen.

Lead is ductile, the softest of the common metals and is highly malleable which allows it to be shaped with ease. Lead recrystallises during cold working and this factor means that work hardening does not take place to any great extent. Low tensile strength, softness and lack of elasticity give lead a tendency to 'creep', but this can largely be overcome by the application of correct installation procedures. The term 'creep' must not be used to describe the slipping of lead down a pitched roof due to inadequate fixings failing to provide support. Fatigue cracking does occur but is usually caused by bad design or poor workmanship in using oversized sheets and not making sufficient allowance for expansion and contraction.

Milled lead sheet is available in standard rolls 2.4 m wide in lengths of 3, 6 or 9 m. These sheets are very heavy, difficult to handle and are only recommended where large areas are to be covered. For a small extra charge rolls of smaller width (75 mm up to 600 mm) known as strip can be supplied, these being more economic in use and save time in cutting on site. The thickness of lead is designated by a BS specification code number, by the thickness in millimetres and by an identifying colour, these factors are shown in Table 9.3.

The thicknesses of lead for various situations as recommended by the Lead Development Association are given in Table 9.4.

The choice between the thicknesses of sheet given in Table 9.4 depends on the quality of

Table 9.3 Indentification of sheet lead.

BS specification Code No.	*Colour*	*Thickness (mm)*
3	Green	1.25
4	Blue	1.80
5	Red	2.24
6	Black	2.50
7	White	3.15
8	Orange	3.55

Table 9.4 Thickness of sheet lead for various fixing positions.

Fixing position	*BS specification Code No.*
Small flats with no pedestrian traffic	4 or 5
Large flats with or without traffic	5, 6 or 7
Gutters — parapet, box or tapering valley	5 or 6
Dormer cheeks and roofs	4 or 5
Chimney flashings	4 or 5
Soakers	3 or 4
Cornices	5 or 6
Valleys, hip, ridge, and cover flashings	4 or 5
Vertical cladding	4 or 5
Pipe weathering	4 or 5
Cornice weathering	4, 5 or 6
Damp-proof courses	3, 4 or 5

building, length of life required, roof design and shape of individual panels. The initial cost is however, an important consideration so the thinnest lead to suit the fixing position will usually be specified.

Aluminium

Chemical symbol	: Al
Colour	: Bluish-white
Density	: 2,705 kg/m^3
Melting point	: 660 °C
Coeff. of linear expansion	: 0.0000234 °C
Tensile strength	: 90–150 MN/m^2

BS specification No. 1470
BSCP 143, Part 15

Aluminium is obtained from the ore bauxite, large reserves being found in the USA, Guyana, Central Europe, the USSR, Australia, Ghana and India. The metal is young in comparison with lead and copper, the first ingot being shown at the Paris Exhibition of 1855. Commercial production did not commence until the end of the nineteenth century, when its potential as a weathering material was quickly appreciated. Before the turn of the century roofs were being covered, one example that still displays the original metal is the church of San Gioacchino in Rome which was weathered before 1897 using 98.28 per cent pure aluminium sheet. It was not until the second half of the twentieth century that aluminium became competitive in price with the older roofing materials.

Aluminium is malleable and ductile giving when in a soft or annealed condition similar working properties to those of copper. The metal is unaffected by 'creep' but will suffer work hardening and fatigue cracking if cold worked.

Annealing is easily achieved but due to the low melting point care must be taken to avoid overheating. A suitable temperature can be judged by heating until a matchstick drawn across the surface leaves a black charcoal line. When this occurs the metal should be quenched in water or allowed to cool naturally in the air.

In BS 1470 five grades of aluminium are specified but only two are recommended for roofwork being S1 and S1B. Details are given in Table 9.5.

Table 9.5 Grades of sheet aluminium.

Nominal composition (%) showing maximum permissible impurities	*Grade*		
	S1	*S1B*	
Al%	99.99	Al%	99.5
Cu%	not more than 0.01%	Cu%	0.05
Fe%		Fe%	0.4
Si%		Si%	0.3
Mn%		Mn%	0.05
Temper available	Soft		Soft

Due to an almost 100 per cent aluminium content, the Grade S1 is known as super purity aluminium. Both S1 and S1B can be specified for all roof weathering and flashing applications. Two thicknesses are available, these being 0.6 mm (standard) for use as a flashing material and 0.8 mm (heavy duty) which is suitable for covering large areas of bays.

Sheet aluminium can be obtained in standard 8 m rolls in widths of 150, 300, 600 and 900 mm. Profiled apron flashing (PAF) is also obtainable, this having an apron shape to fit into the contours of corrugated asbestos sheets. PAF is available in 1.200 m lengths of 300 mm width and 0.8 mm thickness to fit over 75 mm and 150 mm corrugated sheets.

Zinc

Chemical symbol	: Zn
Colour	: Greyish-white
Density	: 7,200 kg/m^3
Melting point	: 416 °C
Coeff. of linear expansion	: 0.000029 °C
Tensile strength	: 105–150 MN/m^2

BS specification No. 849
BSCP 143, Part 5

Zinc is obtained from several ores, the main sources being zinc blende or sphalerite and calamine. Among the major mining countries are Canada, the USA, the USSR and Australia.

Early use of the metal as a weathering material is recorded at Liège in Belgium where the Church of St Barthelemy and the Cathedral of St Paul were covered in 1811 and 1812 respectively. In

Great Britain an example of zinc roofing which was fixed in 1845 can still be seen at Kemble Station in Gloucestershire. Many early roof coverings were carried out using zinc in the form of interlocking tiles.

Commercial zinc sheet has a low ductility and is the least malleable of roofing metals. These factors combined with the fact that it is also hard, make manipulation difficult and working (bossing) an impracticable operation. Zinc is also affected to some extent by creep but this and certain other disadvantages have been largely overcome by the development of zinc alloys.

Zinc alloys have a reduced creep (strength), good ductility, a tensile strength of 170–220 Mn/m^2 and a linear expansion rate of less than two-thirds that of commercial-quality sheet. Two alloys are currently available, namely zinc/titanium and zinc/lead, these being marketed under the trade name of Metizinc and Metiflash respectively, Metizinc sheet is for covering large roof areas while Metiflash, which is malleable and has a greater ductility, is for flashings and small details such as canopies, dormers and bay window tops.

Commercial zinc is available in sheets 2.438 m × 914 mm with a recommended thickness of 0.6 mm for flashings and 0.8 mm for large roof areas.

Zinc/titanium can be obtained in sheets up to 3 m long by 1 m wide (standard sheet is 2.438 m × 914 mm) and in coiled strip up to 1 m wide (standard widths are 610 mm, 686 mm and 1 m). Various thicknesses can be supplied from 0.2 mm to 2 mm, the usual thickness being the same as those for commercial zinc.

Zinc/lead is 0.6 mm thick and is supplied in 10 m coils in widths of 150, 240, 300, 480 and 600 mm. Rolls of 900 mm width and 6 m length are also available.

Zinc sheet has a grain which runs down the length of the sheet, and although this occurs in all rolled sheet metals it is of particular significance in zinc, making it more difficult to obtain a sharp turn along the length of the sheet than across its width, i.e. across the grain. For this reason when zinc work is set out, it should be arranged so the majority of turns or folds are made across the sheet, typical examples being cap flashings or the apron pieces fixed round the edge of a zinc-covered flat roof. In cold weather zinc becomes brittle and unless warmed slightly when being prepared for fixing, it will split or crack when folded. A note of caution is also essential when applying heat as too much heat will alter the crystal structure and make the metal brittle – warmth to the touch of the hand is sufficient.

Temper is not usually referred to when specifying this material but as a general guide, commercial zinc is hard, zinc/lead is soft and zinc/titanium alloy is between the two.

Behaviour under working conditions

When the surface covering material has been selected, other factors must be considered in regard to behaviour under differing conditions. The result of these considerations will influence the method of application and could identify the need for precautions to ensure a long working life of the material.

Thermal expansion has already been mentioned and it is essential to make provision for this by limiting sheet size and providing fixings that allow movement.

Corrosion is another important detail which must not be overlooked or failure of the metal will occur. Before discussing the application of corrosion to roofing metals some explanation of the basic process is necessary.

Corrosion

Corrosion is a chemical or electrochemical action which causes the decay or 'eating-away' of metals. Chemical corrosion is the attacking of metal by acid, although it must be appreciated that alkalis and other agents have a similar effect. Electrolytic corrosion, another form of chemical corrosion, is not so readily appreciated.

Electrolytic corrosion

A simple experiment demonstrating electrolytic corrosion is illustrated in Fig. 9.5, where the placing of the two metals in a water-filled beaker is shown to produce a simple electric cell or

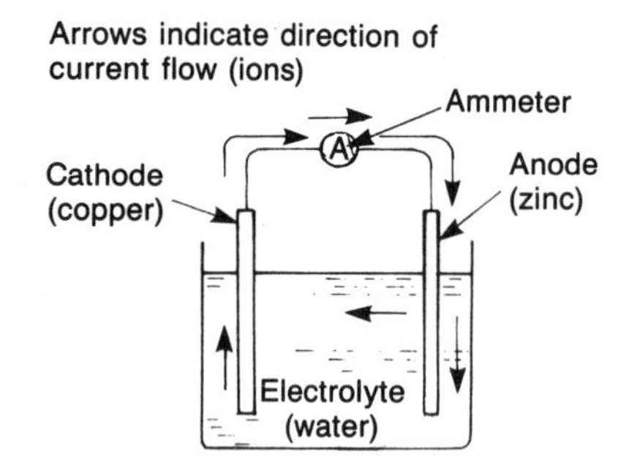

Fig. 9.5 Electrolysis. The flow of electrically charged ions from the anode to the cathode eventually results in the complete destruction of the anode. A similar situation can exist if two dissimilar metals are used together in weathering or water supply work. Water acts as the electrolyte and the anode is destroyed.

battery. Electrically charged particles called ions flow from cathode to anode through the electrolyte producing a current that will register on the ammeter. This action eventually causes the destruction of the anode.

The rate of electrolytic corrosion is increased if the electrolyte is acidic or the two metals are widely separated in the list of metallic elements shown in Table 9.6. The higher placed metal always forms the positive electrode (anode) and the lower the negative electrode (cathode).

Oxidation

Oxidation is a form of corrosion that involves the combination of the oxygen in the air with another substance and can be seen when freshly cut metal is exposed to the atmosphere. The cut sample

Table 9.6 Electrochemical series in metals used in plumbing.

HIGH	Magnesium
↓	Aluminium
	Manganese
	Zinc
	Chromium
	Iron
	Nickel
	Tin
	Lead
LOW	Copper

Table 9.7 Patina coloration of metals.

Metal	*Natural colour*	*Patina colour*
Copper	Reddish-brown	Green
Lead	Blue-grey	Silver-grey
Aluminium	Bluish-white	Grey
Zinc	Greyish-white	Silver-grey

quickly 'dulls' due to the action of oxygen producing a surface coating which is an oxide of the metal involved, e.g. copper forms copper oxide. The action of moisture and certain chemicals such as carbon dioxide and sulphur dioxide on the metal leads to the formation of a 'skin' that is commonly referred to as a *patina*.

Patina coloration is different to the natural colour of the metal as is shown in Table 9.7.

With non-ferrous metals, i.e. those that do not contain iron, the patina adheres to the metal forming a protective coating which is effective in preventing further attack.

Owing to the effects of electrolytic corrosion, the porous 'skin' (rust) on ferrous metal flakes off leaving a new unprotected metallic surface beneath. Further consideration is given in later chapters to this important topic which results in the complete conversion of steel to rust if it is left unprotected.

The practical effects of using, for example, copper and zinc together in the presence of an electrolyte will result in the rapid corrosion of the latter. For this reason careful consideration is necessary in the choice of materials for sheet roofing fixings, e.g. nails and screws made of galvanised iron, or aluminium alloys must not be used for fixing sheet copper or they will be destroyed by electrolysis if water is present too. Further reference is made to this subject in relation to roofwork in Chapters 10 and 11.

Electrolytic corrosion also affects the materials used for water supplies (see Chapters 5 and 6).

One further point should be noted at this stage. It will be seen from Table 9.6 that although tin and iron are not as far removed from copper, as for instance, zinc or aluminium, they are also anodic to copper and should not be used in conjunction with it in wet or damp conditions.

Occurrence of corrosion
Students may be puzzled as to how corrosion can affect roof coverings as, if similar metals are used and the air is dry, there will be no corrosive chemicals and decay will not occur. However, an electrolyte – water – is always present in the form of condensation, mist or rain.

Carbon dioxide and sulphur dioxide are normal constituents of an industrial atmosphere and when dissolved in water, form weak carbonic and sulphurous acids; similar corrosive agents develop in a marine environment. The moss and lichen which grow on roofs can form a harmful acid in the same way as some timbers, e.g. red cedar, fir, oak, teak and chestnut. The use of timber preservatives such as copper sulphate provides the cathode that will have a deleterious effect on any anodic metal placed in contact with it. Alkaline attack from cement, lime and plaster will lead to corrosion of some metals, particularly in permanently damp conditions.

Protection against corrosion
Protection against corrosion is required as indicated in Table 9.8.

The effects of industrial and salt-laden air depend on both the thickness of the zinc and the roof pitch. To avoid corrosive depositions becoming lodged in crevices or forming puddles, a minimum fall of 5° is recommended on roofs weathered with this material.

Table 9.8 Metals requiring protection against corrosion.

Corrosive agent	*Zinc*	*Lead*	*Copper*	*Aluminium*
Organic acid	✓	✓	–	✓
Industrial air	–	–	–	–
Marine air	–	–	–	–
Alkalis	–	–	–	✓
Zinc and water	–	–	–	–
Lead and water	–	–	–	✓
Copper and water	✓	–	–	✓
Aluminium and water	–	–	–	–

Comparative costs

It is not possible to provide market prices for materials as they are constantly varying due to supply and labour costs. What can be stated with reasonable certainty is that the comparative costs, including laying of roofing materials, show aluminium as the cheapest followed by zinc, copper and lead. Other considerations such as maintenance, thicknesses used, long life and appearance vary considerably according to the situation and usage and are quite beyond simple comparison.

Further reading

The following, which contain much useful information, are obtainable from HM Stationery Office:

BS 6915: Specification for design and construction of fully supported lead sheet roof and wall coverings.
CP 143:Part 12: Copper sheet roof and wall coverings.
CP 143:Part 15: Aluminium sheet for building.
CP 143:Part 5: Sheet zinc roof and wall coverings.
CP 143:Part 2: Lead sheet roofing.

Further information is available from:
Aluminium Federation, Broadway House, Calthorps Road, Fiveways, Birmingham, B15 1TN.
Copper Development Association, Orchard House, Mutton Lane, Potters Bar, Herts, EN6 3AP.
Lead Sheet Manuals, 1, 2 and 3.
L.D.A., St Johns Road, Tunbridge Wells, Kent, TN4 9XA.
Metra, Non-ferrous Metals Ltd, Pinder Road, Hoddesdon, Herts, EN11 0DE.
Zinc Development Association, 42 Weymouth Street, London, W1N 3LQ.

Self-testing questions

1. List the type, thickness and temper of sheet materials recommended for use as flashings.
2. Name the roofing material that is affected by alkali attack.
3. Describe the process used to soften 'work hardened' copper sheet.

4. Identify which of the following pairs of metals would be most affected by electrolytic corrosion:

   (a) lead and zinc
   (b) aluminium and copper
   (c) copper and lead

5. Define the term 'coefficient of linear expansion'.
6. Calculate the mass of a square metre of zinc/titanium alloy which is 0.8 mm thick. (Density of alloy is 7,200 kg/m^3.)
7. Make a table showing the density, melting point and coefficient of linear expansion of each of the roofing materials discussed.
8. Select (a) the sheet roofing material least likely to be affected by thermal expansion; (b) the most dense material, and (c) the material with the lowest melting point.
9. Describe why lead is more liable to 'creep' than other roofing materials.
10. Describe the difference between fatigue cracking and creep in relation to metal roofing.

10 Sheet weatherings – working processes

After completing this chapter the reader should be able to:

1. Detail preparatory work needed prior to applying roof-covering materials.
2. Describe preventative and remedial measures to protect roof coverings from the influence of wind, corrosion and thermal movement.
3. State the recommended falls for effective roof drainage.
4. Select suitable jointing methods for sheet weatherings.
5. Select suitable fixing methods for sheet weatherings.
6. State the principles involved in forming, shaping and jointing details in roofing materials.

The fixing of each weathering material will be considered in this chapter but it is first necessary to deal with certain physical problems and technical knowledge which must be understood before any practical operations can commence.

Capillarity and fatigue cracking

The make-up of matter

Just as a large block of flats is merely a complicated arrangement of bricks and mortar, cement, glass, wood and metals, so an analysis of any substance shows that it is made from one or more of about 92 different basic materials. These are called *elements* and can be in the form of a gas, a liquid or a solid, examples being oxygen (gas), mercury (liquid) and iron (solid). Imagine 92 bottles, each containing one of these elements and from these bottles, everything known is composed.

All known substances are either pure elements or are obtained by combining different elements to form mixtures or compounds. Examples of these substances are lead (element), solder (mixture of the elements tin and lead), and water (compound of the elements hydrogen and oxygen).

Every substance is composed of many tiny parts which are too small for the eye to see, these being termed *molecules*. Although these are the smallest particles of a substance, they can be further broken down into *atoms* of the same element or atoms of two or more different elements. The atoms of a molecule of the element copper would be copper but the atoms of a molecule of the compound water would be hydrogen and oxygen.

It is advisable to study this topic in greater detail and there are many books on building science that will provide the necessary information.

Cohesion and adhesion

It has been established that every substance is composed of many molecules and it is of interest how these are held together. Although the full scientific explanation is complicated, the force between molecules can be thought of as being similar to the attraction between a magnet and steel. The force is, however, very much more powerful than a magnet and a simple term to describe it is *cohesion*.

An example of cohesion can be seen in Fig. 10.1(a) which shows a glass overfilled with water. The liquid has not overflowed as the molecules of water are attracted to each other and those at the surface are pulled together and downwards, thus creating *surface tension* which acts in the same manner as a surface 'skin'.

There is another force or attraction between the water molecules and the side of the vessel in

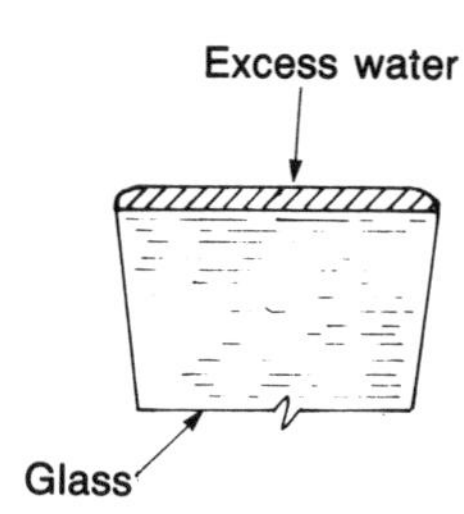

(a) Cohesion

The surface tension of the water molecules is capable of holding a body of water above the edges of the vessel containing it.

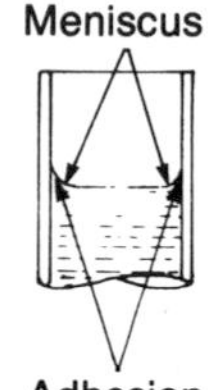

(b) Adhesion

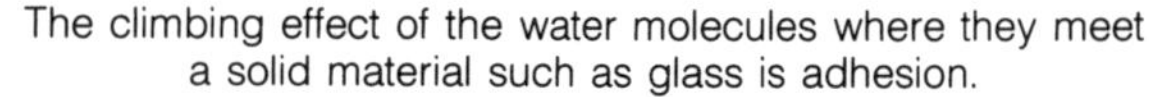
The climbing effect of the water molecules where they meet a solid material such as glass is adhesion.

Fig. 10.1 The two forces causing capillarity

which they are contained. This is called *adhesion*, an example of which is given in Fig. 10.1(b). If a clean glass tube is placed in the water, the water will adhere to its sides and be raised above the true water level in the tube, thus creating a dish-shaped water level in the tube, known as a *meniscus*. When reading the water level in a glass tube the lower part of the meniscus is taken as being the correct level.

Capillarity

The two forces of cohesion and adhesion combine when capillarity occurs. The molecules of water by the action of adhesion between the water and the solid 'haul' themselves upwards wherever there is contact. By the simultaneous action of cohesion the main body of water is lifted higher, thereby permitting adhesion to take place again to achieve a yet higher level. These processes are both continuous and simultaneous and cease only when the water column becomes too heavy or when the surface area is too large to be supported by adhesion to the sides of the vessel.

The illustration in Fig. 10.2 shows this quite clearly, the water in the glass tubes of larger diameter having a larger surface area, is unable to reach the height attained by that in the tubes of smaller bore. It also indicates the approximate height that a column of water will achieve in a given tube diameter; the smaller the tube bore, the higher will be the column of water capable of support by capillarity.

Water will find its way into buildings due to capillarity if lapped surfaces are close together, so the most practicable method of prevention is to widen the gap between the surfaces. A typical example of both the problem and its solution can be seen in Fig. 10.3.

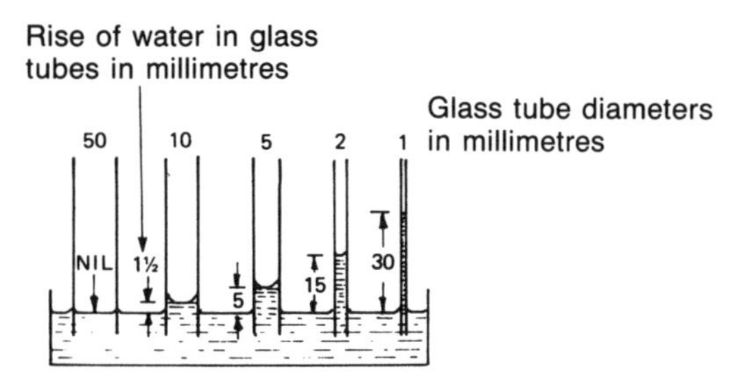

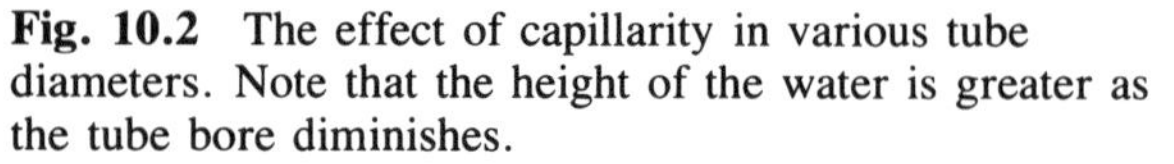
Fig. 10.2 The effect of capillarity in various tube diameters. Note that the height of the water is greater as the tube bore diminishes.

Water enters between the splash lap and undercloak at A. Water rises between the two surfaces and enters the building at B

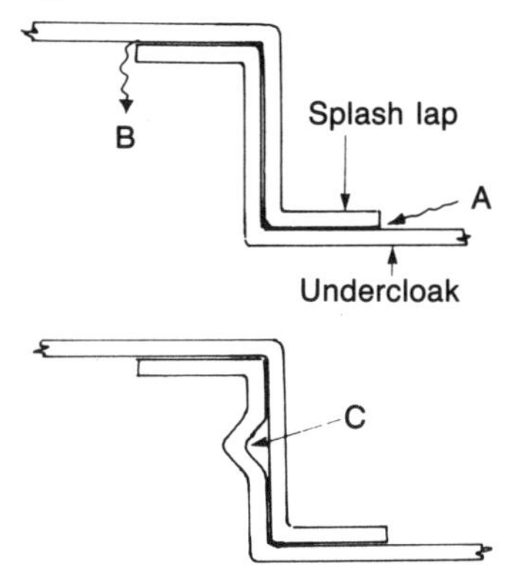

Fig. 10.3 Preventing the entry of water into a building by capillarity via a drip in a lead gutter. (For further details of drips in lead gutters, see Fig. 10.8). The formation of an anti-capillary groove shown at C prevents any water rising above it.

Fatigue cracking
In Chapter 9 work hardening was considered together with the resultant effects of reduced ductility which lead to fatigue cracking. This problem can occur when forming sheet metal, but it can also be caused by the use of designs and fixing methods that ignore two important natural factors, wind effect and temperature change.

Wind effect
The effect of the wind can cause metal sheets which have a free edge, e.g. cover flashings, to 'flap' about and even continually hit against adjacent surfaces. This 'cold working' resulting from the bending or the blows can eventually lead to the fracture of the metal. Another result of wind effect is the continuous rise and fall of covering materials when subjected to alternating negative and normal air pressures, and this will also lead to fatigue cracking. This is illustrated in Fig. 10.4 where a low-pressure zone is created by the wind blowing over the surface — a similar effect can be seen when paper and leaves are 'drawn up' behind a moving motor vehicle.

Temperature change
Temperature change can also produce fatigue cracking similar to that shown in Fig. 10.4. The subject of linear expansion has been discussed and it is easy to imagine that if no provision is made to compensate for this, the increase in material size will cause 'bulging'. On cooling the material contracts and the continual 'up and down' movement will eventually result in fatigue cracking.

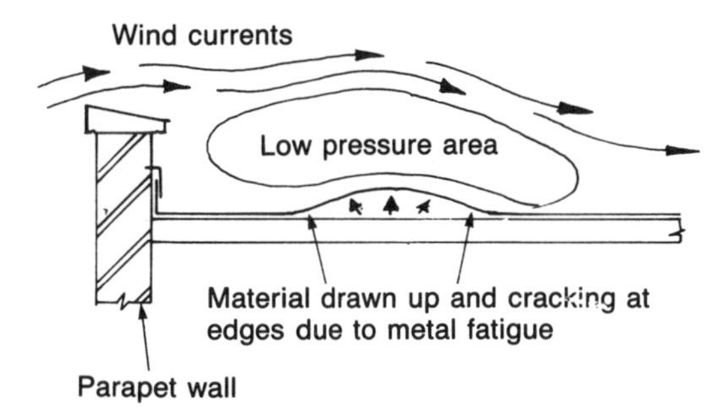

Fig. 10.4 Causes of fatigue cracking of light sheet roofing materials

Fatigue cracking can be limited by the use of:

(a) the correct gauges of material as specified in Chapter 9
(b) recommended maximum bay or panel sizes
(c) correct jointing methods
(d) adequate fixings

Roofing terms

Prior to considering the operations involved in applying sheet weathering materials, two important roofing terms, roof pitch and maximum bay sizes, must be clearly appreciated.

Roof pitch
Roof pitch is the angle between the sloping roof surface and the horizontal plane as shown in Fig. 10.5. A roof is generally referred to as *flat* when the pitch is less than 10° and *pitched* when it is 10° or more. The minimum satisfactory pitch or fall is $2\frac{1}{2}°$, as anything less will result in a slow rainwater run-off that could cause flooding and allow entry of water into the building.

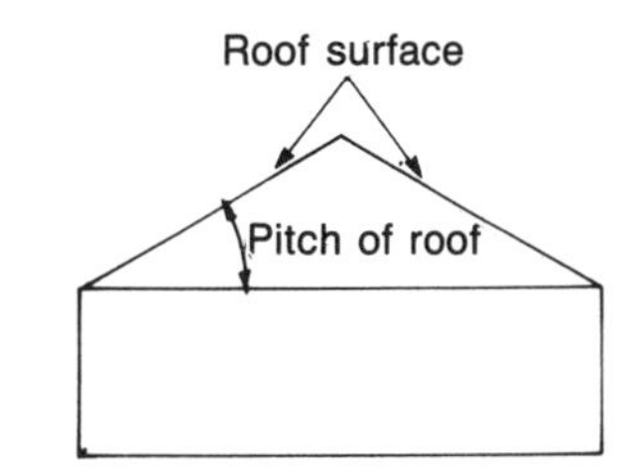

Fig. 10.5 Angle of roof pitch

Maximum bay sizes
Maximum bay or panel sizes are required to avoid excessive thermal movement, reduce 'creep' and limit problems created by wind pressure. Large roof surfaces are divided into a series of smaller areas, the sizes depending on type and gauge of material and roof pitch. Areas in relation to these factors are specified in CP 143 and its recommendation should be strictly adhered to in order to ensure a trouble-free and weathertight roof.

The application of sheet weatherings can be conveniently divided into three operations, jointing, fixing and the formation of details such as corners, flashings and gutters. First, however, it will be appropriate to discuss the type of support necessary for the various materials.

Preparation of decking

Some roofing materials such as corrugated steel, rigid aluminium cladding and non-asbestos corrugated sheet, have sufficient natural strength to maintain their shape and form, so these require little additional support in the form of decking. It should be noted that non-asbestos corrugated sheet is a cement-bonded fibrous material now marketed as a substitute for cement asbestos. However, the sheet metals discussed in Chapter 9 are for application to a substructure which will provide them with complete support over the whole area, i.e. they are fully supported roof coverings. The type of material used for decking should conform to the relevant British Standard and is usually one of the following:

(a) tongued and grooved or square-edge boards
(b) blockboard
(c) chipboard
(d) compressed straw slabs
(e) plywood

It is not within the activities of the plumber to lay roof decking but it is important for him to be aware of the requirements for suitable decking which are as follows.

(a) The roof undersurface is to be flat and without indentations that could lead to the formation of puddles of water. Stagnant water can be a hazard to health and can cause the accumulation and concentration of undesirable acids formed by organic and atmospheric pollutants.
(b) In order to prevent warping and springiness, the boards are to be not less than 25 mm nominal thickness and laid either diagonally or with the fall of the roof.
(c) All nails are to be punched down below the board surface, screws countersunk and all grit or debris removed to leave a smooth surface. Sharp edges of any decking materials which could damage the covering are to be removed, especially if lead is the covering material.
(d) Wood or wood preservative which could cause corrosion must not be used.
(e) Where appropriate, precautions must be taken against the alkaline nature of concrete and mortar which affect certain materials.
(f) Adequate falls are to be incorporated into the roof structure to allow the quick removal of rainwater.
(g) An insulating felt underlay is to be used with sheet metals.

Most of these factors are self explanatory or have been mentioned in Chapter 9, but the use of felt underlays requires further consideration.

Felt underlay

The recommendation that a suitable underlay should be used on the prepared decking when sheet metal coverings are to be used is made for the following reasons:

(a) To provide some measure of thermal insulation, i.e. to prevent solar heat unduly affecting the building.
(b) To insulate the building against the sound of rain and wind.
(c) To allow the metal to move freely when thermal expansion takes place and to prevent abrasion between the weathering material and decking.
(d) To prevent electrolytic corrosion arising from the reaction between metal fixings used on the decking and the sheet roofing metal.
(e) To prevent corrosion due to contact between the roof metal and any organic alkaline or other chemical agent present in the decking material.

For wooden decking two materials are in common use as underlay, BS 747 (Type 4A, Brown No. 1 inodorous felt) and waterproof building paper to BS 1521 – Class A. Felt is more generally specified as its thickness helps to overcome uneven surfaces, while building paper is not recommended for use with aluminium as it may not be copper-free and could cause corrosion.

The fixing of underlay must be done with care as galvanised or copper nails and brass or steel screws can lead to electrolysis depending on the roof covering used. Waterproof paper should be laid with 50 mm lapped joints while felt sheets should be butted together. On no account should bituminous sheet, i.e. ordinary roofing felt, be used as it will soften due to heat, adhere to decking and weathering material and thus restrict the movement caused by thermal expansion. The felted surfaces should be covered immediately by the final metal roof coverings to avoid trapping any moisture.

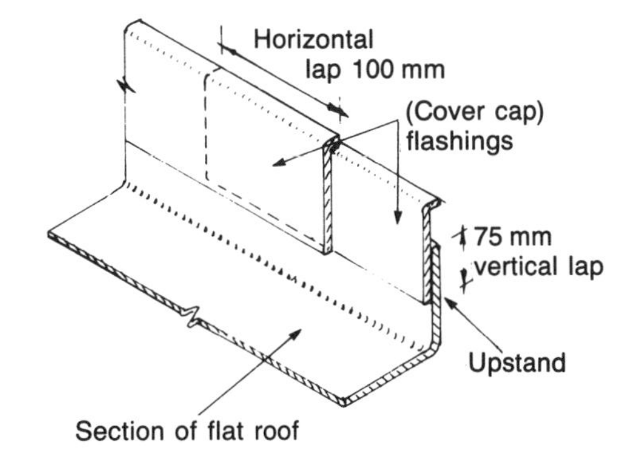

Fig. 10.6 Recommended cover for laps on cap or cover flashings

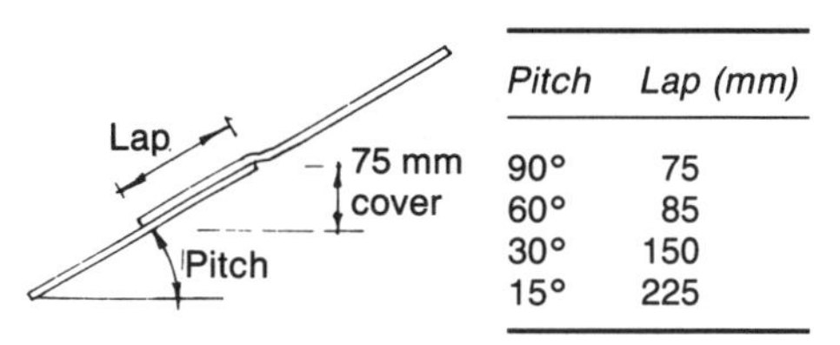

Pitch	*Lap (mm)*
90°	75
60°	85
30°	150
15°	225

Fig. 10.7 Effective cover for laps on pitched roofs

Jointing methods

All joints *must* be watertight, although the choice of jointing method is dependent on the following considerations.

(a) position of the joint in relation to rainwater flow, i.e. 'with' or 'across' the fall of the roof surface
(b) roof pitch
(c) type of roofing material
(d) accessibility by pedestrian traffic, i.e. will the roof be regularly walked on

In order to avoid contamination (electrolytic corrosion) tools which have been previously used with copper or lead must be well cleaned before use with aluminium and zinc. Some plumbers keep a separate set of wood tools for hard metal work, as these materials cut into and deface such tools, which if used on lead sheet would result in a lot of unsightly marks on its surface.

Laps

Laps are the simplest joints and are made by laying one sheet edge over another, the illustration in Fig. 10.6 shows typical details.

To avoid water entry due to capillary attraction a minimum lap of 100 mm is recommended for laps on cap flashings and 75 mm for those in the vertical position. For sloping surfaces the lap should correspond to an equivalent vertical height of 75 mm as can be seen from Figs. 10.6 and 10.7. When in any position other than vertical, laps must only be used across the fall of the roof. The edges of these joints must be securely fixed to prevent the problems encountered through wind effect. Clips of a suitable material that will not cause electrolytic corrosion must be used.

Drips

Drips are positioned across the fall on very low pitched roofs, i.e. under 15°, where a lap joint would allow the entry of water. The drip is basically a step formed in a flat roof or gutter, the depth being related to the type of material as indicated in Table 10.1 to prevent capillarity taking place.

Drips formed in these materials are shown in Fig. 10.8. A minimum drip depth of 40 mm is permissible for lead drips provided an anti-capillary groove (see Fig. 10.8(a)) is formed in its step. The splash lap shown on the lead drip is intended to stiffen the free edge of the lead sheet and to assist in keeping the overcloak in position.

Table 10.1 Depth of drip according to type of material.

Material	*Drip depth (mm)*
Zinc	50
Aluminium	50
Copper	65
Lead	50

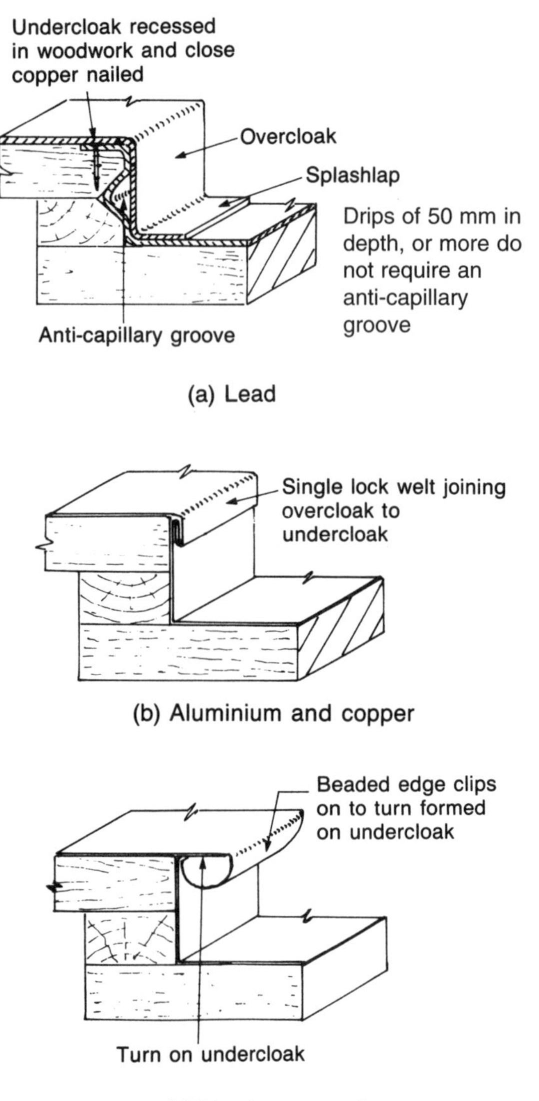

Where soft temper zinc is employed the method used is similar to that of copper and aluminium.

Fig. 10.8 Sections through drips in sheet metal weathered wooden gutters

Rolls

Rolls are used to form the joints that run in the direction of the flow of water or 'with' the fall of the roof. The majority of these joints are made over a wooden core, sections of which are shown in Fig. 10.9. One type of roll, used only for pitched roofs covered with sheet lead, has no core, and for this reason is called a *hollow roll*.

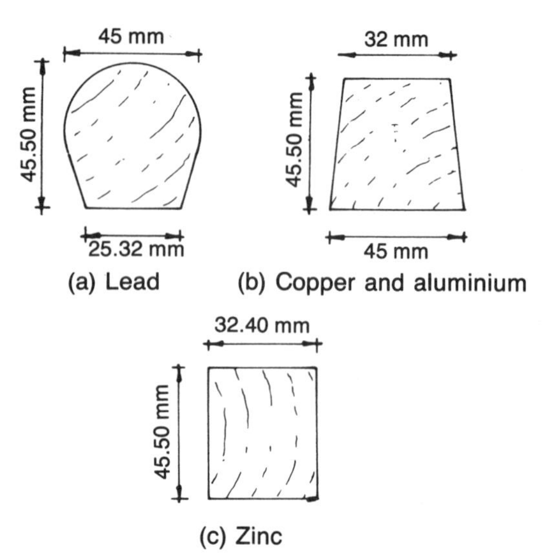

Fig. 10.9 Wooden cores for rolls for various materials

Typical lead roll A typical lead covered roll is shown in Fig. 10.10. One fault which can occur with this joint if the roof has a low pitch is that if water stands on the surface it may be drawn up between overcloak and undercloak by capillary attraction. This would, in fact, be a design fault as the roof slope should be sufficient to prevent water standing, and rolls formed in this way are most suitable for pitches up to 30°. Above this pitch it is preferable not to use a splash lap but to secure the free edge by using a 50 mm wide clip, in which case a 5 mm clearance is left between the roof surface and edge of the overcloak to prevent the formation of a vertical space that could encourage capillary attraction. The splash lap serves the same function on rolls as it does when

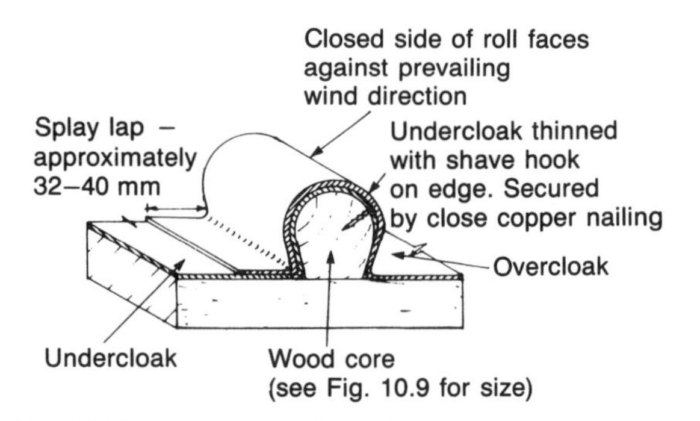

Fig. 10.10 Lead-covered wood-cored roll

used on a drip and it should be positioned on the side of the roll that will be least affected by wind. It should be noted that the undercloak is taken about two-thirds around the wood core and the edge chamfered using a shave-hook. This prevents the formation of a sharp edge which would show through the overcloak after dressing into position.

Hollow rolls Hollow rolls are suitable for steeply pitched roofs where they will not be walked on, and are formed over a wooden core or bending spring which is withdrawn afterwards. The stages in forming this type of roll are shown in Fig. 10.11. The use of hollow rolls is especially useful for roof work details having curved surfaces such as domes or bay window weatherings.

Rolls for hard metals The wood core shown in Fig. 10.9(b) and (c) is termed a batten roll and is for use with hard metals. When the covering metal is aluminium or copper the roll is formed as in Fig. 10.12.

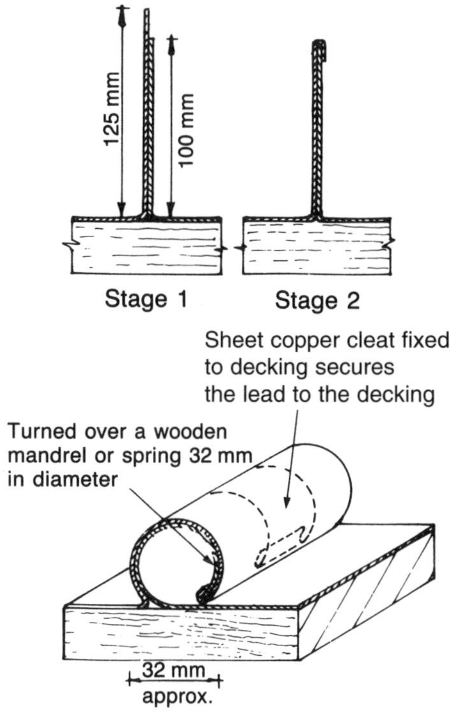

Fig. 10.11 Formation of hollow rolls

'Roll cap' system for zinc Because zinc is stiff and springy a procedure known as the *roll cap* system was developed to overcome the practical difficulties that are encountered when weathering with this material. Figure 10.13 shows a typical

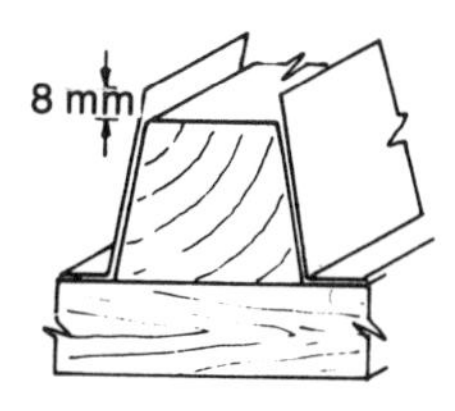

Stage 1
Upstand turned up to roll allowing approximately 8 mm for undercloaks to welt

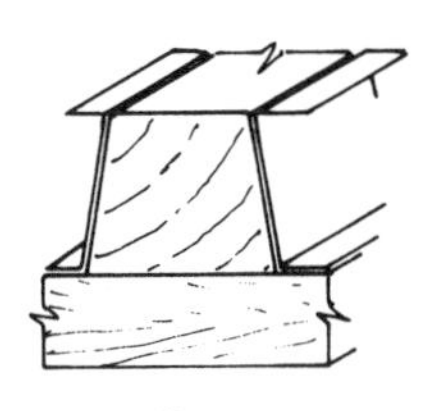

Stage 2
Allowance for welt turned out

Stage 3
Prepared capping slid over turns

Stage 4
Welts completed by turning through 90° to lock. It is suggested that a holding-down block is placed on the capping during this operation to help achieve clean sharp turns

Fig. 10.12 Stages of weathering copper or aluminium rolls

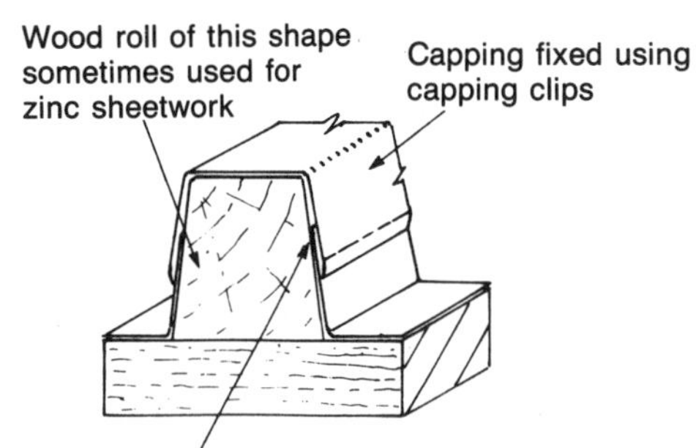

Fig. 10.13 Section through a zinc roll

joint where it can be seen that the edges of the roll capping are set-in to ensure a close contact with the upstand. The difficulties outlined are particularly relevant to commercial zinc sheet, but the development of the zinc alloys discussed in Chapter 9 has largely eliminated these problems. The roll cap system using the batten roll is still the traditional longitudinal jointing procedure for hard-temper zinc.

Welts

Welts are formed by folding sheet metal, the number of folds suggesting the name of the joint so that a single-lock welt will be folded once while a double-lock welt is folded twice. Both single- and double-lock welts are used as transverse joints, i.e. across the fall of the roof. Single-lock welts are recommended for pitched roofs over 45° or on vertical surfaces, while the double-lock welt is used on pitches of 6° and up to 45°. The forming of a single-lock welt is shown in Fig. 10.14(a), the double-lock welt (Fig. 10.14(b)) being made in a similar manner, except that the heights of the upstands are greater and the welt is given an extra turn.

On copper roofs of under 20° the edges forming the double welt should be painted with boiled linseed oil or a non-hardening mastic before folding. The reason for this treatment is that the welt could become covered with water and capillary action would cause water entry. This capillary action would be increased if the seam was filled with water as syphonic action (see Chapter 7) would then occur. It is essential to

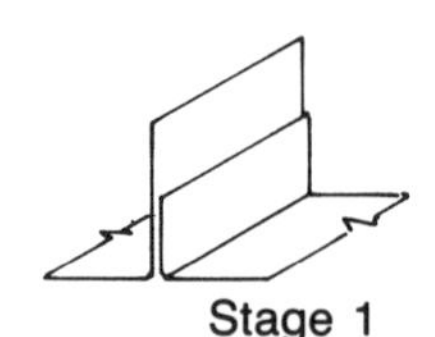

Stage 1
Turn upstands to required dimensions, usually 25–27 mm for the overcloak and 12 mm for the undercloak

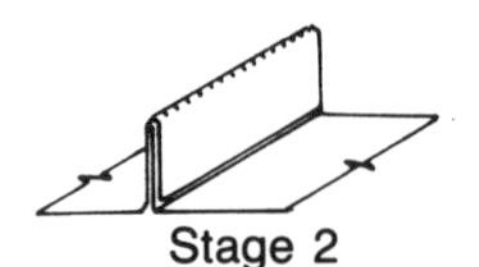

Stage 2
Turn the overcloak on to the undercloak

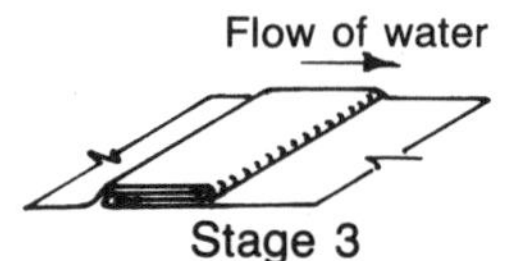

Stage 3
Turn welt through 90° to complete

(a) Forming single-lock welt

Welts are normally used for joints across a pitched roof weathered with copper or aluminium sheet. The arrow indicates direction of flow of water.

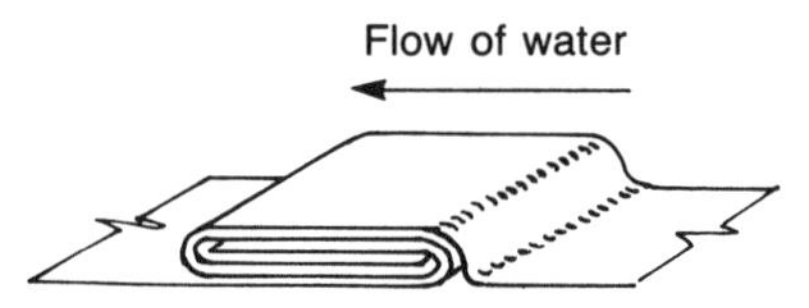

(b) Section through a double-lock welt

Used for joints across a pitched roof of less than 45°.

Fig. 10.14 Single- and double-lock welts

position welts used as transverse joints so that the water run-off is in the direction of the arrow shown in Fig. 10.14; this allows water to run over the closed side of the joint.

In addition to joining the edges of panels or bays the single-lock welt is utilised in the formation of a wide variety of roofing details, some of which are shown in Chapter 11.

Folded edges The folded edge shown in Fig. 10.15 is *not* a joint but is used to stiffen the exposed free edges of aluminium, zinc or copper flashings and to provide a 'safe edge', as the edges of hard sheet metals are very sharp after being cut.

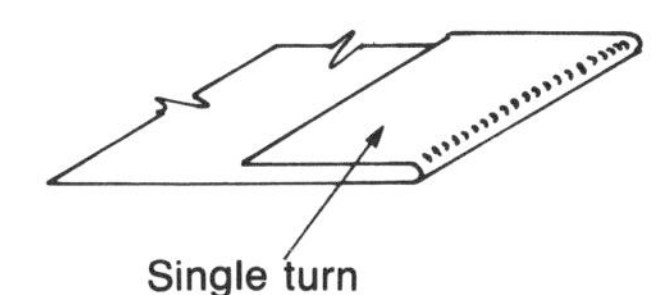

Fig. 10.15 Folded edge. Used to stiffen the edge of copper, aluminium and zinc.

Standing seams

Standing seams should never be used in positions where there is a possibility that people will walk on the roof surface as the seam is easily flattened. This will cause the roof to look unsightly and may allow water to enter as the joint is basically a single-lock welt which has been raised above the roof surface. The standing seam is an alternative to a wood-cored roll for roofs having pitches in excess of 6°. Figure 10.16 shows the stages in making the seam, two purpose-made forming blocks edged with steel being used for the hard sheet metals (see Fig. 10.17).

Dog-eared corners

The stages for forming dog ears are shown in Fig. 10.18. Figure 10.19 shows a forming block which is used to make the first turn. The block illustrated is dowelled to prevent it twisting and has a steel face. The saw cut, approximately 1 mm wide, is made to a slightly greater depth than the height of the upstand, i.e. for an upstand of 100 mm the saw cut should be about 105 mm deep.

First one side of the corner is turned up at 90°, then the block is placed over the upstand as shown in Stage 1 and held in position as the other upstand is pulled up. This will be made easier by 'setting in' on the dotted line with a chase wedge or plumber's hammer, the straight pein type being ideal for this purpose. To complete the corner, the dog ear is simply turned at 90° as in Stage 2.

So that the upstand lies tightly against the abutment, it is 'set in' as shown.

The methods of folding, welting and forming dog-eared corners are used for copper, aluminium and zinc, traditionally called hard metals. Lead is normally 'bossed' into the various shapes required but it can be welted in cases where narrow strips are to be joined, i.e. covering of parapets and stone cornices. Copper and aluminium can be

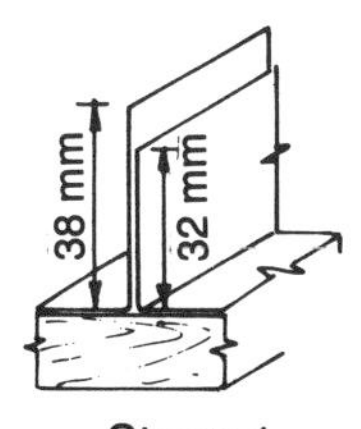

Stage 1
The under- and overcloaks are turned to the required dimensions

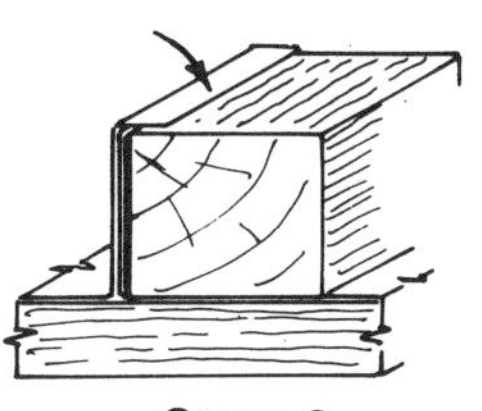

Stage 2
The first turn is made over the thicker of the two forming blocks

Stage 3
Place the block on the other side and close the first turn

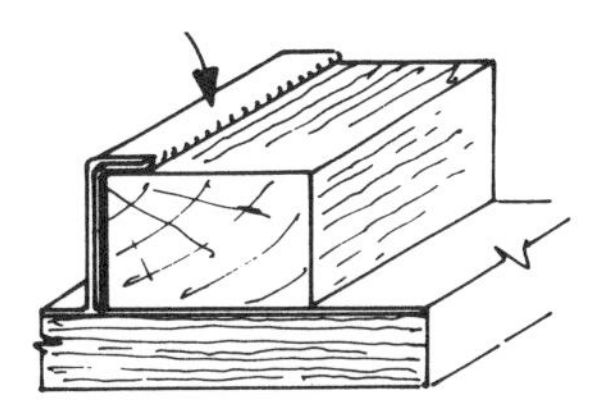

Stage 4
Select the thinner block and fold the first turn through 90°

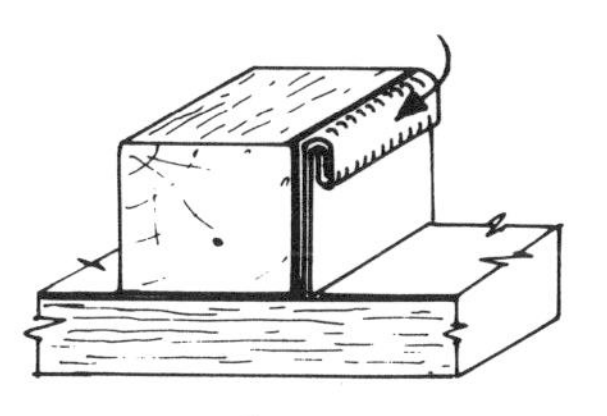

Stage 5
Reverse the block and close the second turn. The standing seam is now completed

Fig. 10.16 Forming standing seams for sheet copper and aluminium roof work

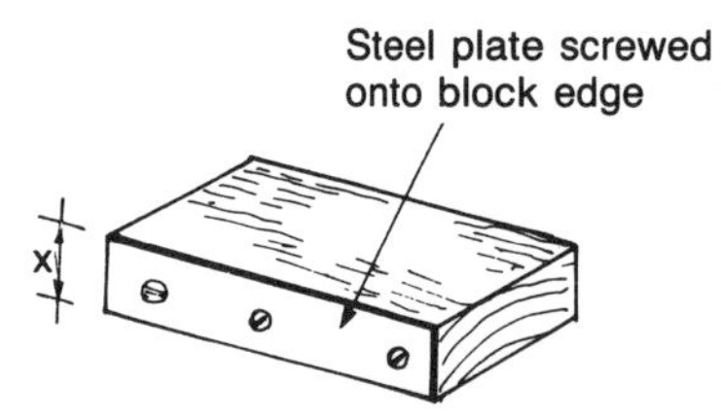

Two blocks are necessary. The thickness x is 32 mm (for first turn) and 25 mm (for the second turn

Fig. 10.17 Forming blocks for standing seams

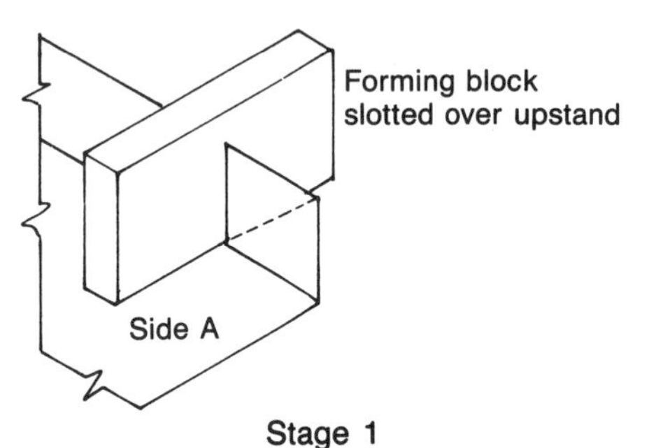

Stage 1

Mark out and turn an upstand on one side of the corner. Place block over upstand and hold in position while dressing up side 'A', taking care that the fold in the corner occurs on the dotted line

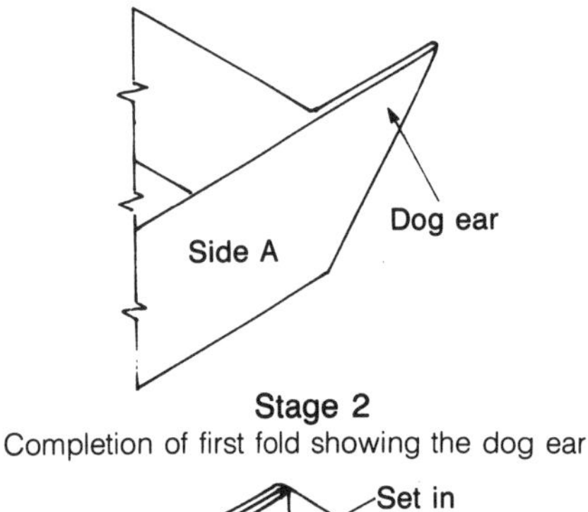

Stage 2

Completion of first fold showing the dog ear

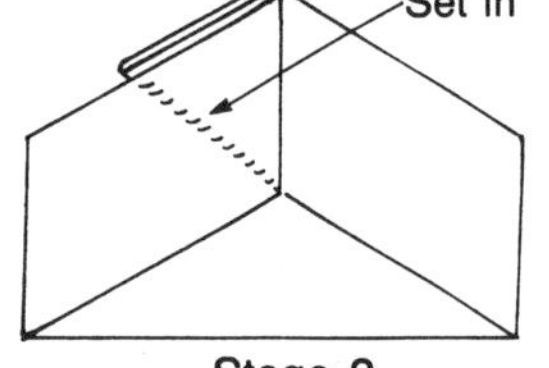

Stage 3

The block should now be placed over side 'A' and held in position while the dog ear is turned through 90° and dressed flat against the upstand. A neat finish will be obtained if the corner is 'set in' with a chase wedge as shown

Fig. 10.18 Forming dog-eared corners in hard metals

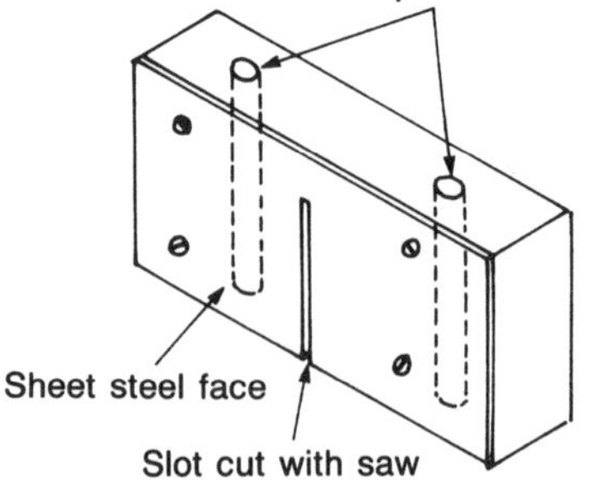

Fig. 10.19 Dog-ear formation block. The use of this simple tool enables dog-ear corners to be made quickly and accurately.

bossed to a certain extent if frequently annealed as these metals work harden, but the process is a long one, and is not generally accepted as a commercial proposition.

Soft-soldered joints

Soft soldering is a process of joining two metals together by the use of lead/tin alloy (solder) as distinct from copper or silver/tin alloys which must be used for water supplies. Soft solder has a lower melting point than the metals being jointed. The process is different to welding where the edges of the metals to be jointed are brought to melting point and then fused together. In soldering the filler material flows between the unmelted metal surfaces to which it chemically adheres combining to form a compound 'intermetallic' layer as indicated in Fig. 10.20. Tin is a metal that is capable of combining with certain other metals such as copper, lead and zinc to form an intermetallic compound and this fact forms the basis of soft-soldering alloys.

The three preliminary stages in forming a soldered joint are:

(a) thoroughly clean the surfaces
(b) apply a flux
(c) heat the joint to the desired temperature

It is vital for each of these three stages to be carried out precisely so they are described in detail below.

(a) Cleaning is achieved by the use of steel wool,

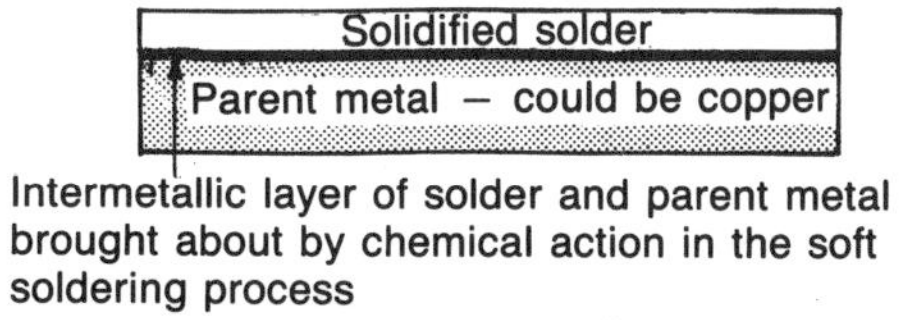

Solidified solder
Parent metal – could be copper
Intermetallic layer of solder and parent metal brought about by chemical action in the soft soldering process

Fig. 10.20 Detail showing the adhesion of soft solder to parent metals

wire brush, file or, in the case of lead sheet and pipe, a shave-hook. The aim is to mechanically remove all grease, dirt, traces of corrosion and oxide film, as solder will not adhere to non-metallic substances.

(b) Applying the flux has four basic functions as follows.
   (i) To assist in the 'wetting' process in which the solder is made to flow easily over the surfaces being joined, allowing bonding or alloying with the parent metal at every point. If 'wetting' does not take place the solder will remain on the surface as beads or globules, and even if the surface 'held together' the joint would not be sound or watertight.
   (ii) To chemically clean the surface of the parent metal by removing or reducing any oxide film that remains after mechanical cleaning. (Reduction is the reverse chemical process to oxidation, i.e. it is the removal, rather than the addition, of oxygen.)
   (iii) To prevent metal surfaces from oxidising while they are being heated to soldering temperature when they are more reactive towards oxygen.
   (iv) To float away any foreign matter from the soldering operation.

(c) Heating the joint to the required temperature is achieved on pipe joints by the use of oxy-fuel gas, propane or natural gas blowpipes. Propane blowpipes are also utilised for certain soldering applications on lead sheet, but it is more usual for a copper bit or soldering iron to be used for sheet zinc work. Solder is applied to lead sheet using a blowlamp and it is then 'wiped' into position using a wiping cloth.

The application of solder to zinc sheet is achieved by means of a copper bit which is chosen because it has a high thermal conductivity, i.e. a capacity to transfer heat quickly to the job. The quantity of heat available in a copper bit depends on its size or mass and this will be dependent on the work being undertaken. For soldering zinc seams a large heat capacity is required and a 1 kg bit is recommended.

It should be noted that gas-heated soldering bits are available from most manufacturers of portable gas equipment. The combined burner and copper bit are readily adaptable to the hand-held shank, and where extensive soldering operations are carried out they usually save both time and fuel.

Prior to commencing the soldering operation it is necessary to tin the soldering 'iron' as it is often called, and this is done by heating it to the required temperature (when it will melt the solder), then cleaning the faces by filing, then rubbing it on sal ammoniac crystals (or some other flux) and applying it to a stick of solder. This should result in the end of the bit being completely covered with a layer of solder.

A diagrammatic close up of the soft-soldering operation using a copper bit is shown in Fig. 10.21. The action of the boiling flux is shown removing the oxides from the metal to expose an oxide-free surface enabling the solder to 'take' – or 'wet' – the parent or base metal and form a thin surface alloy with it.

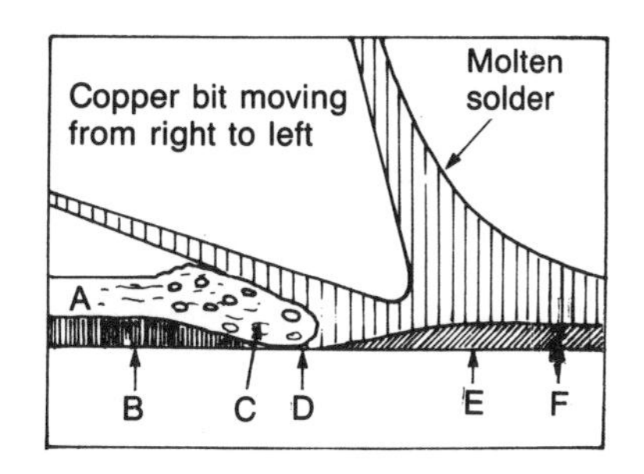

Fig. 10.21 Soft soldering with a copper bit. A: Flux on the oxidised metal. B: Oxidised metal. C: Boiling flux removes oxides. D: Bare metal exposed. E: Surface alloy of tin and base metal (intermetallic layer). F: Solidified solder.

Soft solders

Soft solder, used for purposes other than water supply work, is manufactured to BS 219 and is an alloy of tin and lead, although for some solders small amounts of antimony are also included. Tin is very expensive and to reduce costs antimony is used as a substitute, but this can only be included in amounts of up to 6 per cent of the tin content or the properties of the alloy are affected. Antimony must never exceed 0.5 per cent of the tin content in solders to be used for zinc and alloys of zinc such as brass. A wide range of solders are available but those detailed in Table 10.2 are the most common types used by the plumber, Grade D being used for the repair of lead gutters, flats and wiped solder dots and Grade F for general purpose copper-bit work.

When solder is in a liquid state both tin and lead are molten. On cooling, crystals of the excess metal begin to form in the molten solder and this provides a consistency between liquid and solid states that is referred to as 'pasty' or 'plastic'. This plastic condition is essential for the formation of wiped joints where the plumber has to have sufficient time to shape the solder. In other solders, such as those used for zinc, shaping is not necessary and the alloy is required to set as soon as it is deposited on the parent metal. The cooling periods can be seen in the equilibrium diagram shown in Fig. 10.22, the temperature range between liquid and solid being called the 'plastic range'. It can easily be seen that a wide plastic range is necessary for wiping solder but for copper bit work a solder is needed which will set as the copper bit moves forward.

Equilibrium diagrams are used to record the changes that occur during the solidification and subsequent cooling of alloys. In Fig. 10.22 'D' indicates the proportion of lead and tin in Grade D wiping solder. It should be noted that it remains plastic over a wider temperature range than Grade F solder ('F' in the diagram). An increase in the proportion of tin in the alloy reduces its plastic range. The *eutectic point* in an alloy indicates two things:

(a) its lowest melting point
(b) the point at which it solidifies without going through a plastic range.

It will be noticed from a study of the diagram that the *eutectic point* has two important characteristics.

(a) An alloy of 63 per cent tin and 37 per cent lead has no plastic range; it both melts and solidifies at the same temperature.
(b) The lowest temperature at which an alloy of these two metals will melt is 183 °C. (The word *eutectic* does in fact mean 'easily melted'.)

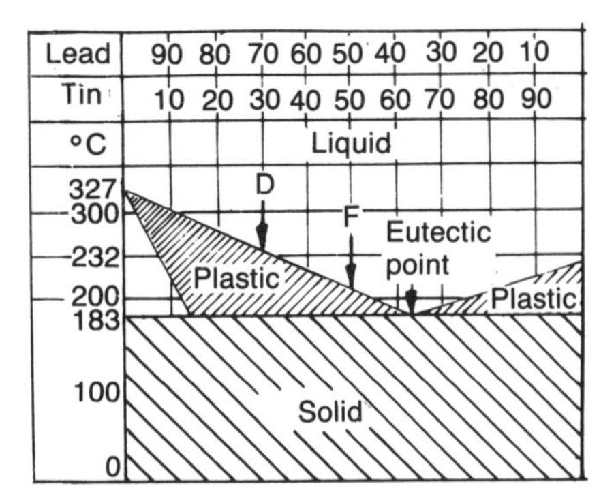

Fig. 10.22 Equilibrium diagram for lead/tin alloys

Table 10.2 Common types of solder used by the plumber.

BS 219 grade	*Melting point (°C)*		*Composition (%)**		
	Solid	*Liquid*	*Sn*	*Pb*	*Sb*
D (Wiping solder)	185	248	30	68.3	1.7
F (Bit solder)	183	212	50	49.5	0.5

*Sn = tin; Pb = lead; Sb = antimony

Fluxes

Fluxes are classified under two general categories depending on their ability to attack the oxide layer on metals – *corrosive* (or active) and *non-corrosive* (or inactive). All fluxes are to some extent acidic as this is essential for dissolving and removing metallic oxides. Active fluxes such as zinc chloride will attack metals at any temperature but inactive fluxes such as tallow are not very reactive at room temperature, even if left in contact for a long

Table 10.3 Fluxes in general use in plumbing.

Flux	*Type*	*Function*	*Origin*
Tallow	Inactive	Soldering lead	Organic fat of cattle containing oleic acid
Resin*	Inactive	Tinning brass and copper	Gum from pine tree bark that contains abiatic acid. Used in powdered form or mixed with spirit or formed into a paste.
Manufactured paste fluxes	May be active or inactive	Tinning brass or copper	Various deoxidants bound in Vaseline or similar pastes that remain fluid at soldering temperatures
Zinc chloride†	Active	Zincwork	Zinc dissolved in hydrochloric acid until acid is 'dead'. Commonly called 'killed spirits'.

*Resin-based fluxes are used in flux-cored solder for soldering electrical components.
†Zinc chloride, although still used, has been superseded by liquid fluxes specially formulated for zinc work.

period; they are, however, sufficiently active at soldering temperatures. Corrosive fluxes must be removed after soldering is completed as the residues absorb water readily from the atmosphere and form acids that will continue to attack the metal. Removal is affected by washing in hot, diluted, soda solution followed by a further hot water wash. Because of the difficulty of avoiding corrosion with zinc chloride it must never be used for electrical work or any other operation which cannot be washed effectively. Non-corrosive fluxes do not require to be removed due to their low reactivity at normal temperatures.

Details of the fluxes in general use in the plumbing industry are given in Table 10.3.

It should be noted that while resin was traditionally used as a flux for copper and brass it has been superseded by manufactured paste fluxes of which there are many types now produced.

Soldering, brazing and welding

The use of soft solder as a jointing method for sheet copper is bad practice due to the difference in expansion rates that could lead to cracking of the joint. Zinc sheet can be satisfactorily soft soldered as the coefficients of linear expansion of both metals are similar; typical joints are illustrated in Fig. 10.23. It must be stressed, however, that neither soldering process can be used for jointing the parts of an entire roof covering as this would not make any allowance for expansion to take place due to atmospheric temperature changes.

Welding and brazing processes require a high flame temperature and this is provided by the use of a special blowpipe requiring compressed oxygen and a fuel gas, usually dissolved acetylene. A detailed description of the equipment used and the processes involved is dealt with in Volume 2 of this series.

Welding requires the melting or fusion (i.e. fusion welding) of the two sheets being joined, whereas brazing is a similar process to soldering and is often referred to as hard soldering. The jointing material for welding is basically rods made from the parent metal plus certain additives. Brazing requires the use of filler rods which have a lower melting point than the sheet material upon which they are used.

Lead can be fusion welded using either a butt or lap seam as illustrated in Fig. 10.24. This is a better jointing method for sheet roofing than soldering as the welded joint has the same coefficient of linear expansion as the lead sheet. By ensuring that expansion of roof material and jointing material are the same, failure due to cracked joints is avoided.

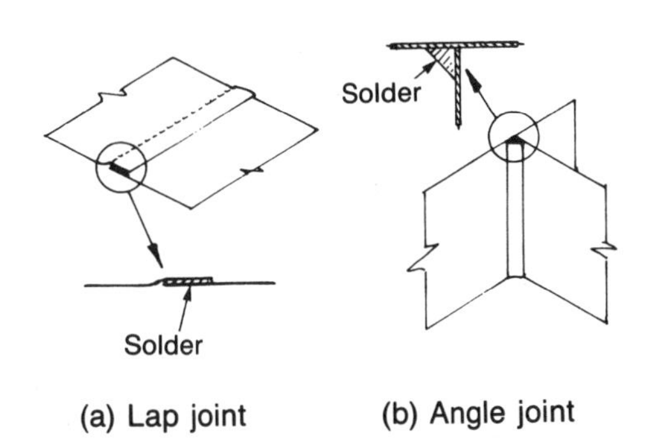

Fig. 10.23 Typical soft-soldered joints in sheet zinc

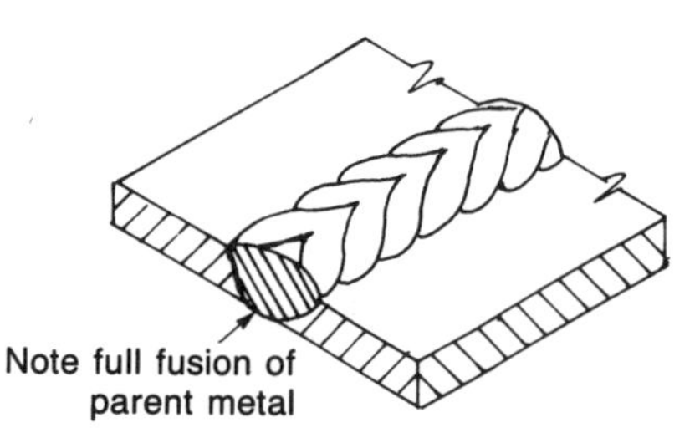

(a) Butt joint

Used mainly for off site fabrications, i.e. chimney aprons and gutters.

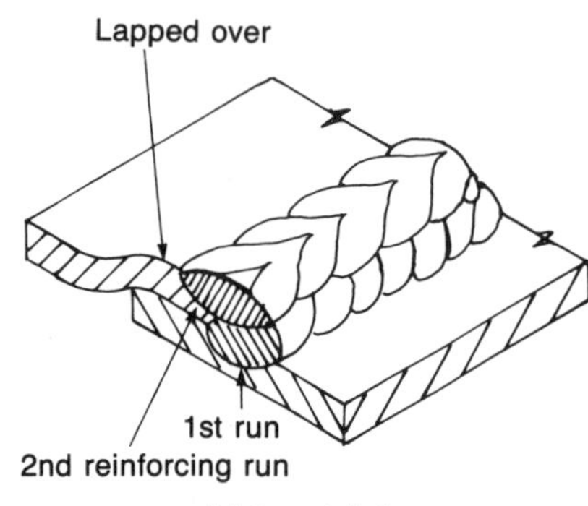

(b) Lap joint

Used for both fabrication and on-site work. Note that the underside of the lead is not melted, enabling this method to be used for repairs to lead roofs with little danger of fire risk to the timber decking.

Fig. 10.24 Lead-welded joints

Copper sheet can be joined by bronze welding, and aluminium sheet by welding or brazing, but these processes are rarely, if indeed ever, used in modern weathering practice. Silver brazing is particularly useful in the fabrication of small weathering details such as rainwater chutes and cesspools. It must never be used as the principal means of jointing on a roof as no provision would be available to allow for linear expansion to take place and distortion would occur.

Fixing methods

Reference was made earlier in this chapter to the influence of wind effect and temperature change on roof coverings. It is of primary importance to consider carefully the fixing methods to be employed because, while all fixings must hold materials in position, some must also allow for the movement caused by thermal expansion. Another factor to be taken into account is that electrolytic corrosion can occur if clips, screws or nails are of a different material to the roof sheet – reference should be made to the electrochemical series shown in Table 9.6.

Lead sheet fixings

The materials used for fixings should be as follows:

Nails Large-headed copper having barbed shanks not less than 25mm long. 10 SWG (approx 3 mm) diameter, and conforming to BS 1202, Part 2, Table 2.

Screws Brass or stainless steel not less than 25 mm long or 10 SWG shank diameter and conforming to BS 1210.

Cleats Not less than 0.6 mm thick copper sheet of $\frac{1}{4}$ hard temper and conforming to BS 2870; or stainless steel, not less than 0.375 mm thick and complying to the relevant B.S.

Solder (for dots) Either Grade D or Grade J conforming to BS 219. (The only difference between these grades is the antimony content.)

The usual practice for fixing the undercloaks of drips or wood-cored rolls is to position nails at 50 mm centres. Using nails at 75 mm centres is satisfactory for fixing the upper edges of sheets on pitched roofs that are to be covered by a lap or cover piece which will prevent water entering the nail holes. Nailing is only suitable if the decking, i.e. the undersurface, is timber. Nailing can be used with a concrete substrate if a soft-wood batten impregnated with preservative has been inset, but where this is not the case screws using suitable plugs and brass washers are utilised.

The edges of sheets and joint positions are usually retained in place by the use of 50 mm wide clips of either copper or lead sheet. The applications of clips are too numerous to detail individually but some typical examples are shown in Fig. 10.25. Clips are usually positioned at a maximum of 500 mm intervals.

Lead wedges positioned at about 500 mm as in Fig. 10.26 are the method of securing the edges of lead flashings into brickwork joints.

A method of fixing lead to stone which may be

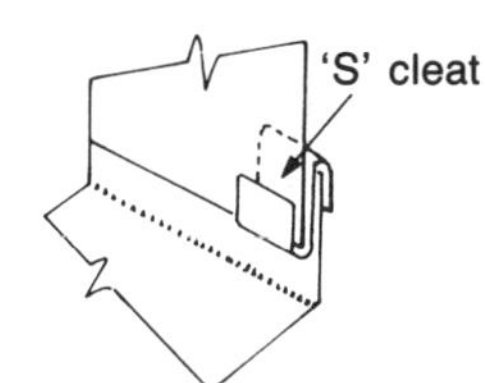

(a) Method of securing 'cap' flashing

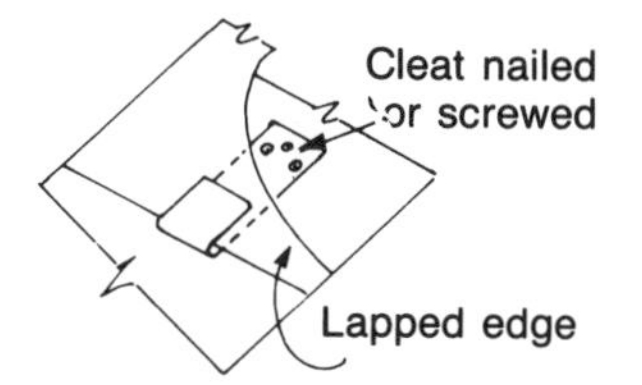

(b) Securing the free edge of a lap joint

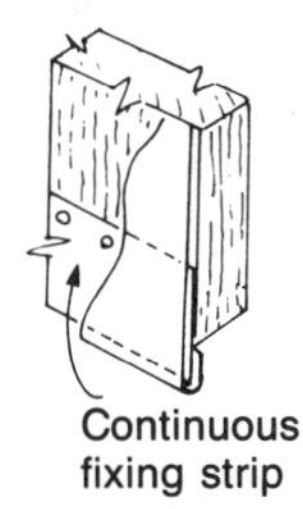

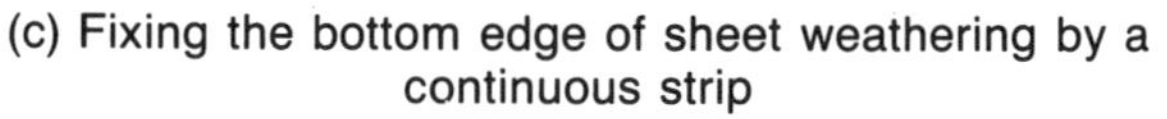
(c) Fixing the bottom edge of sheet weathering by a continuous strip

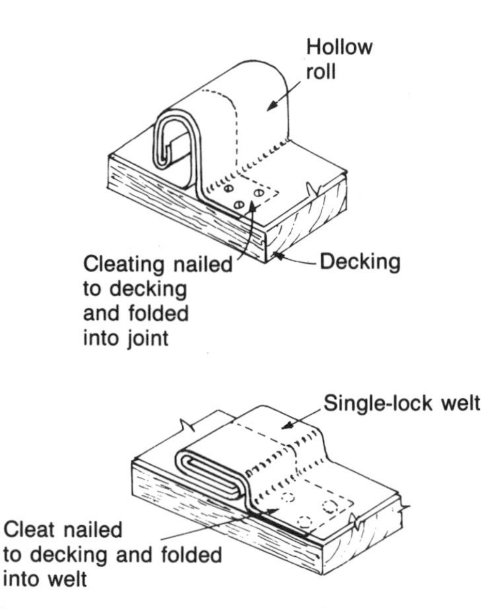

(d) Holding down cleats nailed to the decking of a roof and folded into the joint as it is made

Fig. 10.25 Use of clips and cleats for lead sheetwork

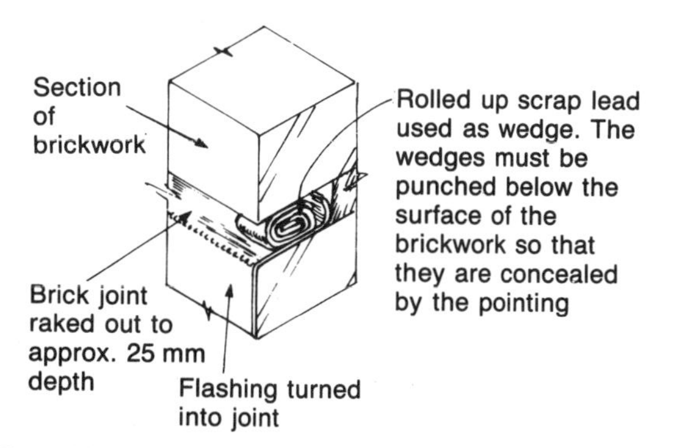

Fig. 10.26 Securing weatherings into brickwork joints

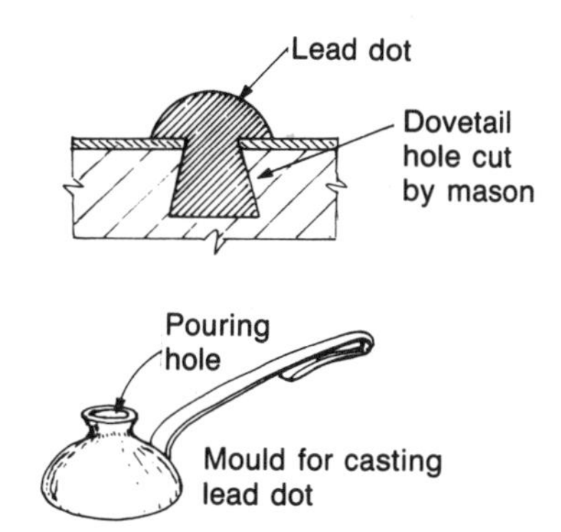

Fig. 10.27 Lead dot method of fixing lead sheet to masonry

seen in older buildings is the *lead dot*. This required a dovetail hole to be cut in the stone which was then filled with molten lead using a dot mould (see Fig. 10.27). Figure 10.28 shows another method of fixing using screws and lead-welding techniques.

Intermediate fixings Fixings which do not occur at the edges of sheets are known as intermediate fixings and can either be *visible* or *secret*. The secret method requires the welding of a sheet lead tack to the back of a panel of sheet lead. This tack is then passed through a slot in the timber backing and secured as in Fig. 10.29 using round-headed brass screws and washers.

The traditional type of visible intermediate fixings are illustrated in Fig. 10.30(a) and (b) and require the use of either a lead-welded cap or a soldered dot over the brass wood screw to prevent water entering through the screw hole.

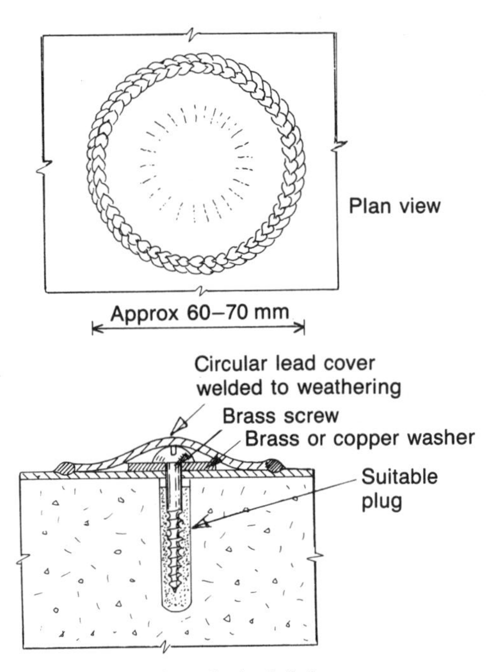

Fig. 10.28 Screwed method of fixing to masonry

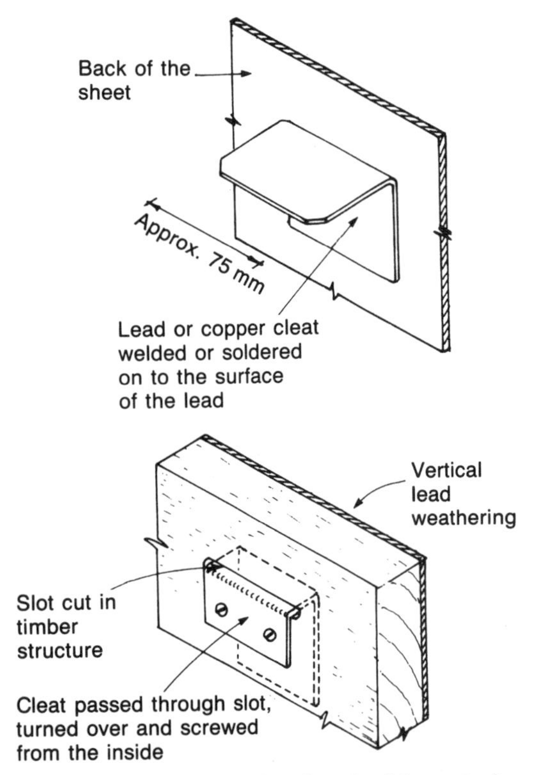

Fig. 10.29 Secret fixings for sheet lead in vertical positions such as dormer cheeks

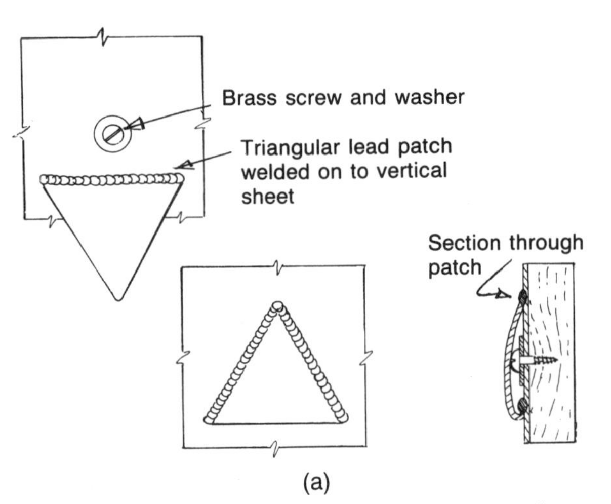

(a)

Triangular patch turned upwards through 90° concealing the screw and washer prior to completing lead weld on sides of patch

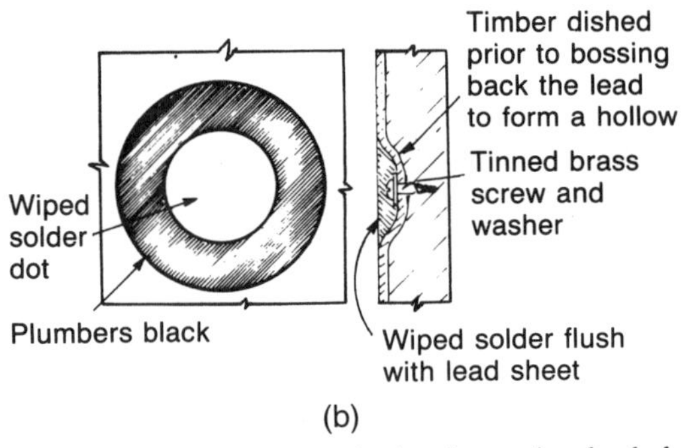

(b)

Fig. 10.30 Alternative methods of securing lead sheet in a vertical position

Copper sheet fixings

The materials used for fixings should be as follows:

Nails Copper or copper alloy having barbed shanks not less than 25 mm long or 2.6 mm diameter, weighing not less than 1.5 kg per 1,000, having a head diameter of not less than 6 mm and conforming to BS 1202, Part 2.

Screws Brass, not less than 25 mm long and not less than 2.6 mm shank diameter. Screws used for securing wooden battens can be of steel provided they are countersunk below the surface and the exposed steel either coated with paint or bitumen or covered with a felt ring or washer.

Clips Copper sheet clips should be the same thickness as the roof sheeting being used.

Both intermediate and edge fixings for copper sheet utilise copper clips or cleats. Clips are either 38 mm or 50 mm wide depending on fixing position and are placed at centres as follows:

Standing seams 380 mm max.
Batten roll and upstands 460 mm max.
Drips, eaves and verges 300 mm max.
Cross welts two per bay (single lock) or one per bay (doubled lock)

Two copper nails or brass screws are positioned close to the turn-up when fastening clips to the understructure. Fixing to concrete substructures can be achieved by a similar procedure to that described for sheet lead. Examples showing specific applications to copper or aluminium sheet are given in Fig. 10.31(a) and (b).

Wedges of rolled-up copper are used to fix copper sheet in brick- and blockwork in a similar way to that used with lead sheet, as indicated in Fig. 10.26.

Aluminium sheet fixings

The materials used for these fixings must not include copper or copper alloys such as brass or electrolytic corrosion will occur. It is also essential that where aluminium is to be embedded in brickwork, as for example cap or cover flashings, a coat of bituminous paint must be applied to the surfaces in contact with mortar.

Aluminium, zinc or good-quality galvanised steel nails and screws of similar sizes to those stated for copper are recommended, in that order of preference.

The types of fixing are identical to those for copper but as aluminium is a soft metal it is essential to turn cleats back over the nails or screws as in the standing seam detail in Fig. 10.31 to protect the covering from damage by the nail or screw heads. All fixing clips and wedges must be of aluminium.

Zinc sheet fixings

The materials used for fixings must not include copper or copper alloys.

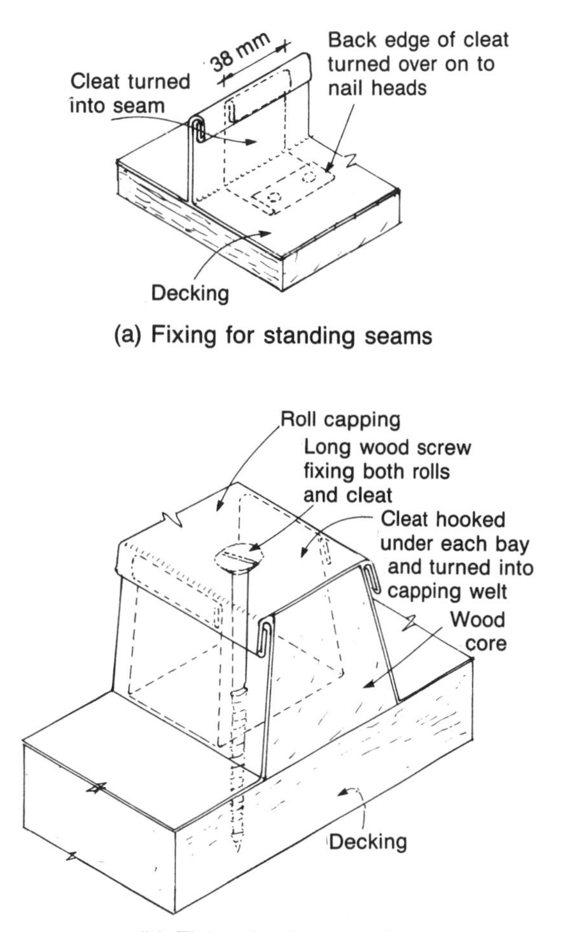

Fig. 10.31 Intermediate fixings for copper and aluminium sheet

Nails Galvanised steel or aluminium alloy of sufficient length to allow penetration of the substructure to a depth of at least 22 mm and conforming to BS 1202.
Screws Details as for nails, but conforming to BS 1210.

Details of fixing procedures are similar to those specified for copper and aluminium but as zinc is a hard metal there is no requirement to cover nail or screw heads used for clips; they can be left as shown in Fig. 10.25(e). Fixing clips must be formed from sheet zinc but wedges for fixing into raked brick joints can either be of zinc or lead.

Forming roofing details

Roofwork details are the positions where it is necessary to form the material into a certain shape such as internal or external corners, roll ends and the abutment of a standing seam to a vertical surface.

In Chapter 9 it was seen that the characteristics and physical properties of the materials used for roofing vary considerably. These characteristics and properties must be taken into account when deciding on the technique used for forming the details.

Lead sheet working

Probably the best-known characteristic of lead, apart from its weight, is its malleability, a property which enables it to be worked or 'bossed' without the need for annealing. The term 'working' lead is often used by plumbers, this merely being another way of describing the process of bossing. Bossing is the traditional method for forming the various details in sheet leadwork. An alternative procedure widely used by the modern plumber is to cut the sheet, remove or insert areas of lead and then to join them using the oxy-fuel gas-welding process.

Bossing When bossing is to be used it is essential to decide how the detail is to be formed, i.e. how is the shaping operation to be undertaken. A study of Fig. 10.32 will provide an indication of these considerations. Figure 10.32(a) shows the setting out of an external corner and it can be seen by comparing this with the finished detail in Fig. 10.32(b) that the shaded area of lead is superfluous to requirements. It is this area which must be removed.

An internal corner is illustrated in Fig. 10.33 and a study of the sketch will show that lead has to be gained from elsewhere to allow the corner to be formed. The shaded area in Fig. 10.33(a) must be at least equal to that shown in Fig. 10.33(b), but in practice more than the minimum requirement is provided to facilitate ease of

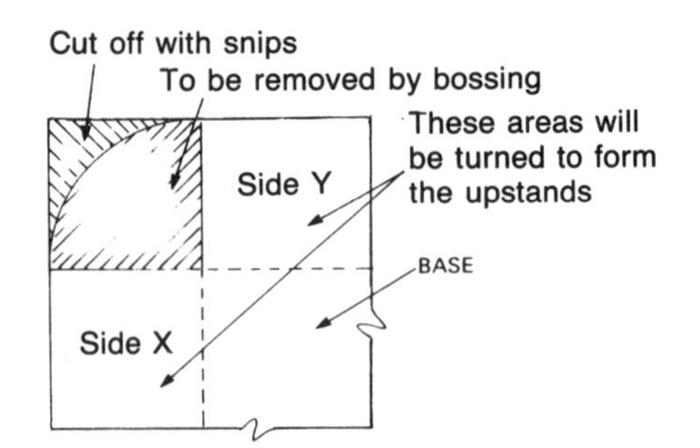

(a) Corner marked out before bossing commences

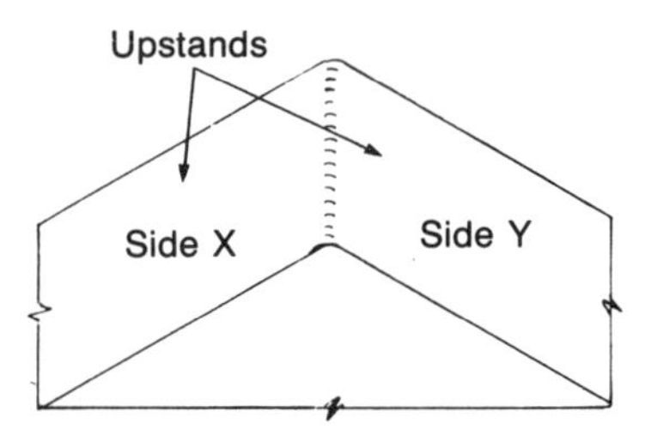

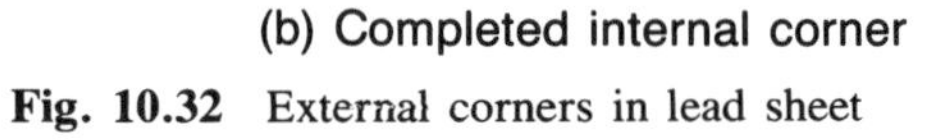

(b) Completed internal corner

Fig. 10.32 External corners in lead sheet

working by increasing the allowance for the upstand where a corner of this type occurs.

These two operations, i.e. the *removal of surplus lead* or *the moving of lead from one place to another*, form the basis of all sheet lead bossing. Examples of the first operation include roll ends, undercloaks to drips and external corners. The second operation is required for such details as aprons, back gutters and internal corners.

Once the working method has been decided the marking out can proceed using a felt-tip pen or a chalk line. (Harder metals such as zinc and copper can be marked using a pencil but this is not recommended for lead. Sheet material of any kind should never be marked out using anything that will scratch or score, as this could cause failure such as cracking along the marked line.) The use of a chalk line is shown in Fig. 10.34. Chalk is rubbed into the line which is then stretched tightly across the work to coincide with the measurement marks. The line is picked up in the centre and then released allowing it to snap back on to the sheet where it leaves a sharp, clearly defined line without damage to the surface of the lead. The next stage is to remove any excess metal with

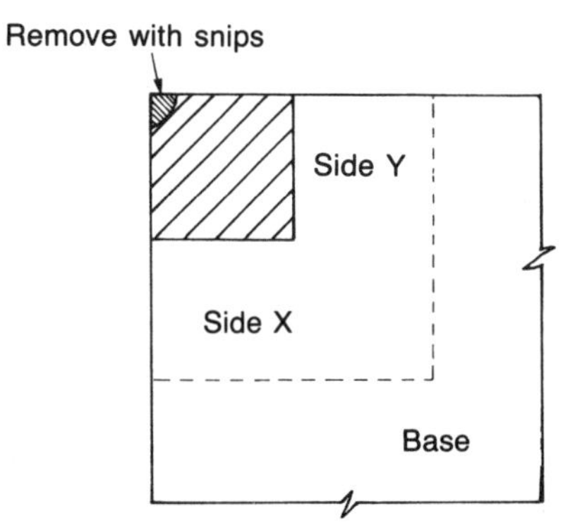

(a) Corner marked out before bossing commences

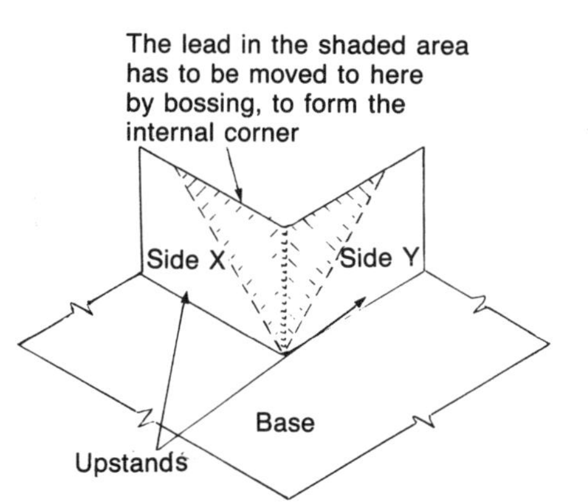

(b) Completed internal corner

Fig. 10.33 Internal corners in lead sheet

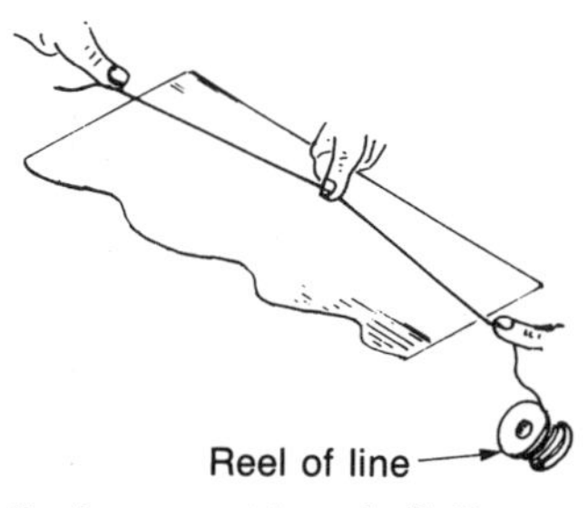

Fig. 10.34 Setting out with a chalk line

tinsnips so that the bossing process is simplified. Examples are shown in Figs 10.32 and 10.33.

Bossing an external corner One of the more common operations involving lead bossing is the formation of an external corner. It should be marked out in a similar way to that shown in Fig. 10.32(a), any excess metal then being trimmed off. The upstands are next turned up 90° using a timber former, a piece of softwood 600 mm long, 100 mm wide and 50 mm thick is ideal for this purpose. The angles are then 'set in', using a setting-in stick, a tool similar to a dresser but having a wedge-shaped face like a chase wedge. This edge is placed in contact with the angle formed between the upstand and the base, and a sharp blow is applied to the back of the tool with a mallet. This setting-in process fixes the position of the upstands preventing any tendency for them to wander while bossing is in progress. To ensure a good square base to the corner and fix its position, a slight groove or 'belly' is made on the base of the sheet. Figure 10.35(a) shows the corner at this stage. Surplus lead is then removed by bossing as illustrated in Fig. 10.35(b), the first blows being diverted inwards to stiffen the base of the corner. The direction in which the blows are struck is most important as that is the direction in which the lead is being driven.

The actual bossing procedure shown is not always used. The internal support is usually provided by a mallet or a dummy but the external

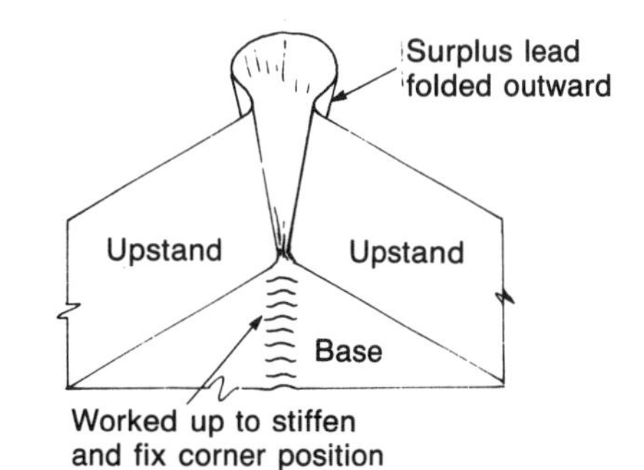

(a) Preparing the external corner prior to bossing

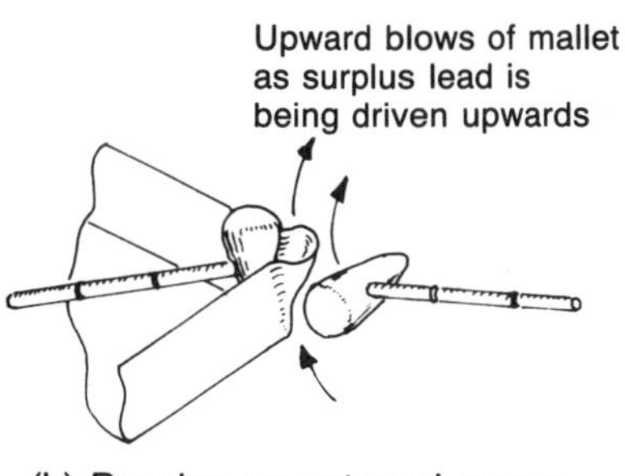

(b) Bossing an external corner

Fig. 10.35 External corners in lead sheet

shaping is carried out by many plumbers using a bossing stick. When the excess material has been bossed out and the corner formed to the required angle the detail is trimmed to the required height. Bear in mind the objective with all bossing operations is to maintain a uniform thickness in the whole of the area of working.

Bossing an internal corner An example of a detail where lead has to be gained, i.e. an internal corner, can be seen in Fig. 10.36 where one side of a chimney front apron is shown. The area of lead required to be worked into the corner is indicated by shading. First the upstand is folded to the angle of the roof pitch and the return ends turned to the angle of the chimney breast, which is usually 90° although other shapes including circular chimneys may be encountered. This will cause a hump to occur that is then worked into the desired position using a mallet, bending or bossing stick. The procedure is illustrated in Fig. 10.37(a).

The illustration shown in Fig. 10.37(b) is another example showing how lead may be worked from one area to another, in this case to fabricate one side of a chimney back gutter.

Problems encountered bossing lead Once the skills of gaining or removing lead have been mastered they can be applied to the shaping of any roofing detail. Several important factors must, however, be borne in mind when bossing sheet lead.

When surplus lead is being removed there is a tendency for it to thicken and crease. These creases must not be allowed to form or they will result in cracking. When forming internal corners there is always a danger of stretching the lead into position, i.e. creating a greater area by thinning. This may reduce the thickness to an unacceptable level for the application concerned and could cause splitting to occur. From these factors it can be seen that the objective in bossing is to maintain the original thickness over the whole worked area and this requires considerable practice. The plumbing student can assess his own ability by cutting sections from completed exercises and examining

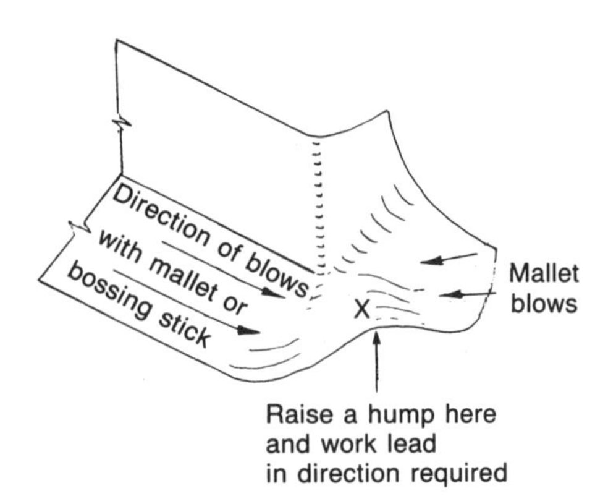

(a) Working round one end of a chimney apron
A chimney apron is an internal corner there being a shortage of lead at the point marked 'X'. Lead is driven from both directions, as shown, to provide material needed at 'X'.

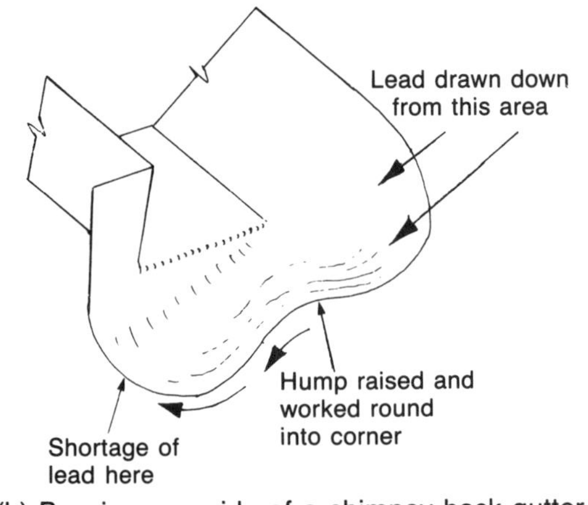

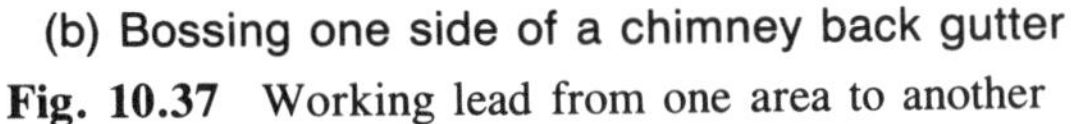
(b) Bossing one side of a chimney back gutter

Fig. 10.37 Working lead from one area to another

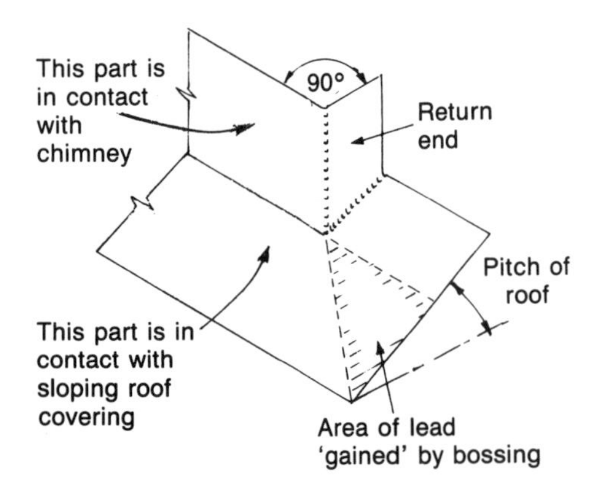

Fig. 10.36 Area of lead to be 'gained' for a chimney apron

them to check for even thickness. It is better to strike many blows lightly when bossing, rather than fewer and heavier blows.

Even if the forming of lead by means of bossing is undertaken by a competent craftsman, it is still a slow and laborious process, although it has the advantage of requiring few tools and so is often very convenient. The use of the 'cut and weld' technique is much quicker and if the welding is carried out competently it ensures that the lead remains of a uniform thickness throughout. This procedure requires a knowledge of simple geometry as it consists basically of cutting out or inserting pieces of lead to achieve the desired shape. The individual pieces are then welded or fused together using one of the seams illustrated in Fig. 10.24. Examples of the 'cut and weld' procedure are given in Fig. 10.38 where internal and external corners are shown formed by cutting and turning or inserting a gusset.

A comprehensive range of details formed by the methods described are given in the publication *Lead Sheet in Building* which is available from the Lead Development Association.

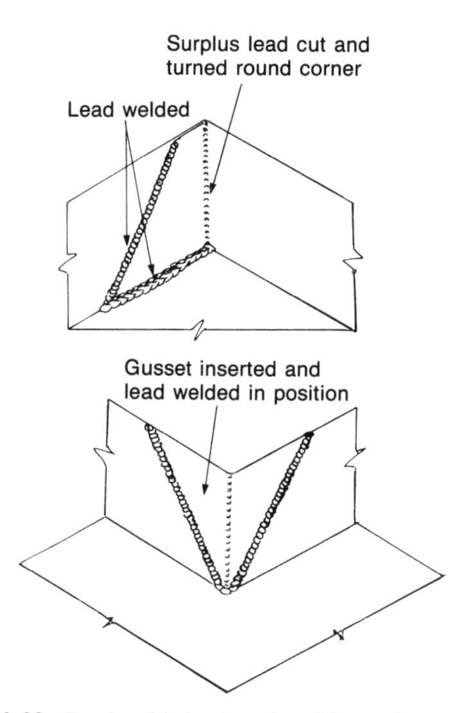

Fig. 10.38 Lead-welded external and internal corners

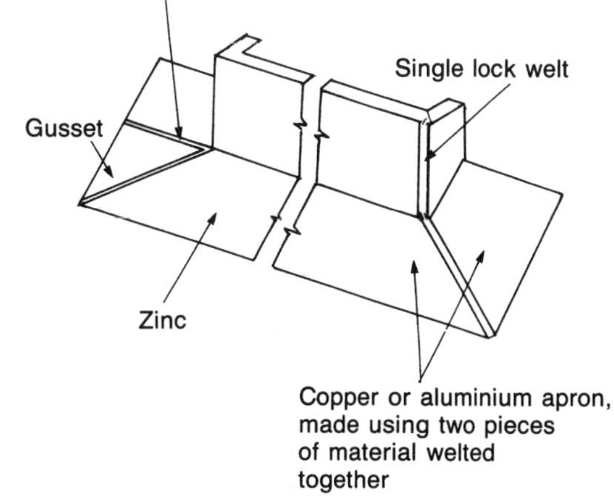

Fig. 10.39 Fabrication of front aprons using zinc, copper or aluminium

Copper, aluminium and zinc sheet working

Copper, aluminium and zinc alloy are hard metals and much less malleable than lead, and are not normally formed by the bossing method. An elementary knowledge of geometry is again needed for these working techniques which consist basically of folding, cutting and jointing by means of the methods previously described, an example being the chimney apron shown in Fig. 10.39. Fabrication details can be set out using a soft pencil but crayons and felt-tipped pens are also satisfactory. (Further examples of these details and methods of setting out are described in Chapter 11.)

Further reading

Detailed practical information in respect of jointing, fixing and forming procedures for the various roofing materials is provided in the literature recommended for further reading in Chapter 9.

Self-testing questions

1. Describe the preparation of a roof surface prior to laying metal sheet.

2. Name two materials that are used for underlay and state their special qualities.
3. (a) Describe the causes of fatigue and 'creep' in sheet-metal roofing.
   (b) State how these two defects can be avoided on a well-designed and fixed metal roof.
4. State why soft soldering is not recommended for joining sheet copper.
5. Describe two techniques for forming details in sheet lead.
6. Select and describe two methods for fixing lead sheet to steeply pitched or vertical surfaces.
7. Explain the use of a 'continuous fixing strip'.
8. State the minimum recommended pitch or fall of a sheet-metal roof surface. Describe what would happen if a roof pitch were below the minimum.
9. (a) What is the main reason for the long working life of metal roof coverings?
   (b) Name two factors which enable metal weatherings to withstand the effects of thermal movement.
10. State the function of an anti-capillary groove.

11 Sheet weatherings — simple applications

After completing this chapter the reader should be able to:

1. Identify the ways in which dampness can enter a building.
2. Select appropriate methods of weathering for all types of roof covering.
3. Describe the methods of employing bossing, welting and gusseting to fabricate simple roofing details.
4. Calculate the amount of material required for a given job.
5. Select appropriate tools for given roofwork details.

Some idea of the working properties and techniques of sheet-weathering materials will have been gathered from the previous chapters. One of the purposes of this chapter is to show how some of the more common of these processes are applied in actual on-site situations. In roofwork the plumber is involved mainly with preventing the entry of water into the building, and his work consists of flashing and weathering flat roofs, dormers and chimneys.

However, in addition to entering a building from above, water can also enter a building from below ground level by capillary action through the brick, stone or concrete used in construction, and for this reason damp-proof courses are used.

Damp-proof courses

Most of the clayware and concrete products used for structural work are porous and as such provide the ideal medium through which moisture can pass by capillary action. Unless steps are taken to prevent the ingress of water in this way, dampness will penetrate a building causing damage to the structure and creating a cold, damp atmosphere which could also endanger the health of the occupants.

The method used to overcome this problem is the installation of a damp-proof course (DPC). Any material used for such work must be impervious to moisture and must be built into the building structure to provide a waterproof barrier between various components. For many years slate was used to prevent the moisture in the earth rising above the ground-floor level, but due to its high cost and scarcity asphalt, plastic strip or engineering bricks are used in most modern buildings. Although engineering bricks are made of clay, the methods of manufacture result in a high-density brick which, for all practical purposes, is water resistant.

All these materials are fixed by a bricklayer as they are actually built into joints in the building structure. The main disadvantage of the materials named is that any settlement of the building will result in a crack in the DPC which may allow the passage of moisture at that particular point.

Lead or copper sheet, although expensive, is sometimes used as an alternative to the materials already mentioned, as their strength and other qualities make them far more likely to resist damage by settlement faults and they maintain a waterproof barrier under all circumstances. If sheet lead is used, it should be coated with a bituminous solution to resist the alkali present in the cement mortar in which it is laid. Code No. 4 or 5 lead is considered to be satisfactory for DPCs or in the case of copper, 0.45–0.6 mm thickness. The method of jointing for each material is a simple lap joint with not less than 100 mm overlap.

One of the most common causes of dampness is the moisture in the earth seeping into the footings of the building and rising by capillarity through the lower area of the walls. This is commonly called 'rising damp'. To prevent this dampness rising to the ground-floor rooms of the building, a DPC is built into the wall not less than 150 mm from the ground level (see Fig. 11.1). Older buildings were often built without DPCs and, in such cases the walls at ground-floor level are damp, sometimes to a height of one metre or more.

Building components exposed to the weather at higher levels of a building also absorb dampness and can bypass the flashing as shown in Fig. 11.2. A typical example of this occurs in parapet walls, and a sheet metal DPC fitted as shown in Fig. 11.3 prevents the downward movement of dampness. It is also arranged to act as the cover flashings over any weathering such as a gutter or flat roof on the inside of a parapet.

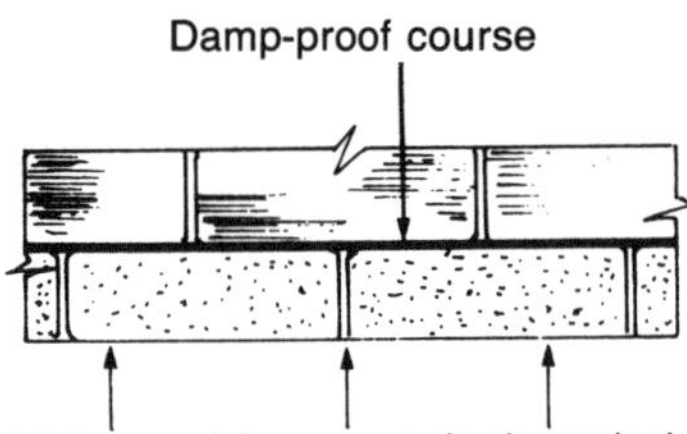

Fig. 11.1 Damp-proof course

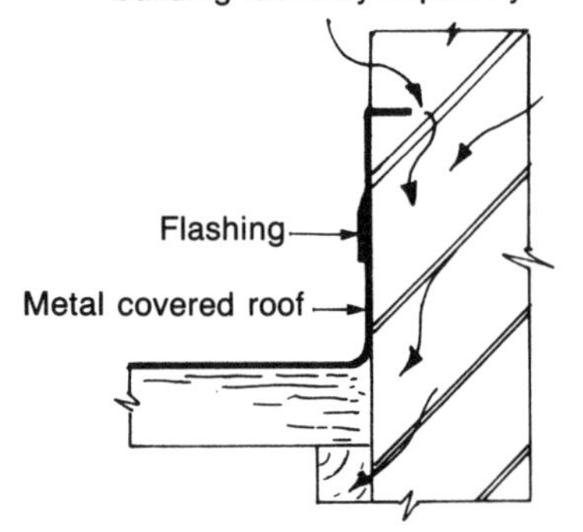

Fig. 11.2 Water can be absorbed by masonry in very exposed positions and gain access into the building

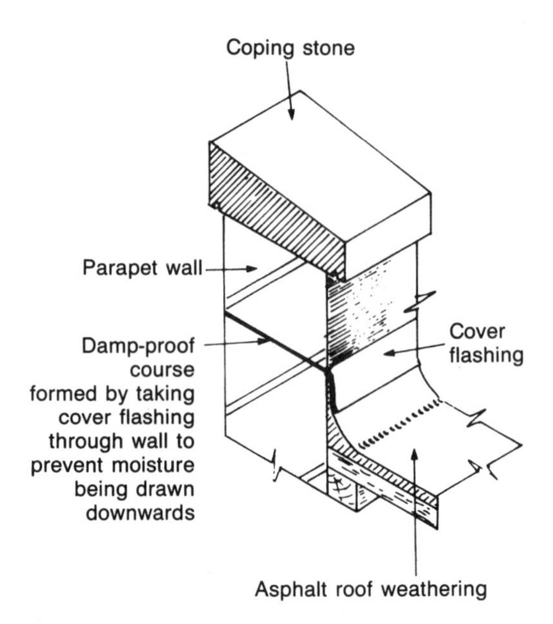

Fig. 11.3 Damp-proof course through parapet wall

DPCs to chimney stacks that penetrate pitched roofs are unusual except in very exposed conditions. Any moisture absorbed by the fabric of a chimney normally dries out in the area of the roof space before it can penetrate to a habitable room. In the case of a chimney penetrating a flat roof the situation is quite different, there being no roof space in which the moisture can be dissipated. In such circumstances a sheet lead DPC built into the chimney brickwork is absolutely necessary, and Fig. 11.4 shows how this should be fitted. The

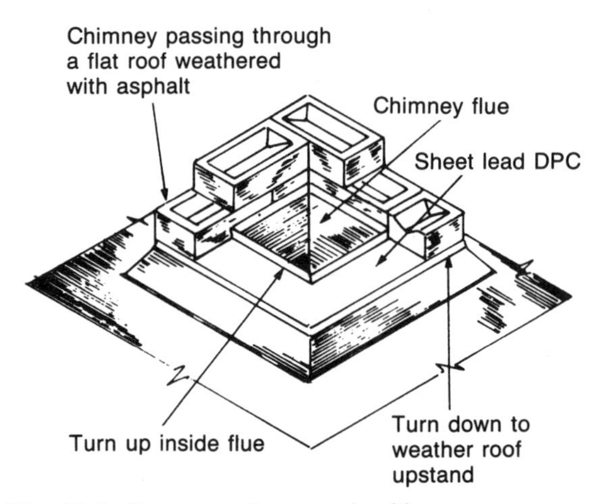

Fig. 11.4 Damp-proof courses in chimneys

turned down portion over the upstand of the flat roof can be bossed or lead welded.

There are many other different examples of damp-proofing the building structure with sheet metal, such as the weathering of openings and lintels in cavity walls. Only the more common examples have been included in this book, but further details can be found in the further reading suggested at the conclusion of Chapter 9.

Simple external weatherings

Probably one of the areas in which sheet metal weatherings will never be surpassed for their properties of long life, flexibility and durability is in the weathering of chimneys, small canopies and vent pipes. Modern developments and techniques have introduced prefabricated (non-metallic) units for weathering vent pipes where they pass through a roof, but these are not always suitable for all pipe materials, and while some types embody a degree of flexibility to accommodate a wide range of roof pitches, they are not always adaptable for all roof coverings or for the weathering of all sizes of pipe diameters.

Weathering chimney stacks

The weathering of a typical chimney where it passes through a pitched roof is shown at Fig. 11.5. Before the weatherings are fitted, the first step is to mark out and rake all the joints to receive the turn in from the flashings. They should be raked to a depth of 25 mm to ensure a good fixing and to enable the wedges securing the flashings to be driven home below the surface of the finished brickwork. If ordinary plain tiles or slates are used, the point at which they abut the chimney is weathered by soakers (unless step and cover flashings are used, soakers being unnecessary with this type of flashing).

Soakers

The length of a soaker will vary according to the type of roof covering with which it is used. Its width should not be less than 175 mm which allows an adequate amount of material to lie under the tile or slate (100 mm), and for the upstand against the brickwork (75 mm). The length of a soaker is normally found by calculation based on the gauge and lap of the roof covering used. Figure 11.6 illustrates these terms.

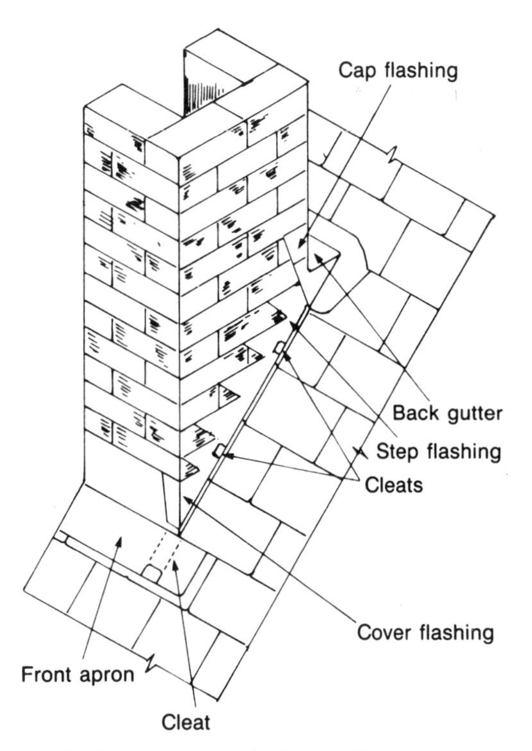

Fig. 11.5 Chimney stack flashing with lead sheet

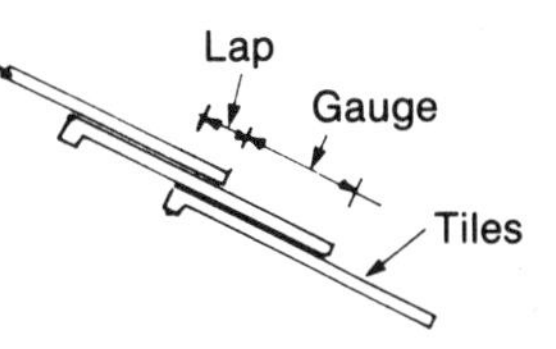

Fig. 11.6 Illustrating the terms 'gauge' and 'lap'

The gauge is the actual amount of the tile or slate that can be seen when the roof is finished, while the lap relates to the cover given to a tile by the overlap of the tile next but one above it. Using plain tiles on a roof pitched at approximately 40°, the gauge is usually 100 mm, with a lap of 50 mm. These measurements vary with different roof pitches, for example, a lower pitch requires a longer lap. The following formula is used to calculate the length of soakers:

$$\text{Length of soaker} = \frac{\text{Tile length} - \text{Lap}}{2} + \text{Lap} + 25\text{ mm for fixing.}$$

Example A slate, 500 mm in length is laid with a gauge of 200 mm and a lap of 100 mm. Calculate the length of the soakers required.

$$\begin{aligned}\text{Length of soaker}\\ &= \frac{\text{Slate length} - \text{Lap}}{2} + \text{Lap} + 25\text{ mm}\\ &= \frac{500-100}{2} + 100 + 25\text{ mm}\\ &= \frac{400}{2} + 100 + 25\text{ mm}\\ &= 200 + 100 + 25\\ &= 325\text{ mm}\end{aligned}$$

A soaker 325 mm in length will be required.

To determine the number of soakers needed for a given job, it is necessary to know:

(a) The length of the roof to be weathered
(b) The gauge to which the roof covering is laid

the former simply being divided by the latter thus:

$$\frac{\text{Length of roof}}{\text{Gauge}}$$

Example Assuming the gauge to be 200 mm and the length of the abutment 2.4 m, calculate the number of soakers required.

$$\begin{aligned}\text{No. of soakers} &= \frac{\text{Length of roof}}{\text{Gauge}}\\ &= \frac{2.4}{0.2}\\ &= 12\end{aligned}$$

Twelve soakers will be required to weather this length.

The soakers are fixed in position by the tiler as the tiles or slates are laid and nailed on to or turned over a batten. Figure 11.7 shows a detail of the soakers after the tiler has completed his work. It will be seen that although the edge of each tile has been weathered, the rainwater running down the brickwork could enter the building down the back of the soaker. It will be shown later in this chapter how step flashings are used to cover the top edge of soakers and so complete the weathering of the side of the abutment.

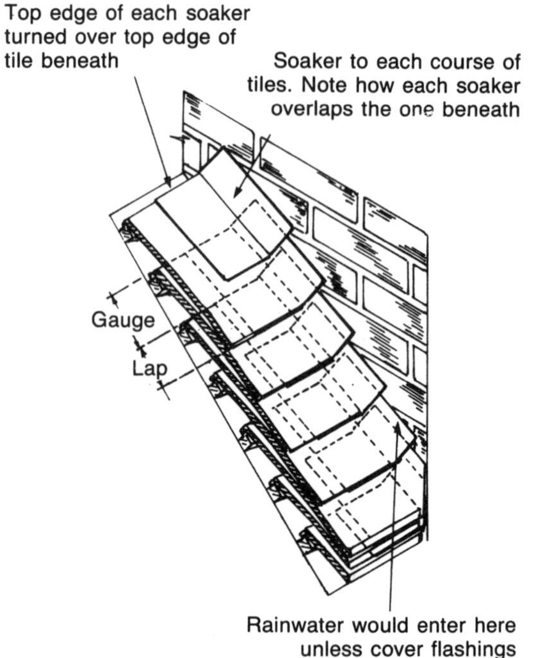

Fig. 11.7 Fixing soakers in position

Chimney back gutters

The normal way of weathering a roof is to start at the bottom and work upwards to ensure the various pieces of weathering overlap in the correct manner. It is often more convenient, however, to fix the completed back gutter of a chimney first so it can be fitted in with tiles or slates as they are laid. The chimney apron and back gutter shown in Fig. 11.5 are made of lead and have been bossed, using traditional methods. Obviously the basic shape of the weatherings will be the same whatever type of material is used, only the methods of fabrication will differ.

Figure 11.8 shows the method of setting out and forming one end of a back gutter in copper or aluminium sheet. The use of a fillet block avoids a sharp corner at the intersection of the sole board and brickwork. No measurements are given as these would be taken from the actual job, but as a general guide the upstand at the back of the stack is normally 100 mm, the sole 150 mm, and the allowance under the tiles or slates ranges from 200 mm to 225 mm depending on type and pitch.

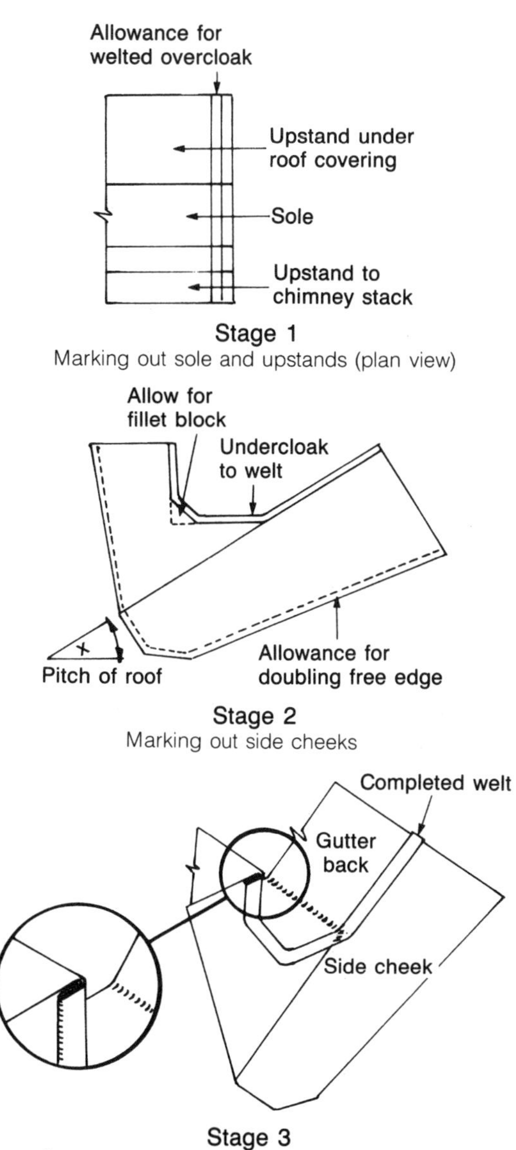

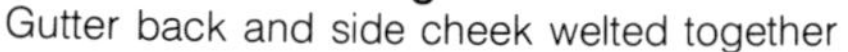

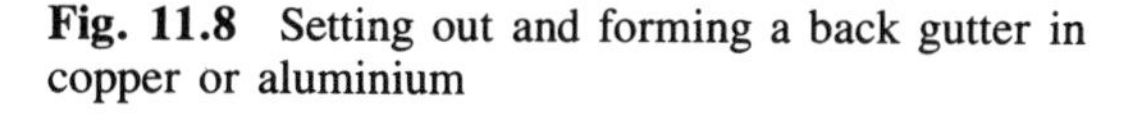
Fig. 11.8 Setting out and forming a back gutter in copper or aluminium

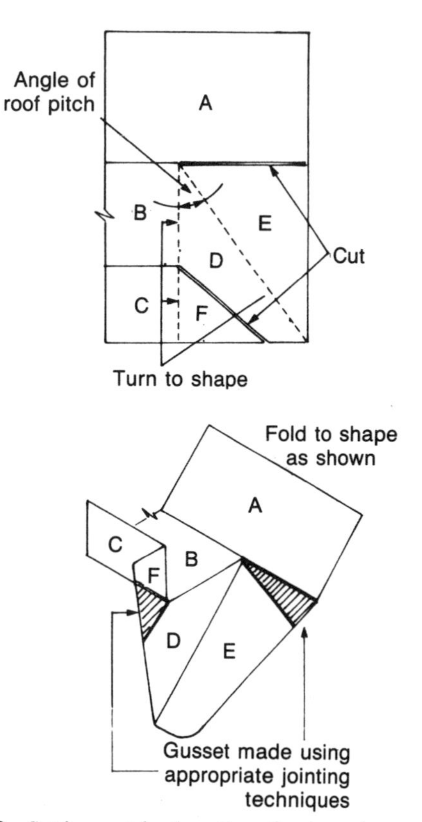

Fig. 11.9 Setting out back gutters for inserting gussets

Usually a back gutter has water outlets at each end, and thus is made in three pieces which are joined together by means of single-lock welts. Note the doubled edge indicated by the dotted line which has the effect of stiffening the free edges and giving a neat finished appearance. It also reduces the risk of cut fingers when fixing. It is good practice to double all free edges of thin materials for the reasons given, not only on the back gutters, but on all roofing details.

Figure 11.9 shows how the same weathering is fabricated when zinc- or lead-welding techniques are used, the material being cut and gusseted to form the required shape. With zinc the gussets are soldered with a copper bit. If lead is used and welding techniques are employed, the gussets are marked out as shown.

It is not good practice to turn the upstand of a back gutter directly into a brickwork joint, as this restricts movement brought about by thermal expansion. The correct procedure is to use a cap flashing as shown in Fig. 11.10 which allows a watertight joint to be made, yet still permits movement of the back gutter.

Assuming the roof covering is complete and the back gutter is in position, the next step is to fabricate the front apron. When using sheet copper or aluminium, the apron is cut and folded for

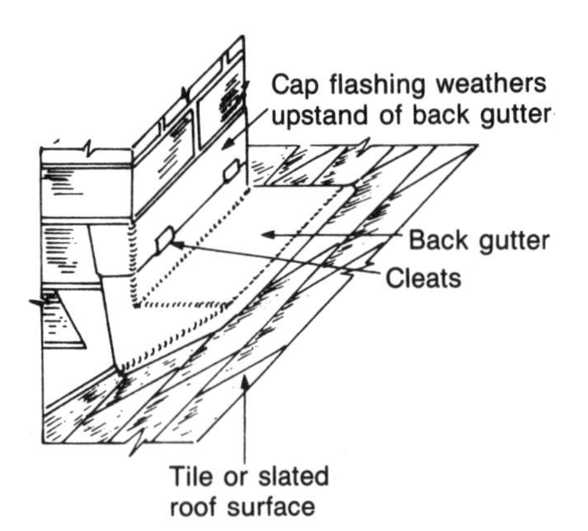

Fig. 11.10 Back gutter in position

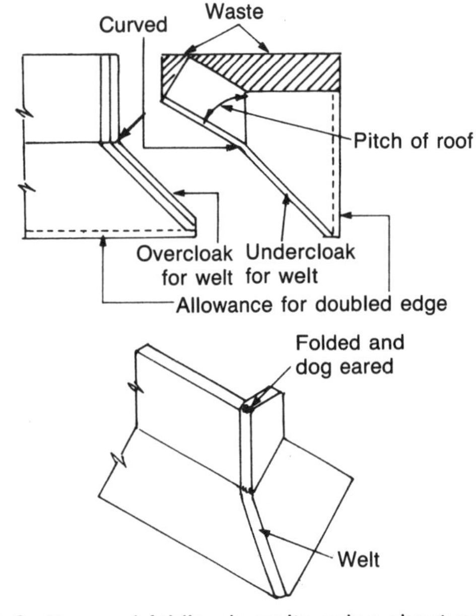

(a) Cutting and folding in welts using sheet copper or aluminium

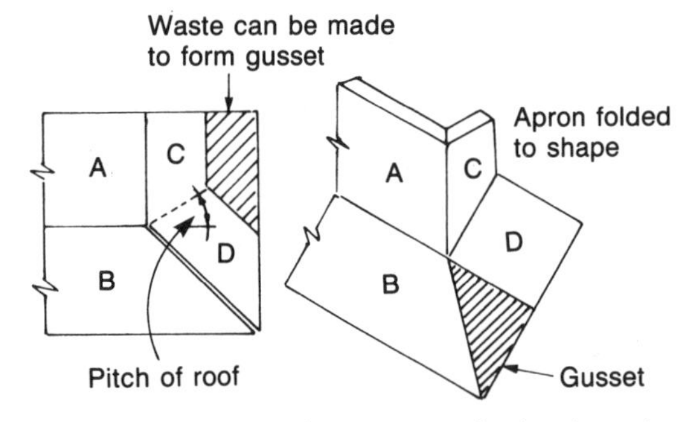

(b) Cutting and inserting gussets for lead or zinc

Fig. 11.11 Fabricating the front apron

welting as shown in Fig. 11.11(a), or, if zinc or lead welding techniques are to be used, it is made by cutting and inserting gussets as in Fig. 11.11(b).

The two spots marked 'curved' on Fig. 11.11(a) should be noted. It is important that the metal is shaped to a curve with a file at these points, not a sharp angle, or the sheet will split as the welt is formed. (A further point to be observed when cutting sheet metal with snips is not to allow the blade to close fully, otherwise a sharp point is produced each time the blades close. The blades should be closed until the ends almost meet and the cutting stroke repeated.)

Prior to positioning the apron, the cleats (see Fig. 11.5) which secure the bottom edge and prevent any tendency to creep, are fixed. They are made of two strips of lead or copper sheet approximately 50 mm wide, either wedged into the stack or turned over and nailed to a convenient tile batten. The top edge of the apron is then turned into a brickwork joint approximately 75–100 mm above the roof line and secured by lead wedges. If soakers are used they must be fitted over the top of the apron.

One very important point that cannot be stressed enough is the need for accuracy when chimney aprons and back gutters are set out. This is particularly true with those materials that employ the use of gussets or welts as a means of fabrication. Unlike lead, when metals such as copper and aluminium are turned and welted it is almost impossible to alter them in any way. An error in setting out or a carelessly taken measurement may result in the whole job being scrapped. A good practical hint is to set out the work on stiff paper and make a pattern, this can easily be checked against the actual job, and if found to be satisfactory its shape can easily be transferred by marking round it on to the sheet metal. If the job is to be repeated, the same pattern can be used for the whole batch.

Flashings

Step flashing The term 'step flashing' is given to the flashing which weathers a pitched roof to brickwork, its name being derived from its step-like appearance. There are two main types of flashing, the first being that shown on the chimney stack in Fig. 11.5, known simply as step or 'skeleton' flashing and always used in conjunction with soakers. A more detailed illustration is shown in Fig. 11.12(a). Where the use of soakers is not a practical possibility, as in the case of interlocking tiles and corrugated sheeting, the second type, step and cover flashings, are used (see Fig. 11.12(b)).

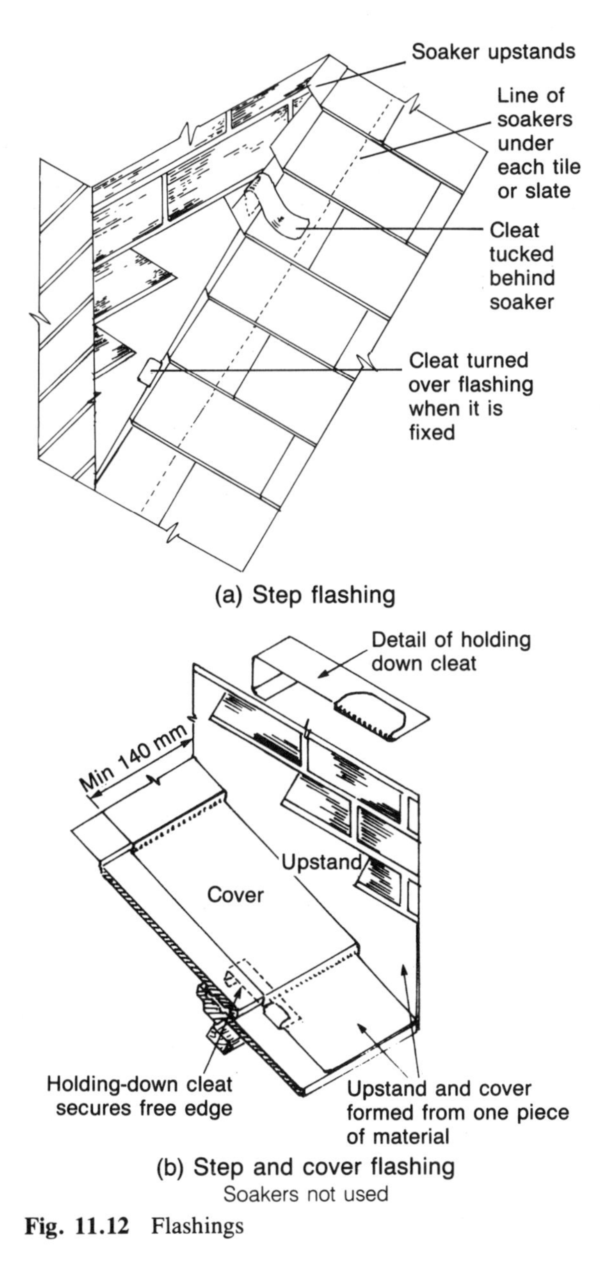

(a) Step flashing

(b) Step and cover flashing
Soakers not used

Fig. 11.12 Flashings

It will be seen that the upstand on which the steps are marked out is the same in both cases, the difference being the free edge of step and cover flashing which is extended outward over the roof covering. Sheet lead is generally accepted as the best material for this type of flashing as it can easily be dressed down on to the somewhat awkward shape of most interlocking tiles. The disadvantage of a cover flashing is its tendency to lift in high winds, but this can be overcome when a sufficient number of cleats are used, these being firmly fixed at joints in the roof covering. Sufficient lead should also be left on the chimney apron to allow it to be turned over to form a cleat and hold down the lower end of the cover flashing as shown in Fig. 11.13.

The type of flashing so far dealt with is for use with materials such as brick, where the horizontal joints are equally spaced. In areas where buildings are constructed of stone, slightly differing techniques of flashing are used as illustrated in Fig. 11.14(a) and (b). Where the stones are laid to regular horizontal courses, called coursed stonework, it is possible to use step flashing, but due to the uneven appearance of the steps, the method that is commonly used is shown at Fig. 11.14(a), having separate pieces of lead for each step. Where random stonework is used the best method is to cut a groove in the stone into which the lead is turned as shown in Fig. 11.14(b) – a disc cutter is essential for this work.

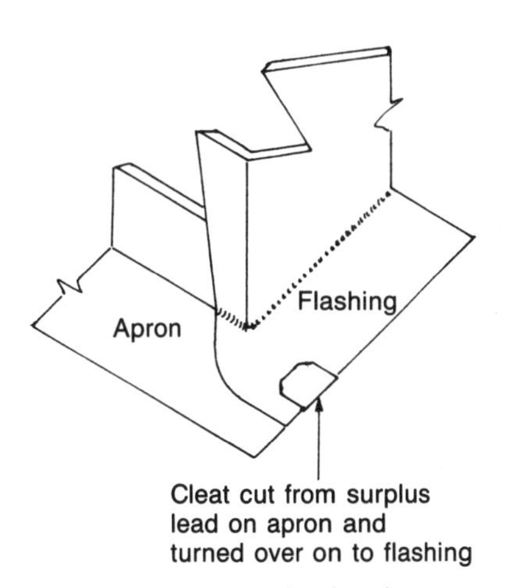

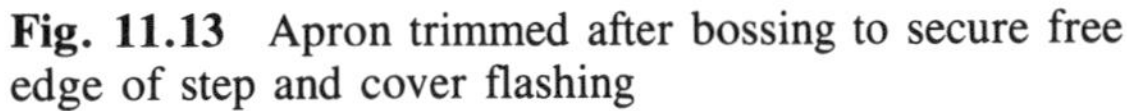

Fig. 11.13 Apron trimmed after bossing to secure free edge of step and cover flashing

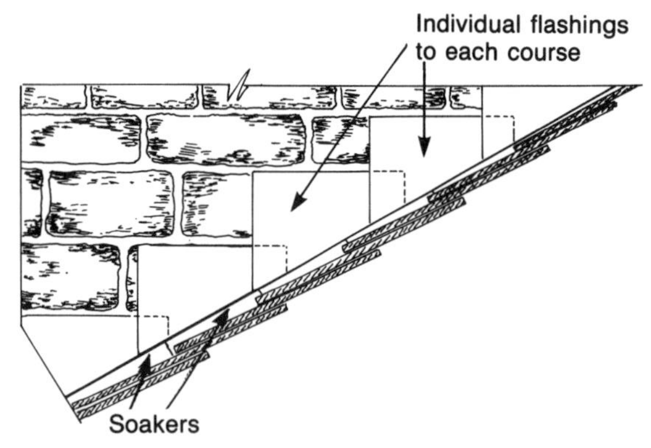

(a) Flashings to coursed stonework

Note that the horizontal courses are regular but due to the differing thicknesses of stone the widths of each course varies.

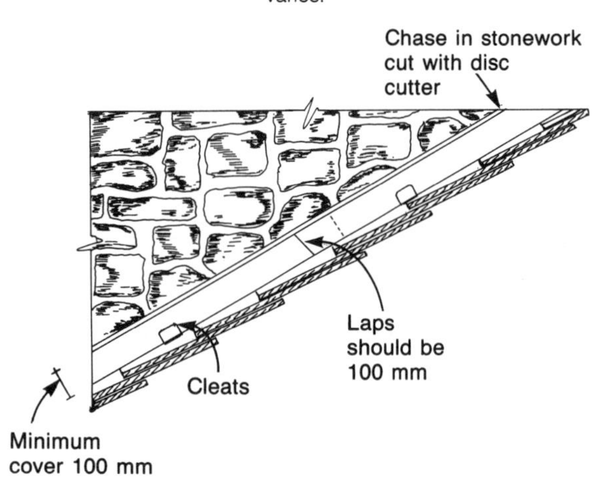

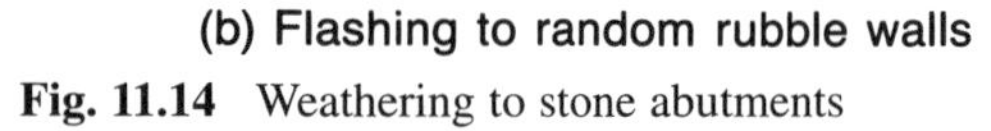

(b) Flashing to random rubble walls

Fig. 11.14 Weathering to stone abutments

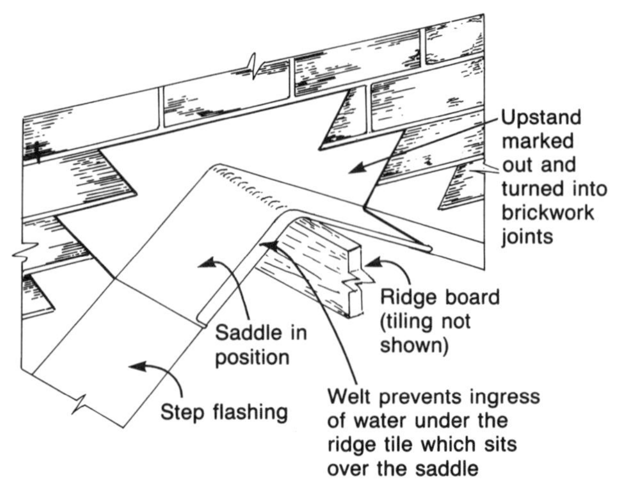

Fig. 11.15 Lead saddle

Where a chimney passes through the ridge of a building, or when a double-pitched roof abuts a wall, as shown in Fig. 11.15, a saddle piece will be required to weather the tops of the side flashings. These can be bossed or lead welded as shown in Fig. 11.16(a), (b) and (c). Figure 11.16(a) shows the technique of bossing this detail, the lead being worked from the end towards the middle to form a bulge. When the ends are pushed down to form the angle, the bulge will flatten. This operation will have to be repeated several times before the correct angle is achieved. The absolute minimum size of lead required for this job is 350 × 450 mm, but measurements taken from the job will be more accurate. Allowances must be made for final trimming after the bossing operation is completed. Figure 11.16(b) illustrates one method of forming a saddle using the lead-welding technique. Whichever method is employed when the saddle is formed, the steps are marked out to correspond with the brickwork joints in the same way as marking out step flashings. When the saddle is in position, with the steps firmly wedged, the tiler can bed the ridge tile adjacent to the abutment over the saddle.

Marking out step flashings Assuming the sides of a brick chimney are to be weathered with step flashing, a strip of the flashing material 160 mm wide by the length of the roof slope alongside the stack is cut. An allowance of approximately 75 mm should be added to this length to permit the bottom of the flashing to be turned around on to the front of the apron. A water line of 65 mm is marked off on the material before placing it in position on the side of the chimney. Using a bevel or folding rule, the brickwork joints are then marked on the flashing as shown in Fig. 11.17. This marking out must be accurate and the marks made parallel with and about 12 mm above the bottom of the brick joint. The stage by stage illustration in Fig. 11.18 shows how the steps are set out from the marks taken from the brickwork joints. Snips are used to make the cuts shown by the heavy lines (not

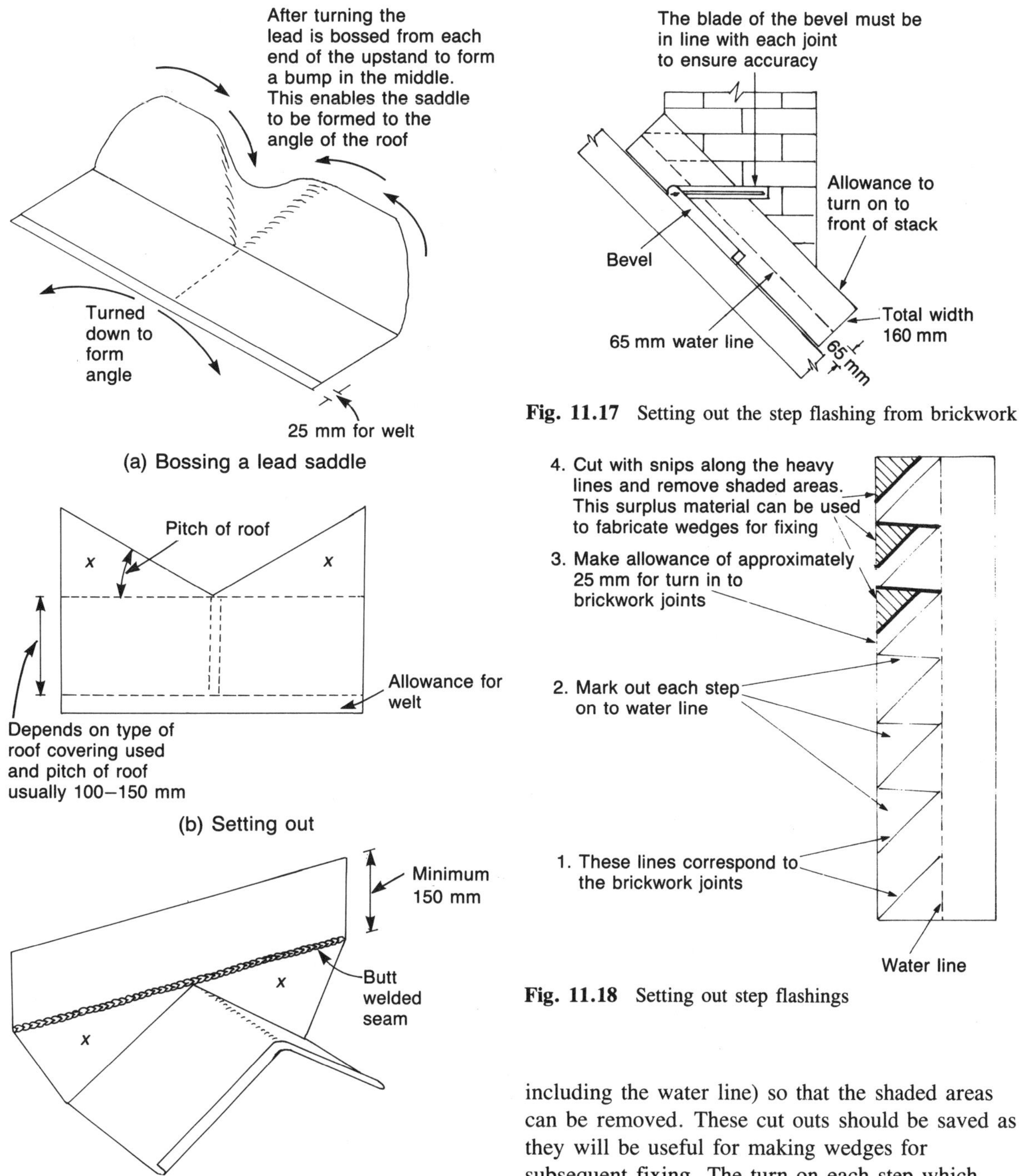

Fig. 11.17 Setting out the step flashing from brickwork

Fig. 11.18 Setting out step flashings

(c) The saddle turned with extension piece welded on

Steps marked out from job.

Fig. 11.16 Fabricating a lead saddle using lead-welding techniques

including the water line) so that the shaded areas can be removed. These cut outs should be saved as they will be useful for making wedges for subsequent fixing. The turn on each step which forms the fixing on the brickwork is made with the step turner described in Chapter 2. On low pitched roofs it may be necessary to use two wedges to fix each step securely as the turn in on each step is longer.

Step and cover flashings are set out using the same techniques, the only difference being that wider strips of material are required to allow for the cover on the roof. There is no hard and fast rule for the width of the cover, but it is usual to cut the flashing strips from material of 310 mm width, allowing approximately 160 mm for the step and the upstand, this allows 150 mm cover on the roof which is adequate for most applications.

Pipe flashings It is often necessary to weather ventilating or flue pipes where they pass through a roof. The traditional name for such a flashing is a 'lead slate' because for many years only lead was used. Since the use of other materials for this purpose is increasing, the term 'pipe flashing' more accurately describes this type of weathering. Figure 11.19(a) shows a typical flashing made of a lead sheet which has been fabricated by lead welding a purpose-made pipe to a base.

If copper or aluminium is used, the pipe section of the flashing is made from sheet in the same way as the lead, but joined by a single lock welt (see Fig. 11.19(b)). This section is then welted to a base as shown. It should be noted that this welt is quite bulky as there are four thicknesses of material in a single lock welt, and unless sufficient space is allowed the flashing will be too tight to fit comfortably over the pipe to be weathered. To overcome this problem, the pipe section should be made with a slight taper.

It is possible to obtain purpose-made pipe flashings. These incorporate a fold which permits the angle at which the pipe section fits on the base to be varied to suit a variety of roof pitches.

Pipe flashings made with sheet lead or zinc are prepared in a similar way where it is intended to use soldering or lead welding techniques as shown in Fig. 11.19(c).

When the flashing has been fitted over the pipe, and the base is weathered into the roof, it is important to ensure that water does not penetrate into the building between the pipe and the top edge of the upstand. If the pipe terminates only a short distance above the roof surface, the flashing can be dressed over and into the top of the pipe as seen in Fig. 11.20(a). An alternative is to weather the edge of the flashing by a socketed pipe turned

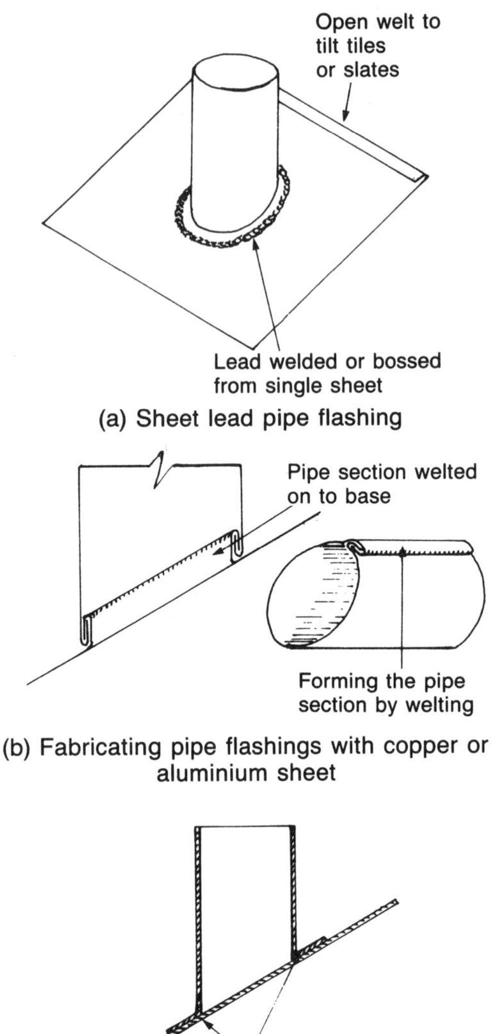

(a) Sheet lead pipe flashing

(b) Fabricating pipe flashings with copper or aluminium sheet

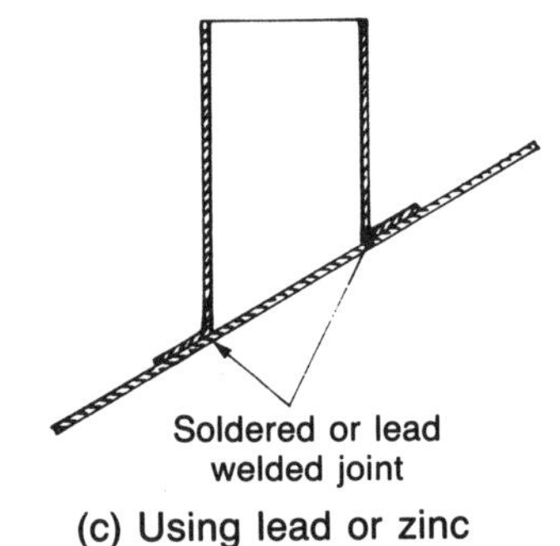

(c) Using lead or zinc

Fig. 11.19 Fabrication of pipe flashings with sheet lead, copper, aluminium and zinc

upside down as in Fig. 11.20(b), but the difficulty with this method lies in making an effective joint to the pipe in this position. However, it is a convenient method to use when weathering gas flues where a terminal can be used instead of a length of pipe. In the event of these two methods being unsuitable, as for instance in the case of a flag pole or long length of pipe fitted above the

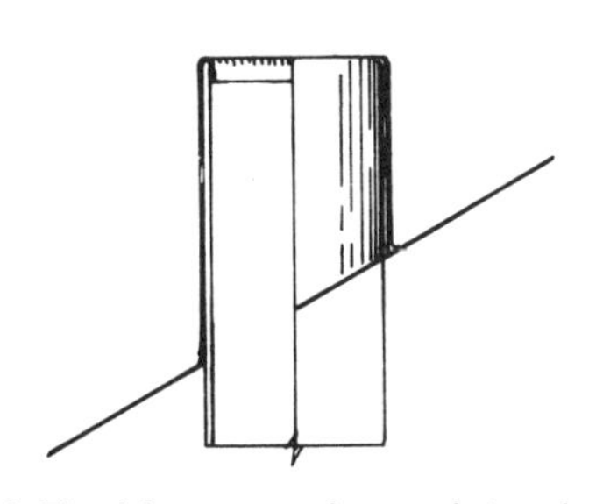

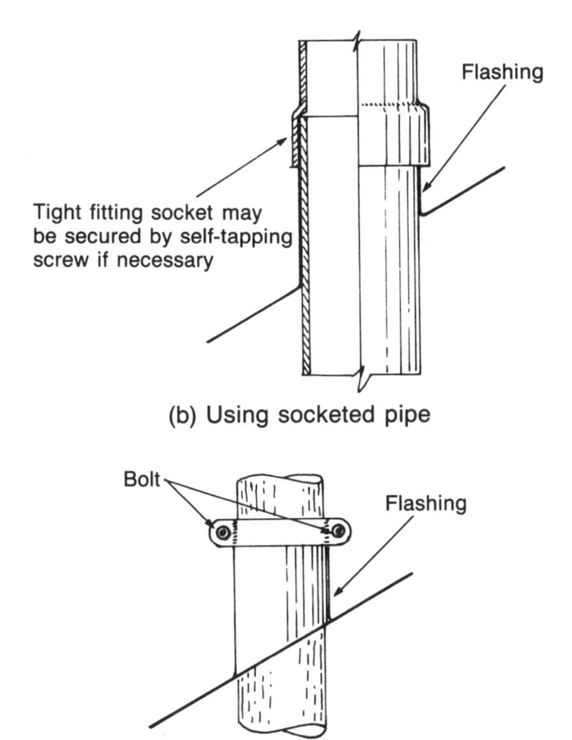

Fig. 11.20 Weathering flashings to pipe

surface of a roof, a clamp made of non-ferrous metal can be used (see Fig. 11.20(c)). A suitable mastic should be applied around the top edge of the flashing before the clamp is tightened to ensure that the joint is weathertight.

Weathering small canopies and flat roofs

It is impossible within the confines of this book to deal with all the possible variations of weathering details using all the available materials with which the plumber is likely to come into contact. The object within the following pages is to show typical examples of the basic principles of weathering small canopies and flat roofs using some of the materials used in the plumbing industry to carry out the work.

Metal sheet weatherings are a far more suitable and permanent alternative to bituminous felt for covering small roofs, especially in the case of 'one-off' jobs. The plumber has to have a working knowledge of the methods used to weather small areas of this type with all the materials previously described. For low cost work, aluminium or zinc can be used, while for better-class work lead or copper is more usual.

Weathering a porch canopy in sheet lead

A small double-pitched canopy is shown in Fig. 11.21, and as its superficial area is not likely to be too large the method used shows it made in one piece. To make this weathering, an upstand of approximately 200 mm is first turned on edge to abut the wall. A bulge is formed at the apex in the upstand in a similar way as that when forming a saddle (see Fig. 11.16(a)). Before fitting the canopy, some cleats or a continuous fixing strip should be nailed round the front and side edges of the canopy. This will allow the front and side edges to be secured and prevent the lead being lifted in high winds, this type of fixing is shown in detail in Fig. 10.25(c). When the canopy is fitted and the two sides dressed down, the lead forming the bulge will flatten and can be dressed back to the abutment.

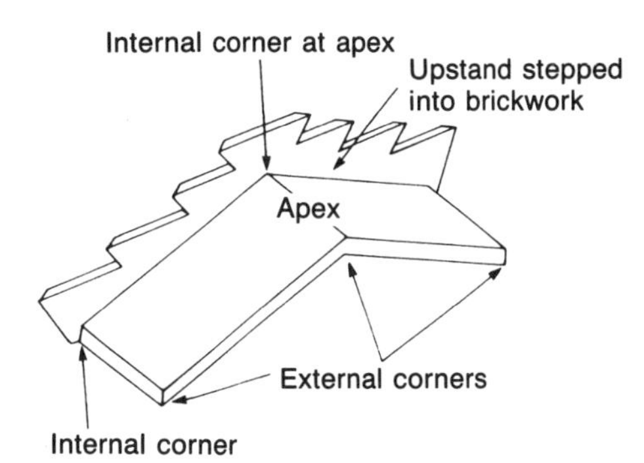

Fig. 11.21 Small lead canopy

The next step is to work down the bottom edge adjacent to the abutment. This is really an internal corner and a gain in material is necessary. The method used in shown in Fig. 11.22(a) and is not unlike that used to form the overcloak to a drip. A bulge is made in the upstand which is bossed downward in the direction of the arrow, taking care not to crease the lead in the corner. When the bossing operation and trimming has been completed, the lead should appear as seen in Fig. 11.22(b).

To turn down the front edge of the canopy, the three external corners shown have to be bossed, and should be worked down using the methods described in Chapter 10. If these corners are worked down in position the lead should be fixed in place by suitable clamps to prevent it easing forward during this last operation. Should this be allowed to happen a gap will occur between the lead upstand and the brickwork which will be difficult if not impossible to correct.

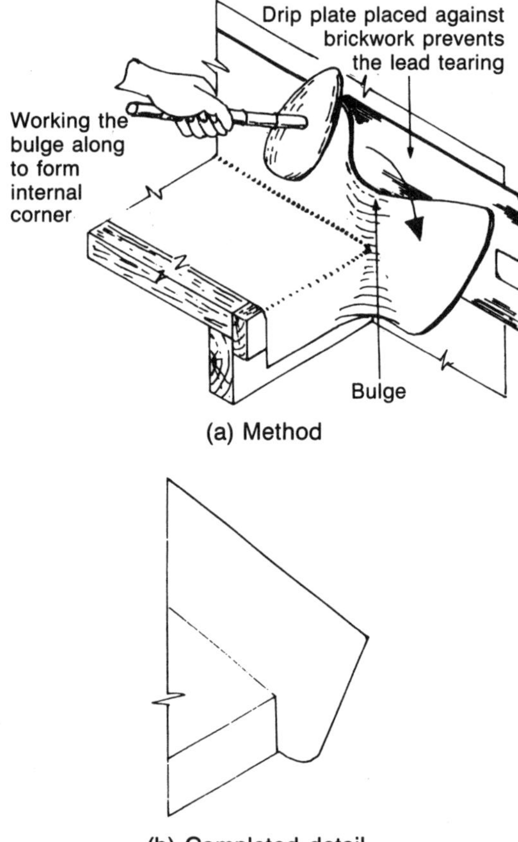

Fig. 11.22 Working the internal corners on a lead canopy

The next operation is to mark out the flashing to the upstand for turning into the brickwork in the same way as step flashing before it is fitted to the decking. The canopy is finally secured by wedging the steps into the brickwork and turning the front and side edges under the continuous fixing strip. If larger canopies of this type are weathered it may be necessary to use an expansion joint at the apex. This can take the form of a welt or a wood-cored roll. It will also be necessary to use separate flashings, as only on very small canopies is it permissible to turn the upstand directly into the abutment.

Sheet zinc work

Two types of zinc are available, zinc lead alloy sold in strips of various widths, and titanium zinc which is marketed in standard sheets of 2.438 m × 0.914 m. These are the metric conversions of the imperial 3 ft × 8 ft sheets. The main use of zinc lead alloy is for flashings, because, being softer, it can more easily assume the sometimes complex details required. When it is used for this purpose, the techniques of working are similar to those of lead sheet when lead welding is employed, only instead of welding, the gussets are soldered in. This material, like other roofing metals, can be used effectively for cladding large areas where a technique employing standing seams is used. This, however, is outside the scope of this book and the method of weathering using the traditional roll cap system with titanium/zinc is described here.

This material has no malleability whatsoever from a practical point of view and the method of jointing takes this into account. It is worth mentioning at this stage that all the details shown can be made up using stiff card and a pair of scissors. This is useful as they can be adapted to most of the situations one meets in weathering using this material, and for the reader who is not familiar with sheet zinc, practising on paper is a lot cheaper than making mistakes on the actual job.

Like other thin sheet metals, zinc can be particularly vicious to the unwary; in short, it can be the cause of some nasty cuts, so watch out for sharp corners and either file or cut them off when they occur. Always have a first aid kit at hand!

Weathering a small canopy or dormer top A bench having a steel angle on one edge will be found useful, not only for zinc but for all other hard metal sheetwork. Some pieces of timber will also be required for use as holding-down blocks. The first step to take after preparing the decking is to make and fix the edging or drop apron. This is shown in Fig. 11.23(a), (b), (c) and (d). The 'T' section dimensions shown will usually be found to be satisfactory. Because zinc is easier to turn across its grain, the usual practice is to cut strips of the required width across the sheet. First turn a stiffening edge of approximately 10–15 mm then mark off the vertical section of the apron shown at y in Fig. 11.23(b) to the required dimension. Figure 11.23(c) illustrates the four steps necessary to form the apron. The purpose of the protruding edge section, 25 mm wide, is to secure the beaded edge of the bays when they are fitted. Where joints in the apron are required, the turns on one end are open slightly, the next piece being slipped into it to about 20–25 mm in depth – this is illustrated in Fig. 11.23(d). To alter the direction of the apron, as for example when a corner is made, simply cut across the top of the 'T' section and turn it to the required angle. Where it meets the abutment it should be cut and turned in a similar way, opening at the 'T' section in contact with the abutment as shown in Fig. 11.23(a). It will be seen that a diagonal cut is made on the top section so that the welted gusset shown in Fig. 11.28(b) can be accommodated when the bay is positioned. This is another of the examples where a little practice with cardboard and scissors will be helpful.

The apron sections should now be fixed using suitable galvanised or stainless steel nails at 1.25 mm centres. The batten roll should also be fixed at this stage using corrosion-resistant nails or screws.

Figure 11.23(a) shows the cleats under the roll which, when fixed, should be turned up either side of it. They will be turned over the roll upstands to secure the bays when they are fitted. It should be noted that a gap of about 10 mm is left between the end of the roll and the abutment to accommodate the dog ears on each bay. Similarly the front end, splayed at about 60°, should be fixed at approximately 20 mm short of the front edge of the apron, which must also be cut away here to accommodate the half stops.

Preparation of the bays The next step is to measure and set out the dimensions of the bays. Figure 11.24 illustrates this and it will be noted that four different details are shown which are dealt with separately. It is recommended that the shaded areas should be cut out before any forming is carried out. At this stage it should be noted there is a sequence in forming these zinc bays, the rule being that the beaded edges should be formed first. Note also that measurements x and y are taken from the outside of the 'T' section apron, not the actual decking. This is because the apron overhangs the decking on the front and side by 25 mm. Detail (1) is shown in Fig. 11.25(a) and shows how the corner of the beaded edge is formed. It is first set out with a compass as shown, and if this is done accurately the cut-out can be used as a template for other corners. The 40 mm allowances for the beaded edge are then dressed up through 90° simultaneously, knocking back the section shown as x–x to form a semi-internal dog ear. This is a tricky operation and a little practice will be necessary to get it just right. The corner should now look as shown in Fig. 11.25(b). The beaded edge can now be completed by dressing it over a steel tube or bar about 18 mm in diameter (a piece of the old $\frac{3}{8}''$ steel tube about 1 m long is just right for this job). Figure 11.25(c) shows this operation and Fig. 11.25(d) shows the completed corner. An alternative method of forming this detail is to mitre the corner and join the mitres using solder, but the method shown saves time and is a stronger job. Detail (2) shown in Fig. 11.26(a), (b) and (c) illustrates the formation of the half stop; this is the section that occurs where the beaded edge adjoins the upstand to the batten roll. Figure 11.26(a) shows how it is set out. The bay is then laid on

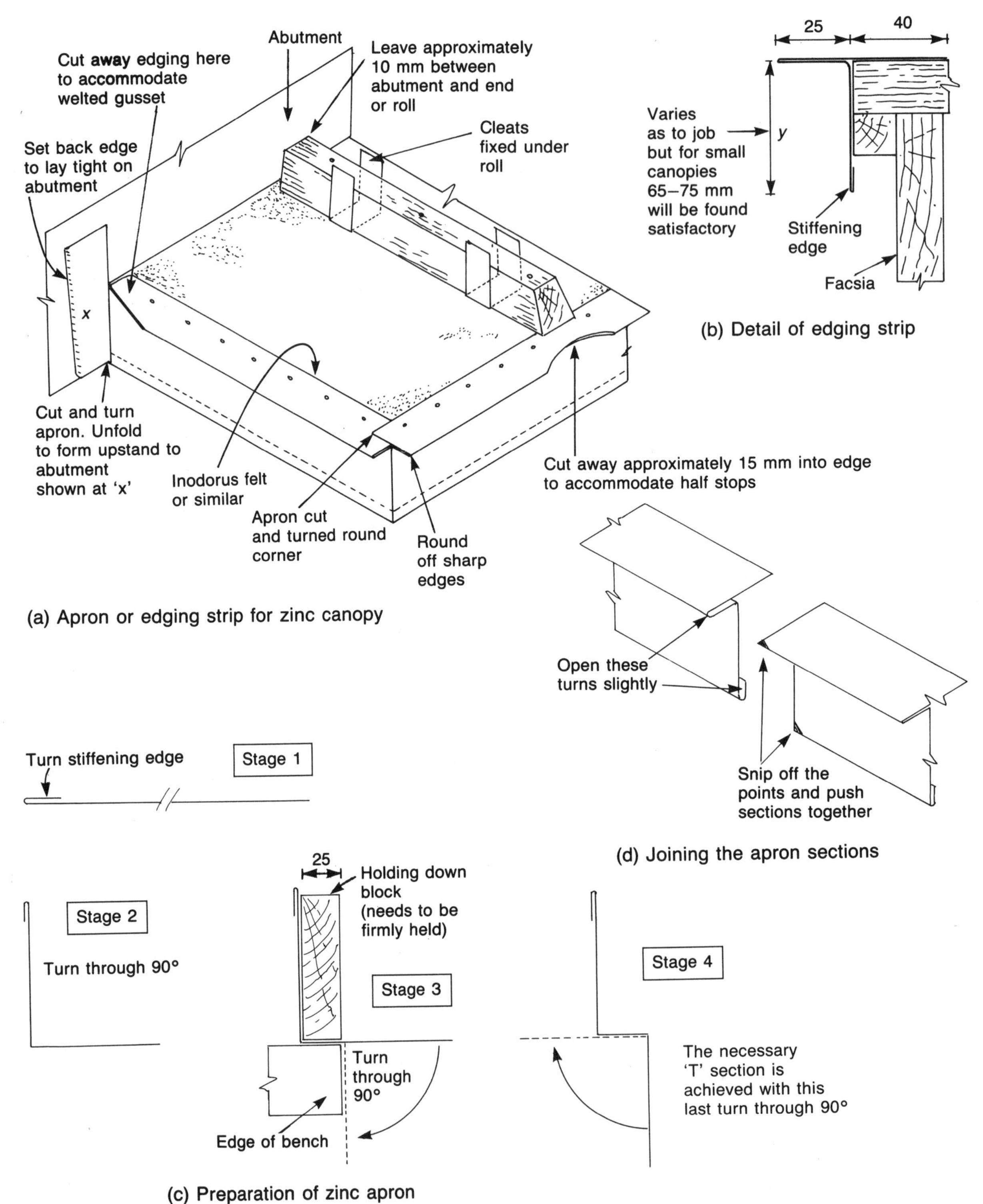

Fig. 11.23 Edging strips

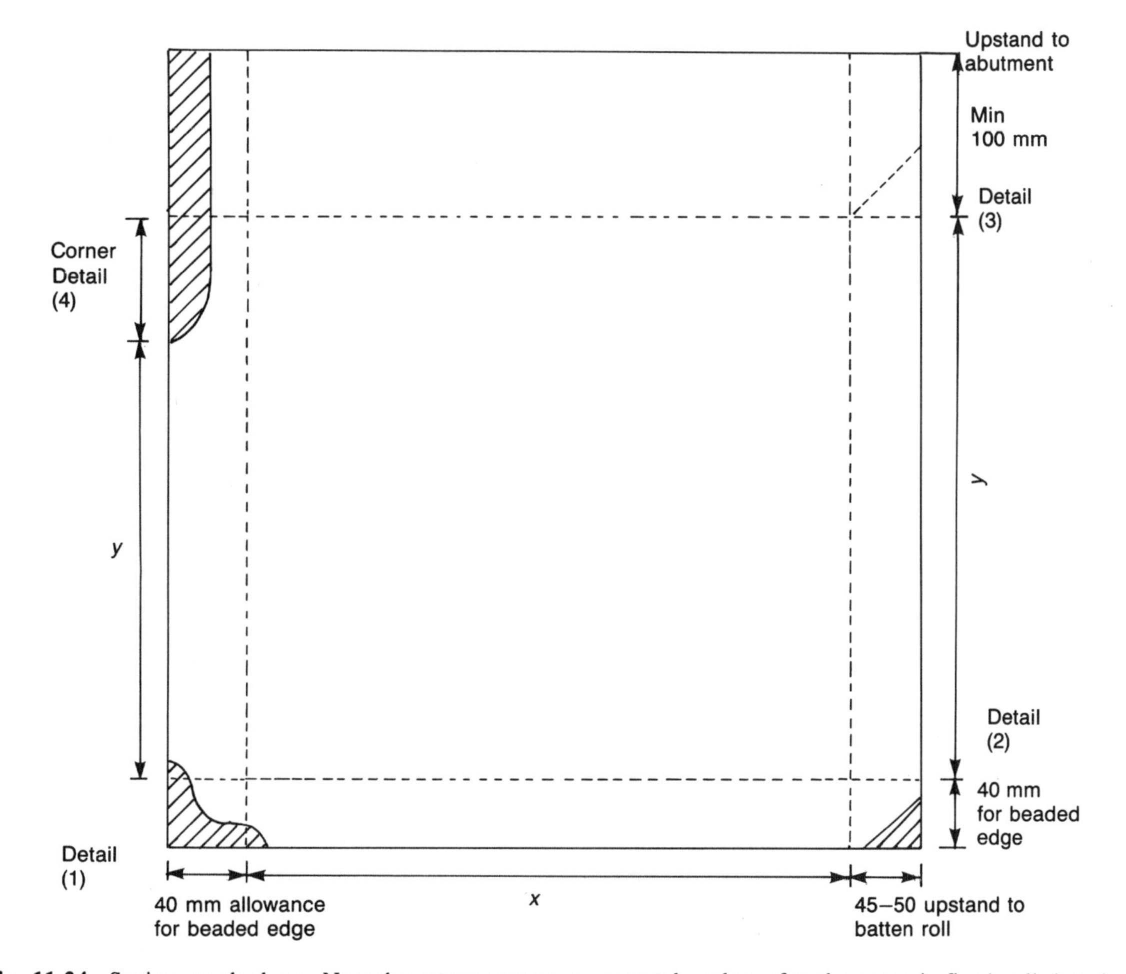

Fig. 11.24 Setting out the bays. Note that measurements $x-y$ must be taken after the apron is fixed. All dotted lines indicate folds or turns.

the edge of the bench, and holding it down with a wooden block (see Fig. 11.26(b)), the upstand to the roll is dressed down. This has the effect of pulling in the beaded edge and it is for this reason the semi-circular cut-out is made on the apron in front of the roll. Figure 11.26(c) illustrates the completed detail and again a little practice with card and scissors will be helpful here.

The dog ear shown in Fig. 11.27 needs little explanation. It is made using a forming block as shown in Fig. 10.19. It should be noted that these dog ears are not turned back on themselves as is normal, but left open so they can be fitted in the 10 mm gap between the end of the roll and the abutment. The next step is to form the detail where the edge of the bay abuts the wall. To do this a gusset piece is marked out and the shaded area removed to the dimensions shown in Fig. 11.28(a) and turned over to form the undercloak of a welt. The cut-away area should be dressed down so that the beaded edge, which has already been turned, is dressed down so that it gradually merges into a welt overcloak. The gusset may then be positioned and welted in. If a seaming tool suitable for setting in the welt is not available, a wooden forming block having an appropriately sized groove cut in it provides a suitable alternative. To prevent movement the gusset can be locked into

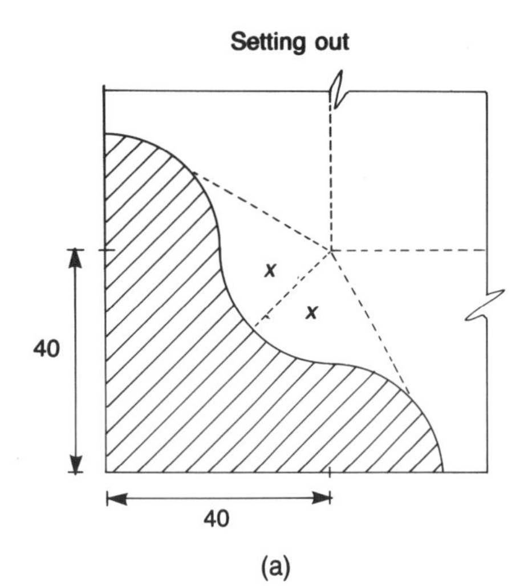

(a)
Set compass to 20 mm to scribe arcs. If set out accurately the cut out can be used as a template for other corners.

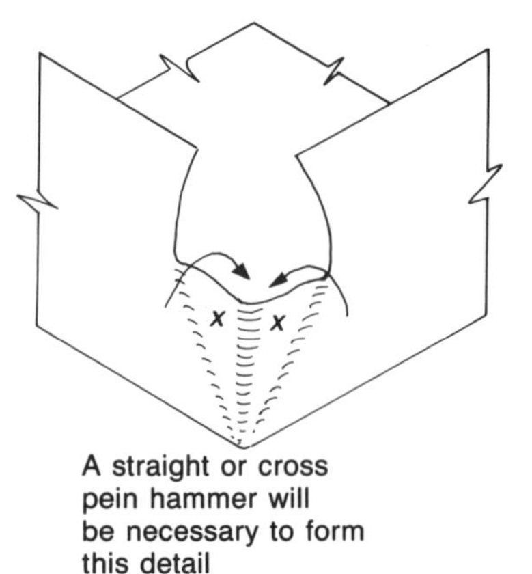

(b)
The two sides forming the beaded edges are turned so that an internal dog ear is formed. The two upstands are turned over a steel bar, as shown, prior to closing up the corner with a tinsmith's mallet.

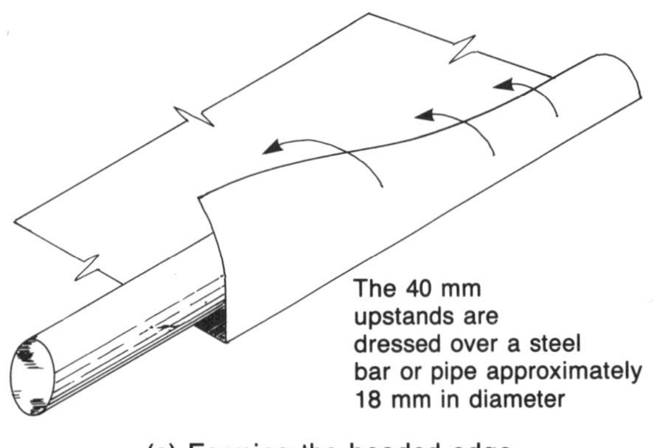

(c) Forming the beaded edge

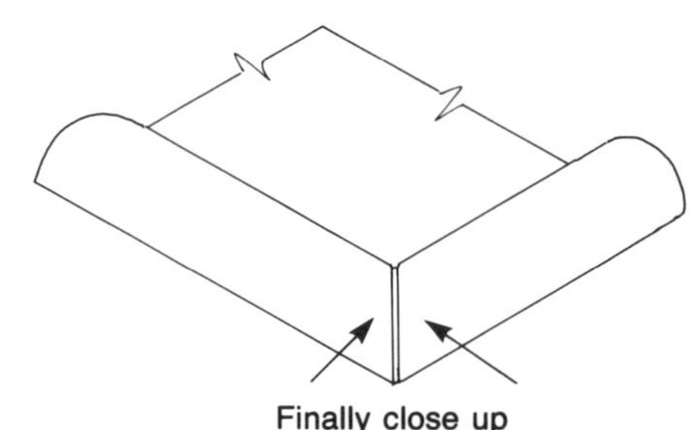

(d) The finished corner

Fig. 11.25 Forming the corner (Detail (1))

place with a centre punch, or alternatively soldered on the back. The detail should now look as shown in Fig. 11.28(b). Laying the bay on the bench and using a wooden block to hold it down, the upstand to the abutment can now be dressed up as shown in Fig. 11.28(c). Finally turn down the rounded edge of the gusset to form a drip edge and set back the side as shown. This causes the gusset to lie tightly against the wall. It is not usual to double the edges of zinc as with copper or aluminium, as being a stiffer material it is, to some degree, self supporting.

The completed bay can now be positioned on the canopy as shown in Fig. 11.29, the protruding edge of the apron being tucked into the beaded edge of the bay which holds it down. The cleats are also turned over the upstand of the roll at this stage.

Roll capping Assuming the other bay has been prepared and fixed, the next step is to form the roll capping. This is best obtained ready made as it cannot be easily formed by hand. A section of roll capping is shown in Fig. 11.30. Note the set-in edges causing it to lie tightly against the roll upstands. The termination of the roll at the upstand is shown in Fig. 11.31(a) and is formed as shown in Fig. 11.31(b), (c) and (d). The end of the roll

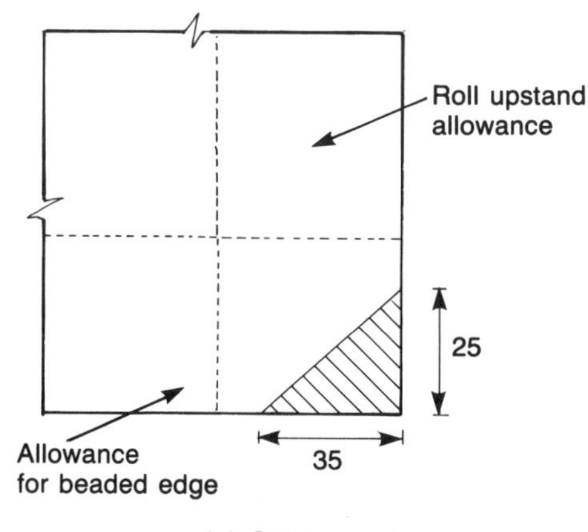

(a) Setting out

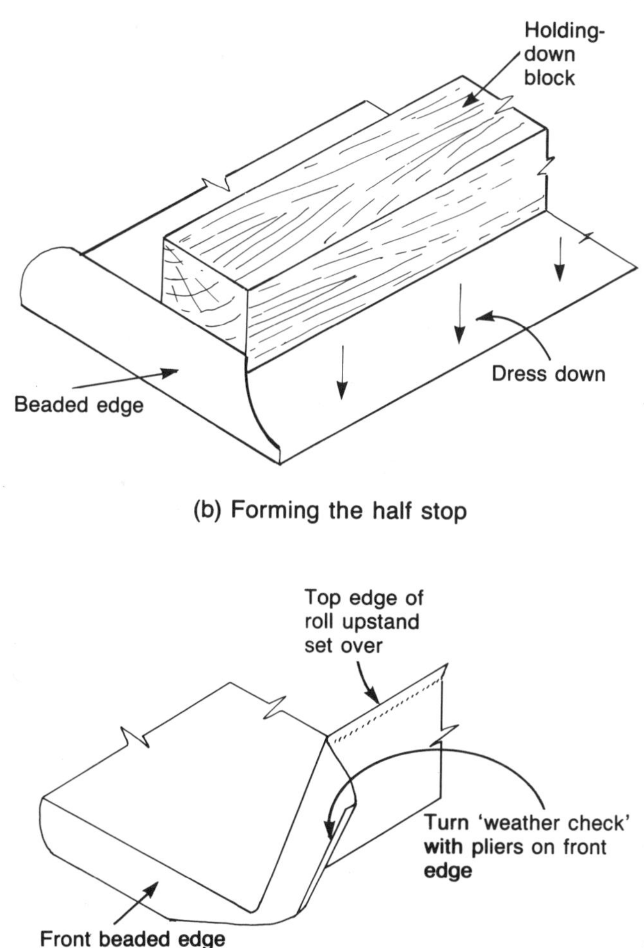

(b) Forming the half stop

(c) Completed half stop

Fig. 11.26 Half stops (Detail (2))

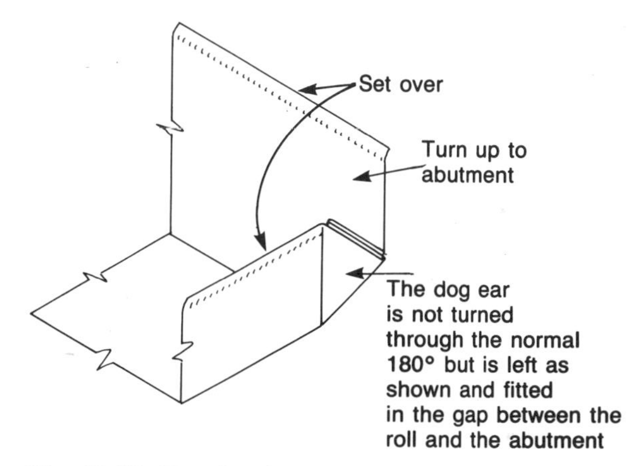

Fig. 11.27 Forming the dog-ear corner (Detail (3))

is flattened for a distance of about 150 mm, depending on the height of the upstand and marked out as shown in Fig. 11.31(b). The broken lines indicate the doubling which must be made to form the capping as shown in Fig. 11.31(c). It is then placed over a wooden former (a length of batten roll is ideal for this job), pressing down the sides when it will be seen that the doubling begins to open. At this stage a hammer blow directed at B will start the turn shown by the broken line (see Fig. 11.31(d)). The capping can now be removed from the wooden block and formed up to assume the shape shown in Fig. 11.31(a); a pair of pliers will be helpful here.

The roll end at the eaves is made on a forming block. These blocks can be made of hardwood for the one-off job, but those made of cast iron or steel are still available and will last a lifetime. To make this detail a length of capping is placed in the block with an overhang of approximately 150 mm. This measurement may vary slightly depending on the height of the batten roll and the angle to which it is cut. The end is opened out in a similar way to that when forming the upstand. This operation is shown in Fig. 11.32(a). The overhang is then pressed down, simultaneously forcing the sides back over the front of the forming block so that, in effect, two internal dog ears are produced. These are indicated by the broken lines seen in Fig. 11.32(b), which shows the completed roll end with the bottom edge shaped so that it clips under the beaded edge.

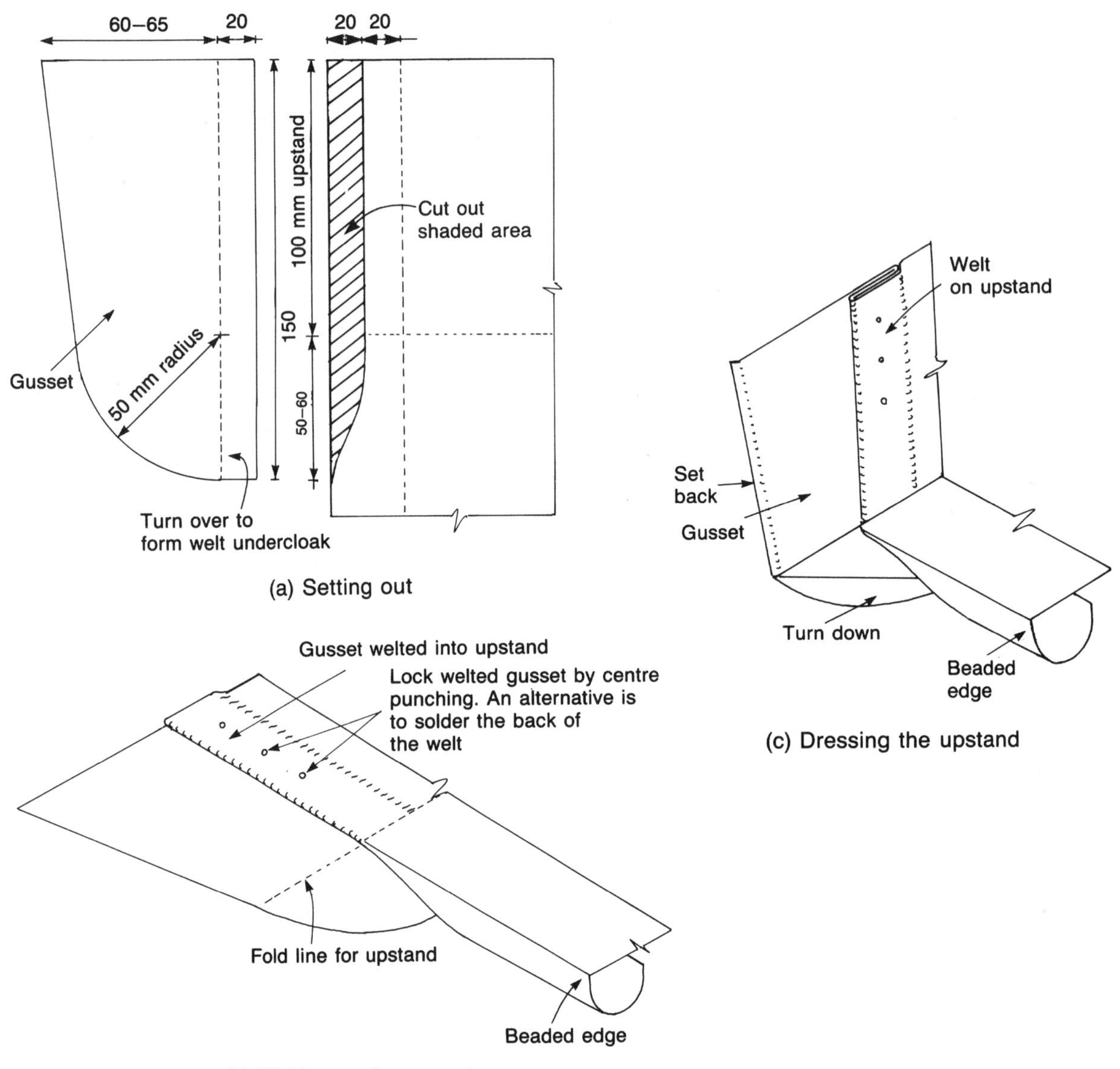

Fig. 11.28 Forming (Detail (4))

Holding-down clips The main purpose of these is to provide a fixing for lengths of roll capping, but they can also be used to join together two short lengths of capping, thus reducing waste. They are made from offcuts of zinc sheet to the dimensions shown in Fig. 11.33(a). If possible arrange for the grain of the zinc to run lengthways through the clip, as this makes it easier to form. The front edge of the clip is turned back on itself as shown in Fig. 11.33(b) prior to placing it on a section of batten to dress the sides down. Figure 11.33(c) shows the completed clip and how it is fixed.

Cap flashing The final operation is to weather the upstand against the abutment using cap flashing. As with the apron sections it is usual to cut the

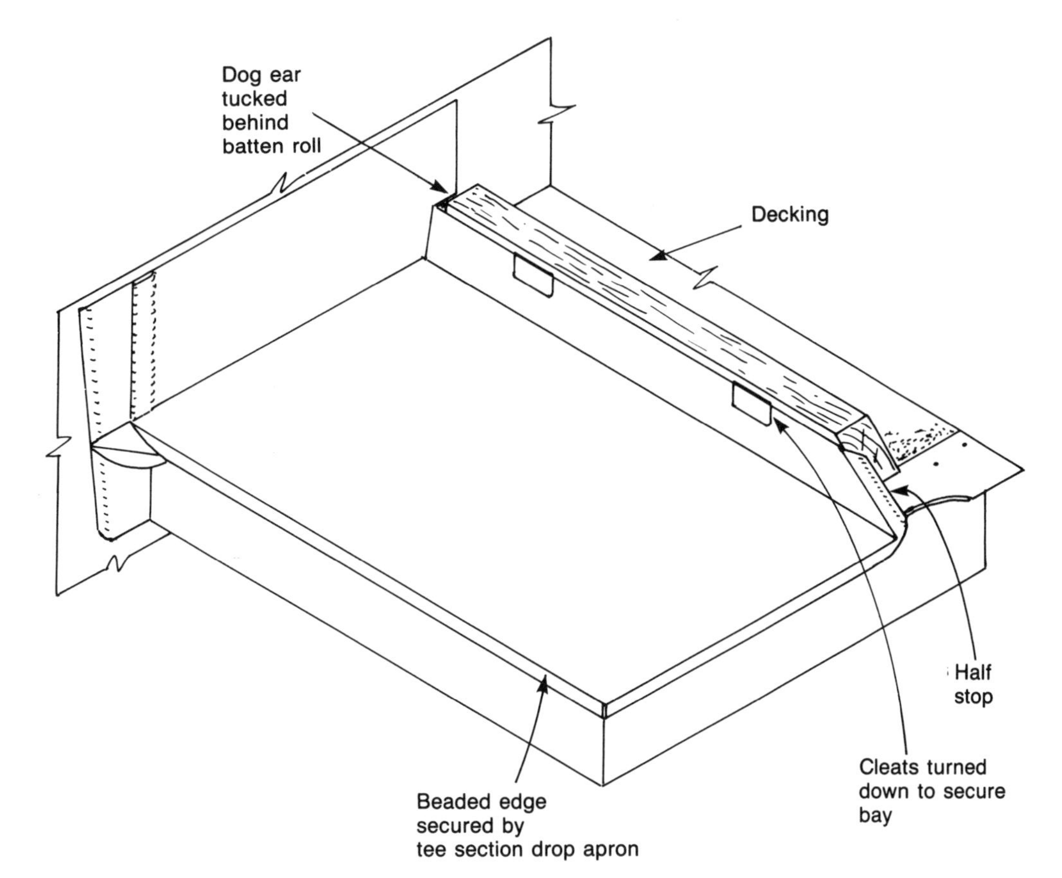

Fig. 11.29 One bay fitted, indicating the relative positions of all the details

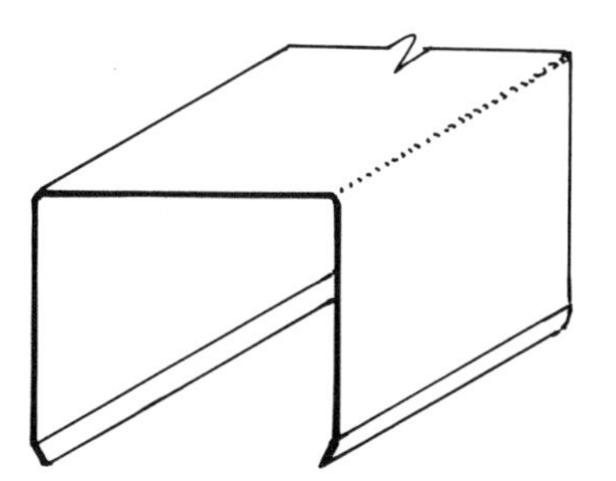

Fig. 11.30 Section through roll capping

length of flashing across the sheet, doubling the lower edge to afford some degree of stiffening, the top edge being turned into the abutment 25 mm. Each section of flashing should be lapped 100 mm and fixed with rolled up zinc or lead wedges. Cleats should be provided at approximately 600 mm centres to secure the flashing to the upstand of the bays. A section of the completed canopy is shown in Fig. 11.34.

Weathering flats and canopies with sheet copper and aluminium

The weathering of small areas in these materials will present no difficulty if basic principles are observed. As stated earlier in this chapter, it is necessary to be very precise when measuring and setting out with these materials. Figure 11.35 shows a canopy in an angle of a building weathered in copper or aluminium using traditional methods.

Before any sheet metal is fixed fillet blocks should be fitted as shown in Fig. 11.36(a). They are used to avoid the difficulties of forming 90° angles where different surfaces of the sheet intersect. As they have to withstand a considerable amount of hammering when the apron welt is

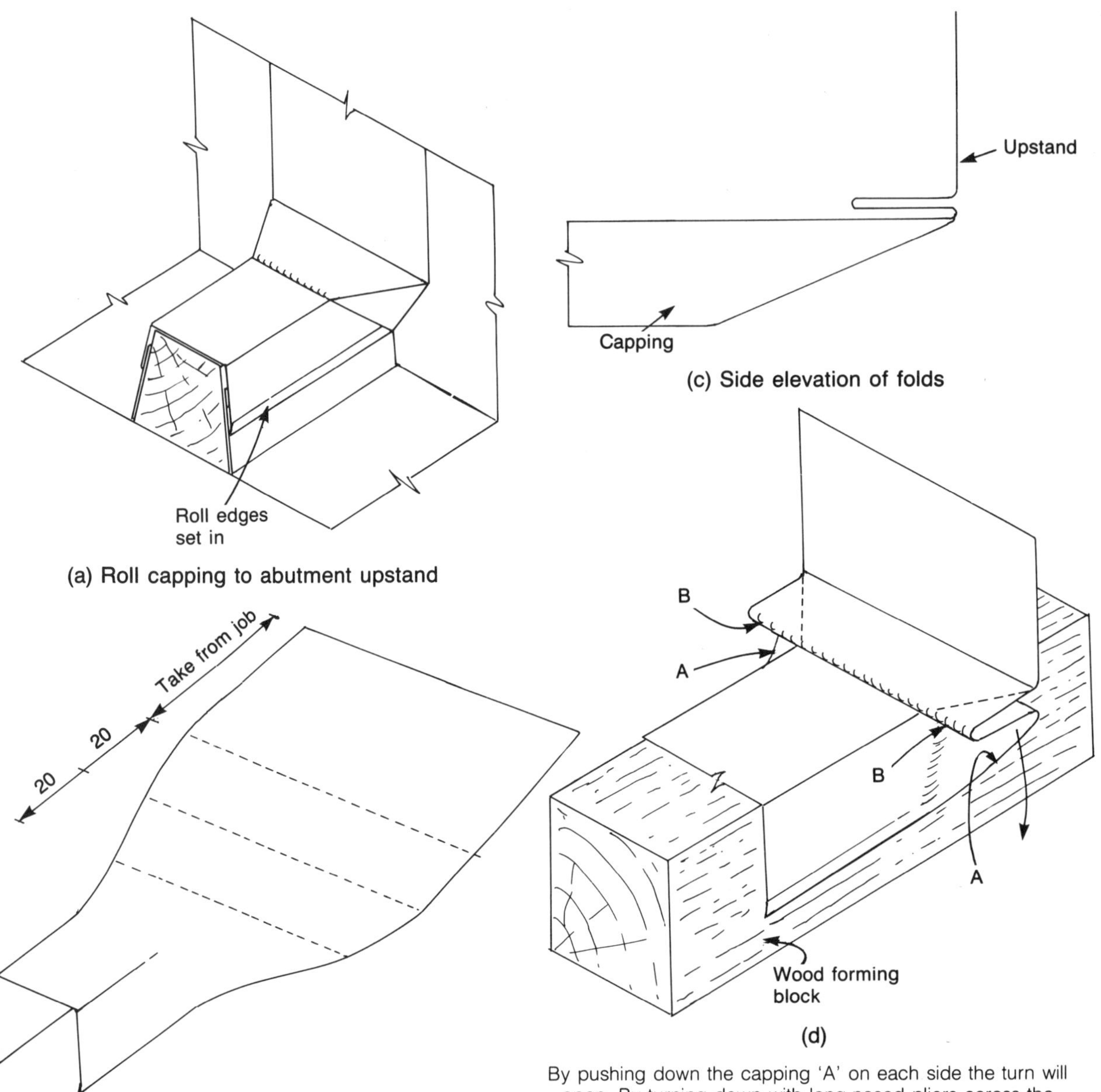

(a) Roll capping to abutment upstand

(c) Side elevation of folds

(b) Setting out the opened up section of capping for forming the upstand to the abutment

(d)

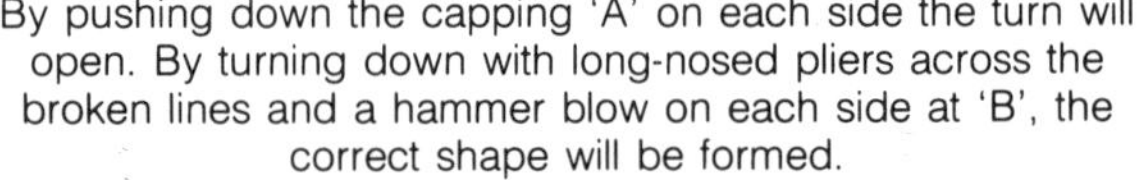
By pushing down the capping 'A' on each side the turn will open. By turning down with long-nosed pliers across the broken lines and a hammer blow on each side at 'B', the correct shape will be formed.

Fig. 11.31

made over them, they are best made of hardwood. It is usual both to glue and screw them in position for the same reason.

The apron should now be fixed as shown in Fig. 11.36(b). The ends of the apron are welted to the wings (this name being derived from their shape) where they abut the wall. They should be set-out as shown in Fig. 11.36(c) and are roughly 'L' shaped. As two are normally required for each job they may be marked out on one piece of sheet to avoid waste. It is important to remember that the wings are turned for both right- and left-hand

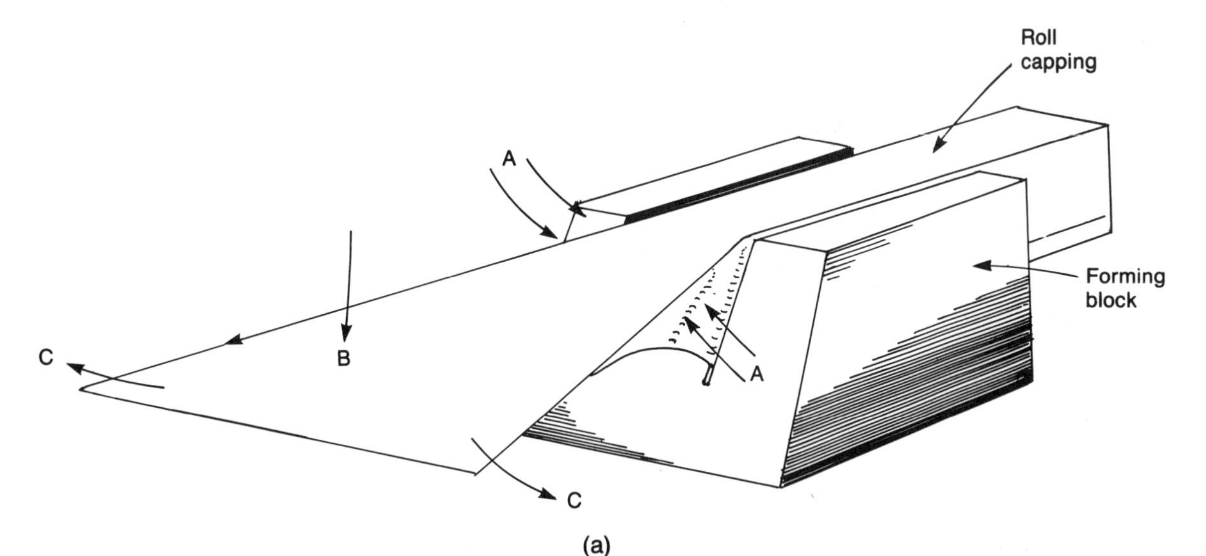

(a)
With the capping lying on the forming block with the front end protruding by about 150 mm, apply blows inward as shown at 'A' using a straight pein hammer, simultaneously pushing down at 'B'. When the angle thus formed is approximately 135° open out the bottom edge 'C' before finally dressing it down on the forming block. This gives the traditional splayed roll end that is characteristic with roll cap roofing.

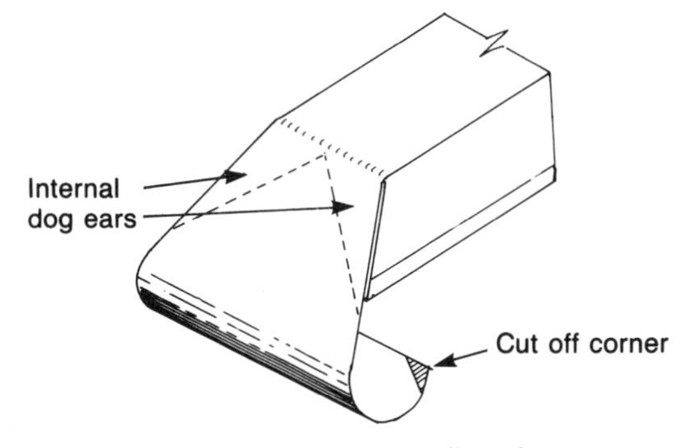

(b) The completed roll end

Fig. 11.32 Forming the front end of the capping

sides – it is both wasteful and time consuming if it is found that two 'right-hand' or two 'left-hand' pieces have been formed. Figure 11.36(d) shows a left-hand wing prepared for welting to the apron.

The bay sheets are then prepared by first turning the upstands for the walls and for the standing seams. Remember to allow enough material for the welt which secures the front and side edges of each bay to the apron shown in Fig. 11.36(b) although the welt must not be made until the standing seam has been completed. The curve or saddle in the welt on the standing seam where it merges with the upstand is marked out using the two purpose made tools shown in Fig. 11.37. They can be fabricated from angle iron and are used to make the two turns required to form the welt. The three stages of forming this detail prior to making the final welt are shown and described in Fig. 11.38.

The only other special tool needed is that used to merge the standing seam into the apron at the eaves which is made from a 25 mm square bar, one end of which is tapered to an angle of approximately 15° (see Fig. 11.39). It is most important when working with copper or aluminium to know where the undercloak can be safely cut to avoid the build up of an unnecessary and unwieldy thickness of material where welted joints intersect. This thickness presents great difficulties when welts are being formed and there are many instances in copper or aluminium roofing where it is necessary and quite safe to cut away some of the superfluous thickness of metal, but it would be impossible to detail all of them. Practical experience with these materials and a little forethought will enable the student to overcome most of the problems likely to be encountered. When in doubt it is essential to check by making the detail on a piece of scrap material first, remembering the joint must be weatherproof. It is

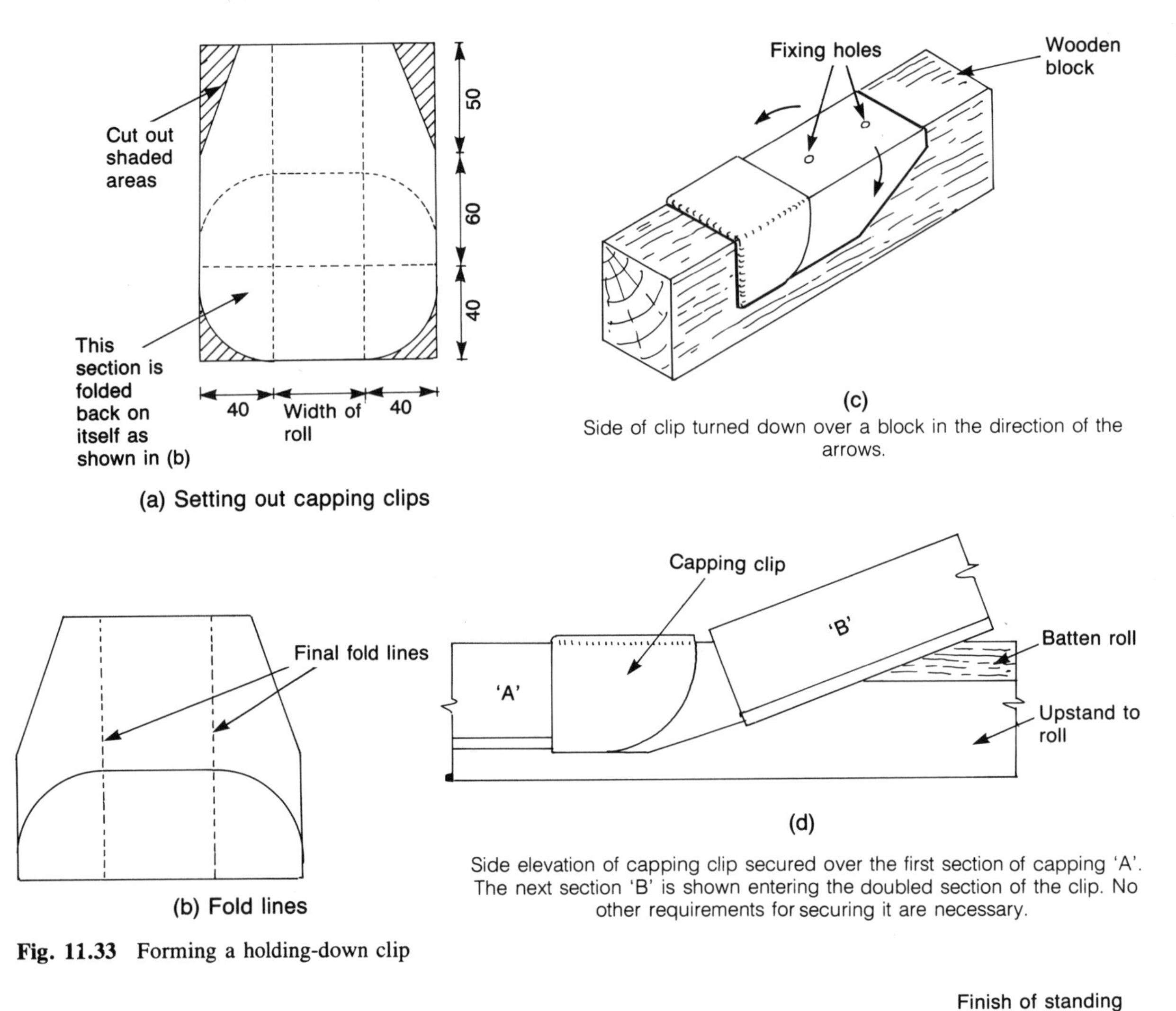

Fig. 11.33 Forming a holding-down clip

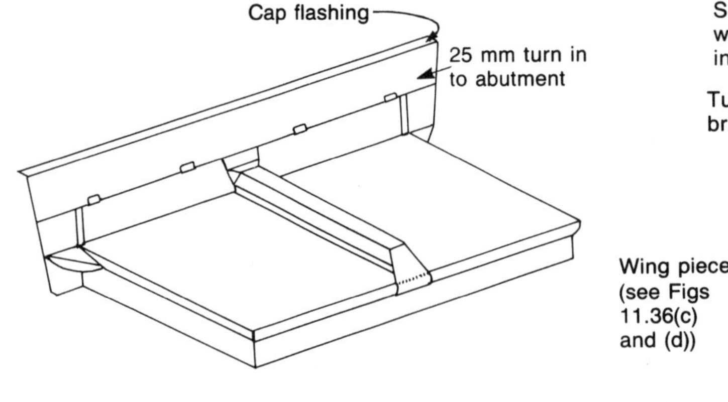

Fig. 11.34 Zinc canopy with all details in position

Fig. 11.35 Small copper-covered canopy

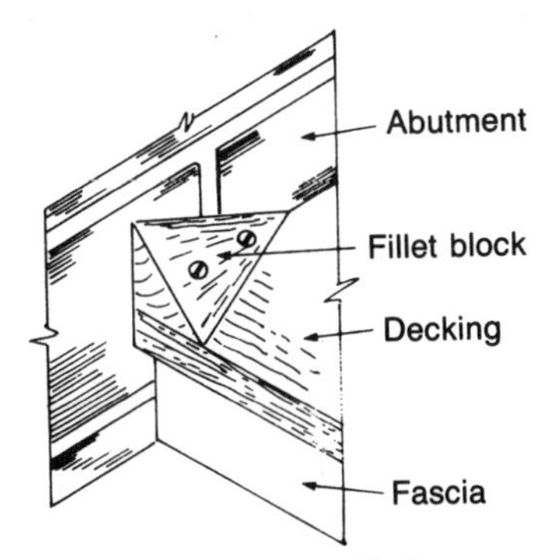

(a) Fixing fillet blocks to avoid sharp corners at the abutment

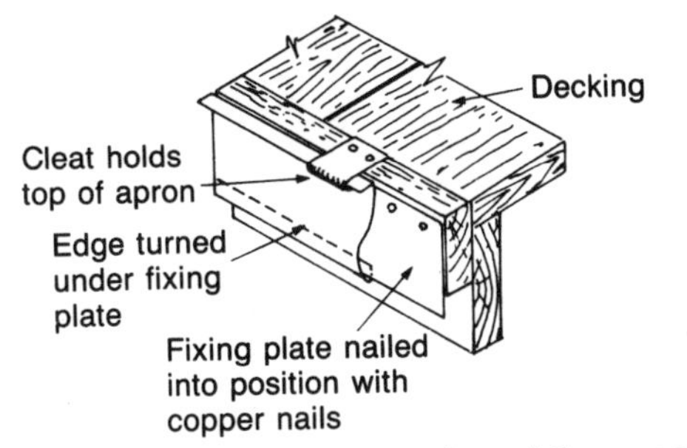

(b) Fixing the apron to the edge of the canopy

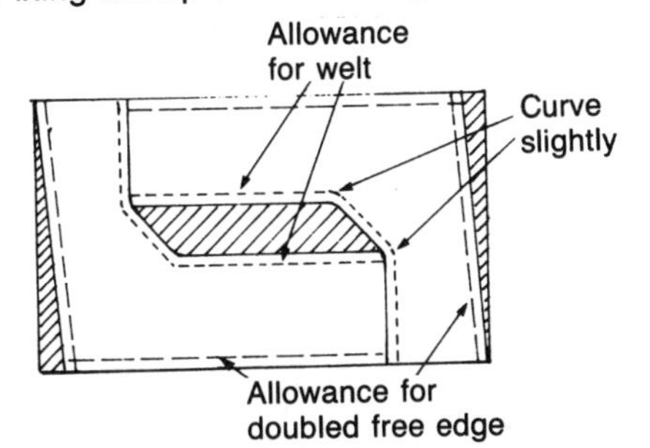

Note shaded area indicates surplus material to be cut away

(c) Setting out the wings to avoid waste

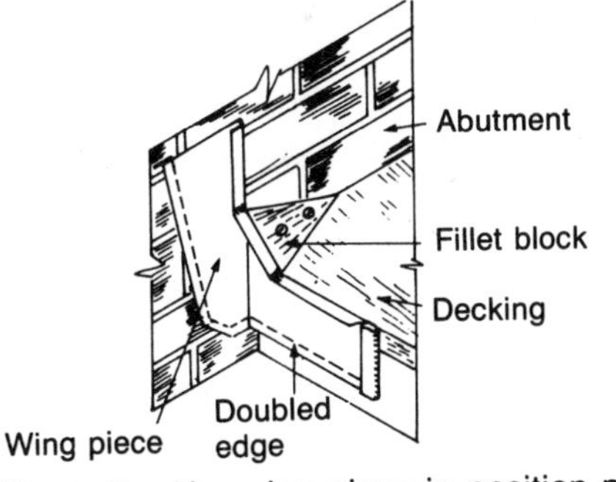

(d) Illustrating the wing piece in position prior to completing the apron

Fig. 11.36 Copper-covered canopy fixings

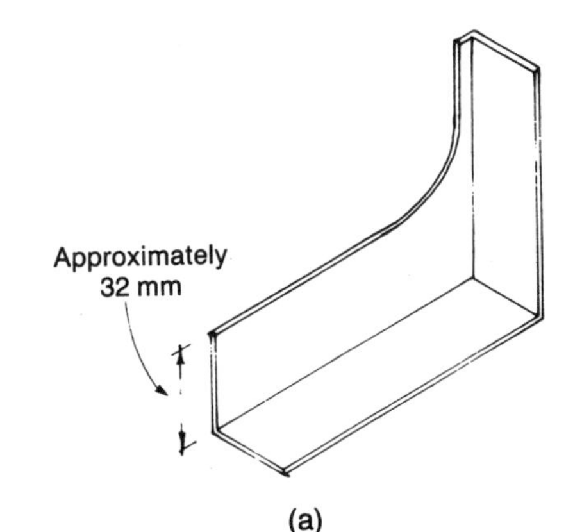

(a)

Purpose-made tools for forming the upstand (saddle) of a standing seam to the butment.

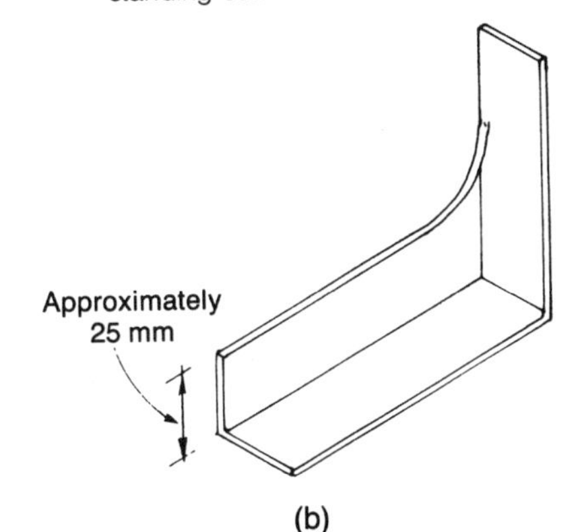

(b)

Tools are made of light angle iron, cut and welded to form corner. Two are required for each turn of the standing seam welt.

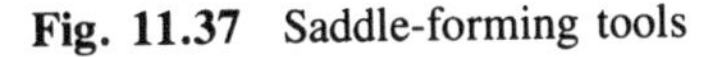

Fig. 11.37 Saddle-forming tools

also useful to remember that many of these details can be simulated off the site with a piece of cartridge or brown paper.

A typical example of removing unnecessary thicknesses, or 'thinning' as it is sometimes called, is shown in Fig. 11.40 where the undercloak of the standing seam merges with the apron welt. First mark and cut out the shaded area prior to the formation of the standing seam. This has the effect of reducing the thickness of material at the end where it turns into the apron welt from five thicknesses to three. When this has been completed, the welt with the apron at the edge of the canopy can be formed. It is closed with a mallet prior to turning down, using the purpose-

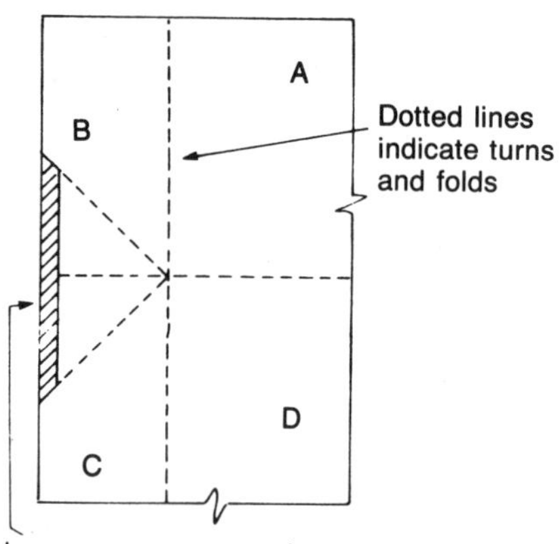

Stage 1
Marking out

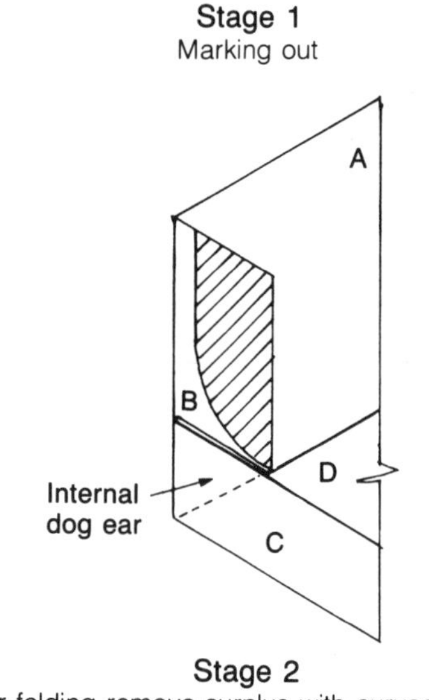

Stage 2
After folding remove surplus with curved snips to form saddle

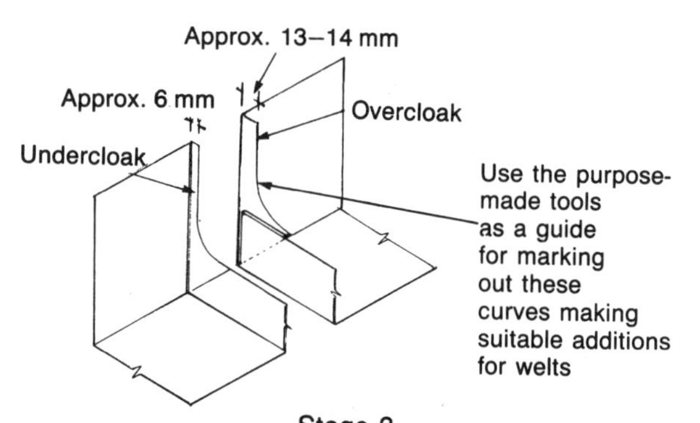

Stage 3
The two upstands are brought together and welted using the purpose-made tools shown in Fig. 11.37

Fig. 11.38 Marking out and cutting for saddle on standing seam

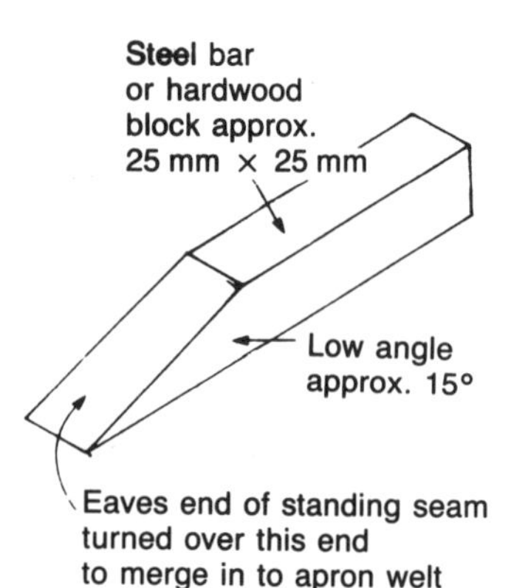

Fig. 11.39 Standing seam-forming tool

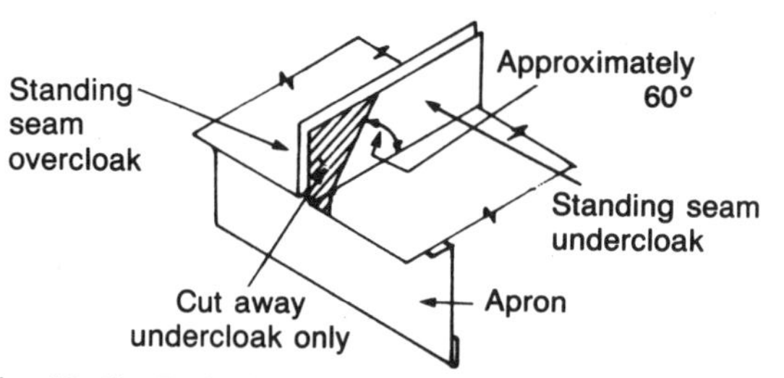

Fig. 11.40 Reducing the thickness of the standing seam welt prior to turning over at the eaves

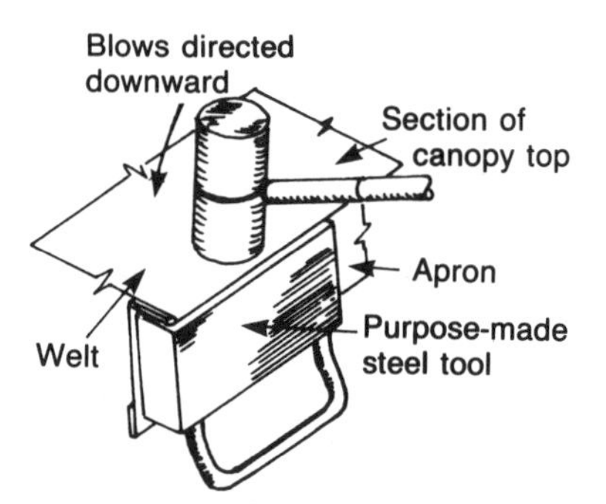

Fig. 11.41 Closing the apron welt

made tool to support the underside as shown in Fig. 11.41.

Next, the corner on the apron welt is formed in the manner shown in Fig. 11.42. Stage 1 shows the method of setting out allowing for the welt and the removal of surplus material which is indicated by the shaded area. Stage 2 illustrates the first turn of the welt being locked on to the apron, while Stage 3 shows one side of the welt with the second turn completed. The curved area of the corner should now be driven under the other side of the

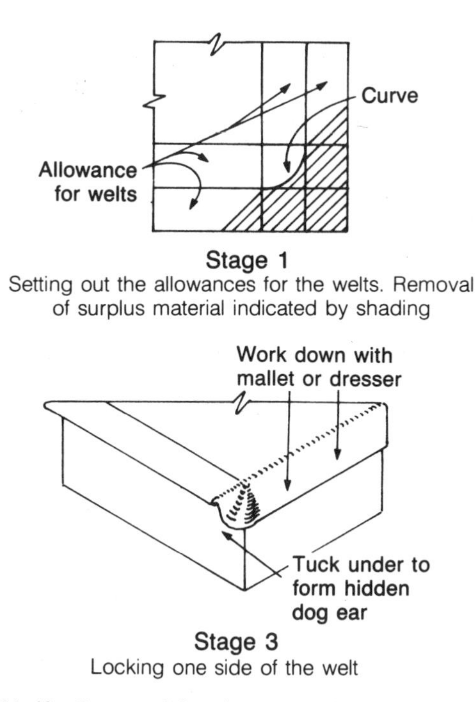

Stage 1
Setting out the allowances for the welts. Removal of surplus material indicated by shading

Stage 3
Locking one side of the welt

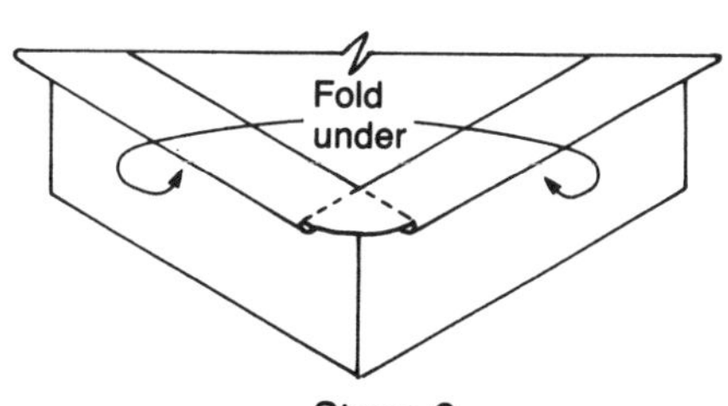

Stage 2
Forming the first stage of the welt

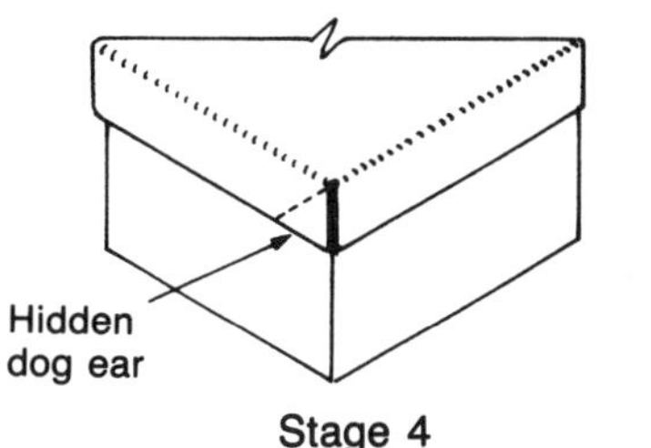

Stage 4
Both sides of the apron welt locked in position

Fig. 11.42 Stages of forming external corners at the apron

welt forming an internal dog ear as it is dressed down. The completed corner is shown at Stage 4, the broken line indicating the internally folded dog ear.

Valley gutters

Tapering valley gutters are those formed between the junction of the eaves of two roofs, and are of the open type being wide enough to allow a person to walk along them if necessary to effect repairs.

Smaller gutters, simply called valleys, occur when a gable intersects with a pitched roof. Figure 11.43 shows in diagram form both types and their relative positions on a roof.

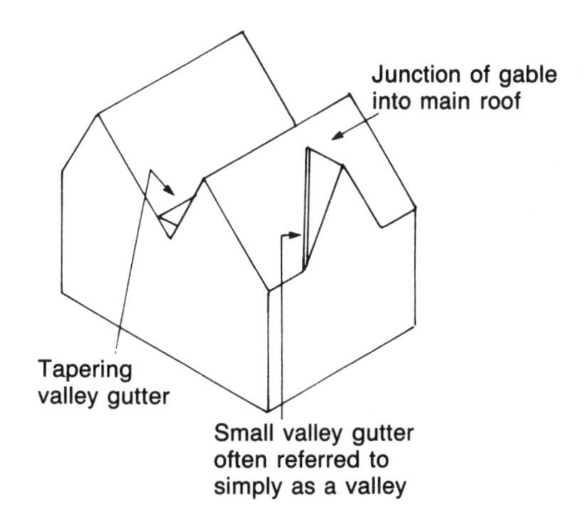

Fig. 11.43 Types of valley gutters

Small valleys can be weathered in a number of ways. For example, when plain tiles are used, it is possible to dispense with metal altogether as specially shaped valley tiles can be used. These are illustrated in Fig. 11.44. The only weathering required in this case will be to cap the intersection of the ridge with the sloping roof in a similar way to that shown in Fig. 11.45 where a metal valley gutter is illustrated. Special tiles are also made to form valleys for roofs weathered with interlocking tiles. However, sheet metal is always used in conjunction with slates and often on tiled roofs where, for one reason or another, it is more suitable than valley tiles.

Figure 11.46(a) illustrates a section through a metal covered open valley with a distance of approximately 50 mm left between the edges of the tiles, while Fig. 11.46(b) shows a secret valley gutter. While the external appearance of the secret gutter is more attractive, they have a tendency to

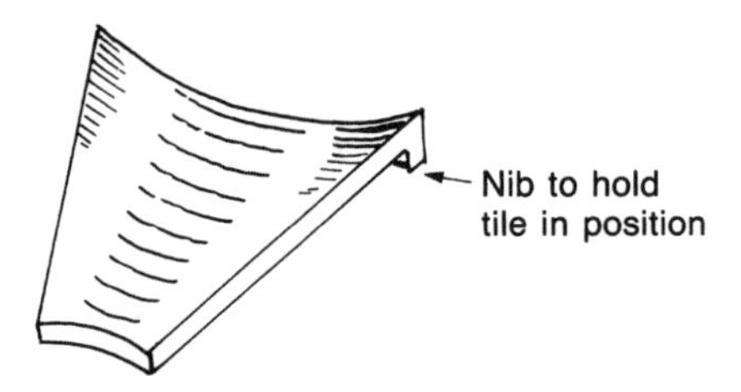

Fig. 11.44 Valley tile. These tiles are shaped to form the valley; each tile being laid as the roof tiling proceeds.

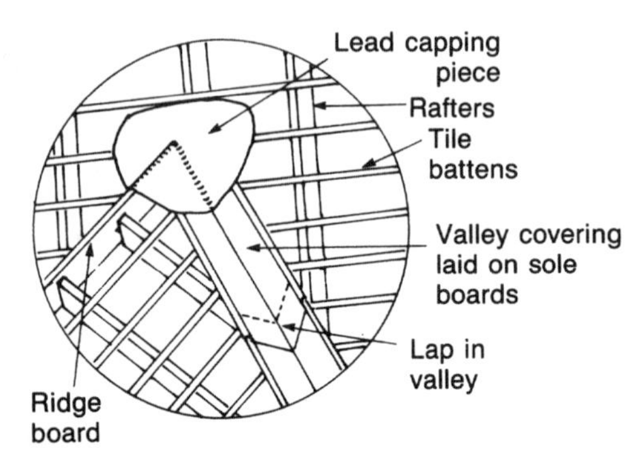

Fig. 11.45 Detail of weathering at the top of valley gutter. Finish with a capping piece

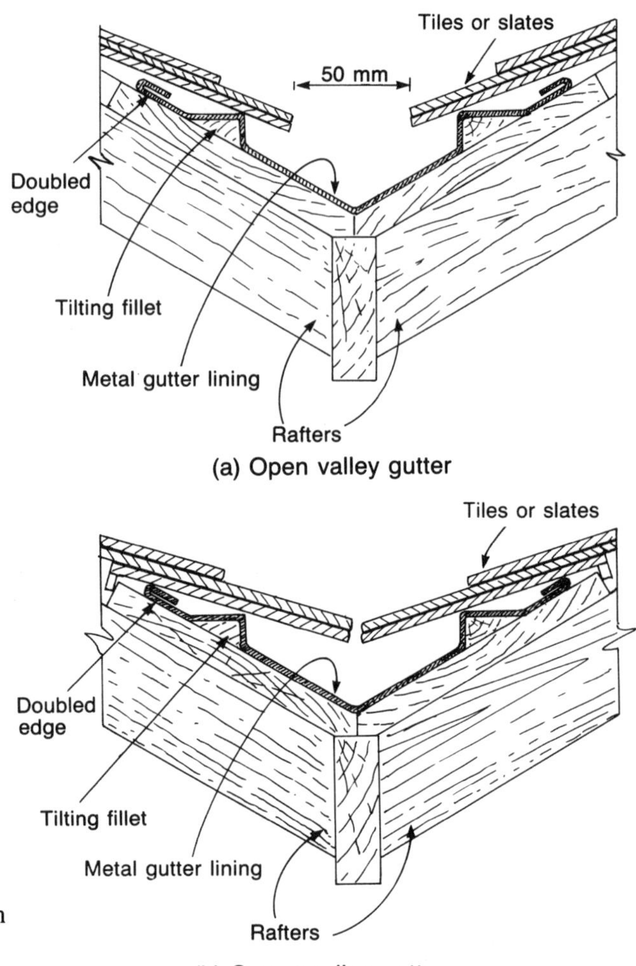

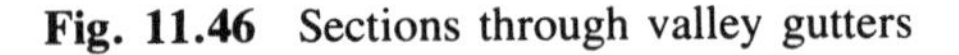

Fig. 11.46 Sections through valley gutters

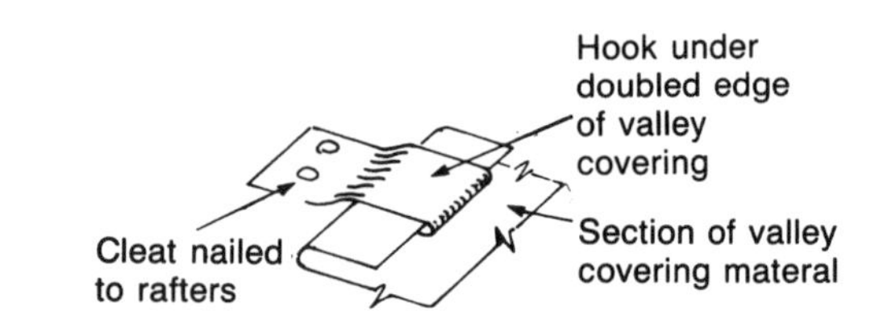

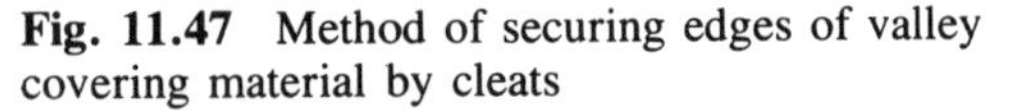

Fig. 11.47 Method of securing edges of valley covering material by cleats

collect debris washed off the roof and often become blocked, causing water to enter the building.

The method of laying valleys is the same in both cases, the covering material being formed or dressed into shape and laid on wooden sole boards, the top edge of each length being secured by nailing. There is no hard and fast rule relating to the length of each section, apart from a recommended maximum of 1.5 m. The turned edges at the sides serve to prevent water flooding over the sides of the gutter and at the same time provide a good method of intermediate fixing while also allowing for longitudinal movement. The edges are held down with cleats, shown in Fig. 11.47 which permit longitudinal expansion. The edges must *not* be nailed down as this will restrict the movement of each length.

Tilting fillets

Sometimes called 'springing', this term relates to fixing a batten under the slates or tiles at the eaves or where the courses are broken by a chimney or flat roof. This batten tilts each slate or tile upwards (hence the name tilting fillet), so the gap occurs at each overlap preventing capillary attraction. Figure 11.48(a) shows how the top edge

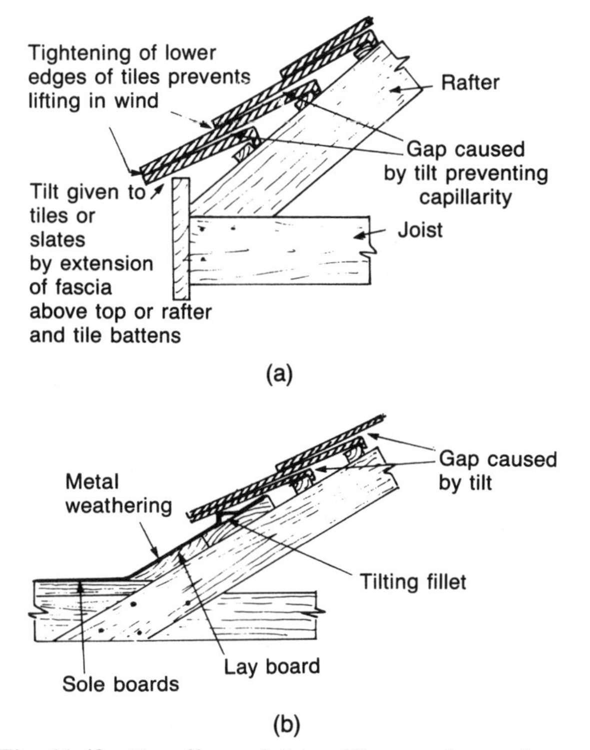

Fig. 11.48 The effects of tilting fillets on tiles or slates to prevent capillary attraction and to maintain a tightening of the lower edges to prevent lifting by wind

of the fascia board is fixed to provide a tilt to the tiles or slates, and Fig. 11.48(b) shows the section of a tilting fillet where the roof is broken by a flat roof or chimney.

Roofing calculations

The plumber should know the procedures for measuring and ordering the materials for a specified roofing job. In order to do this effectively, some knowledge of the methods of measuring and calculating area is necessary. The basic unit of area in the SI system is the square metre shown in Fig. 11.49 and is written as $1\,m^2$. When it is necessary to calculate the area of a surface, the length has to be multiplied by the breadth, and some examples of this follow.

Example 1 Ignoring the welts and turn downs, calculate the surface area of a small canopy measuring 2.0 m by 1.5 m:

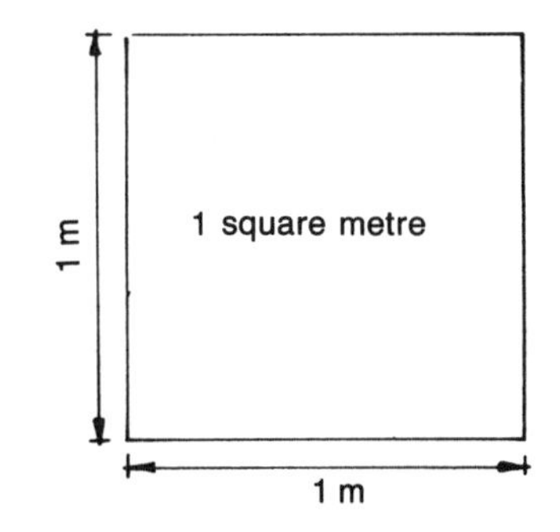

Fig. 11.49 The basic unit of area: 1 square metre ($1\ m^2$)

$$\begin{aligned}\text{Area} &= \text{Length} \times \text{Breadth}\\ &= 2.0 \times 1.5\\ &= 3.0\end{aligned}$$

The answer is 3 square metres or $3\,m^2$.

The square metre is rather a large unit and in many cases a component may have an area of less than this as the next example shows.

Example 2 Calculate the area of a piece of lead sheet measuring 1.0 m × 0.75 m:

$$\begin{aligned}\text{Area} &= \text{Length} \times \text{Breadth}\\ &= 1.0\,m \times 0.75\,m\\ &= 0.75\,m^2\end{aligned}$$

The area will be $0.75\,m^2$ and its relationship to 1 square metre is shown by the shaded area in Fig. 11.50. (In this example the breadth measurement is in fact 750 mm, but may be expressed as 0.750 m. It is necessary to express it in this way as the

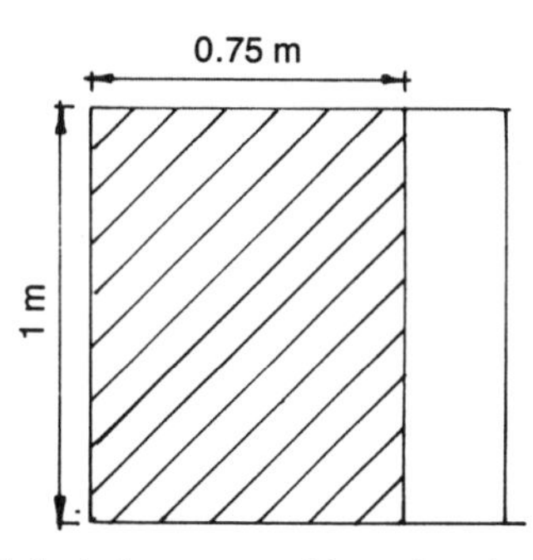

Fig. 11.50 Calculating areas of less than $1\ m^2$ (see text for details)

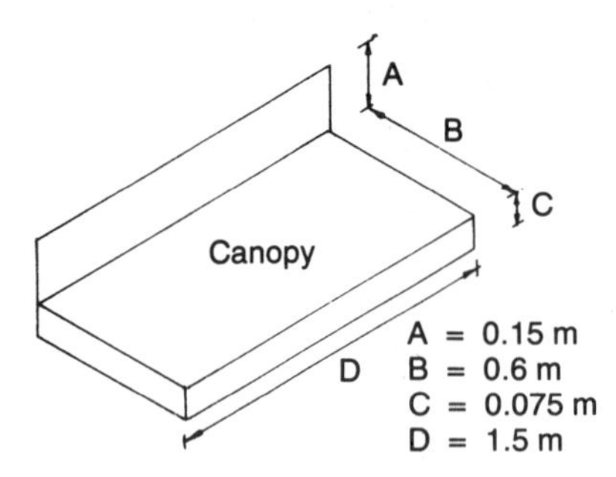

Fig. 11.51 Calculating the area of a small canopy

answer is required in square metres, and metres cannot be multiplied by millimetres.)

From the information given it should now be possible to calculate the total area of material required to cover the small canopy shown in Fig. 11.51. The total width of the canopy is found by adding measurements A, B and C, and multiplying this by the total length, found by adding D, C and C (C will occur twice here as there will be a turn down on both sides):

$$\begin{aligned} \text{Width} &= 0.15 + 0.6 + 0.075 = 0.825\text{ m} \\ \text{Length} &= 1.5 + 0.075 + 0.075 = 1.65\text{ m} \\ \text{Area} &= \text{Length} \times \text{Width} \\ &= 0.825 \times 1.650 \\ &= 1.36125\text{ m}^2 \end{aligned}$$

The area required is 1.36125 m^2, but as usual an approximation is made to the first three places to the right of the decimal point to give 1.361 m^2.

It is sometimes necessary to calculate the area, or partial area, of a circular surface in connection with sheet weatherings. To calculate the area of sheet material required to weather the top of a semicircular bay window, the formula described in Chapter 4 is used, where the area of a full circle is found by using the formula πr^2. In this example only half the area of a full circle is required so the formula must be modified to:

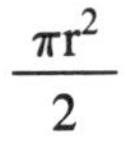

$$\frac{\pi r^2}{2}$$

Example 3 Calculate the area of the semicircular canopy shown in Fig. 11.52 using the measurements given.

Using the modified formula:

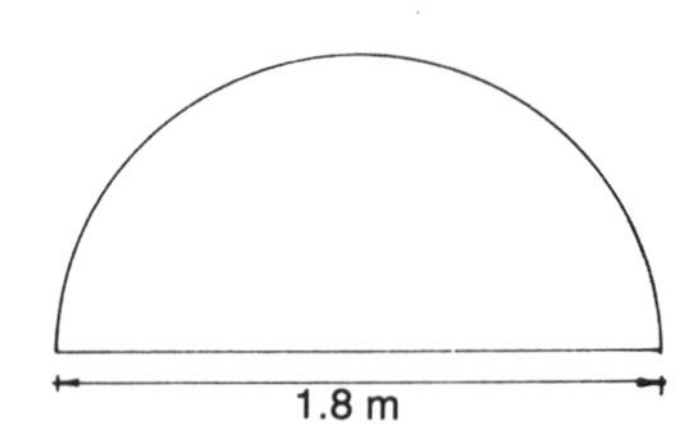

Fig. 11.52 Calculating the area of a semicircular canopy

$$\frac{\pi r^2}{2}$$

The diameter of the canopy is 1.8 m, so the radius will be 0.9 m (radius = $\frac{1}{2}$ diameter). Therefore:

$$\begin{aligned} r^2 &= 0.9 \times 0.9 \\ &= 0.81\text{ m}^2 \\ \pi \times r^2 &= 3.142 \times 0.81 \\ &= 2.54502\ (2.545\text{ approx.})\text{ m}^2 \\ 2.545 \div 2 &= 1.2722\text{ m}^2 \end{aligned}$$

The approximate answer is 1.272 m^2, which is the amount of material required to cover the flat part of the canopy, to which must be added allowances for rolls, welts or seams, depending on the material chosen.

In the next example it is necessary to use the perimeter or circumference of a circle, which is the line that forms its outer edge (see Fig. 11.53). Sometimes it is essential to determine the length of a circumference when the plumber is dealing with circular shapes. Typical examples in weathering occur when the circumference or perimeter of a pipe is to be measured in order to make a pipe flashing, or when the length of an apron surrounding a circular canopy needs to be found.

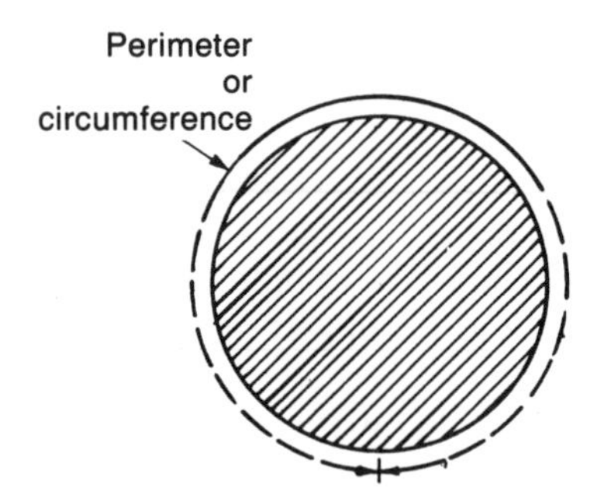

Fig. 11.53 Circumference of a circle

The formula given in Chapter 4 for the calculation of the circumference of a circle is

Diameter × π.

Example 4 Assume an apron having a width of 0.075 m is required for the same circular canopy shown in Fig. 11.53 as only half the circumference is required, the formula can be modified thus:

$$\frac{\pi D}{2}$$

$$\pi \times D = 1.8 \times 3.142$$
$$= 5.6556 \text{ (5.656 approx.) m}$$
$$5.656 \div 2 = 2.828 \text{ m}$$

The length of the apron required is therefore 2.828 m. To obtain the area of sheet needed for the apron, the length must be multiplied by the width of the apron:

$$2.828 \times 0.075 = 0.2121 \text{ m}^2$$

The total area will be approximately 0.2121 m^2.

The plumber is also called upon to weather areas that are triangular in shape, typical examples being the weathering of dormer window cheeks and small canopies over bay windows. Triangular shapes fall into three categories as shown in Fig. 11.54. A point to note is that the angles of any triangle always add up to 180°, the same number as that contained in a semicircle. The same method of calculating the area of triangular figures is adopted whatever its type. The formula may be stated as:

Length of base of triangle × $\frac{1}{2}$ Vertical height

$$\text{Area of triangle} = \frac{\text{Height}}{2} \times \text{Base}$$

Example 5 Calculate the area of the triangle shown in Fig. 11.55:

Length of base = 1.4 m
Vertical height = 2.2 m
(note that this is *not* the sloping height)

$$\text{Area of triangle} = \frac{\text{Height}}{2} \times \text{Base}$$
$$= \frac{2.2}{2} \times 1.4$$
$$= 1.4 \times 1.1$$
$$= 1.54 \text{ m}^2$$

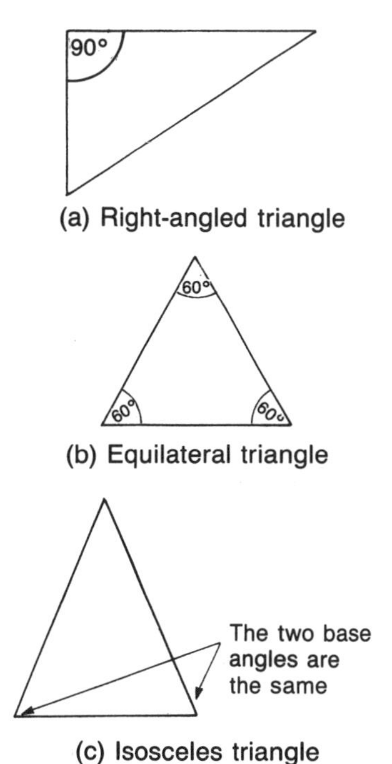

Fig. 11.54 Types of triangle

Another practical application of measurement which integrates calculations involving triangular figures, may be seen in Fig. 11.56 which represents the plan of a small, five-sided canopy over a bay window. The lines show how it can be divided up into five equal traingles, and by calculating the area of one triangle and multiplying by five, the total area can be found thus:

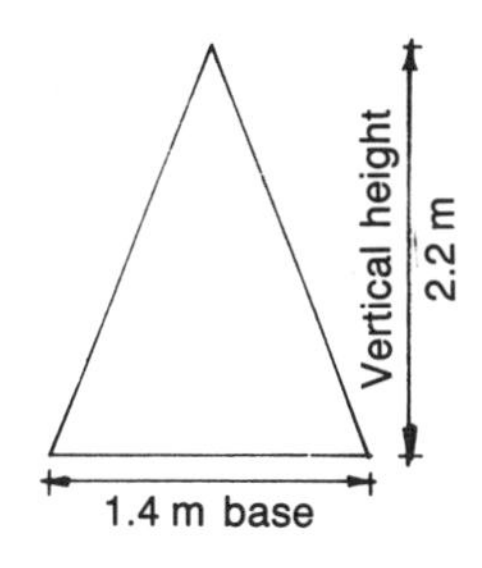

Fig. 11.55 Calculating triangular areas

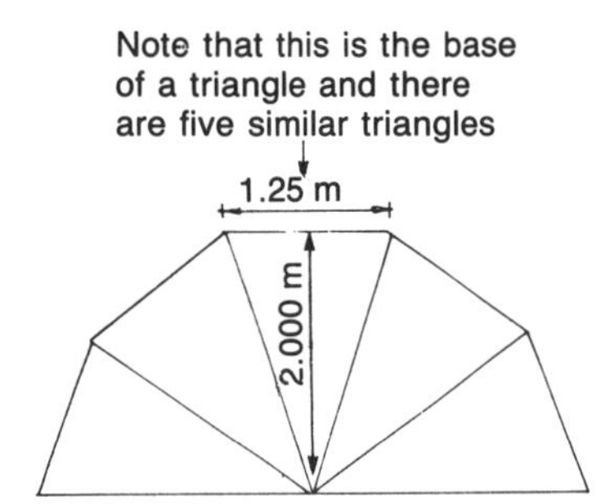

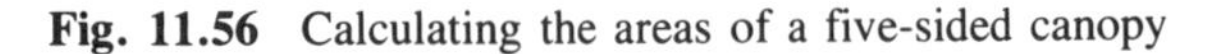

Fig. 11.56 Calculating the areas of a five-sided canopy

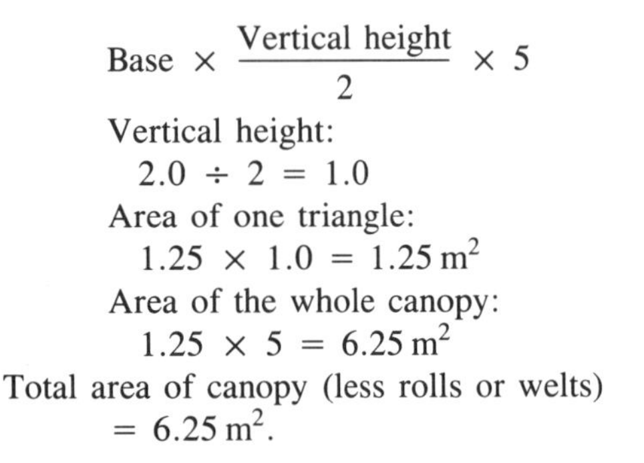

$$\text{Base} \times \frac{\text{Vertical height}}{2} \times 5$$

Vertical height:
$2.0 \div 2 = 1.0$
Area of one triangle:
$1.25 \times 1.0 = 1.25\,\text{m}^2$
Area of the whole canopy:
$1.25 \times 5 = 6.25\,\text{m}^2$
Total area of canopy (less rolls or welts)
$= 6.25\,\text{m}^2$.

Percentages

Percentages are used in calculating prices and costs in plumbing and are a convenient way of expressing the relationship of one quantity or price with another; the percentage symbol is %. If the original cost of an article is £100 and it increases by £50, it can be said that the increase is one half of its cost or 50%. To give another example using the same article costing £100. If it is subject to 15% value added tax, the actual cost will be £115. The foregoing is very simple and can be seen at a glance but it becomes a little more difficult if the original quantity is not 100. Assume a large flat roof is to be weathered in metal sheet and the measured area is $35\,\text{m}^2$. The plumber would be very foolish if he submitted a price on this quantity because a reasonable allowance must be made for waste; let us say in this case 7%. To relate this to the original area, a simple calculation must be made and this is done by dividing 35 into 100 equal parts which will indicate 1%. Thus:

$$\frac{35}{100} = 0.35$$

$0.350\text{ m} = 1\%$ of $35\,\text{m}^2$

To find 7% of $35\,\text{m}^2$

$$0.350 \times 7 = 2.45$$

The sum 2.45 m is 7% of 35 m and this is the quantity of sheet that should be allowed in this case for waste.

Example The cost of a quantity of lead sheet is £78 + $17\frac{1}{2}$% VAT. What will be the cost to the plumber?

$$\frac{78}{100} \times \frac{17.5}{1} = 13.65$$

£78 + £13.65

The plumber will pay £91.65 for the lead. It should be noted that 100 is a very easy number with which to divide, all that is necessary is to insert a decimal point and by moving it two places to the left we have effectively divided by 100. Taking the number 59% or 1/100 of 59 can be expressed as 0.59. A further example of this using a number already containing a decimal point is shown as follows: 1% or 1/100 of 174.6 = 1.746. It is perhaps worth noting the following; to **divide** by 10 the decimal point is moved 1 place to the left, to **multiply** by 10 or 100 it is moved one or two places to the right respectively. The occasion might arise when the price increase is given as a percentage increase and the ability to convert this to a sum of money is essential if a cost comparison is to be made. Assume two suppliers have increased the cost of a certain type of sheet material per m^2. Supplier (a) originally quoted £4.64 but has increased his cost by 5%. Supplier (b) original cost was £4.75 but his increase is only 2%. Which of the two can supply at the lowest price?

Supplier (a) $\frac{4.64}{100} \times 5 = 0.23$ (23 pence)

New price £4.64 + 23p = £4.87

Supplier (b) $\frac{4.75}{100} \times 2 = 0.095$ (approx. 10p)

New price £4.75 + 10p = £4.85

Supplier (b) is the cheapest.

Further reading

Refer to the literature recommended in Chapter 9.

Self-testing questions

1. State the essential properties of a damp-proof course.
2. State the advantages of copper or lead damp-proof courses.
3. Describe the difference between 'step' and 'step and cover' flashings.
4. Explain why the free edges of copper, aluminium and zinc weatherings should be doubled.
5. Sketch a chimney stack using double lines for mortar joints. On the slope of the abutting roof, draw a piece of lead of suitable length, showing how single-stepped flashings would be marked out, cut and turned.
6. Specify the material you would recommend for flashings and weatherings on interlocking tiles and state the reasons for your choice.
7. Describe two methods used to weather the top of a pipe flashing.
8. Sketch and describe three special purpose-made tools that are required for copper and aluminium roofing.
9. State the formula used for calculating the area of triangular figures.
10. Taking the formula for determining the perimeter of a circle to be πD, calculate the amount of material required to make a pipe flashing for a flat roof to the dimensions given in Fig. 11.57.

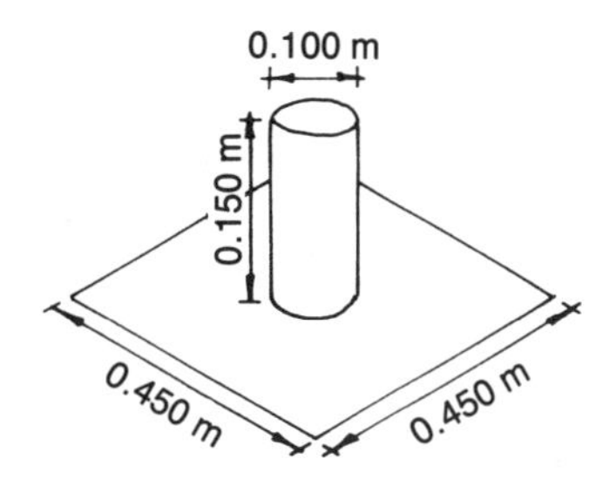

Fig. 11.57 Question 10

11. Explain the reasons for using fillet blocks with copper and aluminium sheet weatherings.

12 Eaves gutters

After completing this chapter the reader should be able to:

1. Enumerate the principal materials used for eaves gutters, stating their advantages and disadvantages.
2. State suitable methods for jointing and fixing for each material.
3. Sketch and describe methods of supporting gutters in various circumstances.
4. Decide upon suitable falls and positioning of outlets for eaves gutters.
5. Specify the materials used for rainwater pipes and describe methods of fixing them.

Introduction

The eaves may be defined as the lower edges of a pitched roof. Gutters around the eaves of the building provide the means of conveying rainwater which falls on the roof to the underground drainage system or, in some cases, soakaways.

The cross-sectional shape of these gutters varies considerably depending to some extent on the type of material used. Sections are illustrated in Fig. 12.1, the most commonly used being half round, square and ogee.

Gutter fittings

Some of the more common brackets and fittings are shown in Figs 12.2 and 12.3. Unless the contrary is specifically stated, they are common to gutter systems of all materials.

Materials for eaves gutters

PVC (BS 4576:1970, Part 1)

This is probably the most commonly used material for smaller premises and domestic work and has much to commend it. It is light, cheap and flexible, requiring no maintenance such as painting. It is currently made in white, brown, black and various shades of grey as these colours offer the greatest resistance to fading in strong sunlight. PVC has a smooth internal surface which offers a better flow rate than other materials. Its biggest disadvantage is its high expansion rate, but providing it is properly fixed according to the manufacturer's recommendations this does not present any problems. It softens in hot weather but at low temperatures becomes very brittle and therefore ladders must not be rested against it when one is working on a roof. It is best to remove a length of gutter to reduce the possibility of any damage to it and to avoid potentially dangerous working conditions. As an alternative, special rests are available to hold the ladder off the wall giving sufficient clearance for the gutter, a typical example being shown in Fig. 12.4. It is important for the usual safety precautions to be carried out too, i.e. the ladder must be fastened securely top and bottom, with the bottom of the ladder on firm ground.

Cast iron (BS 460:1964)

The use of this material has declined considerably since PVC became popular. It is very strong but requires constant painting both inside and out to maintain it in good condition. It is still used, however, especially in cases where a moulded profile is required, a common example being the ogee shape which is difficult to produce, and lacks strength, if made in PVC. Note that when ordering

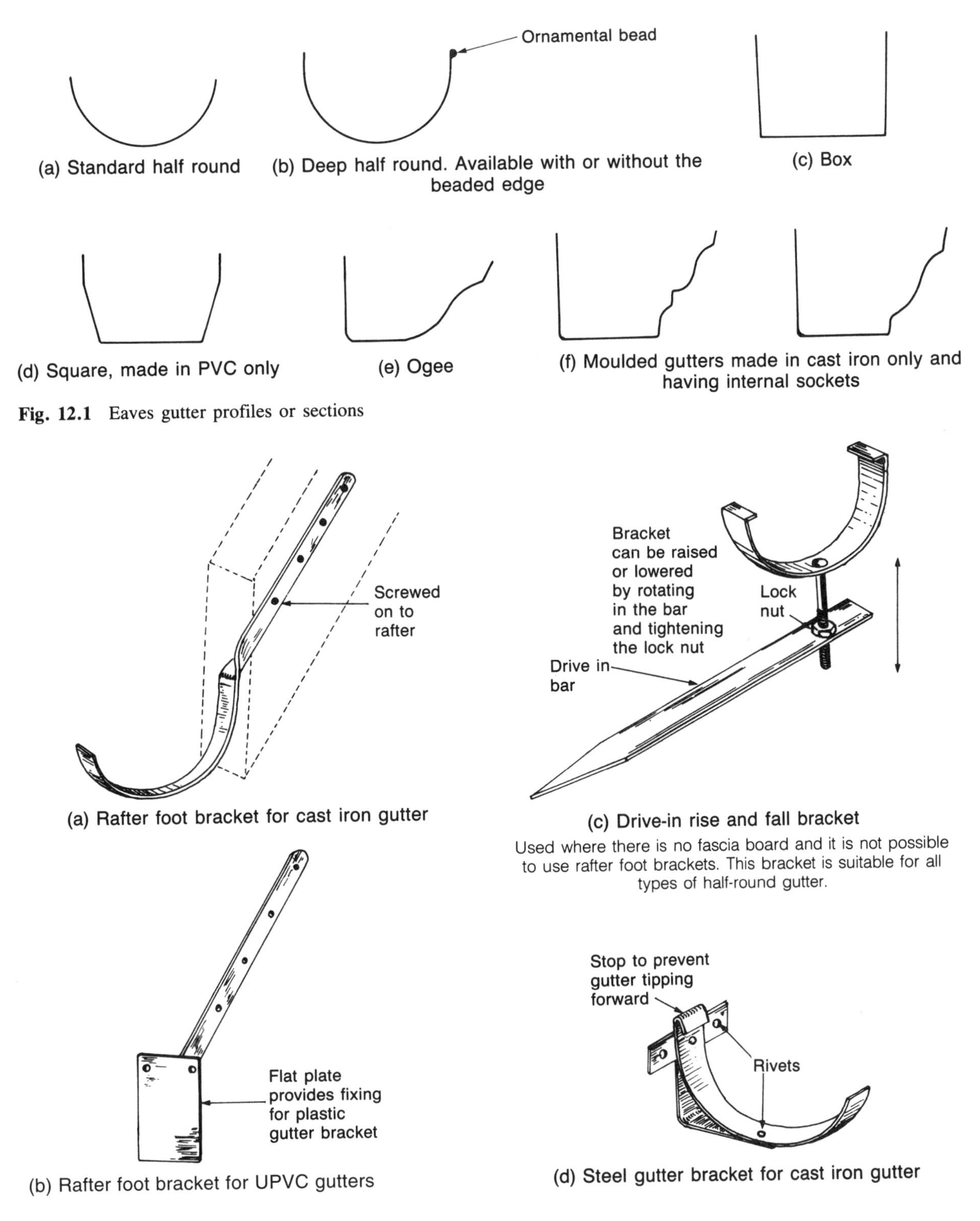

Fig. 12.1 Eaves gutter profiles or sections

Fig. 12.2 Gutter brackets

(a) Stop end

May be external or internal. The former is more useful as it does not require a socket on the gutter, thus cut unsocketed lengths may be used.

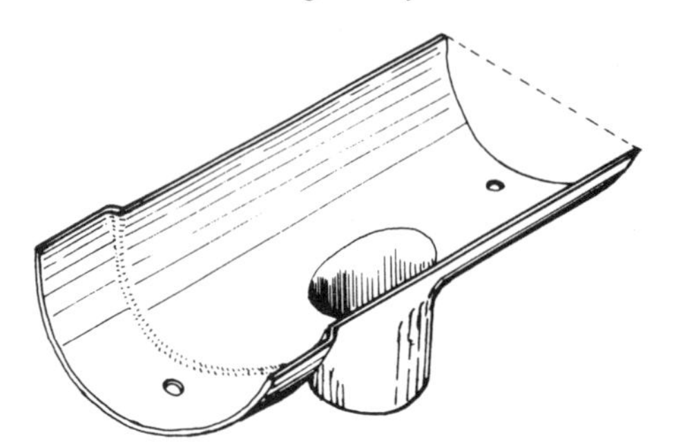

(b) Gutter outlet in half-round cast iron

The type illustrated has a socket at one end only but outlets can be obtained with sockets at each end, enabling short, unsocketed lengths of gutter to be used. Stop end outlets are also available.

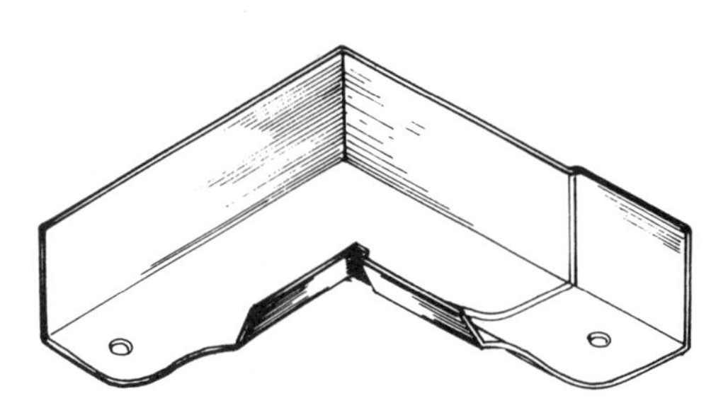

(c) Internal ogee gutter angle

When ordering, internal or external angles and right-hand or left-hand sockets must be specified.

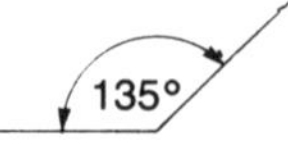

(d) Standard angles for all types of gutter

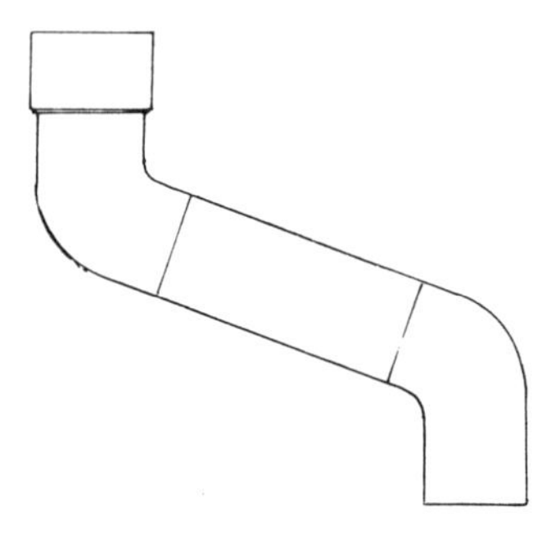

(e) Swan neck

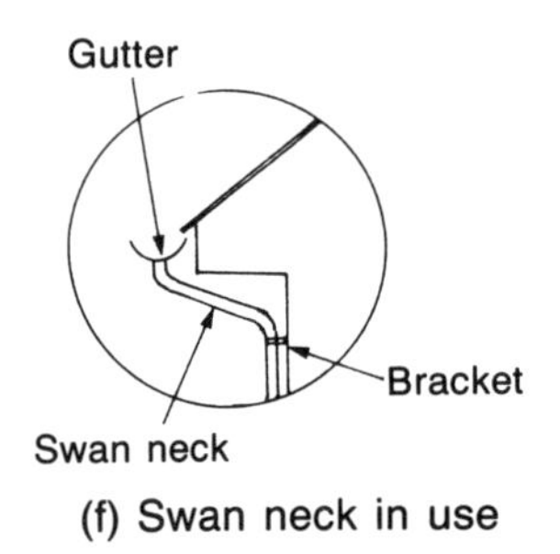

(f) Swan neck in use

Fig. 12.3 Gutter fittings

ogee gutter angles, internal or external should be specified as, unlike angles for other types of gutter, they are not interchangeable.

Sheet steel (BS 1091:1963)

Like cast iron, the use of this material has declined, especially for stock sizes of gutter. It is mainly used now in cases where non-standard gutter is required; and it is made to order by sheet metal workers. It is manufactured in several thicknesses and should be galvanised to protect it from corrosion. The methods of painting and fixing are the same as those described for cast iron.

Asbestos cement

Asbestos cement gutters were used extensively as an alternative to those of cast iron prior to the introduction of gutters made of PVC. As they are no longer made, any repair work cannot be

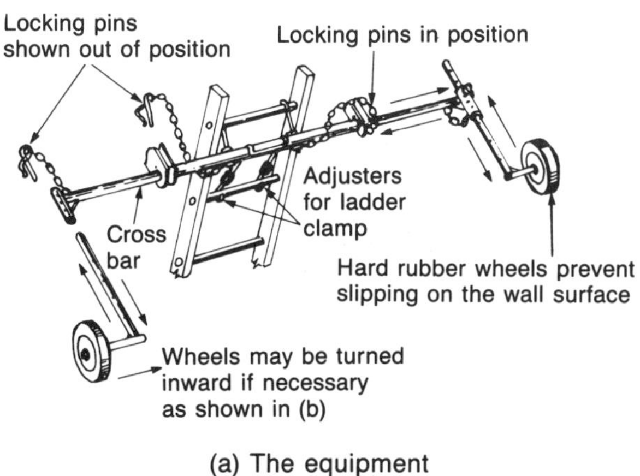

(a) The equipment
Note that the arrows indicate the 'in' and 'out' movements of each part of the equipment

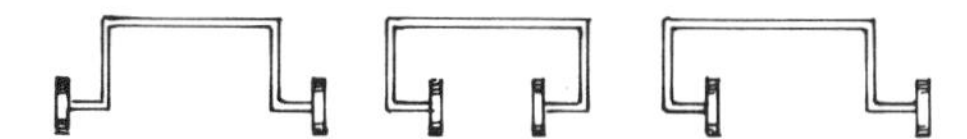

(b) Alternative arrangements for fitting wheels to the cross-bar make this a very versatile tool

Fig. 12.4 Typical equipment for holding ladders off gutters and eaves

undertaken and the complete guttering system will have to be replaced. Before its removal the local environmental officer should be contacted, as some authorities have stricter regulations relating to this material than others. Generally speaking it is not considered to be as dangerous as, for example, asbestos insulation, as the cement binds the asbestos fibres. Always remember, however, that any asbestos products should be treated with caution.

Sheet zinc (BS 1431:1960)

This material was popular in some parts of the country during the period 1900–40. Except when required to meet a special order, it is always stocked in ogee section, which gives it greater rigidity. Its use is rather labour-intensive, each length being soldered to its neighbour, while all angles, stop ends and outlets have to be fabricated from the gutter itself. To resist the weight of ladders leaning against it, it is also necessary to insert stays at 450 mm centres, the stays being made of bell tube or sheet zinc angle pieces soldered in place. One of the big advantages of zinc gutter is that it can be fabricated to suit small bay windows having non-standard angles. Such a job would be difficult to carry out using other materials.

Purpose-made gutters

These are made from sheet metal by specialist sheet metal companies which manufacture both gutters and fittings from copper, aluminium, zinc and low carbon steel sheet, the latter being galvanised as protection from corrosion. Such gutters are obviously expensive and careful ordering is necessary to ensure that no waste occurs. Gutters made of these materials should conform to the following British Standards:

Aluminium	BS 2997:1958
Copper and zinc	BS 1431:1960
Low carbon steel	BS 1091:1963

Methods of cutting, jointing and fixing gutters

Eaves gutters of cast iron and galvanised low carbon steel are produced with a socket cast or made on to one end of each length. Jointing is carried out by filling the socket with a suitable mastic or linseed oil putty. The spigot end of the gutter is bolted to the socket, using a galvanised mushroom-headed screw. The surplus jointing medium must then be removed from both inside and outside the joint. It should be noted that whenever linseed oil putty is used as a jointing medium, the surfaces must first be painted to form a suitable key.

Sheet steel and cast iron gutters are cut with a hacksaw, although in the case of the latter it is usual to make a cut of only approximately 1 mm in depth around the perimeter and to complete the cut as shown in Fig. 12.5. Note that this method can be used successfully on cast iron pipe in the event of suitable pipe cutters not being available. Due to the brittle nature of cast iron, eye protection must be worn when this operation is carried out. When a half round cast iron gutter is to be cut, an alternative method is to make a hacksaw cut right through its base then give it a sharp tap on a hard surface such as concrete. However, this method does not always produce such a clean cut, and it

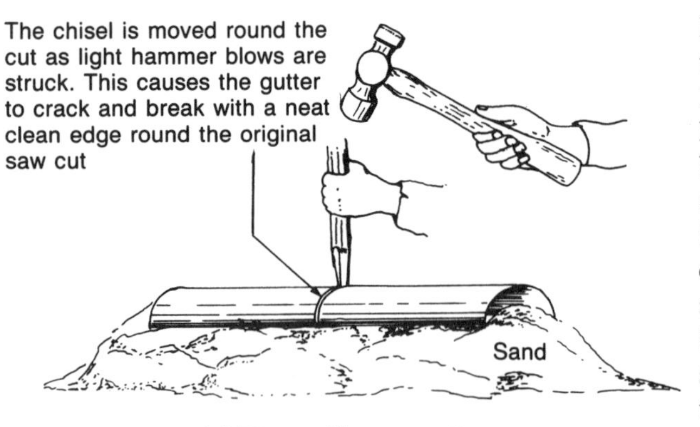

(a) The cutting operation
Shallow cut made with hacksaw round circumference of gutter (see section in (b))

(b) Section
The cut has the effect of removing the hard skin formed on the surface of all cast iron products and exposes the grey iron for cutting with a hammer and chisel

Fig. 12.5 Cutting cast iron gutters

can result in a cracked length. Its main advantage is that it is quick, but it is advisable to practise on some offcuts first. A cut having been made, a hole must now be made to line up with that in the socket. To do this accurately, lay a socketed length of gutter over the cut end using the hole in the socket as a guide for the drill.

Gutters of half round or square section are fixed to a fascia board using purpose-made brackets, but when buildings have no fascia, rafter foot or drive-in brackets are used. All brackets and screws used for gutter fixing should be galvanised as, once fixed, they are very difficult to paint. It is not unknown for brackets made of low carbon steel to corrode causing gutters to become unsafe. Ogee gutters, although giving a pleasing exterior to a building, do have one big disadvantage in so far as once they are fixed it is impossible to paint the back. For this reason it is advisable to give the gutter at least three coats of paint before fixing. Do not use bituminous paint as it will run into any synthetic paint used on the gutter face causing the painter endless trouble. Ogee gutters are not normally fixed using brackets as the back of the gutter has cast-in holes through which is passed a large-headed galvanised gutter screw, similar to that shown for fixing zinc gutters, but only 32–38 mm long. Fixings for moulded and deep half round cast iron gutters must be very robust because these gutters are heavy. They may be supported on brackets that are also made of cast iron, or in some cases the brickwork is corbelled out to support the gutter throughout its length. It should be noted that moulded gutter is made with internal sockets so that a smooth unbroken line is shown on the face of the building. Unfortunately, this allows the bottom of the internal surface of the gutter to become a collecting ground for silt, reducing the flow rate and causing corrosion. Because of this fact, such gutter is seldom installed in modern buildings.

Forming and fixing

Zinc gutters

These are made from preformed zinc sheet of 0.6, 0.8 or 1.0 mm thickness. No fittings need be purchased as it is more convenient to make them from the gutter itself using the soldering technique described in Chapter 10. Due to the difficulty of forming a neat finish on any soldered angles if a lap joint is used, edge to edge jointing is recommended. The lack of strength in such a joint is compensated for by soldering in strengthening pieces where required. The gutter is fixed by screwing through the 'soldered in' strengthening pieces with long galvanised gutter screws made specially for this purpose. Outlets and stop ends are fabricated in a similar way to angles. Rainwater pipes are manufactured but they can be made on site by rolling a strip of zinc sheet around a suitable piece of steel pipe and soldering or welting the seam. Rainwater shoes or offsets can be made by cutting, mitring and soldering. Do not forget to wipe off any surplus flux as it is very corrosive. Figure 12.6 illustrates details of zinc gutter work which, although involving a good deal of labour, can be used with advantage where it is difficult to use other materials effectively.

PVC gutters

PVC gutters are usually of the half round or square pattern. Most manufacturers provide

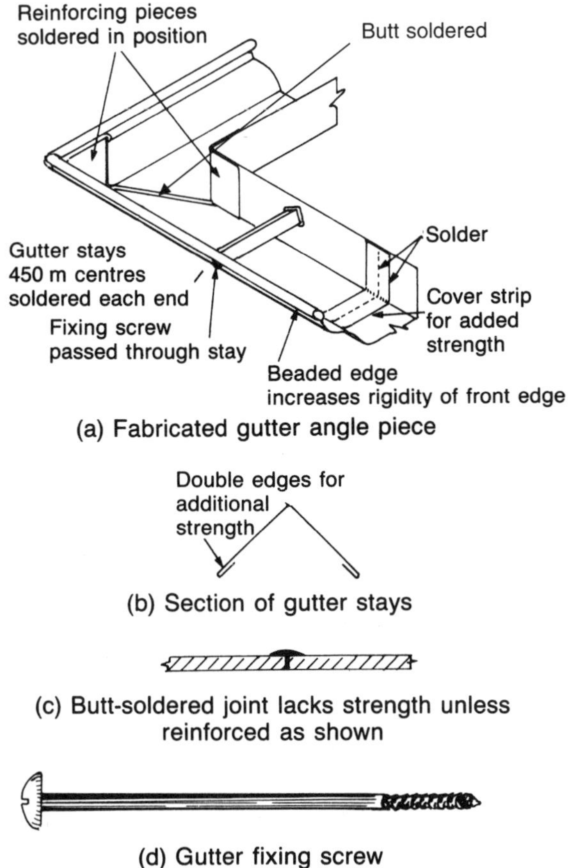

(a) Fabricated gutter angle piece

(b) Section of gutter stays

(c) Butt-soldered joint lacks strength unless reinforced as shown

(d) Gutter fixing screw
Long galvanised screw passes through gutter stays

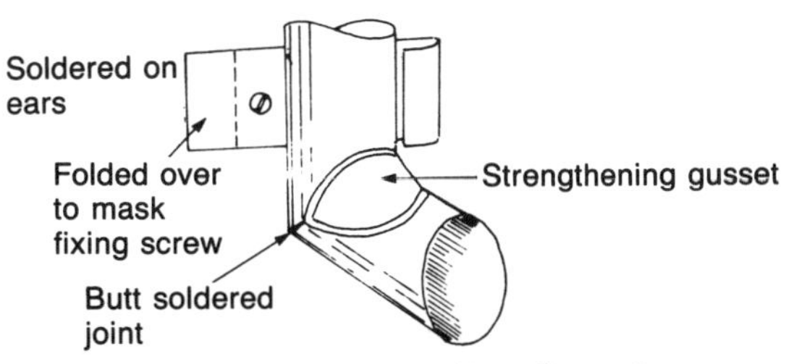

(e) Fabricated pipe and fittings for a zinc gutter

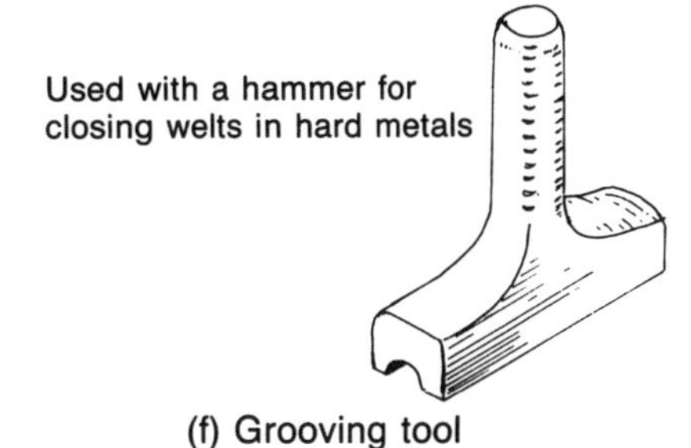

(f) Grooving tool
Used with a hammer for closing welts in hard metals

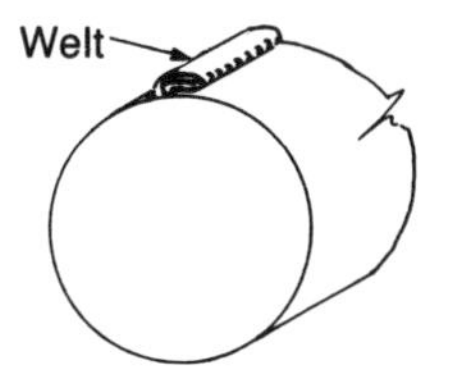

(g) Section of zinc pipe made from sheet zinc with a welted seam
A suitably sized piece of steel pipe held in a vice can be used as a support while the welt is made

Fig. 12.6 Zinc gutter details

detailed instructions for their own products and these should be carefully studied prior to installation. PVC, like most synthetic plastics, has a high rate of expansion and approximately 3 mm must be allowed per 1 m run. The allowance for expansion must be made in each socket; thus when a 2 m length is fitted, a 6 mm gap must be allowed. If greater lengths are employed then the gap must be correspondingly larger. Some systems employ a method of fixing where the brackets are independent of the jointing process. Others use two types of brackets (see Fig. 12.7): one simply supports the gutter, the other not only supports it but acts as a joint between two lengths. While a saving of some support brackets can be effected with systems of this type, more care has to be taken in setting out the jointing brackets to ensure that the correct expansion gap is left at each end of the gutter. The jointing system with all types of PVC is simple. A synthetic rubber pad is secured during manufacture to the inside of each socket by an adhesive; the spigot end is made into the socket by firmly pushing it down until it slips under the moulded clips formed on the socket. It is usually best to tuck the back of the gutter under the rear of the bracket before the front is clipped home. Figure 12.8 illustrates the recommended procedure. A little soap solution or silicone grease on the rubber seal makes jointing much easier, and in very cold weather a little heat applied with a blowlamp will make the gutter more flexible. However, do not overheat it as it easily chars.

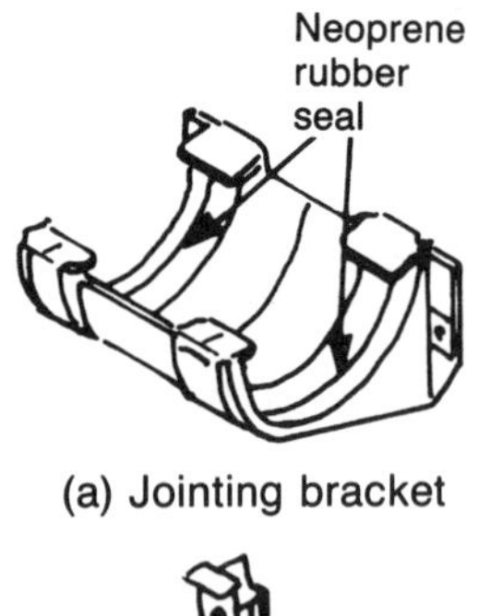

(a) Jointing bracket

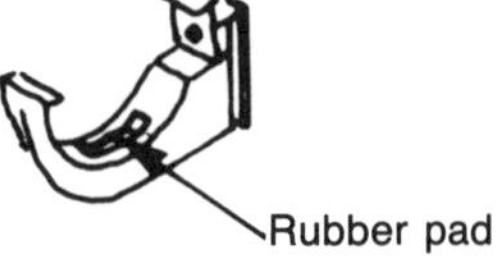

(b) Support bracket

Fig. 12.7 Brackets for PVC gutters

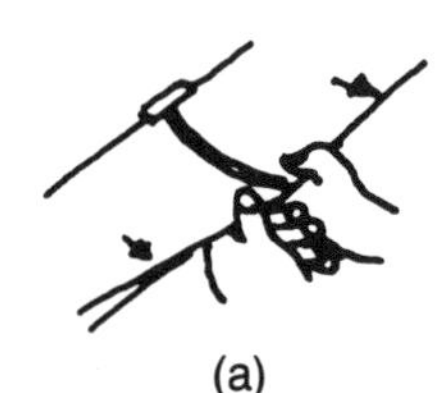

(a)
Pull gutter and brackets outward in direction of the arrows

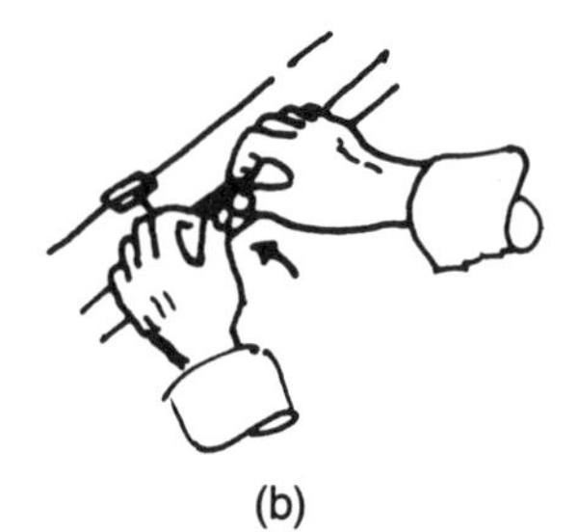

(b)
When gutter engages under the clip, push upward and forward to secure the gutter.

Fig. 12.8 Snapping PVC gutter into brackets

Before the joint is made wipe the rubber seal clean, as failure to do this may result in a leaking joint.

Support brackets should be 1 m apart unless the manufacturer's instructions suggest otherwise. If the roof is steeply pitched, it may be advisable to reduce this spacing to about 750 mm as after a heavy snowfall extra weight on the gutter can cause it to collapse.

In cases where PVC gutters are to be connected

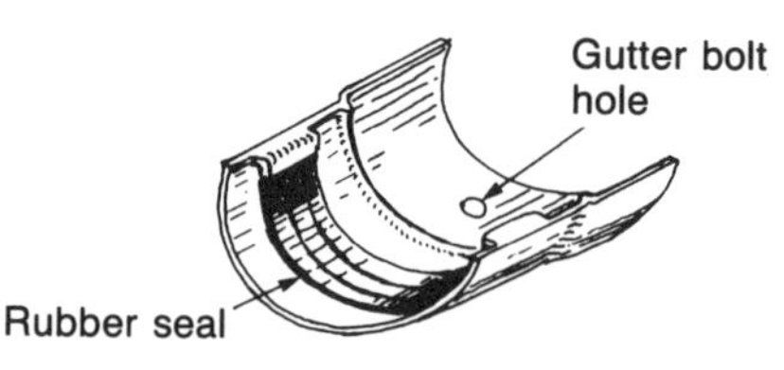

(a) Half-round traditional gutters to half-round PVC

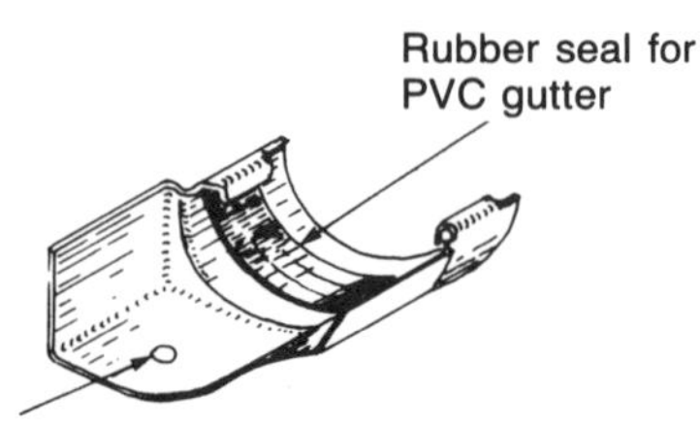

(b) Ogee to PVC half-round gutter, ordered right or left hand as required

Fig. 12.9 Gutter adaptors for connecting PVC to existing sheet steel or cast iron gutters. Both adaptors are made of aluminium alloy to enable a traditional putty or mastic joint to be made to the existing gutter.

to existing gutters of other materials, a suitable adapter should be used to ensure that the joint will be trouble-free; see Fig. 12.9. Manufacturers of PVC gutter systems all produce adapters suitable for making joints with other gutter materials and sections. Do not be tempted to try to joint the PVC directly to, for instance, a cast iron gutter using a putty joint. Putty contains linseed oil, and oil of any kind should not be allowed to come into contact with plastic materials as it will cause them to soften and degrade.

Setting out gutter falls

The brackets should be fixed as high as possible to avoid the run off from the eaves being blown back behind the gutter: see Fig. 12.10. Setting out must be done very carefully. Consider not only the bracket spacings, but the amount of fall given to the gutter, as this and the positioning of the outlets have a considerable influence on the gutter's capacity. A centrally mounted outlet doubles the capacity, and a gutter with a fall of 1 in 600 discharges one third as much water again than a gutter fixed level. Table 12.1 serves to illustrate

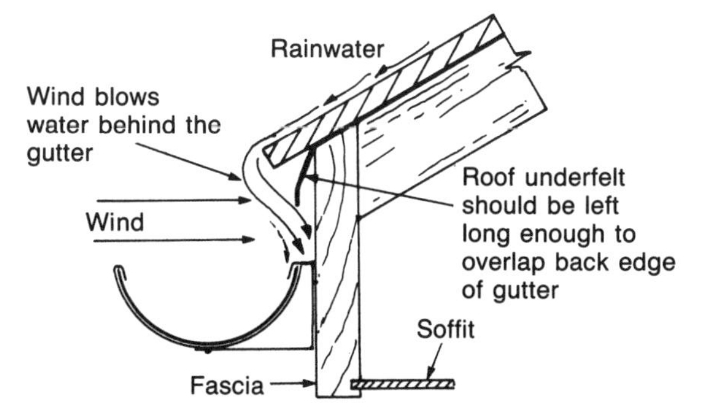

Fig. 12.10 Effect of fixing gutters too low on fascia

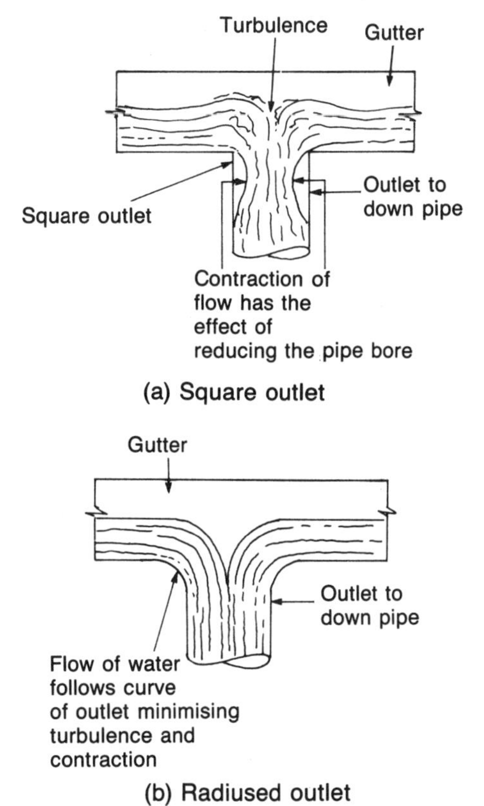

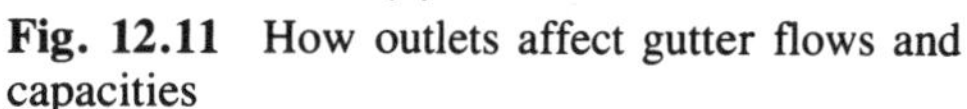

Fig. 12.11 How outlets affect gutter flows and capacities

this. Although only one size is shown, the carrying capacity of all gutter sizes is increased by roughly the same proportion. It will be seen that by carefully choosing the outlet position and apportioning a suitable fall, gutters can be made to remove rainwater quickly from the roof while overflowing during storms and periods of heavy rainfall is avoided.

The influence of angles and outlets on the flow of water is also very important, as sharp changes of direction reduce the flow rate of the water as shown in Fig. 12.11. There is, of course, an aesthetic consideration relating to falls. While it has been shown that a good fall is desirable, some thought must be given to the appearance of the job, and nothing looks worse than a gutter with one end tight underneath the tiles and the outlet on the bottom of the fascia. While the fascia should be perfectly level, in practice this is not always so, and should it fall in the opposite direction to the gutter, it is sometimes difficult to achieve a good fall without giving the gutter a lopsided appearance. Unless the position of the gutter outlets can be changed, it may not be possible to give as much fall as would be desirable. In extreme cases it may even be necessary to use a larger gutter. This is something a plumber must decide for each individual job.

Table 12.1 Discharge of water from gutters.

Gutter size and section	*Litres discharged per minute*			
	Gutter level		*Gutter having 1 in 600 fall*	
	Central outlet	*Outlet at one end*	*Central outlet*	*Outlet at one end*
120 mm diameter half-round section	104	52	146	73

The following procedure is adopted when setting out the gutter brackets to determine the highest and lowest points. Assume a length of gutter of 9 m is required with a fall of 1 in 600; overall fall on the gutter will be 15 mm. With a spacing of one bracket every metre, ten brackets will be required.

The highest bracket should be fixed on the fascia as tight under the tiles as possible, and to achieve the desired fall the one at the other end is fixed 15 mm lower. A line is stretched across the distance between the brackets, and the intermediate brackets are then carefully fixed to the line as shown in Fig. 12.12, the bottom of each bracket just touching the line without distorting it. Ogee

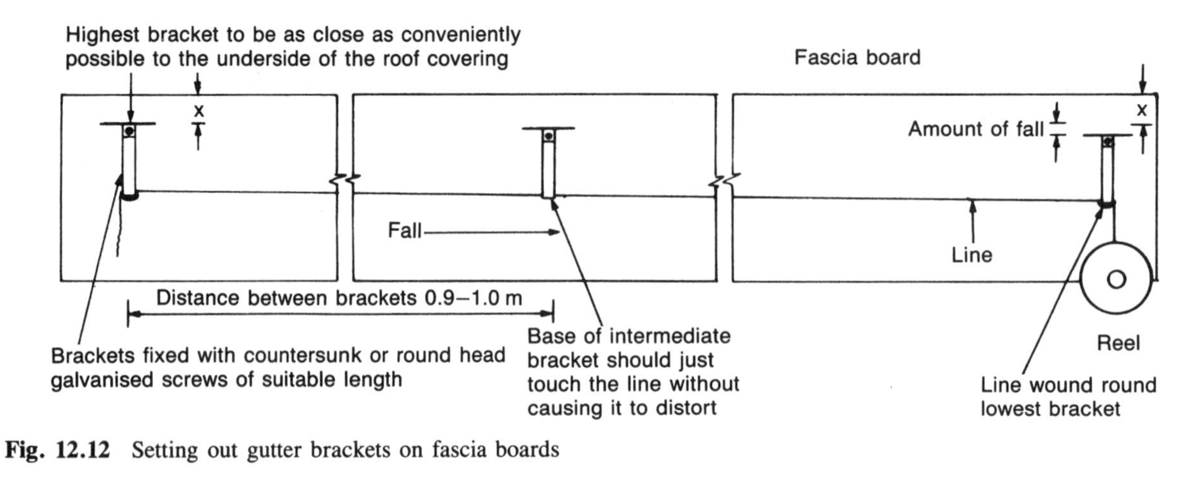

Fig. 12.12 Setting out gutter brackets on fascia boards

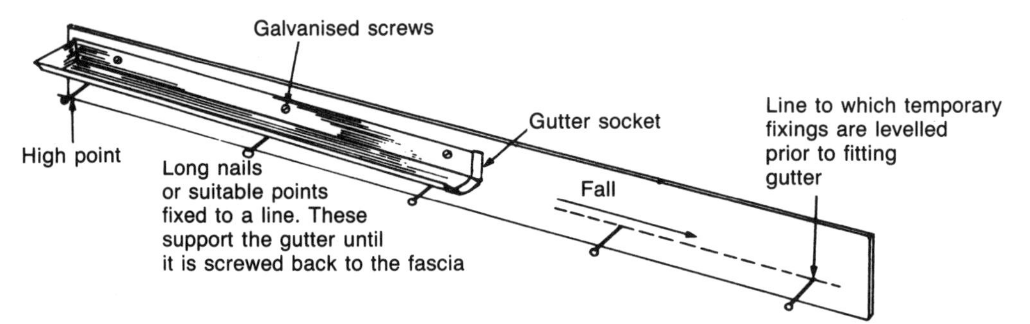

Fig. 12.13 Fixing ogee or moulded guttering

gutters, having no brackets, are fixed as shown in Fig. 12.13, the gutter resting on fixing points or long nails while it is jointed and fixed.

The line level shown in Fig. 12.14 can be used to establish a level prior to setting out the falls in a gutter. It should never be assumed that fascia boards, even when fitted, are absolutely level due to slight imperfections in the timber.

Standard sizes of various gutters are shown in Table 12.2. The sizes given are approximate as there is a slight variation, usually of only 1 or 2 mm, between different materials.

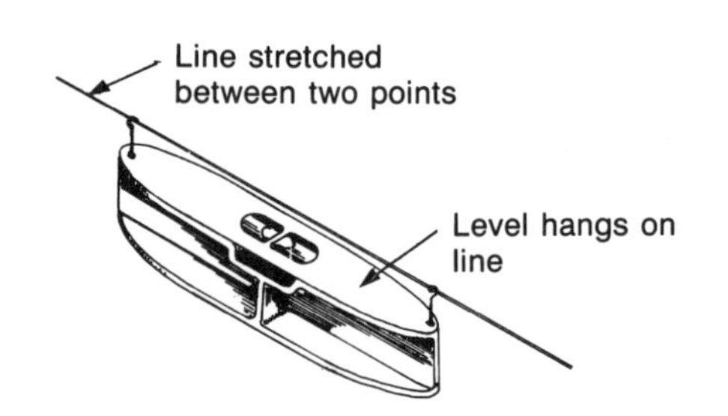

Fig. 12.14 Spirit level. A small lightweight spirit level which can be hung on a line to take levels over long distances. It is a convenient tool to use when setting out gutter falls where it might not be possible to use an ordinary level. It is important that a good quality line is used and it must be stretched as tightly as possible.

Rainwater pipes for eaves gutters

It is essential, for the sake of appearance that rainwater pipes or, indeed, any vertical pipe fixed externally should be perfectly upright. It is possible to check each length with an upright level before fixing, but it is usually much quicker to strike a chalk line on the face of the wall, especially if several lengths are to be fixed. The method used is similar to marking out sheet lead with a chalk line, but in this case a plumb bob is used as illustrated in Fig. 12.15.

Table 12.2 Standard gutter sizes for common materials.

Gutter materials	*Nominal dimensions of profile*		*Notes*	*BS No.*
	Half round (mm)	*Ogee (mm)*		
Cast iron, also sheet steel up to 100 mm only	75 100 114 125 150	114 125 150	Gutters also formed in square or moulded profiles	Cast iron 460:1964 Steel: 1091:1963
PVC	75 100 115 125 150	Not produced in this material	Nominally square sections are also produced to similar dimensions	4576:1970, Part 1
Copper and zinc sheet	75 100 114 125	75 100 114 125	Nearly always made in ogee section to give greater rigidity	1431:1960

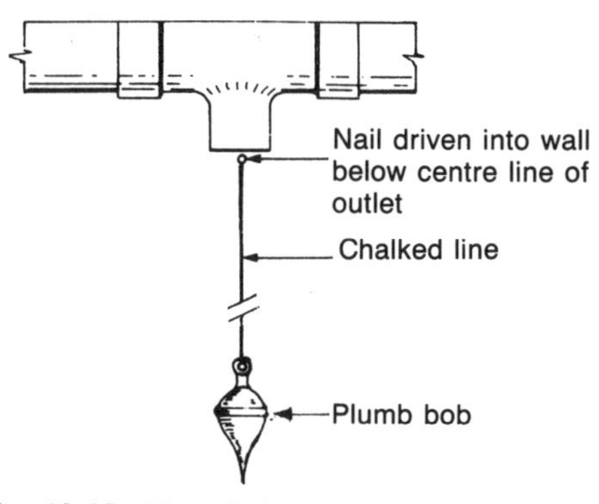

Fig. 12.15 Use of plumb bob for fixing vertical pipes. When the plumb bob is stationary pluck the line, which will leave an easily removable mark on the wall surface, serving as a guide for the centre line of each pipe fixing.

Rainwater pipes are made of the same material as the gutter and normally require no special jointing, the spigot simply entering the socket. Details of joints and fixings are shown in Fig. 12.16. When working with PVC rainwater pipes, be sure to leave a gap of about 6–8 mm for each 2 m length of pipe between the spigot end of the pipe and the bottom of the socket to allow for expansion. Rainwater pipes terminate in a variety of ways depending on the method of disposal. The traditional method is to use a rainwater shoe discharging over a trapped gulley as often found in older properties. Two such shoes are shown in Fig. 12.17. It should be noted that the anti-splash type is only made in the cast iron range of rainwater goods. The usual method in current use is to use a back inlet gulley having the rainwater pipe connected directly to it, or if a separate rainwater drain is available, no gulley is normally necessary and the rainwater pipe may be connected to the underground drain. Figure 12.18(a) and (b) illustrates these two methods. The two adaptors shown in Fig. 12.19(a) and (b) show how direct connections may be made between all types of underground drain and rainwater pipes.

Special offsets called 'plinth pieces' shown in Fig. 12.20 are available to overcome difficulty when fixing rainwater pipes to walls which are built with a projection at some point. This is common in very old properties where the width of the brickwork footings were extended above ground level. Plinths are sometimes built into modern buildings as an architectural feature.

Rainwater collection in buildings having large areas of flat roof

The method employed will depend largely on the material used for the roof covering. If sheet metal

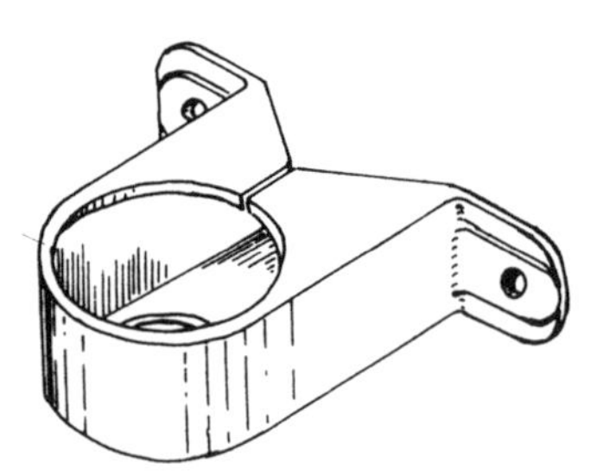

(a) This fixing is made to open and clamp round rainwater pipe connector

To and fro movement

Clip

(b) Adjustable pipe clip

The hole in the back fixing bracket is slotted to permit limited movement of the clip to accommodate any variations of the fixing surface

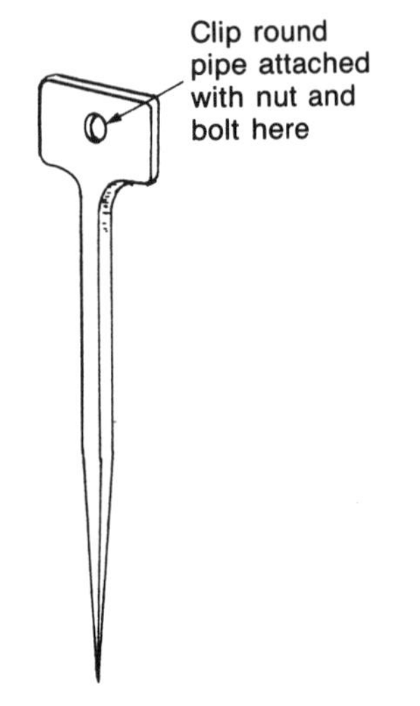

(c) Drive-in fixing for walls with uneven surfaces

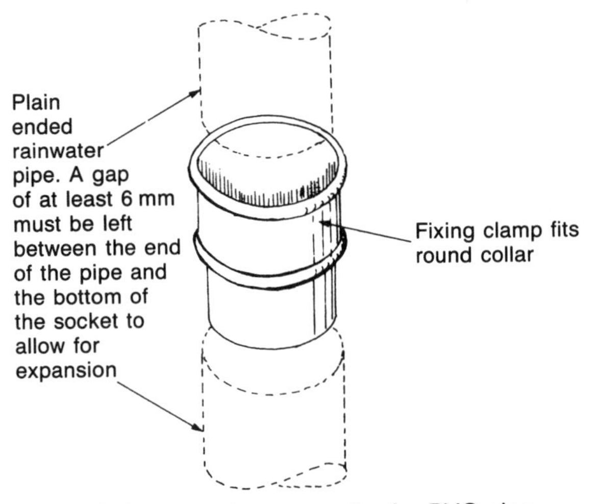

(d) Rainwater pipe connector for PVC pipe

Fig. 12.16 PVC rainwater pipe fixings

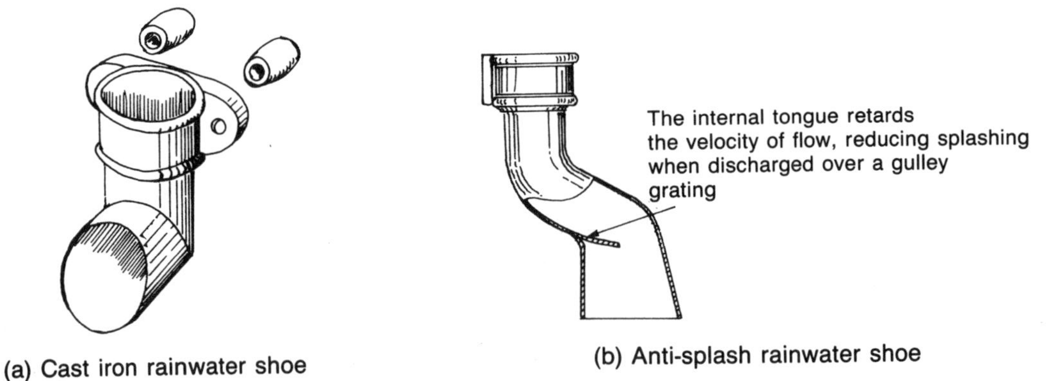

(a) Cast iron rainwater shoe

Purpose-made cast iron 'bobbins' or spacers are fitted behind the fixing plate of cast iron rainwater components enabling them to be spaced off the wall. This allows access to the back of the pipe for painting. Short pieces of steel pipe can be used as an alternative

(b) Anti-splash rainwater shoe

Fig. 12.17 Discharge of rainwater pipes to underground drainage system

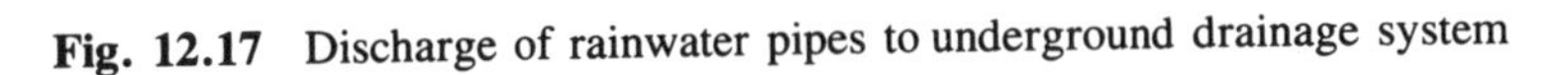

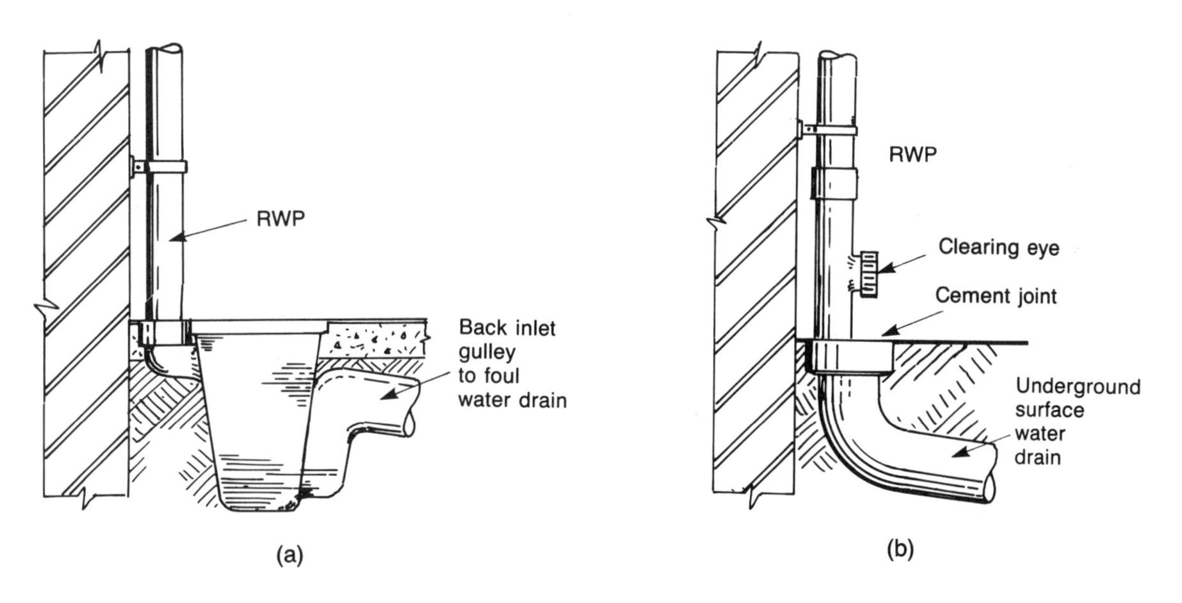

Fig. 12.18 Methods of discharging surface water into underground drains

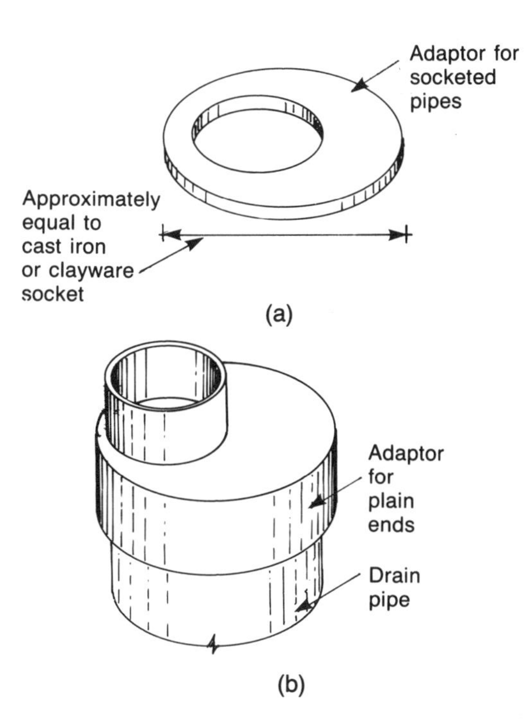

Fig. 12.19 Adaptors for reducing clayware and cast iron pipes to a smaller diameter rainwater pipe

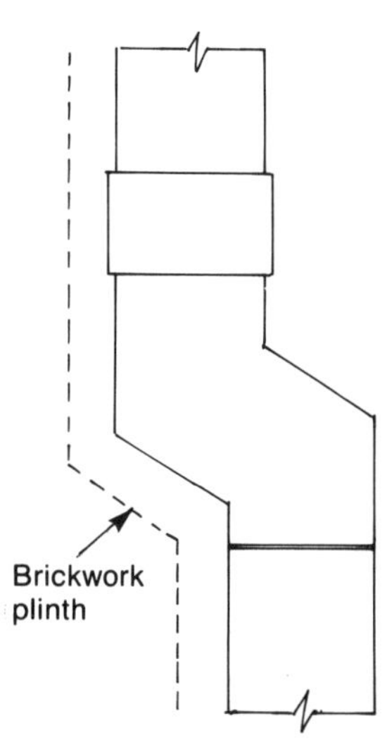

Fig. 12.20 Plinth offset made by some producers of cast iron and plastic rainwater fittings

is used, the gutters are usually made of the same material and will in effect become part of the roof covering. Details of gutters of this type are dealt with in Book 2 of this series as they are more closely allied to sheet roof work than any other subject. In modern buildings it is common to use materials such as asphalt or built-up felt roofing for weathering large flat roofs, as the initial costs are less than that of sheet metal. In many cases the roof is surrounded by a parapet wall and its drainage may be accomplished in one of two ways. If the rainwater pipe is fitted on the face of the building a chute is fitted through the parapet wall and discharges into a hopper-head in a similar way to that of a sheet metal gutter. If the rainwater pipe is to be fitted internally, a roof outlet of the type shown in Fig. 12.21 is used. This type is made of cast iron or PVC. The main feature is the domical grating which prevents it becoming clogged with leaves or other debris. It is important that the roof covering is keyed into the outlet as shown to ensure a permanently watertight joint. Rainwater pipes inside buildings are made of copper, PVC or cast iron, and are jointed and fixed using the same techniques as those used for sanitary pipework. Good workmanship is essential as any defect could flood the building.

Rainwater collection from large areas of pitched roof

In most cases mass-produced eaves gutters are not big enough to remove water from large industrial and commercial buildings. It is possible, however, to obtain large-section gutters purpose-made by specialist producers. These are made of the same materials as eaves gutters and the methods of jointing are normally similar. A selection of such gutter sections is shown in Fig. 12.22. It should be noted that, when ordering, the angle of the roof pitch should be given to the manufacturer as once the angles of the gutter are formed it is difficult, often impossible, to alter them.

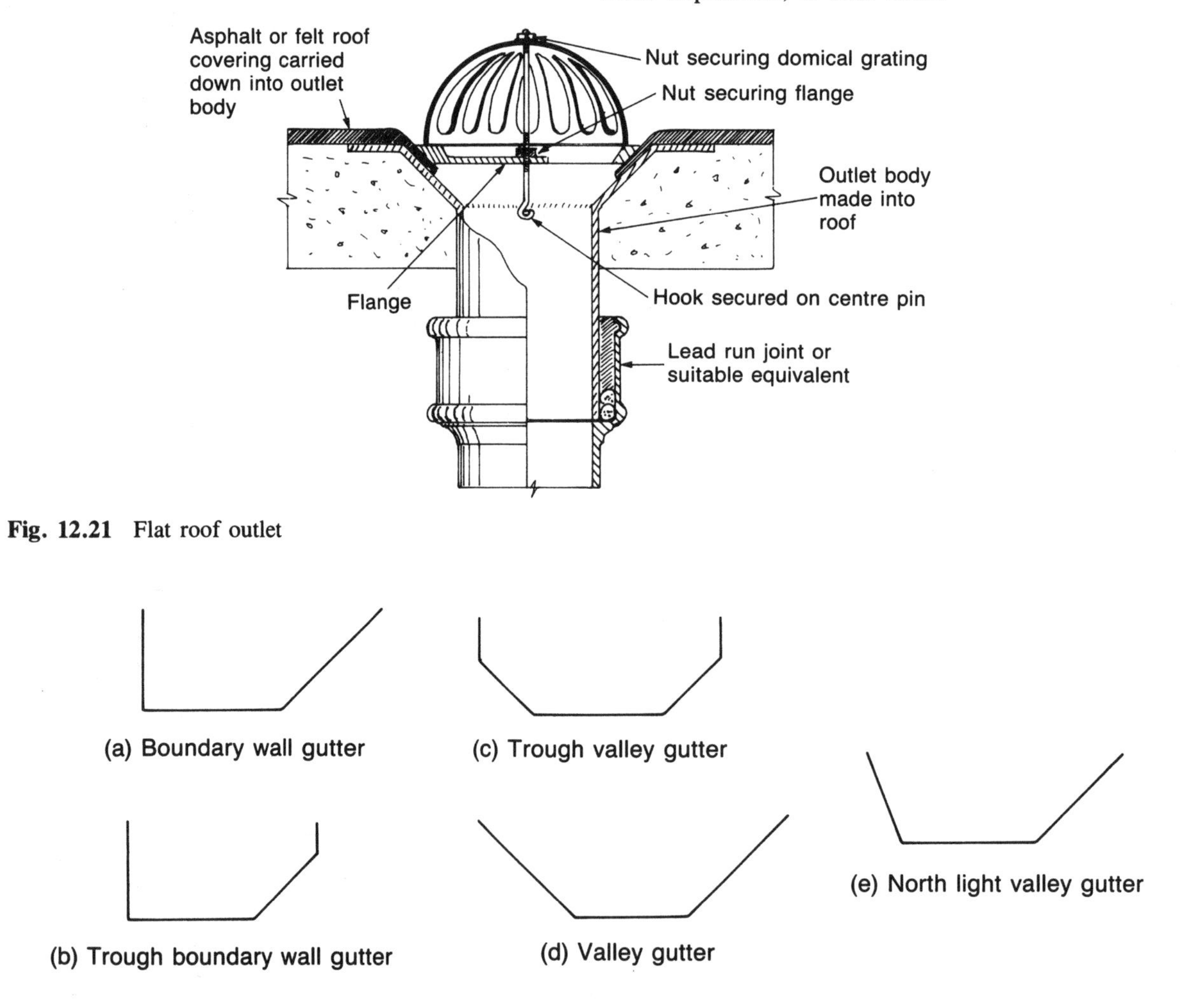

Fig. 12.21 Flat roof outlet

Further reading

Much useful information can be obtained from the following companies who produce rainwater goods.

Caradon Terrain Ltd, Aylesford, Kent, ME20 7PJ.
Marley Plumbing, Dickley Lane, Lenham, Maidstone, Kent, ME17 2DE.
Hunter Building Products, London, SE28.
Osma Plastics, Wavin Building Products Ltd, Parsonage Way, Chippenham, Wiltshire, SN15 5PN.

Self-testing questions

1. List five materials used for eaves gutters.
2. State why ladders should not be rested on PVC gutters and describe a method of avoiding this.
3. Describe how to make a joint in cast iron or sheet steel gutter. What special provision is necessary when jointing such gutters to those made of PVC?
4. What are the recommendations for painting ogee gutters before fixing? State why they are necessary.
5. (a) Describe how provision for expansion is made when installing PVC gutters.
   (b) Explain the procedure for setting out the fall in a gutter.
6. State the generally recommended maximum distance between brackets in a run of eaves gutter.
7. (a) What are the advantages of fitting the outlet in the centre of a gutter rather than at one end?
   (b) In what way does the design of angles and outlets influence a gutter's discharging capacity?
8. (a) List the tools required for fixing half round eaves gutters.
   (b) If a gutter is 25 m long and has an overall fall of 30 mm, what will be the fall per 1 m run?
9. Describe a method of setting out to ensure that rainwater pipes are fixed in an upright position.
10. Make a sketch illustrating the use of a swan neck.

Appendix A: Assignments

Cold water assignment

An existing underground lead service pipe (Fig. A.1) laid at a depth of 600 mm is to be replaced with polythene pipes:

1. State the minimum and maximum depths at which the new service should be laid and the reasons for this.
2. Sketch or describe how the new service should be installed to avoid damage due to subsidence or shrinkage of the subsoil in dry weather.
3. During the course of carrying out the work it is discovered that the wooden floor and its supports are in a very poor condition, and it is decided to replace it with a concrete floor finished with a 50 mm cement screed. Describe two methods you could use, in compliance with the water bylaws, to re-run the service pipe from its point of entry into the building to the consumer's existing stop cock which is 2 m from the point of entry into the building.
4. A few days after completing the work the customer complains that the screw-down tap over the kitchen sink has developed a persistent drip. State a likely cause and remedy for this.

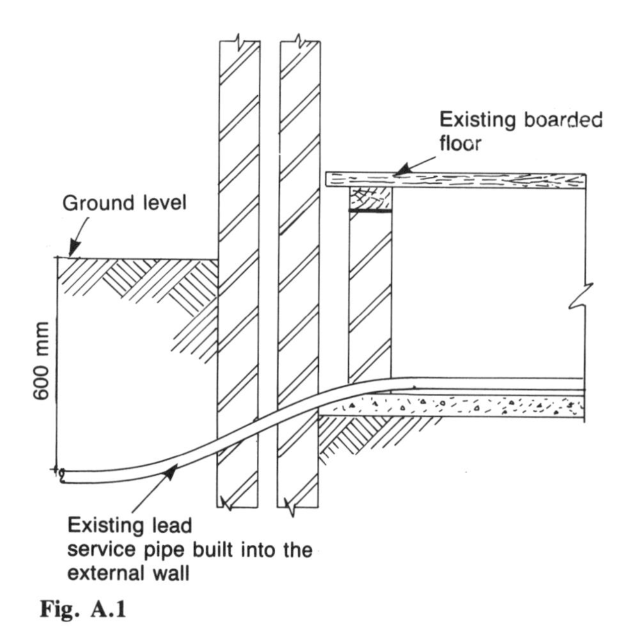

Fig. A.1

Hot water assignment

A plumber is called to identify the cause of several defects in a direct hot water system having a secondary circulation. The system is shown in Fig. A.2 and the complaints made by the customer are listed as follows:

1. (a) Noises are heard in the boiler and circulating pipes where the boiler operates.
   (b) When running a bath, after an initial flow of hot water from the hot tap, the water temperature drops so that only lukewarm water is drawn off.
   (c) The hot water draw-off splutters and discharges with a mixture of water and air.
   (d) Although the water in the storage vessel is hot when the boiler shuts down at night, it is found that the temperature drop is unacceptable when hot water is required on the following morning.

From a study of the illustration, list the causes of the foregoing defects and your proposals for their rectification.

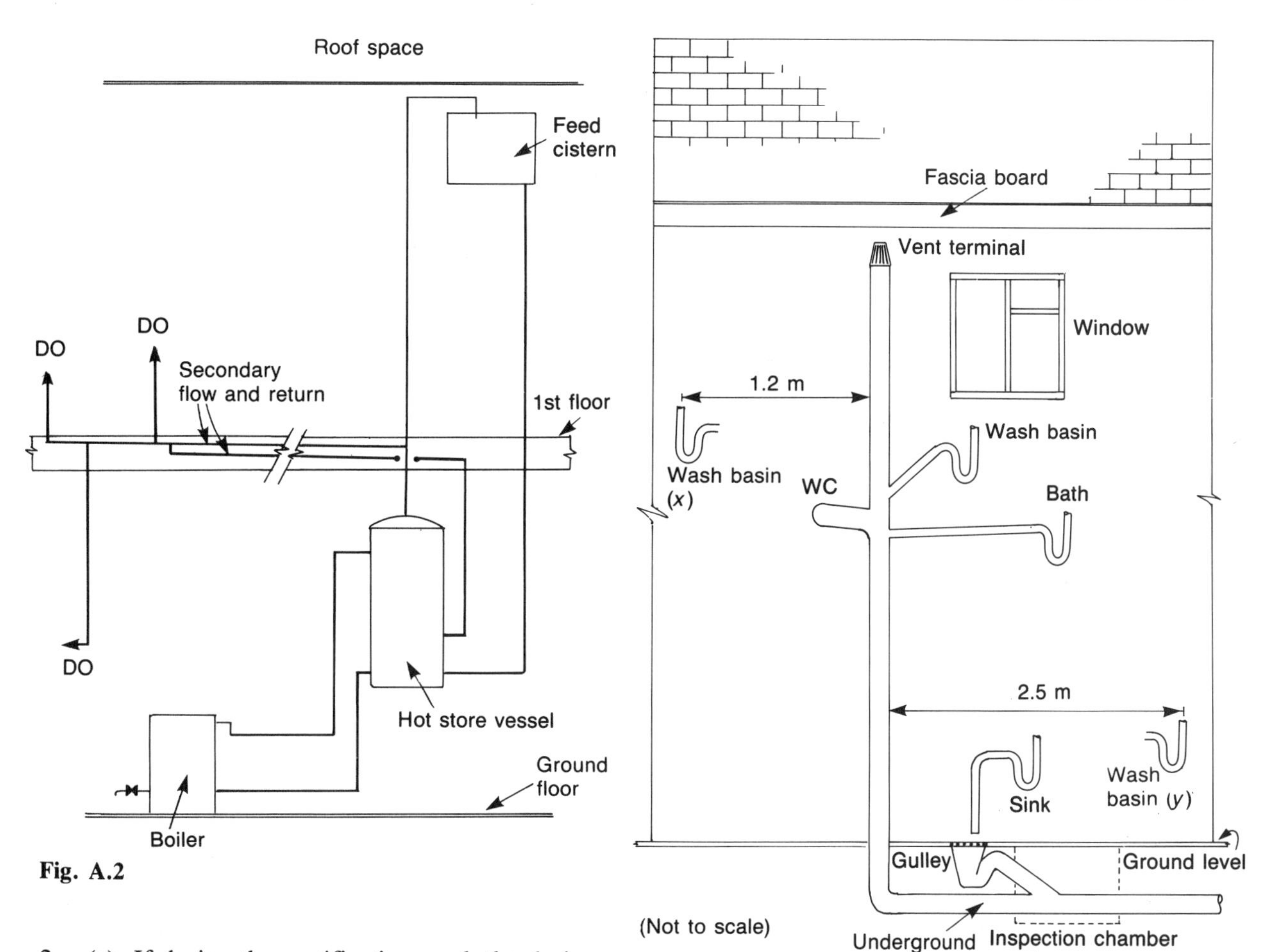

Fig. A.2

Fig. A.3

2. (a) If during the rectification work the drain-off cock is operated, state how you would ensure that it is in working order and will not let by when the system is refilled.
   (b) On refilling the feed cistern it is noted there is a very poor flow through the float-operated valve. List two possible causes for this.
   (c) When the boiler is fired after the system has been rectified, it is noted that a one-pipe circulation is taking place in the vent pipe. Show by means of a sketch how you would prevent this loss of heat.

Sanitary pipework and appliance assignment

You have been asked by a client to prepare a report on the discharge pipework in an existing property he is considering buying. His surveyor has advised him the system is defective as shown in Fig. A.3. The pipework is constructed of PVC.

1. Identify the faults in the system and in each case, by means of a drawing, show how they should be rectified.
2. Assuming your client buys the property and has the remedial work carried out, state the type of scaffolding you would employ and list any safety requirements that must be observed in its use.
3. The flushing cistern serving the WC, which is of the wash-down type, will not flush. Give two possible causes for this problem.
4. (a) Your client has asked you to fix two extra wash basins, x and y, to be supported on

towel-rail brackets in the position shown in the figure. It is found, however, that the walls are constructed of very soft bricks with lime mortar joints, and it is not considered suitable for reliable fixings using screws. Name an alternative type of towel-rail bracket that could be used and list the steps you would take to ensure a satisfactory fixing.

(b) Show on your amended drawing how you intend to connect the discharge pipes from the two new wash basins to the existing drainage system. From the graph in Chapter 8 determine the correct fall of the pipe from basin *x*. In connection with basin *y* the discharge pipe may be connected to the main discharge stack or discharged into the underground drainage system. Sketch and describe the requirements to comply with good practice in each case.

Chimney stack weathering assignment

A client has requested that you investigate the cause of dampness in the roof timbers of his home. The affected timbers are adjacent to a chimney stack passing through a pitched roof weathered with plain tiles. On examination it is found that the cement fillets weathering the stack to the tiles have cracked away from the brickwork due to the shrinkage of the roof timbers, allowing water to seep into the structure. The sole of the back gutter also reveals a number of fatigue cracks due to the fact it was incorrectly fixed, as illustrated in Fig. A.4. You have recommended to your client that the chimney is re-weathered using lead sheet flashings.

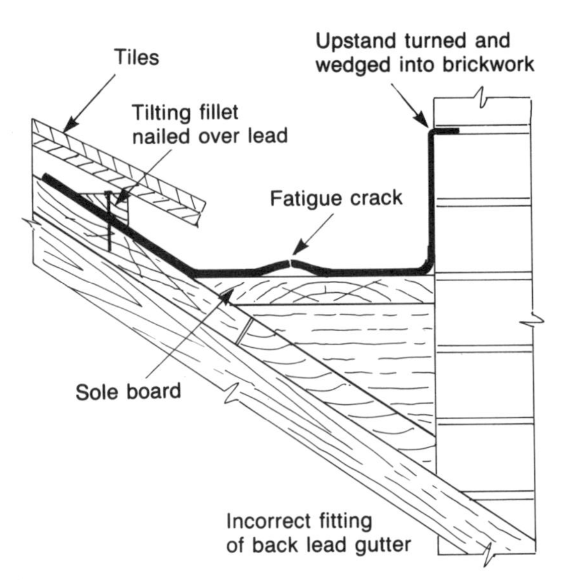

Fig. A.4

1. (a) A small tower scaffold is available to obtain access to the eaves of the building. State how you would ensure safe working practices in its use, especially bearing in mind the outward thrust to which it will be subjected due to the fact that the base of the ladder used to gain access to the chimney will be resting on the working platform.
   (b) List the personal safety requirements that must be observed when cutting away the existing cement fillets and raking the joints. Indicate also any precautions and notices that may be required at ground level to warn the client and his family that work is going on overhead.
2. List the tools necessary to carry out the work.
3. Specify a suitable thickness of lead sheet suitable for the new weatherings and indicate its colour code.
4. Assuming the apron has been prepared and is ready for fixing, state or sketch how you intend to secure its bottom edge.
5. (a) Name the type of side flashing necessary if single-lap interlocking tiles were used instead of the plain tiles shown.
   (b) Make a sketch showing the dimensions of the water line on the side flashings and indicate how the steps are set out.
6. (a) The new back gutter is to be fabricated using welding techniques. Make a sketch showing how you would set out the lead sheet before forming it and welding in the gussets.
   (b) Make a sectional sketch indicating the correct way in which the back gutter should be fitted to avoid future defects.

Appendix B: National Vocational Qualifications

NVQs in mechanical engineering services—plumbing

National Vocational Qualifications (NVQs) are a new system of certification which have been introduced in England and Wales. They have replaced the long-established method of assessing awards by external examinations, which has been accepted by industry for many years, in recognition of the achievement of a craft qualification. NVQs are based on continuous assessment rather than examinations and, like their Scottish equivalent, Scottish Vocational Qualifications (SVQs), reflect standards that have been agreed by the plumbing industry. They have been developed jointly by the British Plumbing Employers Council, plumbing employers, organisations and the Joint Industrial Board (JIB) for the Plumbing Industry.

In England and Wales two levels of NVQ in plumbing have been introduced: levels 2 and 3 in Mechanical Engineering Services – Plumbing. Level 2 replaces the craft certificate and level 3 the advanced craft certificate which were awarded by the City and Guilds of London Institute.

Structure

Each level has a specified number of units of competence, which are divided into elements of competence. Each element of competence has performance criteria and range statements. The element defines the standards which have to be met and the range defines the circumstances in which the standard must be applied.

Assessment

The syllabus (which summarises what has to be assessed) is dictated by the element of competence, the performance criteria and the range statements. Generally, the assessment for each unit comprises:

- practical observations
- oral questions
- written questions

Written evidence is required for the practical observations.

There is no need to assess all elements of a unit at once. Each element can be assessed separately. When a candidate has satisfied all the performance criteria for each element in a unit the candidate will be credited with that unit. The candidate must meet all performance criteria successfully for each element and satisfy the knowledge requirements of the element.

There is no requirement for units to be completed in a particular order. Candidates can complete separate elements in several units before completing any one unit. Units and elements can also be assessed in any order. In addition, units do not have to be completed within a certain time.

Work-based assignments

An integral part of the course is the continuous work-based assessment of the candidate's work which can be carried out in the workplace or at an appropriate centre. Assessments are made on a range of practical activities and on documentary evidence provided by the candidates to prove their competence. The evidence may be in the form of photographs or written statements by employers or clients. Where assessment in the workplace is not possible, it can be carried out through simulated activities and tasks in an approved centre.

Activity record

An activity record is a written description of the activities that have been undertaken by the candidate during the completion of a unit. The record lists details of the following:

- Activities undertaken, including production techniques used accompanied by appropriate sketches.
- Any problems or difficulties encountered and how they were resolved.
- Knowledge of what was done and of any relevant legislation and company policies that applied to the activities.
- How the job was planned, including any liaison with other personnel, co-contractors and customers.
- All safety precautions taken.
- Materials and equipment used.

Related knowledge and understanding

A candidate's performance may be demonstrated directly. However, in many cases satisfactory performance alone is not enough and the candidate must show an understanding of the task being done. Standard assessments have been prepared by the awarding body to allow candidates to show that they have this knowledge and understanding.

The assessments comprise a series of questions which require a short written answer or a sketch. Sometimes this will need to be supplemented by oral questions, which will be asked by the assessor. Centres may sometimes provide their own assessments.

Accreditation of prior learning (APL)

Candidates can use evidence from work or other activities undertaken before starting the NVQ in the assessment of a unit. The evidence must satisfy the performance criteria and range statements in the same way as evidence gathered while working towards the NVQ.

Portfolio of evidence

The 'portfolio of evidence' is the documentation which the candidate submits to the assessor for assessment of an element or unit. This will usually be a file or folder in which the candidate keeps all their evidence. It should be emphasised that it is the candidate's responsibility to keep the portfolio and add the appropriate evidence.

The portfolio should include a completed record of assessment and the supporting evidence. The evidence can be anything that illustrates the candidate's competence. Forms of evidence can include authenticated photographs, job cards, time sheets, appraisals from line managers or supervisors, testimonials, video and audio tapes and computer disks.

The portfolio must be accessible to the reader. It should be divided into sections which relate to the units and elements of the NVQ. The information must be coherent and include a contents page, dates on all entries and titles and headings to describe the contents. All related entries should be cross-referenced, indicating the relevant units and elements of the standards.

Sources of information

Further information on NVQs and SVQs can be obtained from the following organisations.

British Plumbing Employers Council,
14–15 Ensign Business Centre,
Westwood Business Park,
Westwood Way,
Coventry CV4 8JA
Tel. 01203 470626.

Scottish Vocational Educational Council (ScotVEC),
Hannover House,
Douglas Street,
Glasgow
Tel. 0141 248 7900.

Appendix C: Study guide for NVQ units

This appendix shows how the underpinning knowledge given in this volume of *Plumbing: Mechanical Services* relates to the NVQ units. It follows the **main** performance requirements in the 'type of evidence' sections of the unit specification, indicating the pages or chapters in Book 1 that contain the relevant underpinning knowledge; Book 2 of *Plumbing: Mechanical Services* is referenced where appropriate. The supplementary evidence requirements are not covered.

The requirements are listed unit by unit and element by element. The number of each performance requirement matches the number in the unit specifications.

Unit P2/1 Install and test the components of the system

Element P2/1.1 Interpret the installation requirements of the components of the system

1. Hot and cold water, hot water heating — Chapters 5 and 6
2. Unvented and non-storage hot water systems — *Book 2*
3. Above ground discharge pipework and sanitation systems — Chapter 8
4. Below ground drainage system — *Book 2*
5. Gas supply — *Book 2*
6. Electrical systems — *Book 2*
7. Mains services — Page 56
8. Oil supply — *Book 2*

Element P2/1.2 Prepare sites for installation and testing

1. Checks, health and safety with respect to:
   (a) access equipment — Pages 3–8
   (b) excavations — Pages 8–9
   (c) hazardous conditions — Pages 13, 16
2. Input services — *Not covered*
3. Schedules specifications — *Not covered*
4. Checks on site conditions — *Not covered*

Element P2/1.3 Fabricate, position and fix components

Assessment method I
Bending of materials:
- copper — Pages 77–86
- low carbon steel — Pages 85–88

Assessment method II
Jointing of materials:
- cast iron — Pages 52–53
- copper — Pages 50–52, 66–7
- low carbon steel — Pages 31–2, 48–9, 69–70
- pressure pipe (plastic) — Pages 71–6
- soil and waste systems (plastic) — Pages 75–6

Assessment method III
1. Hot and cold water, hot water heating — Chapters 5 and 6 and *Book 2*
2. Unvented and non-storage hot water systems — *Book 2*
3. Above ground discharge pipework and sanitation systems — Chapter 8
4. Below ground drainage system — *Book 2*
5. Gas supply — *Book 2*
6. Electrical systems — *Book 2*

Element P2/1.4 Connect and test components

Assessment method I
1. Connection to incoming service:
   (a) cold water — Pages 108–12
   (b) gas — *Book 2*
   (c) soil system to drain termination — Pages 194–5
2. Connection of new pipework into an existing gas supply — *Book 2*

Assessment method II

1. Water soundness *Book 2*
2. Gas soundness:
   (a) existing systems *Book 2*
   (b) system after connection *Book 2*
   (c) new installation *Book 2*
3. Soil and waste soundness *Book 2*
4. Drainage *Book 2*
5. Flue:
   (a) draught *Book 2*
   (b) soundness *Book 2*
6. Electrical:
   (a) earth continuity *Book 2*
   (b) polarity *Book 2*
7. Mains water pressure *Book 2*
8. Water flow rate *Book 2*

Unit P2/2 Commission and decommission systems

Element P2/2.1 Carry out checks prior to performance testing

1. Hot and cold water, hot water heating Chapters 5 and 6
2. Unvented and non-storage hot water systems *Book 2*
3. Above ground discharge pipework and sanitation systems Chapter 8
4. Below ground drainage system *Book 2*
5. Gas supply *Book 2*
6. Electrical systems *Book 2*

Element P2/2.2 Monitor and compare the dynamic performance

1. Hot and cold water, hot water heating Chapters 5 and 6
2. Cold water Chapter 5
3. Above ground discharge pipework and sanitation systems Chapter 8
4. Below ground drainage system *Book 2*
5. Gas supply *Book 2*
6. Electrical systems *Book 2*

Element P2/2.3 Decommission systems

1. Hot and cold water, hot water heating Chapters 5 and 6
2. Unvented and non-storage hot water systems *Book 2*
3. Above ground discharge pipework and sanitation systems Chapter 8
4. Below ground drainage system *Book 2*
5. Gas supply *Book 2*
6. Electrical systems *Book 2*

Unit P2/3 Maintain the effective operation of systems

Element P2/3.1 Routinely maintain system components

1. Hot and cold water, hot water heating *Book 2*
2. Unvented hot water systems *Book 2*
3. Non-storage hot water systems *Book 2*

Components:

- Stop cock Pages 101–2
- Gate valve Pages 101–2
- Ball valve Pages 101–2
- Radiator valve (thermostatic) *Book 2*
- Float operated valves (including diaphragm type) Pages 105–108
- Pumps *Book 2*
- Motorised valves *Book 2*
- Pressure relief valves *Book 2*
- Temperature relief valves *Book 2*
- Vacuum relief valves *Book 2*
- Pressure reducing valves *Book 2*
- Mixing valves *Book 2*
- Pressure/storage vessels *Book 2*

Element P2/3.2 Diagnose and rectify the cause of faults

1. Insufficient or no water supply Pages 148–56
2. Air locks Pages 153–4
3. Noise in system Page 132 and *Book 2*
4. Component failure
   Refer to individual manufacturer's maintenance instructions
5. Blockage *Refer to all related chapters*
6. Leakage *Refer to all related chapters*
7. Corrosion of components Pages 157–8, 184–5
8. Loss of trap seal Pages 193–4

Unit P2/4 Maintain the safe working environment

Element P2/4.1 Monitor and maintain one's own health and safety

Assessment method I

1. Checks for suitability:
   (a) step ladders — Pages 5, 7–8
   (b) trestle scaffold — Pages 5–6
   (c) ladders — Pages 5, 7–8
   (d) roof ladders — Pages 5, 7–8
   (e) scaffolds — Pages 3–5
   (f) mobile towers — Pages 4–5
2. Lifting equipment checked — Pages 22–3
3. Appropriate lifting techniques used — Pages 19, 22–3
4. Large heavy object is lifted — Pages 22–3
5. Checks carried out on:
   (a) hand tools — Pages 12–13
   (b) immediate work area — *Not covered*
6. Appropriate precautions:
   (a) combustible/noxious/explosive/dangerous gases — *Book 2*
   (b) hazardous materials — *Refer to COSHH Regulations*
7. Personal protective equipment — Pages 13–4

Assessment method II

Accident procedures:

- Authorised person — Pages 1–3
- Details entered in accident book — Pages 1–3
- Appropriate forms completed — Pages 1–3

Assessment method III

Unsafe working practices:

- Authorised person — Pages 1–3
- Appropriate forms completed — Pages 1–3

Element P2/4.2 Contribute to the limitation of damage

1. First aid/emergency procedures
   (a) minor cuts — *Refer to an approved first aid manual*
   (b) minor burns — *Refer to an approved first aid manual*
   (c) electric shock — Pages 9–10
   (d) shock — *Refer to an approved first aid manual*
2. Assistance — *Not covered*
3. Warning/alerting people — *Not covered*
4. Damage to premises — *Not covered*

Element P2/4.3 Contribute to the limitation of damage

This element is beyond the scope of these books.

Element P2/4.4 Agree and maintain a safe environment

1. Hazards identified — Pages 9, 13
2. Hazards made safe — Page 9

Unit P2/5 Maintain effective working relationships

Element P2/5.1 Establish and develop professional relationships with customers and co-contractors

This element is beyond the scope of these books.

Element P2/5.2 Establish and maintain professional relationships with authorised site visitors

This element is beyond the scope of these books.

Element P2/5.3 Maintain effective working relationships with colleagues

This element is beyond the scope of these books.

Unit P2/6 Contribute to quality development and improvement

Element P2/6.1 Promote the organisation's/industry's image

This element is beyond the scope of these books.

Element P2/6.2 Encourage energy efficiency

This element is beyond the scope of these books.

Unit P2/7 Fabricate, install and check sheet weathering systems components

Element P2/7.1 Interpret the installation requirements

Sheet lead installations:

1. Chimney penetration through pitched roof Pages 237–9
2. Pitched roof to vertical wall Pages 240–43
3. Weathering to a dormer *Book 2*
4. Weathering to a roof penetration Pages 237–9, 244–5
5. Weathering to a canopy (flat) Pages 245–59
6. Weathering to a bay top (flat): Pages 245–59
   (a) system material, material specification, etc. Chapter 9
   (b) schedule of materials and components Page 207

Element P2/7.2 Prepare sites for fabricating, installing and checking

1. Checks: health and safety etc. Chapter 1
2. Surface preparation Pages 216–17
3. Temporary protection of building an contents *Not covered*

Element P2/7.3 Fabricate, position and fix components

1. Marking out sheetwork Pages 216–17
2. Preparatory work Pages 216–17
3. Laying inodorous/proprietary underlays Pages 216–17
4. Fabricating:
   (a) internal corners by bossing Pages 243–6
   (b) external corners by bossing Pages 243–6
   (c) internal corners by lead welding *Book 2*
   (d) external corners by lead welding *Book 2*
5. Fixing sheet lead to vertical brickwork Pages 237–8

Unit P2/8 Maintain sheet weathering systems

Element P2/8.1 Interpret the requirements for routine maintenance

Routine maintenance requirements:

1. Keeping free of debris *Book 2*
2. Checking for defects *Book 2*
3. Checking fixings *Book 2*

Schedule of maintenance requirements *Book 2*

Element P2/8.2 Maintain and rectify faults in sheet weathering systems

This element is covered in Book 2.

Index